About the Authors

Prem C. Consul is professor emeritus in the Department of Mathematics and Statistics at the University of Calgary in Calgary, Alberta, Canada. Dr. Consul received his Ph.D. degree in mathematical statistics and a master's degree in mathematics from Dr. Bhimrao Ambedkar University (formerly Agra University), Agra in India. Prior to joining the University of Calgary in 1967, he was a professor at the University of Libya for six years, the principal of a degree college in India for four years, and was a professor and assistant professor in degree colleges for ten years. He was on the editorial board of *Communications in Statistics* for many years. He is a well-known researcher in distribution theory and multivariate analysis. He published three books prior to the present one.

Felix Famoye is a professor and a consulting statistician in the Department of Mathematics at Central Michigan University in Mount Pleasant, Michigan, USA. Prior to joining the staff at Central Michigan University, he was a postdoctoral research fellow and an associate at the University of Calgary, Calgary, Canada. He received his B.SC. (Honors) degree in statistics from the University of Ibadan, Ibadan, Nigeria. He received his Ph.D. degree in statistics from the University of Calgary under the Canadian Commonwealth Scholarship. Dr. Famoye is a well-known researcher in statistics. He has been a visiting scholar at the University of Kentucky, Lexington and University of North Texas Health Science Center, Fort Worth, Texas.

Prem C. Consul
Felix Famoye

Lagrangian Probability Distributions

Birkhäuser
Boston • Basel • Berlin

Prem C. Consul
University of Calgary
Department of Mathematics and Statistics
Calgary, Alberta T2N 1N4
Canada

Felix Famoye
Central Michigal University
Department of Mathematics
Mount Pleasant, MI 48859
USA

Cover design by Alex Gerasev.

Mathematics Subject Classification (2000): 60Exx, 60Gxx, 62Exx, 62Pxx (primary);
60E05, 60G50, 60J75, 60J80, 60K25, 62E10, 62P05, 62P10, 62P12, 62P30, 65C10 (secondary)

Library of Congress Control Number: 2005936506

ISBN-10 0-8176-4365-6 eISBN 0-8176-4477-6
ISBN-13 978-0-8176-4365-2

Printed on acid-free paper.

Printed in the United States of America. (KeS/MP)

9 8 7 6 5 4 3 2 1

www.birkhauser.com

Joseph Louis Lagrange (1736–1813)

Joseph Louis Lagrange was one of the two great mathematicians of the eighteenth century. He was born in France and was appointed professor at the age of 19. He helped in founding the Royal Academy of Sciences at the Royal Artillery School in 1757. He was very close to the famous mathematician Euler, who appreciated his work immensely. When Euler left the Berlin Academy of Science in 1766, Lagrange succeeded him as director of mathematics. He left Berlin in 1787 and became a member of the Paris Academy of Science and remained there for the rest of his career. He helped in the establishment of École Polytechnique and taught there for some time. He survived the French revolution, and Napoleon appointed him to the Legion of Honour and Count of the Empire.

Lagrange had given two formulae for the expansion of the function $f(z)$ in a power series of u when $z = ug(z)$ (*mémoires de l'Acad. Roy. des Sci. Berlin*, 24, 1768, 251) which have been extensively used by various researchers for developing the class of Lagrangian probability models and its families described in this book.

Lagrange developed the calculus of variations, which was very effective in dealing with mechanics. His work *Mécanique Analytique* (1788), summarizing all the work done earlier in mechanics, contained unique methods using differential equations and mathematical analysis. He created Lagrangian mechanics, provided many new solutions and theorems in number theory, the method of Lagrangian multipliers, and numerous other results which were found to be extremely useful.

To Shakuntla Consul and Busola Famoye

Foreword

It is indeed an honor and pleasure to write a Foreword to the book on *Lagrangian Probability Distributions* by P. C. Consul and Felix Famoye.

This book has been in the making for some time and its appearance marks an important milestone in the series of monographs on basic statistical distributions which have originated in the second half of the last century.

The main impetus for the development of an orderly investigation of statistical distributions and their applications was the International Symposium on Classical and Contagious Discrete Distributions, organized by G. P. Patil in August of 1969–some forty years ago—with the active participation of Jerzy Neyman and a number of other distinguished researchers in the field of statistical distributions.

This was followed by a number of international conferences on this topic in various locations, including Trieste, Italy and Calgary, Canada. These meetings, which took place during the period of intensive development of computer technology and its rapid penetration into statistical analyses of numerical data, served inter alia, as a shield, resisting the growing attitude and belief among theoreticians and practitioners that it may perhaps be appropriate to de-emphasize the parametric approach to statistical models and concentrate on less invasive nonparametric methodology. However, experience has shown that parametric models cannot be ignored, in particular in problems involving a large number of variables, and that without a distributional "saddle," the ride towards revealing and analyzing the structure of data representing a certain "real world" phenomenon turns out to be burdensome and often less reliable.

P. C. Consul and his former student and associate for the last 20 years Felix Famoye were at the forefront of intensive study of statistical distribution—notably the discrete one—during the golden era of the last three decades of the twentieth century.

In addition to numerous papers, both of single authorship and jointly with leading scholars, P. C. Consul opened new frontiers in the field of statistical distributions and applications by discovering many useful and elegant distributions and simultaneously paying attention to computational aspects by developing efficient and relevant computer programs. His earlier (1989) 300-page volume on *Generalized Poisson Distributions* exhibited very substantial erudition and the ability to unify and coordinate seemingly isolated results into a coherent and reader-friendly text (in spite of nontrivial and demanding concepts and calculations).

The comprehensive volume under consideration, consisting of 16 chapters, provides a broad panorama of Lagrangian probability distributions, which utilize the series expansion of an analytic function introduced by the well-known French mathematician J. L. Lagrange (1736–1813) in 1770 and substantially extended by the German mathematician H. Bürmann in 1779.

A multivariate extension was developed by I. J. Good in 1955, but the definition and basic properties of the Lagrangian distributions are due to P. C. Consul, who in collaboration with R. L. Shenton wrote in the early 1970s in a number of pioneering papers with detailed discussion of these distributions.

This book is a welcome addition to the literature on discrete univariate and multivariate distributions and is an important source of information on numerous topics associated with powerful new tools and probabilistic models. The wealth of materials is overwhelming and the well-organized, lucid presentation is highly commendable.

Our thanks go to the authors for their labor of love, which will serve for many years as a textbook, as well as an up-to-date handbook of the results scattered in the periodical literature and as an inspiration for further research in an only partially explored field.

Samuel Kotz
The George Washington University, U.S.A.
December 1, 2003

Preface

Lagrange had given two expansions for a function towards the end of the eighteenth century, but they were used very little. Good (1955, 1960, 1965) did the pioneering work by developing their multivariate generalization and by applying them effectively to solve a number of important problems. However, his work did not generate much interest among researchers, possibly because the problems he considered were complex and his presentation was too concise.

The Lagrange expansions can be used to obtain very useful numerous probability models. During the last thirty years, a very large number of research papers has been published by numerous researchers in various journals on the class of Lagrangian probability distributions, its interesting families, and related models. These probability models have been applied to many real life situations including, but not limited to, branching processes, queuing processes, stochastic processes, environmental toxicology, diffusion of information, ecology, strikes in industries, sales of new products, and amounts of production for optimum profits.

The first author of this book was the person who defined the Lagrangian probability distributions and who had actively started research on some of these models, in collaboration with his associates, about thirty years ago. After the appearance of six research papers in quick succession until 1974, other researchers were anxious to know more. He vividly remembers the day in the 1974 NATO Conference at Calgary, when a special meeting was held and the first author was asked for further elucidation of the work on Lagrangian probability models. Since the work was new and he did not have answers to all their questions he was grilled with more and more questions. At that time S. Kotz rose up in his defense and told the audience that further questioning was unnecessary in view of Dr. Consul's reply that further research was needed to answer their questions.

The purpose of this book is to collect most of the research materials published in the various journals during the last thirty-five years and to give a reasonable and systematic account of the class of Lagrangian probability distributions and some of their properties and applications. Accordingly, it is not an introductory book on statistical distributions, but it is meant for graduate students and researchers who have good knowledge of standard statistical techniques and who wish to study the area of Lagrangian probability models for further research work and/or to apply these probability models in their own areas of study and research. A detailed bibliography has been included at the end. We hope that the book will interest research workers in both applied and theoretical statistics. The book can also serve as a textbook and a source book for graduate courses and seminars in statistics. For the benefit of students, some exercises have been included at the end of each chapter (except in Chapters 1 and 16).

This book offers a logical treatment of the class of Lagrangian probability distributions. Chapter 1 covers mathematical and statistical preliminaries needed for some of the materials in the other chapters. The Lagrangian distributions and some of their properties are described in Chapters 2 and 3. Their families of basic, delta, and general Lagrangian probability models, urn models, and other models are considered in Chapters 4 to 6. Special members of these families are discussed in Chapters 7 through 13. The various generating algorithms for some of the Lagrangian probability models are considered in Chapter 16. Methods of parameter estimation for the different Lagrangian probability models have been included. Tests of significance and goodness-of-fit tests for some models are also given. The treatments and the presentation of these materials for various Lagrangian probability models are somewhat similar.

The bivariate and the multivariate Lagrangian probability models, some of their properties, and some of their families are described in Chapters 14 and 15, respectively. There is a vast area of further research work in the fields of the bivariate and multivariate Lagrangian probability models, their properties and applications. A list of the notational conventions and abbreviations used in the book is given in the front of the book.

There is a growing literature on the regression models based on Lagrangian probability models, such as the generalized Poisson regression models and the generalized binomial regression models. We have deliberately omitted these materials from this book, but not because their study is unimportant. We feel their study, which depends on some other covariates, is important and could be included in a book on regression analysis.

The production of a work such as this entails gathering a substantial amount of information, which has only been available in research journals. We would like to thank the authors from many parts of the world who have generously supplied us with reprints of their papers and thus have helped us in writing this book. We realize that some important work in this area might have been inadvertently missed by us. These are our errors and we express our sincere apology to those authors whose work has been missed.

We are particularly indebted to Samuel Kotz, who read the manuscript and gave valuable comments and suggestions which have improved the book. We wish to express our gratitude to the anonymous reviewers who provided us with valuable comments. We would like to thank Maria Dourado for typing the first draft of the manuscript. We gratefully acknowledge the support and guidance of Ann Kostant of Birkhäuser Boston throughout the publication of this work. We also thank the editorial and production staff of Birkhäuser for their excellent guidance of copyediting and production. The financial support of Central Michigan University FRCE Committee under grant No. 48515 for the preparation of the manuscript is gratefully acknowledged by the second author.

Calgary, Alberta, Canada — *Prem C. Consul*
Mount Pleasant, Michigan — *Felix Famoye*

October 25, 2005

Contents

List of Tables

Abbreviations

ARE	Asymptotic relative efficiency
ASD	Abel series distribution
BITD	Bivariate inverse trinomial distribution
BLBD	Bivariate Lagrangian binomial distribution
BLBTD	Bivariate Lagrangian Boral–Tanner distribution
BLD	Bivariate Lagrangian distribution
BLLSD	Bivariate Lagrangian logarithmic series distribution
BLNBD	Bivariate Lagrangian negative binomial distribution
BLPD	Bivariate Lagrangian Poisson distribution
BMPSD	Bivariate modified power series distribution
BQBD	Bivariate quasi-binomial distribution
cdf	cumulative distribution function
cgf	cumulant generating function
CGPD	Compound generalized Poisson distribution
CI	Confidence interval
cmgf	central moment generating function
CV	Coefficient of variation
DF	Distribution function
EDF	Empirical distribution function
EWRC	Empirical weighted rate of change
FBP	First busy period
FPT	First passage time
GLKD	Generalized Lagrangian Katz distribution
GLSD	Generalized logarithmic series distribution
$GLSD_0$	Generalized logarithmic series distribution with zeros
GNBD	Generalized negative binomial distribution
GP	Generalized Poisson
GPD	Generalized Poisson distribution
GPSD	Generalized power series distribution
GSD	Gould series distribution
i.i.d.	independent and identically distributed

Abbreviations Continued

LDPD	Location-parameter discrete probability distribution
LKD	Lagrangian Katz distribution
MBED	Maximum Bayesian entropy distribution
mgf	moment generating function
ML	Maximum likelihood
MLE	Maximum likelihood estimation
MLD	Multivariate Lagrangian distribution
MPSD	Modified power series distribution
MVU	Minimum variance unbiased
MVUE	Minimum variance unbiased estimation
pgf	probability generating function
pmf	probability mass function
QBD	Quasi-binomial distribution
QHD	Quasi-hypergeometric distribution
QPD	Quasi-Pólya distribution
r.v.	random variable
r.v.s	random variables

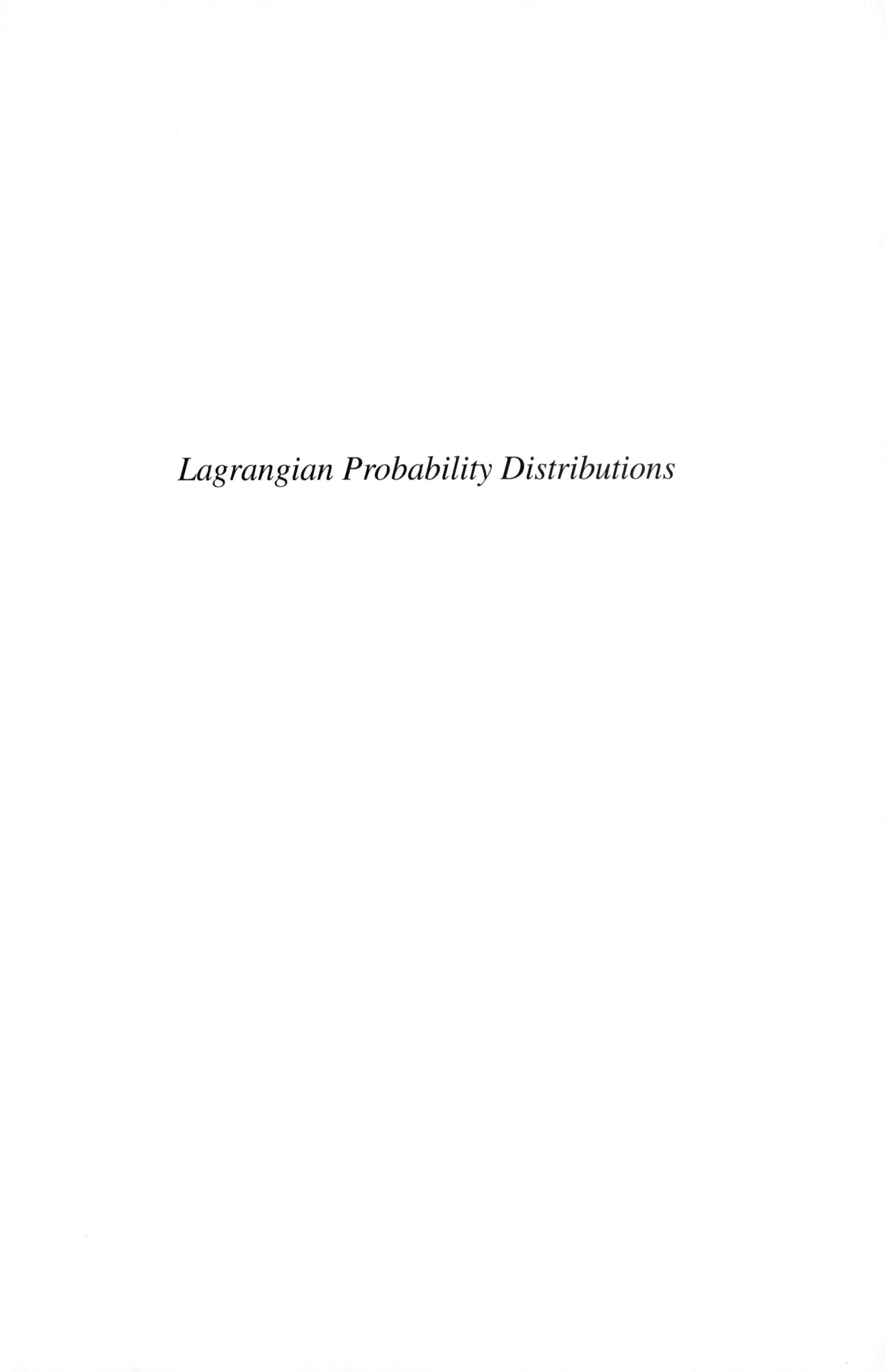

Lagrangian Probability Distributions

1

Preliminary Information

1.1 Introduction

A number of formulas and results are included in this chapter, as they are needed repeatedly in many chapters. The detailed proofs for most of these results have not been provided because they are given in most standard books of mathematics and statistics. Since this book will mostly be used by upper level graduate students and by researchers, we don't need to go into detailed proofs of such preliminary results.

1.2 Mathematical Symbols and Results

1.2.1 Combinatorial and Factorial Symbols

The symbol $n!$ is read as "*n factorial*" and it represents the product of all positive integers from 1 to n. Thus,

$$n! = n(n-1)(n-2)\cdots(3)(2)(1). \tag{1.1}$$

Also, $1! = 1$ and $0! = 1$.

The product of k positive integers from n to $(n-k+1)$ is called the *descending factorial* and is denoted by $n^{(k)}$. Also, $n^{(0)} = 1$ and

$$n^{(k)} = n(n-1)(n-2)\cdots(n-k+1) \tag{1.2}$$

$$= \frac{n!}{(n-k)!}. \tag{1.3}$$

Some authors have used the symbols $n_{(k)}$ and $n^{[k]}$ for the above product of descending factorial. Accordingly, there is no standard notation for this product. Similarly, there is no standard notation for the *ascending factorial* for which the symbol $n^{[k]}$ will be used in this book. Thus $n^{[0]} = 1$ and

$$n^{[k]} = n(n+1)(n+2)\cdots(n+k-1) \tag{1.4}$$

$$= \frac{(n+k-1)!}{(n-1)!}. \tag{1.5}$$

The symbol $n^{[k,s]}$ represents the product

$$n^{[k,s]} = n(n+s)(n+2s)\cdots(n+(k-1)s) \tag{1.6}$$

$$= s^k \cdot \left(\frac{n}{s}\right)\left(\frac{n}{s}+1\right)\left(\frac{n}{s}+2\right)\cdots\left(\frac{n}{s}+k-1\right)$$

$$= s^k \cdot \frac{\left(\frac{n}{s}+k-1\right)!}{\left(\frac{n}{s}-1\right)!} = s^k \left(\frac{n}{s}\right)^{[k]}. \tag{1.7}$$

It may be noted that $\frac{n}{s}$ is not necessarily a positive integer. Also, $n^{[k,1]} = n^{[k]}$ and $n^{[k,-1]} = n^{(k)}$.

The *binomial coefficient* $\binom{n}{k}$ denotes the number of different ways of selecting k items out of n different items and

$$\binom{n}{k} = \frac{n!}{k!(n-k)!} = \frac{n(n-1)\cdots(n-k+1)}{k!} = \frac{n^{(k)}}{k!}. \tag{1.8}$$

Also,

$$\binom{n}{k} = \binom{n}{n-k} \tag{1.9}$$

and

$$\binom{n}{0} = \binom{n}{n} = 1. \tag{1.10}$$

It can be easily shown that

$$\binom{n}{k} + \binom{n}{k+1} = \binom{n+1}{k+1}. \tag{1.11}$$

When n and k are positive integers it is usual to define $\binom{n}{k} = 0$, for $k < 0$ or for $k > n$. However, in an *extended definition of these symbols* for any real value of n and positive integer k, we have

$$\binom{-n}{k} = \frac{(-n)(-n-1)(-n-2)\cdots(-n-k+1)}{k!}$$

$$= (-1)^k \frac{(n+k-1)!}{k!(n-1)!} = (-1)^k \binom{n+k-1}{k}. \tag{1.12}$$

Also,

$$\frac{(-n-1)!}{(-n-k-1)!} = (-n-k)(-n-k+1)(-n-k+2)\cdots(-n-1)$$

$$= (-1)^k(n+1)(n+2)(n+3)\cdots(n+k-1)(n+k)$$

$$= (-1)^k \frac{(n+k)!}{n!}. \tag{1.13}$$

When the exponent m is a positive integer, the *binomial expansion* is

$$(a+b)^m = \sum_{i=0}^{m} \binom{m}{i} a^i b^{m-i} \tag{1.14}$$

$$= b^m + \frac{m^{(1)}}{1!} ab^{m-1} + \frac{m^{(2)}}{2!} a^2 b^{m-2} + \cdots + \frac{m^{(i)}}{i!} a^i b^{m-i} + \cdots + a^m. \tag{1.15}$$

When the exponent m is any real number and $-1 < x < +1$, the *binomial expansion* is

$$(1+x)^m = \sum_{i=0}^{\infty} \frac{m^{(i)}}{i!} x^i = \sum_{i=0}^{\infty} \binom{m}{i} x^i, \tag{1.16}$$

where $\binom{m}{i}$ has the extended definition of (1.12).

Vandermonde's identity

$$\sum_{i=0}^{k} \binom{m}{i} \binom{n}{k-i} = \binom{m+n}{k} \tag{1.17}$$

can be proved by equating the coefficients of x^k in the expansions of $(1+x)^m(1+x)^n = (1+x)^{m+n}$.

The *negative binomial expansion* is a particular case of (1.16). When x and m are both negative, we have

$$(1-x)^{-m} = \sum_{i=0}^{\infty} \frac{m^{[i]}}{i!} x^i = \sum_{i=0}^{\infty} \binom{m+i-1}{i} x^i, \tag{1.18}$$

which gives another *Vandermonde-type identity*

$$\sum_{i=0}^{k} \binom{m+i-1}{i} \binom{n+k-i-1}{k-i} = \binom{m+n+k-1}{k}. \tag{1.19}$$

The *multinomial expansion* is

$$(a_1 + a_2 + a_3 + \cdots + a_k)^n = \sum \frac{n!}{r_1! r_2! \cdots r_k!} a_1^{r_1} a_2^{r_2} a_3^{r_3} \cdots a_k^{r_k}, \tag{1.20}$$

where each $r_i \geq 0, i = 1, 2, \ldots, k$, and the summation is taken over all sets of nonnegative integers $r_1, r_2, \ldots, r_k$ from 0 to n such that $r_1+r_2+r_3+\cdots+r_k = n$. The ratio of the factorials is called the *multinomial coefficient*

$$\binom{n}{r_1, r_2, \ldots, r_k} = \frac{n!}{r_1! r_2! \cdots r_k!}. \tag{1.21}$$

Riordan (1968) has given a large number of formulas between binomial coefficients. Some of them are

$$f_n = \sum_{k=1}^{n} (-1)^{k+1} \frac{1}{k} \binom{n}{k} = 1 + \frac{1}{2} + \frac{1}{3} + \cdots + \frac{1}{n}, \tag{1.22}$$

$$\sum_{k=1}^{n} (-1)^{k-1} \binom{n}{k} f_k = \frac{1}{n}, \tag{1.23}$$

$$\sum_{k=0}^{n}(-1)^{n-k}2^{2k}\binom{n+k}{2k} = 2n+1, \tag{1.24}$$

$$\sum_{k=0}^{n}(-1)^{k}4^{n-k}\binom{2n-k+1}{k} = n+1, \tag{1.25}$$

$$\sum_{k=0}^{n}(-1)^{m+k}\binom{m}{k}\binom{n+k}{k} = \binom{n}{m}, \tag{1.26}$$

$$\sum_{k=0}^{n}(-1)^{m+k}\binom{n}{k}\binom{n}{2m-k} = \binom{n}{m}, \tag{1.27}$$

$$\sum_{k=0}^{n}(-1)^{k}\binom{n}{k}\binom{n}{2m+1-k} = 0. \tag{1.28}$$

1.2.2 Gamma and Beta Functions

The *gamma function* is denoted by the symbol $\Gamma(n)$ and

$$\Gamma(n) = (n-1)! \tag{1.29}$$

when n is a positive integer. For all other positive values

$$\begin{aligned}\Gamma(n) &= (n-1)\Gamma(n-1)\\ &= (n-1)(n-2)\Gamma(n-2)\\ &= (n-1)(n-2)(n-3)(n-4)\Gamma(n-4).\end{aligned} \tag{1.30}$$

The *gamma function* was defined by Euler as an integral

$$\Gamma(x) = \int_0^\infty t^{x-1}e^{-t}dt, \quad x > 0, \tag{1.31}$$

by which the properties (1.29) and (1.30) can be easily proved. Also,

$$\Gamma\left(\frac{1}{2}\right) = \pi^{\frac{1}{2}}, \quad \Gamma\left(n+\frac{1}{2}\right) = \frac{(2n)!\pi^{\frac{1}{2}}}{n!2^{2n}}. \tag{1.32}$$

When n is large, $\Gamma(n+1)$ or $n!$ can be approximated by any one of the two *Stirling's formulas*

$$\Gamma(n+1) \approx (2\pi)^{\frac{1}{2}}(n)^{n+\frac{1}{2}}e^{-n-1}\left(1+\frac{1}{12(n+1)}+\cdots\right), \tag{1.33}$$

$$\Gamma(n+1) \approx (2\pi)^{\frac{1}{2}}(n)^{n+\frac{1}{2}}e^{-n}\left(1+\frac{1}{12n}+\cdots\right). \tag{1.34}$$

The *beta function B(a, b)* is defined by the integral

$$B(a,b) = \int_0^1 t^{a-1}(1-t)^{b-1}dt, \quad a > 0, \ b > 0. \tag{1.35}$$

By putting $t = u(1+u)^{-1}$,

$$B(a, b) = \int_0^\infty u^{a-1}(1+u)^{-a-b} du. \tag{1.36}$$

Also,

$$B(a, b) = B(b, a) = \frac{\Gamma(a)\Gamma(b)}{\Gamma(a+b)}. \tag{1.37}$$

The *incomplete gamma function* is defined by

$$\Gamma_x(a) = \gamma(a, x) = \int_0^x t^{a-1} e^{-t} dt \tag{1.38}$$

$$= \sum_{k=0}^{\infty} \frac{(-1)^k}{k!} \frac{x^{a+k}}{a+k}. \tag{1.39}$$

The ratio $\frac{\Gamma_x(a)}{\Gamma(a)} = \frac{\gamma(a, x)}{\Gamma(a)}$ is more often used in statistics and is also called the incomplete gamma function.

Similarly, the *incomplete beta function*

$$B_x(a, b) = \int_0^x t^{a-1}(1-t)^{b-1} dt, \quad 0 < x < 1, \tag{1.40}$$

$$= \sum_{k=0}^{\infty} \frac{(-b+1)^{[k]}}{k!} \frac{x^{a+k}}{a+k} \tag{1.41}$$

and the incomplete beta function ratio $\frac{B_x(a, b)}{B(a, b)}$ is used in some places.

1.2.3 Difference and Differential Calculus

The symbols E and Δ denote the *displacement (or shift) operator* and the *forward difference operator*, respectively. The shift operator E increases the argument of a function $f(x)$ by unity:

$$E[f(x)] = f(x+1), \quad E[E[f(x)]] = E[f(x+1)] = f(x+2) = E^2[f(x)].$$

In general,

$$E^k[f(x)] = f(x+k), \tag{1.42}$$

where k can be any real number.

The forward difference operator Δ provides the forward difference of a function $f(x)$:

$$\Delta f(x) = f(x+1) - f(x) = E[f(x)] - f(x) = (E-1)f(x).$$

Thus

$$\Delta \equiv E - 1 \quad \text{or} \quad E \equiv 1 + \Delta. \tag{1.43}$$

This symbolic relation in E and Δ enables us to get many interpolation formulas. If h is any real number, then

$$\begin{aligned} f(x+h) = E^h[f(x)] = (1+\Delta)^h f(x) \\ &= f(x) + h\Delta f(x) + (2!)^{-1}h(h-1)\Delta^2 f(x) + \cdots \\ &= \sum_{k=0}^{h} \binom{h}{k} \Delta^k f(x). \end{aligned} \tag{1.44}$$

The symbol ∇ is known as the *backward difference operator* and $\nabla \equiv \Delta E^{-1} \equiv E^{-1}\Delta \equiv E^{-1}(E-1) \equiv 1 - E^{-1}$. Thus,

$$\nabla f(x) = \Delta E^{-1}[f(x)] = \Delta f(x-1) = f(x) - f(x-1). \tag{1.45}$$

By using these three operators one can obtain many relations. If $f(x)$ is a polynomial of degree n, the use of the operator Δ reduces the degree by unity. Thus, for a positive integer n,

$$\Delta^n x^n = n!, \quad \Delta^{n+1} x^n = 0.$$

Also,

$$\begin{aligned} \Delta x^{(n)} &= (x+1)^{(n)} - x^{(n)} \\ &= (x+1)x(x-1)\cdots(x-n+2) - x(x-1)\cdots(x-n+1) \\ &= x(x-1)\cdots(x-n+2)[x+1-(x-n+1)] \\ &= nx^{(n-1)}, \\ \Delta^2 x^{(n)} &= n(n-1)x^{(n-2)} = n^{(2)}x^{(n-2)}. \end{aligned} \tag{1.46}$$

In general when k is a nonnegative integer,

$$\Delta^k x^{(n)} = n^{(k)} x^{(n-k)}, \quad k \le n, \tag{1.47}$$

and for $k > n$,

$$\Delta^k x^{(n)} = 0. \tag{1.48}$$

Also,

$$\frac{\Delta^k 0^n}{k!} = \left.\frac{\Delta^k x^n}{k!}\right|_{x=0} \tag{1.49}$$

is called a *difference of zero*.

The symbol D will be used as a differentiation operator:

$$Df(x) = f'(x) = \frac{df(x)}{dx} \tag{1.50}$$

and

$$D^k x^n = n(n-1)\cdots(n-k+1)x^{n-k} = n^{(k)}x^{n-k}, \quad k \le n. \tag{1.51}$$

The symbol D may also be used as a partial differential operator $\frac{\partial}{\partial x}$. In such use, a subscript will be used to indicate the variate under differentiation. Thus,

$$D_t f(x, t) = \frac{\partial f(x, t)}{\partial t}.$$

If the function $f(x)$ is analytic and successively differentiable, the Taylor's expansion gives

$$f(x+a) = \sum_{j=0}^{\infty} \frac{a^j}{j!} D^j f(x) = \sum_{j=0}^{\infty} \frac{(aD)^j}{j!} f(x) \equiv e^{aD} f(x). \tag{1.52}$$

By comparing (1.52) with (1.44), we get the formal relation

$$E^a \equiv (1+\Delta)^a \equiv e^{aD}, \tag{1.53}$$

which gives

$$e^D \equiv (1+\Delta) \quad \text{or} \quad D \equiv \ln(1+\Delta). \tag{1.54}$$

The formal relations (1.53) and (1.54) can be used to write many formulas for numerical differentiation.

Roman and Rota (1978) have defined an *Abel operator* by $DE^{-\lambda}$, where

$$DE^{-\lambda} f(u) = Df(u-\lambda) = df(u-\lambda)/du. \tag{1.55}$$

Thus,

$$\begin{aligned}(DE^{-\lambda})^x (1+\theta)^n\big|_{\theta=0} &= D^x E^{-\lambda x}(1+\theta)^n\big|_{\theta=0} = D^x(1+\theta-x\lambda)^n\big|_{\theta=0} \\ &= \frac{n!}{(n-x)!}(1-x\lambda)^{n-x},\end{aligned} \tag{1.56}$$

and

$$(DE^{-\lambda})^x e^u\big|_{u=0} = D^x E^{-\lambda x} e^u\big|_{u=0} = D^x (e^{u-x\lambda})\big|_{u=0} = e^{-x\lambda}. \tag{1.57}$$

Also,

$$A_x(\theta;\ \lambda) = \theta(\theta+x\lambda)^{x-1}, \quad x = 0, 1, 2, \ldots \tag{1.58}$$

are called the *Abel polynomials* and the polynomials

$$\begin{aligned}G_x(s;\ r) &= s(s+xr-1)(s+xr-2)\cdots(s+xr-x+1) \\ &= s\frac{(s+xr-1)!}{(s+xr-x)!} = s(s+xr-1)^{(x-1)}\end{aligned} \tag{1.59}$$

for $x = 1, 2, 3, \ldots$ and $G_0(s;\ r) = 1$ were introduced by Gould (1962) and are called *Gould polynomials* by Roman and Rota (1978).

1.2.4 Stirling Numbers

The *Stirling numbers of the first kind* and the *Stirling numbers of the second kind* will be denoted by the symbols $s(n,\ k)$ and $S(n,\ k)$, respectively, a notation used by Riordan (1968). Many other symbols have been used by different authors. An extensive table of such numbers and their properties is given in Abramowitz and Stegun (1965). Charalambides and Singh (1988) provide a very good review and bibliography of these numbers and their generalizations.

A simple definition of the *Stirling numbers of the first kind* can be given by expanding $x^{(n)}$ into a power series of x. Thus

$$x^{(n)} = x(x-1)(x-2)\cdots(x-n+1) = \sum_{k=0}^{n} s(n,\ k)\cdot x^k, \tag{1.60}$$

which gives

$$s(n, n) = 1, \; s(n, n-1) = (-1)n(n-1)/2, \ldots, \; s(n, 1) = (-1)^{n-1}(n-1)!. \tag{1.61}$$

It follows from (1.60) for $k = 0, 1, 2, \ldots, n$ that

$$s(n, k) = (k!)^{-1}[D^k x^{(n)}]_{x=0}. \tag{1.62}$$

Also, for a positive integer n,

$$(x+h)^n = E^h x^n = (1+\Delta)^h x^n = \sum_{k=0}^{\infty} \binom{h}{k} \Delta^k x^n. \tag{1.63}$$

By putting $x = 0$ and $h = x$, the above becomes

$$\begin{aligned} x^n &= \sum_{k=0}^{n} \binom{x}{k} \Delta^k 0^n \\ &= \sum_{k=0}^{n} S(n, k) \frac{x!}{(x-k)!} = \sum_{k=0}^{n} S(n, k) \cdot x^{(k)}. \end{aligned} \tag{1.64}$$

Thus, the *Stirling numbers of the second kind*, $S(n, k)$ for $k = 0, 1, 2, \ldots, n$, are the coefficients of the descending factorials $x^{(k)}$ in the expansion of x^n and

$$S(n, k) = \Delta^k 0^n / k!, \quad \text{for} \quad k = 0, 1, 2, 3, \ldots, n. \tag{1.65}$$

By multiplying (1.60) by $(x - n)$ it can easily be shown that

$$s(n+1, k) = s(n, k-1) - ns(n, k) \tag{1.66}$$

for $k = 1, 2, \ldots, n+1$ and for $n = 0, 1, 2, \ldots$ with initial conditions $s(0, 0) = 1$ and $s(n, 0) = 0, \; n > 0$.

Similarly, by multiplying (1.63) by $x = (x - k) + k$, it is easy to show that

$$S(n+1, k) = S(n, k-1) + kS(n, k). \tag{1.67}$$

Two other important relations are

$$(e^x - 1)^k = k! \sum_{n=k}^{\infty} \frac{S(n, k) x^n}{n!}, \tag{1.68}$$

$$[\ln(1+x)]^k = k! \sum_{n=k}^{\infty} \frac{s(n, k) x^n}{n!}. \tag{1.69}$$

It can be shown that

$$S(n, k) = \frac{1}{k!} \sum_{r=0}^{k} (-1)^{k-r} \binom{k}{r} r^n, \quad k = 0, 1, 2, \ldots, n. \tag{1.70}$$

1.2.5 Hypergeometric Functions

The *Gauss hypergeometric function* is denoted by the symbol ${}_2F_1[a, b; c; x]$ and is defined as a series in the form

$$
\begin{aligned}
{}_2F_1[a, b; c; x] &= 1 + \frac{ab}{c}\frac{x}{1!} + \frac{a(a+1)b(b+1)}{c(c+1)}\frac{x^2}{2!} + \cdots \\
&= \sum_{k=0}^{\infty} \frac{a^{[k]}b^{[k]}}{c^{[k]}}\frac{x^k}{k!}, \quad c \neq 0, -1, -2, \ldots,
\end{aligned}
\tag{1.71}
$$

where $a^{[k]}$ is the ascending factorial. The subscripts 2 and 1 on the two sides of F refer to the number of parameters a, b in the numerator and the one parameter c in the denominator. When a or b is a negative integer then the series terminates as $a^{[k]}$ or $b^{[k]}$ becomes zero for some value of k. When the series is infinite, it is absolutely convergent or divergent according as $|x| < 1$ or $|x| > 1$, respectively. When $|x| = 1$, the series is absolutely convergent if $c - a - b > 0$ and is divergent if $c - a - b < -1$.

The Gauss hypergeometric function occurs very frequently in mathematics and in applied problems. A good reference for numerous results on this function is *Higher Transcendental Functions, Vol.* 1, by Erdélyi *et al.* (1953). When $x = 1$,

$$
{}_2F_1(a, b; c; 1) = \frac{\Gamma(c)\Gamma(c-a-b)}{\Gamma(c-a)\Gamma(c-b)},
\tag{1.72}
$$

where $c - a - b > 0$ and $c \neq 0, -1, -2, \ldots$.

When a and b are nonpositive integers, $a = -n$, $b = -u$, and $c = v + 1 - n$, the hypergeometric function gives the Vandermonde's identity (1.17):

$$
\sum_{k=0}^{n} \binom{u}{k}\binom{v}{n-k} = \binom{u+v}{n}.
$$

For many specific values of a, b, and c the value of ${}_2F_1(a, b; c; x)$ can be expressed in terms of elementary algebraic functions.

Confluent hypergeometric functions (Kummer's Functions), denoted by the symbol ${}_1F_1(a, c; x)$, represent the series

$$
{}_1F_1[a, c; x] = 1 + \frac{a}{c}\frac{x}{1!} + \frac{a(a+1))}{c(c+1)}\frac{x^2}{2!} + \cdots = \sum_{k=0}^{\infty} \frac{a^{[k]}}{c^{[k]}}\frac{x^k}{k!}, \quad c \neq 0, -1, -2, -3, \ldots.
\tag{1.73}
$$

Similar functions are sometimes denoted by $M(a; c; x)$ or $\phi(a; c; x)$ but they will not be used in this book.

The function

$$
H_n(x) = \sum_{k=0}^{[n/2]} \frac{n!x^n}{k!(n-2k)!}\left(\frac{-x^{-2}}{2}\right)^k
\tag{1.74}
$$

is known as the *Hermite polynomial* and it can be expressed as a confluent hypergeometric function:

$$H_{2n}(x) = \frac{(-1)^n (2n)!}{2^n n!} {}_1F_1\left(-n;\ \frac{1}{2};\ \frac{x^2}{2}\right), \tag{1.75}$$

$$H_{2n+1}(x) = \frac{(-1)^n (2n+1)! x}{2^n n!} {}_1F_1\left(-n;\ \frac{3}{2};\ \frac{x^2}{2}\right). \tag{1.76}$$

For special values of the parameters, many hypergeometric functions reduce to elementary functions. Some of them are

$$\begin{aligned}
{}_1F_1(a,\ a;\ x) &= e^x, \\
{}_1F_1(2,\ 1;\ x) &= (1+x)e^x, \\
{}_2F_1(a,\ b;\ b;\ x) &= (1-x)^{-a}, \\
x\ {}_2F_1(1,\ 1;\ 2;\ -x) &= \ln(1+x), \\
2x\ {}_2F_1(1,\ 1/2;\ 3/2;\ x^2) &= \ln(1+x) - \ln(1-x).
\end{aligned}$$

Many well-known polynomials and special functions, like Legendre polynomials, Jacobi polynomials, Chebyshev polynomials, incomplete beta function, Bessel functions, and Whittaker functions, can also be expressed in terms of hypergeometric functions.

1.2.6 Lagrange Expansions

Let $f(z)$ and $g(z)$ be two analytic functions of z, which are infinitely differentiable in $-1 \le z \le 1$ and such that $g(0) \ne 0$. Lagrange (1736–1813) considered the inversion of the *Lagrange transformation* $u = z/g(z)$, providing the value of z as a power series in u, and obtained the following three power series expansions. Two of these expressions are also given in Jensen (1902) and Riordan (1968).

$$z = \sum_{k=1}^{\infty} \frac{u^k}{k!} [D^{k-1} (g(z))^k]_{z=0}, \tag{1.77}$$

$$f(z) = \sum_{k=0}^{\infty} a_k u^k, \tag{1.78}$$

where $a_0 = f(0)$ and

$$a_k = \frac{1}{k!} [D^{k-1} (g(z))^k D f(z)]_{z=0} \tag{1.79}$$

and

$$\frac{f(z)}{1 - z g'(z)/g(z)} = \sum_{k=0}^{\infty} b_k u^k, \tag{1.80}$$

where $b_0 = f(0)$ and

$$b_k = \frac{1}{k!} [D^k (g(z))^k f(z)]_{z=0}. \tag{1.81}$$

If we replace the function $f(z)$ in (1.80) by

$$f_1(z) = \left(1 - z g'(z)/g(z)\right) f(z),$$

the Lagrange expansion in (1.80) reduces to the Lagrange expansion in (1.78). Also, replacing the function $f(z)$ in (1.78) by

$$f_1(z) = \frac{f(z)}{1 - zg'(z)/g(z)},$$

the Lagrange expansion in (1.78) becomes the Lagrange expansion in (1.80). Thus, the two Lagrange expansions in (1.78) and in (1.80) are not independent of each other.

The three expansions (1.77), (1.78), and (1.80) play a very important role in the theory of Lagrangian probability distributions. They also provide a number of useful identities, given by Riordan (1968) and Gould (1972). Some important identities are

$$\sum_{k=0}^{\infty} \frac{a(a+k\theta)^{k-1}}{k!} \frac{[b+(n-k)\theta]^{n-k}}{(n-k)!} = \frac{(a+b+n\theta)^n}{n!}, \tag{1.82}$$

which can easily be proved by comparing the coefficient of y^n in the expansion of $\exp\{(a+b)z\}/(1-z\theta)$ by (1.80) and (1.81) under the transformation $y = z/e^{\theta z}$ with the coefficient of y^n in the product of the expansions of e^{az}, by (1.78) and (1.79), and of $e^{bz}/(1-z\theta)$, by (1.80) and (1.81), under the same transformation

$$y = z/e^{\theta z}.$$

Similarly,

$$\sum_{k=0}^{n} \frac{a(a+k\theta)^{k-1}}{k!} \frac{\{b+(n-k)\theta\}^{n-k-1}}{(n-k)!} = \frac{(a+b)(a+b+n\theta)^{n-1}}{n!}, \tag{1.83}$$

$$\sum_{k=0}^{n} \binom{n}{k} (x+kz)^k (y-kz)^{n-k} = n! \sum_{k=0}^{n} \frac{(x+y)^k z^{n-k}}{k!}, \tag{1.84}$$

$$\sum_{k=0}^{n} \frac{x}{x+kz} \binom{x+kz}{k} \frac{y}{y+(n-k)z} \binom{y+(n-k)z}{n-k} = \frac{x+y}{x+y+nz} \binom{x+y+nz}{n}, \tag{1.85}$$

$$\sum_{k=0}^{n} \frac{x}{x+kz} \binom{x+kz}{k} \binom{y+nz-kz}{n-k} = \binom{x+y+nz}{n}. \tag{1.86}$$

The two Lagrange expansions (1.78) and (1.80) can also be used to provide expansions of the functions $f(\theta)$ and $f(\theta)/\{1-\theta g'(\theta)/g(\theta)\}$, of a parameter θ, in powers of the function $u = \theta/g(\theta)$ as given below:

$$f(\theta) = \begin{cases} f(0), & k = 0, \\ \sum_{k=1}^{\infty} \frac{(\theta/g(\theta))^k}{k!} \{D^{k-1} f'(\theta)(g(\theta))^k\}|_{\theta=0} \end{cases} \tag{1.87}$$

and

$$\frac{f(\theta)}{1-\theta g'(\theta)/g(\theta)} = \sum_{k=0}^{\infty} \frac{(\theta/g(\theta))^k}{k!} \{D^k f(\theta)(g(\theta))^k\}|_{\theta=0}, \tag{1.88}$$

where $D = \partial/\partial\theta$. Since the functions $f(\theta)$ and $g(\theta)$ can be given numerous values, the two expansions (1.87) and (1.88) open a very wide field of study for power series expansions. As examples $f(\theta) = e^{a\theta}$ and $g(\theta) = e^{b\theta}$ provide the expansions

$$e^{a\theta} = \sum_{k=0}^{\infty} \frac{a(a+bk)^{k-1}}{k!} (\theta e^{-b\theta})^k, \tag{1.89}$$

$$\frac{e^{a\theta}}{1-b\theta} = \sum_{k=0}^{\infty} \frac{(a+bk)^k}{k!} (\theta e^{-b\theta})^k, \tag{1.90}$$

and $f(\theta) = (1-\theta)^{-n}$ and $g(\theta) = (1-\theta)^{1-m}$, $m > 1$, provide the expansions

$$(1-\theta)^{-n} = \sum_{k=0}^{\infty} \frac{n}{n+km} \binom{n+km}{k} (\theta(1-\theta)^{m-1})^k, \tag{1.91}$$

$$\frac{(1-\theta)^{-n+1}}{1-m\theta} = \sum_{k=0}^{\infty} \binom{n+km-1}{k} (\theta(1-\theta)^{m-1})^k. \tag{1.92}$$

The Lagrange expansion (1.78), with (1.79), was generalized by Bürmann in 1799 (Whittaker and Watson, 1990, pages 128–132) and this generalization is known as the Lagrange–Bürmann expansion. This generalization has not been used for Lagrangian probability models.

Let $f(z_1, z_2)$, $g_1(z_1, z_2)$, and $g_2(z_1, z_2)$ be three bivariate functions of z_1 and z_2 such that $g_1(0, 0) \neq 0$ and $g_2(0, 0) \neq 0$ and all three functions are successively differentiable partially with respect to z_1 and z_2. Poincaré (1886) considered the bivariate expansion of $f(z_1, z_2)$ in power series of u and v under the Lagrange transformations $u = z_1/g_1(z_1, z_2)$, $v = z_2/g_2(z_1, z_2)$ and obtained the *bivariate Lagrange expansion*

$$f(z_1, z_2) = f(0,0) + \sum_{\substack{h=0 \\ h+k}}^{\infty} \sum_{\substack{k=0 \\ >0}}^{\infty} \frac{u^h v^k}{h!k!} D_1^{h-1} \{ D_2^{k-1} [g_1^h g_2^k D_1 D_2 f$$

$$+ g_1^h (D_1 g_2^k (D_2 f)) + g_2^k (D_2 g_1^h (D_1 f))] \}_{z_1=z_2=0}, \tag{1.93}$$

where $D_i = \partial/\partial z_i$, $g_i = g_i(z_1, z_2)$, $f = f(z_1, z_2)$, and $i = 1, 2$.

If the functions f, g_1, and g_2 are of the form $f(z_1, z_2) = [\phi(z_1, z_2)]^c$, $g_i(z_1, z_2) = [\phi(z_1, z_2)]^{c_i}$, $i = 1, 2$, where c, c_1, c_2 are real constants and $\phi(z_1, z_2)$ is an analytic function in $-1 \le z_i \le +1$, the above formula for the bivariate Lagrange expansion can be written as

$$f(z_1, z_2) = f(0,0) + \sum_{\substack{h=0 \\ h+k}}^{\infty} \sum_{\substack{k=0 \\ >0}}^{\infty} \frac{u^h v^k}{h!k!} \{ D_1^{h-1} D_2^k [g_1^h g_2^k D_1 f)\}_{z_1=z_2=0} \tag{1.94}$$

or

$$f(z_1, z_2) = f(0,0) + \sum_{\substack{h=0 \\ h+k}}^{\infty} \sum_{\substack{k=0 \\ >0}}^{\infty} \frac{u^h v^k}{h!k!} \{ D_2^{k-1} D_1^h [g_1^h g_2^k D_2 f)\}_{z_1=z_2=0}. \tag{1.95}$$

The *multivariate generalization* of the above Poincaré generalization of the Lagrange expansion was obtained by Good (1960) and will be defined and described in Chapter 15 of this book.

1.2.7 Abel and Gould Series

Roman and Rota (1978) have defined and studied the Abel series and the Gould series. Let $A(\theta;\ \lambda)$ be a positive analytic function of a parameter θ which may or may not depend upon another parameter λ and where the parameters have a continuous domain around zero. Let

$$A(\theta;\ \lambda) = \sum_{x=0}^{\infty} a(x;\ \lambda) \cdot A_x(\theta;\ \lambda), \tag{1.96}$$

where $A_x(\theta;\ \lambda) = \theta(\theta + x\lambda)^{x-1}$, $x = 0, 1, 2, \ldots$ are the Abel polynomials defined in (1.58) and the coefficients $a(x;\ \lambda)$ are independent of θ for all values of x. The value of $a(x;\ \lambda)$ can be obtained either by one of the Lagrange expansions or by the formula

$$a(x;\ \lambda) = \frac{1}{x!}(DE^{-\lambda})^x A(u;\ \lambda)|_{u=0}, \tag{1.97}$$

restricting the parametric space to the domain where $a(x;\ \lambda) \geq 0$ for all values of x. The series given by (1.96) is called the *Abel series.*

It may be noted that every positive function $A(\theta;\ \lambda)$ cannot be expressed as an Abel series (1.96). There seem to be only three functions which provide an Abel series, and they are

$$A(\theta;\ \lambda) = e^{\theta}, \quad A(\theta;\ \lambda) = (1+\theta)^n, \tag{1.98}$$

$$A(\theta;\ \lambda) = (1+\theta)(1+\theta+n\lambda)^{n-1}. \tag{1.99}$$

In view of the above, the utility of the Abel series gets diminished.

Let us assume that a positive analytic function $A(n;\ \theta)$ of the parameter n (and possibly another parameter θ) can be expanded into a *Gould series* of the form

$$A(n;\ \theta) = \sum_{x=0}^{\infty} a(x;\ \theta) \cdot G_x(n;\ \theta), \tag{1.100}$$

where $G(n;\ \lambda)$ are the Gould polynomials $n(n + x\lambda - 1)^{(x-1)}$ defined in (1.59) and the coefficients $a(x;\ \theta)$, $x = 0, 1, 2, \ldots$, are independent of the parameter n. The values of $a(x;\ \theta)$ can be obtained either by the use of the Lagrange expansions (1.77) to (1.80) or by some combinatorial identity, based upon one of the Lagrange expansions, and given by Riordan (1968) or Gould (1962).

It may be noted that every positive analytic function $A(n;\ \theta)$ of n cannot be expressed in the form of a Gould series. There seem to be very few functions which can be expressed in the form of such a series. Three such functions are

$$(1-p)^{-n},\ (n+m+\theta k)^{(k)},\ (n+m)(n+m+\theta k-1)^{(k-1)}. \tag{1.101}$$

1.2.8 Faà di Bruno's Formula

Let $G(y)$ and $y = f(x)$ be two functions such that all derivatives of $G(y)$ and of $f(x)$ exist. Thus $G(y) = G(f(x))$ becomes an implicit function of x. Faà di Bruno (1855) considered the problem of determining the nth derivative of $G(y)$ with respect to x and obtained the formula

$$\frac{d^n}{dx^n}G[f(x)] = \sum \frac{n!}{k_1!k_2!\cdots k_n!}\frac{d^pG(y)}{dy^p}\left(\frac{f'}{1!}\right)^{k_1}\left(\frac{f''}{2!}\right)^{k_2}\cdots\left(\frac{f^{(n)}}{n!}\right)^{k_n}, \tag{1.102}$$

where the summation is taken over all partitions of n such that

$$\begin{aligned} p &= k_1 + k_2 + \cdots + k_n, \\ n &= k_1 + 2k_2 + \cdots + nk_n, \end{aligned} \tag{1.103}$$

and f', f'', ..., $f^{(n)}$ represent the successive derivatives of $f(x)$. The formula can easily be proved by induction. This formula is quite useful in obtaining the moments and cumulants of some probability distributions.

1.3 Probabilistic and Statistical Results

1.3.1 Probabilities and Random Variables

Let $E_1, E_2, E_3, \ldots, E_n$ be n events defined on the outcomes of a sample space S. The compound event $(E_1 \cup E_2)$ denotes "either E_1 or E_2 or both" and $(E_1 \cap E_2)$ denotes "both events E_1 and E_2." Similarly, $(E_1 \cup E_2 \cup \cdots \cup E_n) = (\cup_{i=1}^n E_i)$ represents "at least one of the events $E_1, E_2, \ldots, E_n$" and $(\cap_{i=1}^n E_i)$ means "all the events $E_1, E_2, \ldots, E_n$."

The symbols $P(E_i)$, $P(E_1 \cup E_2)$, $P(E_1 \cap E_2)$ denote "the probability of event E_i," "the probability of events E_1 or E_2 or both," and "the probability of both events E_1 and E_2," respectively.

The *conditional probability of an event E_2, given that event E_1 has taken place*, is denoted by $P(E_2|E_1)$ and

$$P(E_2|E_1) = \frac{P(E_2 \cap E_1)}{P(E_1)}, \quad P(E_1) > 0. \tag{1.104}$$

In general,

$$P(E_1 \cap E_2) = P(E_2|E_1) \cdot P(E_1) = P(E_1|E_2) \cdot P(E_2) \tag{1.105}$$

and

$$P(E_1 \cup E_2) = P(E_1) + P(E_2) - P(E_1 \cap E_2). \tag{1.106}$$

The above results can easily be extended to any number of events; e.g.,

$$P(E_1 \cap E_2 \cap E_3) = P(E_1 \mid (E_2 \cap E_3)) \cdot P(E_2|E_3)P(E_3), \tag{1.107}$$

$$\begin{aligned} P(E_1 \cup E_2 \cup E_3) &= P(E_1) + P(E_2) + P(E_3) - P(E_1 \cap E_2) \\ &\quad - P(E_1 \cap E_3) - P(E_2 \cap E_3) + P(E_1 \cap E_2 \cap E_3). \end{aligned} \tag{1.108}$$

Two events E_1 and E_2 are *independent* if $P(E_1 \cap E_2) = P(E_1) \cdot P(E_2)$ and they are *mutually exclusive* if $E_1 \cap E_2 = \emptyset$. Mutually exclusive events are not independent.

A number of events $E_1, E_2, \ldots, E_n$ are said to be exhaustive if $\cup_{i=1}^n E_i$ equals the total sample space. If A is another event defined on the sample space S and if the events $E_1, E_2, \ldots, E_n$ are mutually exclusive and exhaustive, then the *total probability of the event* A is

$$P(A) = \sum_{k=1}^{n} P(A|E_k) \cdot P(E_k). \tag{1.109}$$

A *random variable* (r.v.) X is a real valued function defined on the elements of a sample space S such that there is an associated probability with every value of X. The values of the random variable X provide a partition of S. All random variables (r.v.s) will be denoted by uppercase letters like $X, Y, X_i, \ldots$. The probability associated with the value x of X will be denoted by $P[X = x]$.

The distribution function (DF) of X (also called the cumulative distribution function (cdf) of X) is defined by $P[X \leq x]$ and is denoted by $F_X(x)$. The function $F_X(x)$ is a nondecreasing function of x and $0 \leq F_X(x) \leq 1$. Also, $\lim_{x\to-\infty} F_X(x) = 0$ and $\lim_{x\to\infty} F_X(x) = 1$.

For discrete probability distributions $F_X(x)$ is a step function with an enumerable number of steps at x_j, $j = 1, 2, 3, \ldots$. If the value of the probability $P[X = x_j] = p_j$, then p_j, $j = 1, 2, 3, \ldots$, is called the *probability mass function* (pmf). When X takes the values $0, 1, 2, 3, \ldots$ the value of $P(X = x)$ may be denoted by P_x.

The function $S_X(x) = P[X > x] = 1 - F_X(x)$ is called the *survival function*. Also, the function

$$h_X(x) = \frac{P[X = x]}{P[X \geq x]} = \frac{P_x}{P_x + S_X(x)} \tag{1.110}$$

is known as the *hazard function* (or *failure rate*) of the distribution. The function $h_X(x)$ is called the *force of mortality* in actuarial work. When there is no ambiguity in the variables, the subscripts for $F_X(x)$, $S_X(x)$, and $h_X(x)$ may be dropped.

Let $\{E_1, E_2, \ldots, E_k\}$ be a set of mutually exclusive and exhaustive events or possible probability models for an experiment and let $\{A_1, A_2, \ldots, A_s\}$ be the set of possible events when the experiment is made. Also, let $P(E_i)$, $i = 1, 2, \ldots, k$, be the probabilities of the events or models before the experiment is done; $P(A_j)$, $j = 1, 2, \ldots, s$, be the probabilities of the respective events; and $P(A_j|E_i)$, $i = 1, 2, \ldots, k$, $j = 1, 2, \ldots, s$, be the conditional probabilities that the events A_j will take place, given that the event E_i has taken place. That is, the model E_i generates the event A_j; then the total probability of the event A_j is given by (1.109) and equals

$$P(A_j) = \sum_{i=1}^{k} P(A_j \cap E_i) = \sum_{i=1}^{k} P(A_j|E_i) \cdot P(E_i).$$

The conditional probability $P(E_i|A_j)$ is the *posterior probability* of the event E_i (or model E_i) given that the event A_j has taken place, and it is given by the *Bayes' theorem* as

$$P(E_i|A_j) = \frac{P(A_j|E_i) \cdot P(E_i)}{\sum_{i=1}^{k} P(A_j|E_i) \cdot P(E_i)}. \tag{1.111}$$

When the events $E_1, E_2, \ldots, E_k$ do not form an enumerable set but are given by the values of a parameter θ in an interval $a < x < b$ and $p(\theta)$ is the probability density of the parameter, the quantity $P(A_j|E_i)$ becomes $p(x|\theta)$ and $P(E_i) = p(\theta)$ and the summation becomes the integral over the domain of $p(\theta)$.

1.3.2 Expected Values

Let $X_1, X_2, \ldots, X_n$ denote n discrete random variables and let $P(X_i = x_i) = p_i$, $i = 1, 2, \ldots, n$, denote the respective probability mass functions for them. If the r.v.s are independent, the joint probability mass function for all of them is

$$P(X_1 \cap X_2 \cap \cdots \cap X_n) = \prod_{i=1}^{n} P(X_i = x_i) = p_1 p_2 \ldots p_n. \tag{1.112}$$

The *expected value* of the r.v.s, X and $g(X)$, denoted by $E[X]$ and $E[g(X)]$, are given by

$$E[X] = \sum_x x P(X = x), \quad E[g(X)] = \sum_x g(x) \cdot P(X = x).$$

If C is a constant, then

$$E[C] = C, \quad E[Cg(X)] = CE[g(X)],$$

and

$$E[c_1 g_1(X_1) + c_2 g_2(X_2) + c_3 g_3(X_3)] = c_1 E[g_1(X_1)] + c_2 E[g_2(X_2)] + c_3 E[g_3(X_3)]. \quad (1.113)$$

In general,

$$E[g(X_1, X_2, \ldots, X_n)] = \sum g(x_1, x_2, \ldots, x_n) \cdot P(X_1 \cap X_2 \cap \cdots \cap X_n), \quad (1.114)$$

where the summation is taken over all possible values of $x_1, x_2, \ldots, x_n$.

Also,

$$E\left[\sum_{j=1}^{s} c_j g_j(X_1, X_2, \ldots, X_n)\right] = \sum_{j=1}^{s} c_j E[g_j(X_1, X_2, \ldots, X_n)]. \quad (1.115)$$

When the r.v.s are mutually independent,

$$E\left[\prod_{i=1}^{k} g_i(X_j)\right] = \prod_{i=1}^{k} E[g_i(X_j)]. \quad (1.116)$$

1.3.3 Moments and Moment Generating Functions

The expected value of the r.v. X^k, where k is an integer, is called the kth *moment about zero* and is denoted by

$$\mu_k^{'} = E[X^k]. \quad (1.117)$$

The first moment $\mu_1^{'}$ is also called the *mean* of the probability distribution and is denoted by μ.

The kth moment about the mean μ (also called the kth central moment) is denoted by μ_k and is defined by

$$\mu_k = E\left[(X-\mu)^k\right] = E\left[(X - E[X])^k\right]$$
$$= \sum_{s=0}^{k} \binom{k}{s} (-1)^s (\mu)^s \mu_{k-s}^{'}, \quad (1.118)$$

which gives

$$\mu_2 = \mu_2^{'} - \mu^2,$$
$$\mu_3 = \mu_3^{'} - 3\mu_2^{'}\mu + 2\mu^3,$$
$$\mu_4 = \mu_4^{'} - 4\mu_3^{'}\mu + 6\mu_2^{'}\mu^2 - 3\mu^4,$$
$$\mu_5 = \mu_5^{'} - 5\mu_4^{'}\mu + 10\mu_3^{'}\mu^2 - 10\mu_2^{'}\mu^3 + 4\mu^5. \quad (1.119)$$

The first central moment μ_1 is always zero and the second central moment μ_2 is called the *variance* of X and is denoted by $Var(X) = \sigma^2$. The positive square root of $Var(X)$ is called the *standard deviation* σ. Also, the ratio σ/μ is called the *coefficient of variation* (CV).

The expected value $E[e^{tX}]$ if it exists for $|t| < T$, where T is a constant > 0, is called the *moment generating function* (mgf) of X (or of the probability distribution of X) and is denoted by $M_X(t)$. This function generates all the moments about zero as indicated below:

$$M_X(t) = E[e^{tX}] = E\left[1 + \sum_{r=1}^{\infty} t^r X^r/r!\right] = 1 + \sum_{r=1}^{\infty} \mu_r' t^r/r!. \tag{1.120}$$

Thus,

$$\mu_r' = D^r M_X(t)|_{t=0}. \tag{1.121}$$

It can easily be seen from the definition that

$$M_{X+k}(t) = e^{kt} M_X(t). \tag{1.122}$$

Also, if the r.v.s X_1 and X_2 are mutually independent, then

$$M_{X_1+X_2}(t) = M_{X_1}(t) \cdot M_{X_2}(t) \tag{1.123}$$

and

$$M_{X_1-X_2}(t) = M_{X_1}(t) \cdot M_{X_2}(-t), \tag{1.124}$$

and for a number of mutually independent r.v.s

$$M_{X_1+X_2+\cdots+X_k}(t) = M_{X_1}(t) \cdot M_{X_2}(t) \cdots M_{X_k}(t). \tag{1.125}$$

The *central moment generating function* (cmgf) is given by

$$E[e^{(X-\mu)t}] = e^{-\mu t} M_X(t). \tag{1.126}$$

The kth *descending factorial moment* of the r.v. X is the expected value of $X^{(k)} = X!/(X-k)! = \sum_{j=0}^{k} s(k, j)X^j$, by relation (1.60), where $s(k, j)$, $j = 0, 1, 2, \ldots$, are the Stirling numbers of the first kind. Denoting the kth descending factorial moment $\mathrm{E}\left[X^{(k)}\right]$ by $\mu_{(k)}'$, we have

$$\mu_{(k)}' = E\left[\sum_{j=0}^{k} s(k, j)X^j\right] = \sum_{j=0}^{k} s(k, j)\mu_j'. \tag{1.127}$$

In particular,

$$\begin{aligned} \mu_{(1)}' &= \mu, \quad \mu_{(2)}' = \mu_2' - \mu, \\ \mu_{(3)}' &= \mu_3' - 3\mu_2' + 2\mu, \\ \mu_{(4)}' &= \mu_4' - 6\mu_3' + 11\mu_2' - 6\mu. \end{aligned} \tag{1.128}$$

Also, by (1.64)

$$X^k = \sum_{j=0}^{k} S(k, j)X^{(j)},$$

and therefore

$$\mu'_k = \sum_{j=0}^{k} S(k,\, j)\mu'_{(j)}, \tag{1.129}$$

where $S(k,\, j),\ j = 0,\, 1,\, 2,\, \ldots,$ are the Stirling numbers of the second kind. Since

$$(1+t)^X = 1 + X^{(1)}t + X^{(2)}t^2/2! + X^{(3)}t^3/3! + \cdots \tag{1.130}$$

the *factorial moment generating function* of a r.v. X is $E[(1+t)^X]$.

1.3.4 Cumulants and Cumulant Generating Functions

If the moment generating function $M_X(t)$ of a r.v. X exists, then its logarithm is called the *cumulant generating function* (cgf) of X and is denoted by $K_X(t)$ or $K(t)$. The coefficient of $t^r/r!$ in the power series expansion of $K(t)$ is called the rth cumulant of X and is denoted by κ_r. Thus

$$K_X(t) = \ln M_X(t) = \sum_{r=1}^{\infty} \kappa_r t^r/r!. \tag{1.131}$$

Also,

$$\kappa_r = D^r K_X(t)|_{t=0}. \tag{1.132}$$

Since

$$K_{X+a}(t) = \ln M_{X+a}(t) = \ln\left\{e^{at}M_X(t)\right\} = at + K_X(t), \tag{1.133}$$

the coefficients of $t^r/r!$, $r \geq 2$, in $K_{X+a}(t)$ and $K_X(t)$ are the same, which implies that the values of κ_r, $r \geq 2$, are not affected if a constant a is added to the r.v. X. On account of this property, the cumulants are called semi-invariants. Also, by substituting $a = -\mu$, the mgf $M_{X+a}(t)$ becomes the central mgf and so the cumulants κ_r, $r \geq 2$, become the functions of the central moments. By using (1.132) it can easily be shown that

$$\begin{aligned} \kappa_1 &= \mu, & \kappa_2 &= \mu_2, \\ \kappa_3 &= \mu_3, & \kappa_4 &= \mu_4 - 3\mu_2^2, \\ \kappa_5 &= \mu_5 - 10\mu_3\mu_2. \end{aligned} \tag{1.134}$$

If X_i, $i = 1,\, 2,\, \ldots,\, k$, are independent r.v.s and $Y = \sum_{i=1}^{k} X_i$, then

$$K_Y(t) = \sum_{i=1}^{k} K_{X_i}(t), \tag{1.135}$$

which shows that the cgf of the sum of a number of r.v.s equals the sum of the cgfs of the individual r.v.s.

1.3.5 Probability Generating Functions

Let X be a discrete r.v. defined over the set of nonnegative integers and let P_x, $x = 0,\, 1,\, 2,\, \ldots,$ denote its probability mass function such that $\sum_{x=0}^{\infty} P_x = 1$.

The *probability generating function* (pgf) of the r.v. X is denoted by $g(t)$ and is defined by

$$g(t) = E[t^X] = \sum_{x=0}^{\infty} t^x P_x. \tag{1.136}$$

The probabilities P_x, $x = 0, 1, 2, \ldots$, provide the pgf $g(t)$ and the pgf provides all the probabilities P_x by the relation

$$P_x = (x!)^{-1} D^x g(t)|_{t=0}, \; x = 0, 1, 2, \ldots. \tag{1.137}$$

Thus, there is a relation of uniqueness between the pgf and the probability distribution of a r.v. X. Also, if the variable t in (1.136) is replaced with e^u, the pgf $g(t)$ gets transformed into the mgf $M_X(u)$. Further, the natural logarithm of $M_X(u)$ provides a power series in u, where the coefficients of the different terms are the cumulants of the probability distribution of the r.v. X.

The factorial moment generating function becomes

$$E[(1+t)^X] = g(1+t) \tag{1.138}$$

and the mgf of the r.v. X is

$$M_X(t) = E[e^{tX}] = g(e^t) \tag{1.139}$$

and

$$K(t) = \ln M_X(t) = \ln g(e^t). \tag{1.140}$$

Thus, the pgf $g(t)$ of a r.v. X provides the mgf, the central mgf, the cgf, as well as the factorial moment generating function.

If X_1 and X_2 are two independent discrete r.v.s with pgfs $g_1(t)$ and $g_2(t)$ and if $Y = X_1 + X_2$, then the pgf of Y is $G(t) = E[t^Y] = E[t^{X_1+X_2}] = E[t^{X_1}]E[t^{X_2}] = g_1(t)g_2(t)$. This property is called the convolution of the two probability distributions. Similarly, if X_i, $i = 1, 2, 3, \ldots, n$, are n mutually independent discrete r.v.s with pgfs $g_1(t), g_2(t), \ldots, g_n(t)$, respectively, then the pgf $G(t)$ of their sum $Y = \sum_{i=1}^{n} X_i$ becomes

$$G(t) = g_1(t) \cdot g_2(t) \cdots g_n(t), \tag{1.141}$$

and the pgf for the difference $X_i - X_j$ becomes $g_i(t).g_j(t^{-1})$. Thus, if $X_1, X_2, \ldots, X_n$ are independent r.v.s and if their pgfs do exist, then their probability distributions do possess the convolution property.

If two independent r.v.s X_1 and X_2 possess similar pgfs $g_1(t)$ and $g_2(t)$, differing in the value of one or two parameters only and if the pgf $G(t) = g_1(t).g_2(t)$ is of the same form as $g_1(t)$ and $g_2(t)$, then the probability distributions of X_1 and X_2 are said to be *closed under convolution* or that the r.v.s X_1 and X_2 are *closed under convolution*. This property is more restricted and very few r.v.s possess this property.

1.3.6 Inference

Let $X_1, X_2, \ldots, X_n$ be a random sample of size n taken from a population with cumulative distribution function $F(X_1, X_2, \ldots, X_n | \theta_1, \theta_2, \ldots, \theta_k)$, which depends upon k unknown parameters. The values of the parameters are estimated by some functions of the r.v.s. $X_1, X_2, \ldots, X_n$, say $T_j \equiv T_j(X_1, X_2, \ldots, X_n)$, $j = 1, 2, 3, \ldots$, which are called *statistics* or the *estimators of the particular parameters*. Numerous such statistics can be defined.

A statistic T_i is said to be an unbiased estimator of a parameter θ_i if $E[T_i] = \theta_i$. If μ and σ^2 are the unknown mean and variance of a distribution, then their unbiased estimators are given by

$$\bar{X} = n^{-1}\sum_{i=1}^{n} X_i \quad \text{and} \quad s^2 = (n-1)^{-1}\sum_{i=1}^{n}(X_i - \bar{X})^2. \tag{1.142}$$

Numerous methods of estimation and the corresponding estimators of parameters have been defined. Some of them are the (i) *method of moments*, (ii) *zero-frequency and some moments*, (iii) *maximum likelihood estimation (MLE) method*, (iv) *chi-square method*, (v) *Bayesian method*, (vi) *minimum variance unbiased estimation (MVUE)*, and (vii) *robust estimators*. Many books are available in which these are defined and studied.

All these estimators have some good and useful properties but none possess all the good properties for every probability distribution. These estimators are compared with each other by considering properties such as unbiasedness, asymptotical unbiasedness, efficiency, relative efficiency, consistency, and so on.

2

Lagrangian Probability Distributions

2.1 Introduction

Discrete probability distributions are fitted to observed counts to find a pattern which may lead an investigator to see if some generating models can be set up for the process under study. As every phenomenon is of a multivariate nature, the task of finding the correct model becomes difficult. Practically all models for biological, psychological, social, agricultural, or natural processes are approximations. Thus, every model is a simplification of real life. Possibly, every observed pattern is the steady state of some stochastic process.

As different researchers in the world are faced with observed counts in their respective fields and they try to find a specific pattern in their observations, their efforts lead them to the discovery of new and more complex probability distributions. A large number of discrete probability distributions are now available which are divided into various classes, families, and generalized univariate, bivariate, and multivariate discrete distributions. An extensive account of these discrete probability distributions and their important properties can be found in the well-known works by Balakrishnan and Nevzorov (2003), Johnson, Kotz, and Balakrishnan (1997), Johnson, Kotz, and Kemp (1992), Patil, et al. (1984), Patil and Joshi (1968), Johnson and Kotz (1969), Mardia (1970), Ord (1972), and many others. As this book was ready to go to press, we were informed that the book by Johnson, Kotz, and Kemp (1992) was under revision.

Discrete Lagrangian probability distributions form a very large and important class which contains numerous families of probability distributions. These probability distributions are very useful because they occur quite frequently in various fields. Most of the research work on these distributions is available in journals only. The generalized Poisson distribution is one important model which has been discussed in detail by Consul (1989a). Johnson, Kotz, and Kemp (1992) have described some families of the class of Lagrangian probability distributions under different titles in five chapters of their book. The prime source of importance in the class of Lagrangian distributions is the Lagrange transformation $z = ug(z)$, given by Lagrange (1736–1813) and used by him to express z as a power series in u and then to expand a function $f(z)$ into a power series of u as described in Chapter 1.

Otter (1949) was the first person who realized the importance of this transformation for the development of a multiplicative process. If $g(z)$ is the pgf of the number of segments from any vertex in a rooted tree with a number of vertices, then the transformation $z = ug(z)$ provides the pgf for the number of vertices in the rooted tree. Otter (1949) showed that the number of vertices after n segments in a tree can be interpreted as the number of members in the nth generation of a branching process and that it can be used in the study of epidemics, spread of rumors, population growth, nuclear chain reactions, etc. He considered some simple examples. Neyman and Scott (1964) used Otter's multiplicative process in their study of stochastic models for the total number of persons infected in an epidemic which was started by a single person. Berg and Nowicki (1991) considered the applications of Lagrangian probability distributions to inferential problems in random mapping theory. They considered random mapping models with attracting center and those based on loop probability models.

Good (1960, 1965) developed the multivariate generalization of Lagrange expansion and applied it to stochastic processes and enumeration of trees. Gordon (1962) applied Good's theory to polymer distributions. However, these papers were so highly concise and concentrated (one-line formulae representing one-page expressions) that they did not generate much interest from other researchers. Jain and Consul (1971), Consul and Jain (1973a, 1973b), Consul and Shenton (1972, 1973a, 1973b, 1975), and their co-workers have systematically exploited the technique provided by the Lagrange transformation $z = ug(z)$ and the Lagrange expansions for deriving numerous Lagrangian probability distributions and studying their properties and applications.

Janardan (1987) considered the weighted forms of some Lagrangian probability distributions and characterized them with respect to their weighted forms.

As stated in chapter 1, Lagrange had used the transformation $z = ug(z)$ for giving the two expansions in (1.78) and (1.80) for functions of z in powers of u. These expansions have been used to provide numerous Lagrangian probability distributions.

2.2 Lagrangian Probability Distributions

The discrete Lagrangian probability distributions have been systematically studied in a number of papers by Consul and Shenton (1972, 1973a, 1973b, 1975). The Lagrangian negative binomial distribution was first obtained by Mohanty (1966) by combinatorial methods as a distribution of the number of failures x to attain $n + mx$ successes in a sequence of independent Bernoulli trials. Takács (1962) and Mohanty (1966) showed its usefulness in a queuing process. Jain and Consul (1971) used the Lagrange transformation $z = ug(z)$ to obtain Lagrangian negative binomial distribution by considering $g(z)$ as a pgf. The Lagrangian Poisson distribution was defined and studied by Consul and Jain (1973a, 1973b). Consul's (1989a) book shows how intensively the Lagrangian Poisson distribution has been studied for its modes of genesis, properties, estimation, and applications under the title of a *generalized Poisson distribution*. The Lagrangian logarithmic distribution was defined by Jain and Gupta (1973). Consul (1981) has shown that $g(z)$ need not be a pgf for obtaining the Lagrangian distributions.

The class of Lagrangian probability distributions can be divided into three subclasses:

(i) basic Lagrangian distributions,
(ii) delta Lagrangian distributions,
(iii) general Lagrangian distributions,

according to their probabilistic structure. They will be discussed accordingly.

Basic Lagrangian Distributions

Let $g(z)$ be a successively differentiable function such that $g(1) = 1$ and $g(0) \neq 0$. The function $g(z)$ may or may not be a pgf. Then, the numerically smallest root $z = \ell(u)$ of the transformation $z = u\,g(z)$ defines a pgf $z = \psi(u)$ with the Lagrange expansion (1.77) in powers of u as

$$z = \psi(u) = \sum_{x=1}^{\infty} \frac{u^x}{x!} \left\{ D^{x-1} (g(z))^x \right\}_{z=0}, \tag{2.1}$$

if $D^{x-1}(g(z))^x \big|_{z=0} \geq 0$ for all values of x.

The corresponding probability mass function (pmf) of the *basic* Lagrangian distribution becomes

$$P(X = x) = (1/x!) \left\{ D^{x-1} (g(z))^x \right\}_{z=0}, \qquad x \in N. \tag{2.2}$$

Examples of some basic Lagrangian distributions, based on values of $g(z)$, are given in Table 2.1.

It may be noted that

(i) the geometric distribution is a special case of the Consul distribution for $m = 1$;
(ii) the Haight distribution is a special case of the Geeta distribution for $m = 2$;
(iii) the Geeta distribution becomes degenerate when $m \to 1$ as $P(X = 1) \to 1$;
(iv) when $0 < \beta = \theta < 1$ and $b = (m-1)\theta$, the Katz distribution reduces to the Geeta distribution and when $\beta < 0$ and $\beta = -\alpha$, it becomes the Consul distribution by putting $b = m\alpha$ and $\alpha(1+\alpha)^{-1} = \theta$. Also, when $\beta = n^{-1}$ and $n \to \infty$, its limit becomes the Borel distribution;
(v) Otter (1948) had given the pgf of model (7) of Table 2.1 in the form

$$z = \ell(u) = \left\{ 1 - \sqrt{1 - 4pqu^2}/2qu \right\},$$

not the probability model as given in (7). Also, this model is a particular case of model (10) for $m = 1$.

Many other basic Lagrangian distributions can be obtained by choosing various other functions for $g(z)$. Of course, the pmfs of these probability distributions will possibly be more complex than those given in Table 2.1. Examples of some other functions for $g(z)$ are $e^{\lambda(z^3-1)}$, $e^{\lambda(z^4-1)}$, $q^{m-1}(1-pz^2)^{1-m}$, $(p+qz)(p+qz^2)$, and $(p+qz^4)^m$, and their probability models will be more complicated. Similarly, many other values can be given to $g(z)$.

Delta Lagrangian Distributions

The *delta* Lagrangian distributions can be obtained from the *basic* Lagrangian distributions by taking their n-fold convolutions. However, this method will involve too much labor in computations. The easiest method of obtaining a delta Lagrangian distribution is to put $f(z) = z^n$ in the general Lagrange expansion (1.78) with (1.79), under the transformation $z = ug(z)$. If $D^i(g(z))^{n+i}\big|_{z=0} \geq 0$ for $i = 0, 1, 2, 3, \ldots$ the pgf of the delta Lagrangian distributions becomes

$$z^n = (\psi(u))^n = \sum_{x=n}^{\infty} \frac{nu^x}{(x-n)!x} \left\{ D^{x-n} (g(z))^x \right\}_{z=0} \tag{2.3}$$

and the pmf of the delta Lagrangian distributions can be written as

$$P(X = x) = \frac{n}{(x-n)!x} D^{x-n}(g(z))^x \big|_{z=0} \tag{2.4}$$

for $x = n,\ n+1,\ n+2,\ \ldots$ and zero otherwise.

Table 2.1. Some important basic Lagrangian distributions

No.	Name	$g(z)$	$P(X=x), x=1,2,3,\ldots$
1.	Geometric distribution	$1-p+pz,\ 0<p<1$	$(1-p)p^{x-1}$
2.	Borel distribution [Borel (1942)]	$e^{\lambda(z-1)},\ 0<\lambda<1$	$\frac{(x\lambda)^{x-1}}{x!}e^{-x\lambda}$
3.	Haight distribution [Haight (1961)]	$q(1-pz)^{-1}$ $0<p=1-q<1$	$\frac{1}{2x-1}\binom{2x-1}{x}q^x p^{x-1}$
4.	Consul distribution [Consul & Shenton (1975)]	$(1-\theta+\theta z)^m$ $0<\theta<1,\ \ m\in N$	$\frac{1}{x}\binom{mx}{x-1}\theta^{x-1}$ $\times(1-\theta)^{mx-x+1}$
5.	Geeta distribution [Consul (1990a,b,c)]	$(1-\theta)^{m-1}(1-\theta z)^{1-m}$ $0<\theta<1, 1<m<\theta^{-1}$	$\frac{1}{mx-1}\binom{mx-1}{x}\theta^{x-1}$ $\times(1-\theta)^{mx-x}$
6.	Katz distribution [Consul (1993)]	$\left(\frac{1-\beta z}{1-\beta}\right)^{-b/\beta}$ $b>0,\ 0<\beta<1$	$\frac{1}{xb/\beta+x-1}\binom{xb/\beta+x-1}{x}$ $\times\beta^{x-1}(1-\beta)^{xb/\beta}$
7.	Otter distribution [Otter (1948)]	$p+qz^2$ $0<p=1-q<1$	$\frac{1}{x}\binom{x}{\frac{x-1}{2}}p^{\frac{x}{2}+\frac{1}{2}}q^{\frac{x}{2}-\frac{1}{2}}$, $x=1,3,5,\ldots$
8.	Felix distribution	$e^{\lambda(z^2-1)},\ 0<\lambda<1$	$\frac{(x\lambda)^{(x-1)/2}}{\left(\frac{x-1}{2}\right)!\,x}e^{-x\lambda}$, $x=1,3,5,\ldots$
9.	Teja distribution	$q(1-pz^2)^{-1}$ $0<p=1-q<1$	$\frac{1}{x}\binom{\frac{3(x-1)}{2}}{x-1}q^x p^{(x-1)/2}$, $x=1,3,5,\ldots$
10.	Sunil distribution	$(p+qz^2)^m,\ m\in N$ $0<p=1-q<1$	$\frac{1}{x}\binom{mx}{\frac{x-1}{2}}p^{mx}(q/p)^{(x-1)/2}$, $x=1,3,5,\ldots$
11.	Ved distribution	$(p+qz^3)^m,\ m\in N$ $0<p=1-q<1$	$\frac{1}{x}\binom{mx}{\frac{x-1}{3}}p^{mx}(q/p)^{(x-1)/3}$, $x=1,4,7,\ldots$

The choice of various specific functions for $g(z)$ will provide various members of such delta Lagrangian probability distributions. Examples of some important delta Lagrangian distributions are given in Table 2.2. (Some of these are from Consul and Shenton (1972).)

Note that the negative binomial distribution is a special case of the delta binomial distribution (for $m = 1$) and the Haight distribution is a special case of the delta negative binomial distribution (for $\beta = 2$). Also, the delta binomial distribution and the delta negative binomial distribution become degenerate when $m \to 0$ and $\beta \to 1$, respectively, and provide $P(X = n) = 1$.

Many other delta Lagrangian distributions can be obtained from (2.4) by choosing other values for the function $g(z)$ such that $g(0) \neq 0$ and $g(1) = 1$. Examples of some other functions for $g(z)$ are $(p+qz^2)^m$, $(p+qz)(p+qz^2)$, $(p+qz)^m(p+qz^2)^m$, $p_{-1}+p_0z+p_1z^2$, where $p_{-1}+p_0+p_1 = 1$ and $0 < p_i < 1$. Many more values for $g(z)$ can be defined in a similar manner.

General Lagrangian Distributions

Let $g(z)$ and $f(z)$ be two analytic functions of z which are successively differentiable with respect to z and are such that $g(0) \neq 0$, $g(1) = 1$, $f(1) = 1$, and

$$D^{x-1}\left\{(g(z))^x f'(z)\right\}\Big|_{z=0} \geq 0 \quad \text{for} \quad x \in N. \tag{2.5}$$

The pgf of the discrete *general* Lagrangian probability distribution, under the Lagrange transformation $z = ug(z)$, is given by (1.78) in the form

$$f(z) = f(\psi(u)) = \sum_{x=0}^{\infty}(u^x/x!)D^{x-1}\left\{(g(z))^x f'(z)\right\}\Big|_{z=0}, \tag{2.6}$$

where $\psi(u)$ is defined as in (2.1) and the pmf of the class of general Lagrangian probability distributions becomes

$$\begin{aligned} P(X=0) &= f(0), \\ P(X=x) &= (1/x!)\,D^{x-1}\left\{(g(z))^x f'(z)\right\}\Big|_{z=0}, \quad x \in N. \end{aligned} \tag{2.7}$$

This class of general Lagrangian probability distributions will be denoted by $L(f(z);\ g(z);\ x)$ or by $L(f;\ g;\ x)$ for the sake of convenience.

When the functions $g(z)$ and $f(z)$ are pgfs of some probability distributions they do satisfy the necessary conditions $f(1) = g(1) = 1$ for generating Lagrangian distributions. However, they need not necessarily be pgfs. There are many functions $f(z)$ which are not pgfs but they satisfy the property of $f(1) = 1$. For example, the two functions $[-\ln(1-\theta)^{-1}]\ \ln[1+\theta z/(1-\theta)]$ and $(1-\theta+\theta z)^m$ for $0 < \theta < 1$, $m > 1$ are not pgfs because their expansions provide series whose terms are alternately positive and negative after a few terms; however, their values at $z = 1$ are unity.

The two functions $g(z)$ and $f(z)$ are called the transformer function and the transformed function, respectively. Each set of values of $g(z)$ and $f(z)$, satisfying the condition (2.5) and the conditions $g(0) \neq 0$, $g(1) = f(1) = 1$ will provide a general Lagrangian distribution. Thus, numerous general Lagrangian distributions can be generated. Examples of some important members of general Lagrangian distributions are shown in Table 2.3.

Table 2.2. Some simple delta Lagrangian distributions

No.	Name	$g(z)$	$P(X=x), x=n, n+1, n+2, \ldots$
1.	Negative binomial distribution	$1-p+pz$ $0<p=1-q<1$	$\binom{x-1}{n-1} q^n p^{x-n}$
2.	Delta-binomial distribution	$(1-p+pz)^m$ $0<p<1,\ 1<m<p^{-1}$	$\frac{n}{x}\binom{mx}{x-n} p^{x-n}(1-p)^{n+mx-x}$
3.	Delta-Poisson or Borel-Tanner dist.	$e^{\lambda(z-1)}$ $0<\lambda<1$	$\frac{n}{(x-n)!x}(\lambda x)^{x-n} e^{-x\lambda}$
4.	Haight distribution	$(1-p)(1-pz)^{-1}$ $0<p<1$	$\frac{n}{2x-n}\binom{2x-n}{x} p^{x-n}(1-p)^x$
5.	Delta-Geeta distribution	$(1-\theta)^{\beta-1}(1-\theta z)^{1-\beta}$ $0<\theta<1, 1<\beta<\theta^{-1}$	$\frac{n}{x}\binom{\beta x-n-1}{x-n} \theta^{x-n}(1-\theta)^{\beta x-x}$
6.	Delta-Katz distribution	$(1-\beta)^{b/\beta}(1-\beta z)^{-b/\beta}$ $b>0,\ 0<\beta<1$	$\frac{bn\beta^{x-n}}{bx+\beta x-\beta n}(1-\beta)^{bx/\beta}\binom{bx/\beta+x-n}{x-n}$
7.	Random Walk distribution	$p+qz^2$ $0<p=1-q<1$	$\frac{n}{x}\binom{x}{(x-n)/2} p^n (pq)^{(x-n)/2}$ $x=n, n+2, n+4, \ldots$
8.	Delta-Teja distribution	$(1-p)(1-pz^2)^{-1}$	$\frac{2n}{3x-n}\binom{\frac{3}{2}x-\frac{n}{2}}{x} p^{\frac{x}{2}-\frac{n}{2}}(1-p)^x$ $x=n, n+2, n+4, \ldots$
9.	Delta-Felix distribution	$e^{\lambda(z^2-1)},\ 0<\lambda<1$	$\frac{n}{(\frac{x-n}{2})!x}(x\lambda)^{(x-n)/2} e^{-x\lambda}$, $x=n, n+2, n+4, \ldots$
10.	Delta-Sunil distribution	$(p+qz^2)^m,\ m\in N$ $0<p=1-q<1$	$\frac{n}{x}\binom{mx}{\frac{x}{2}-\frac{n}{2}} p^{mx}(q/p)^{\frac{x}{2}-\frac{n}{2}}$
11.	Delta-Ved distribution	$(p+qz^3)^m, m\in N$ $0<p=1-q<1$	$\frac{n}{x}\binom{mx}{\frac{x-n}{3}} p^{mx}(q/p)^{(x-n)/3}$, $x=n, n+3, n+6, \ldots$

Table 2.3. Some important general Lagrangian distributions

No.	Name of distribution	Transformed $f(z)$	Transformer $g(z)$	$L(f; g; x), x = 0, 1, 2, \ldots$
1.	Double binomial	$(q' + p'z)^n$ $0 < p' < 1$ $q' + p' = 1$ $n > 0$	$(q + pz)^m$ $0 < p < 1$ $q + p = 1$ $m > 0, mp < 1$	$(q')^n, x = 0$ $\frac{n}{mx+1}\binom{mx+1}{x}(q')^n(\frac{qp'}{pq'})(pq^{m-1})^x$ $\times[{}_2F_1(1-n, 1-x; mx-x+2; \frac{p'q}{pq'})]$, $x \geq 1$
2.	Generalized binomial	$(q + pz)^n$	$(q + pz)^m$	$\frac{n}{n+mx}\binom{n+mx}{x}p^x q^{n+mx-x}, x \geq 0$
3.	Binomial-Poisson	$(q + pz)^n$	$e^{\lambda(z-1)}$ $0 < \lambda < 1$	$q^n, x = 0$ $\frac{(x\lambda)^{x-1}}{x!e^{x\lambda}}(npq^{n-1})$ $\times[{}_2F_0(1-n; 1-x; ; \frac{p}{x\lambda q})], x \geq 1$
4.	Binomial-negative binomial	$(q + pz)^n$	$\frac{(q')^k}{(1-p'z)^k}$ $0 < k < q'/p'$, $0 < p' < 1$ $p' + q' = 1$	$q^n, x = 0$ $\frac{(kx+x-2)!}{x!(kx-1)!}npq^{n-1}(p')^{x-1}(q')^{kx}$ $\times[{}_2F_1(1-n; 1-x; 2-x-kx; \frac{p}{qp'})]$, $x \geq 1$
5.	Poisson-binomial	$e^{\theta(z-1)}$ $\theta > 0$	$(q + pz)^m$ $mp < 1$	$e^{-\theta}, x = 0$ $\frac{(\theta q^m)^x}{e^{\theta}x!}[{}_2F_0(1-x, -mx; ; \frac{-p}{q\theta})], x \geq 1$
6.	Generalized Poisson (GP)	$e^{\theta(z-1)}$ $\theta > 0$	$e^{\lambda(z-1)}$ $0 < \lambda < 1$	$\theta(\theta + x\lambda)^{x-1}e^{-\theta-x\lambda}/x!, x \geq 0$
7.	Restricted GP	$e^{\theta(z-1)}$	$e^{\alpha\theta(z-1)}$	$e^{-\theta}(\theta e^{-\alpha\theta})^x(1+\alpha x)^{x-1}/x!, \ x \geq 0$
8.	Poisson-negative binomial	$e^{\theta(z-1)}$ $\theta > 0$	$\frac{(q)^k}{(1-pz)^k}$ $kp < 1$	$e^{-\theta}, x = 0$ $e^{-\theta}[\theta^x q^{kx}/x!][{}_2F_0(1-x, kx; ; -p/\theta)]$, $x \geq 1$
9.	Negative binomial-binomial	$\frac{(q')^k}{(1-p'z)^k}$	$(q + pz)^m$ $mp < 1$	$(q')^k, x = 0$ $\frac{(p'q^m)^x}{(q')^{-k}}\binom{k+x-1}{x}$ $\times[{}_2F_1(1-x, -mx; 1-k-x; -\frac{p}{qp'})]$, $x \geq 1$
10.	Negative binomial-Poisson	$q^k(1 - pz)^{-k}$	$e^{\lambda(z-1)}$ $0 < \lambda < 1$	$q^k, x = 0$ $[kpq^k(x\lambda)^{x-1}e^{-x\lambda}/x!]$ $\times[{}_2F_0(1-x, k+1; ; -p/x\lambda)], x \geq 1$

Table 2.3. (continued)

No.	Name of distribution	Transformed $f(z)$	Transformer $g(z)$	$L(f; g; x), x = 0, 1, 2, \ldots$
11.	Generalized neg. binomial	$q^n(1-pz)^{-n}$	$q^m(1-pz)^{-m}$ $mp < 1$	$\frac{n}{n+mx+x}\binom{n+mx+x}{x}p^x q^{n+mx}$, $x \geq 0$
12.	Gen. logarithmic series	$\frac{\ln(1+pz/q)}{(-\ln q)}$	$(q+pz)^m$ $1 < m < p^{-1}$	$\frac{1}{mx}\binom{mx}{x}\frac{(pq^{m-1})^x}{(-\ln q)}, x \geq 1$
13.	Logarithmic-Poisson	$\frac{\ln(1-pz)}{\ln q}$	$e^{\lambda(z-1)}$ $0 < \lambda < 1$	$\frac{(x\lambda)^{x-1}}{x!}e^{-\lambda x}\frac{p}{(-\ln q)}$ $\times[{}_2F_0(1-x, 1; ; -\frac{p}{x\lambda})], x \geq 1$
14.	Logarithmic-neg. binomial	$\frac{\ln(1-pz)}{\ln q}$	$q^k(1-pz)^{-k}$ $kp < 1$	$\frac{1}{kx+x}\binom{kx+x}{x}\frac{(pq^k)^x}{(-\ln q)}, x \geq 1$
15.	Rectangular-binomial	$\frac{1-z^n}{n(1-z)}$	$(q+pz)^m$ $mp < 1$	$1/n, x = 0$ $\frac{1}{nx}\sum_{r=0}^{a}(r+1)\binom{mx}{x-r-1}$ $\times p^{x-1-r}q^{mx-x+r+1}, x \geq 1$ $a = \min(x-1, n-2)$
16.	Rectangular-Poisson	$\frac{1-z^n}{n(1-z)}$	$e^{\lambda(z-1)}$ $0 < \lambda < 1$	$n^{-1}, x = 0$ $\frac{e^{-x\lambda}}{nx}\sum_{r=0}^{a}(r+1)\frac{(x\lambda)^{x-r-1}}{(x-r-1)!}, x \geq 1$ $a = \min(x-1, n-2)$
17.	Rectangular-negative binomial	$\frac{1-z^n}{n(1-z)}$	$q^k(1-pz)^{-k}$	$n^{-1}, x = 0$ $\frac{q^{kx}}{nx}\sum_{r=0}^{a}\binom{kx+x-r-2}{x-1-r}$ $\times(r+1)p^{x-1-r}, x \geq 1$ $a = \min(x-1, n-2)$
18.	Generalized Katz	$(\frac{1-\beta z}{1-\beta})^{-\frac{a}{\beta}}$	$(\frac{1-\beta z}{1-\beta})^{-\frac{b}{\beta}}$	$\frac{a/\beta}{(a+bx)/\beta+x}\binom{(a+bx)/\beta+x}{x}$ $\times\beta^x(1-\beta)^{(a+bx)/\beta}, x \geq 0$
19.	Shenton Distribution	$(q+pz^2)^n$	$(q+pz^2)^m$	$\frac{n}{n+mx}\binom{n+mx}{x/2}p^{x/2}q^{n+mx-x/2}$, $x = 0, 2, 4, 6, \ldots$
20.	Modified Felix	$e^{\theta(z^2-1)}$	$e^{\lambda(z^2-1)}$	$\frac{\theta(\theta+x\lambda)^{x/2-1}}{(x/2)!}e^{-\theta-x\lambda}$ $x = 0, 2, 4, 6, \ldots$
21.	Modified Ved	$(q+pz^3)^n$	$(q+pz^3)^m$	$\frac{n}{n+mx}\binom{n+mx}{x/3}p^{x/3}q^{n+mx-x/3}$, $x = 0, 3, 6, 9, \ldots$

It is clear from the examples in Table 2.3 that the formula (2.6) can provide almost an infinite number of discrete Lagrangian probability distributions by various choices of the set of functions $f(z)$ and $g(z)$ satisfying the conditions $g(0) \neq 0,\ g(1) = f(1) = 1$, and (2.5).

Each one of the probability models in Table 2.3 has interrelations with other probability models and with the basic Lagrangian and delta Lagrangian probability models. Some of these relations are described later in this chapter.

It may be noted that if the parameter m is replaced by $(m-1)$ in the probability model (11) in Table 2.3, then the model changes to the probability model (2). Thus, these two probability models are the same and have been studied by the name of *generalized negative binomial distribution.* Similarly, the probability models (12) and (14) are also the same because the parameter k in (14) equals $(m-1)$ in (12).

The probability models with the names (i) generalized Poisson distribution, (ii) generalized negative binomial distribution, (iii) generalized logarithmic series distribution, and (iv) generalized Katz distribution will be studied in more depth in later chapters.

If $f_1(z)$ is another analytical function of z in closed interval $[-1, 1]$ such that $f_1(0) \geq 0$ and $f_1(1) = 1$ and if

$$0 < g'(1) < 1 \text{ and } \left\{D^r (g(z))^r f_1(z)\right\}_{z=0} \geq 0 \text{ for } r = 0, 1, 2, \ldots \tag{2.8}$$

the Lagrangian expansion (1.80) with (1.81) under the transformation $z = ug(z)$ defines a random variable Y having another class of general Lagrangian probability distributions, given by

$$P(Y = y) = \begin{cases} (1 - g'(1)) f_1(0), & y = 0, \\ (1 - g'(1))(y!)^{-1} \{D^y (g(z))^y f_1(z)\} |_{z=0}, & y = 1, 2, 3, \ldots. \end{cases} \tag{2.9}$$

The pgf of this class of general Lagrangian distributions becomes

$$H(u) = \frac{(1 - g'(1)) f_1(z)}{1 - zg'(z)/g(z)} = f_2(z), \quad \text{where} \quad z = ug(z), \tag{2.10}$$

whose power series expansion in u is given by (1.80) multiplied by the factor $(1 - g'(1))$. We denote this class of Lagrangian distributions by $L_1(f_1; g; y)$.

When $f_1(z) = z$, the discrete probability distribution, given by (2.9), is called the *basic distribution* and its pmf becomes

$$P(Y = y) = \frac{1 - g'(1)}{(y-1)!} D^{y-1} (g(z))^y |_{z=0}, \ y = 1, 2, 3, \ldots. \tag{2.11}$$

Also, when $f_1(z) = z^n$, the discrete probability distribution, given by (2.9), is called the *delta distribution* and its pmf becomes

$$P(Y = y) = \frac{1 - g'(1)}{(y-n)!} \left\{D^{y-n} (g(z))^y\right\}\Big|_{z=0}, \ y = n, n+1, \ldots. \tag{2.12}$$

Numerous members of this class of Lagrangian distributions are generated by taking various choices of the functions $g(z)$ and $f_1(z)$ in (2.9). Table 2.4 contains twenty-four discrete probability distributions under three subheadings of (i) basic distributions, (ii) delta distributions, and (iii) general distributions, together with the particular values of the functions $g(z)$ and $f_1(z)$.

Jain (1975a) had studied the linear function binomial and the linear function Poisson distributions and had obtained their moments. Lingappaiah (1986) discussed the relationship of linear function Poisson distribution and the generalized Poisson distribution and showed that they need not be studied separately because one is a weighted distribution of the other. Charalambides (1987) obtained the factorial moments and some other properties of the linear function Poisson, binomial, and negative binomial distributions, given in Table 2.4.

2.2.1 Equivalence of the Two Classes of Lagrangian Distributions

Janardan and Rao (1983) and Janardan (1997) had erroneously assumed that the Lagrange expansion (1.80), under the transformation $z = ug(z)$, was independent of the Lagrange expansion (1.78) and had obtained the class $L_1(f_1; g; y)$ of Lagrangian distributions and called it a new class of discrete Lagrangian probability distributions. Consul and Famoye (2001) had further extended that work. However, Consul and Famoye (2005) have proved the following equivalence theorem.

Theorem 2.1. *Let $g(z)$, $f(z)$, and $f_1(z)$ be three analytical functions, which are successively differentiable in the domain $|z| \leq 1$ and such that $g(0) \neq 0$ and $g(1) = f(1) = f_1(1) = 1$. Then, under the transformation $z = ug(z)$, every member of Lagrangian distribution in (2.9) is a member of the Lagrangian distribution in (2.7); and conversely, every member of Lagrangian distribution given by (2.7) is a member of the Lagrangian probability distribution in (2.9) by choosing*

$$f_1(z) = \left(1 - g'(1)\right)^{-1} \left(1 - zg'(z)/g(z)\right) f(z). \tag{2.13}$$

Proof. We need to show that the value of $f_1(z)$ in (2.13) transforms the expression in (2.9) into (2.7). On substituting this value of $f_1(z)$ in (2.9), we have

$$\begin{aligned}
P(Y = y) &= (y!)^{-1} D^y \left\{(g(z))^y \left(1 - zg'(z)/g(z)\right) f(z)\right\}\Big|_{z=0} \\
&= (y!)^{-1} \left[D^y \left\{(g(z))^y f(z)\right\} - D^y \left\{z\,(g(z))^{y-1} g'(z) f(z)\right\}\right]_{z=0} \\
&= (y!)^{-1} \left[D^{y-1} \left\{y\,(g(z))^{y-1} g'(z) f(z) + (g(z))^y f'(z)\right\}\right]_{z=0} \\
&\quad - (y!)^{-1} y D^{y-1} \left\{(g(z))^{y-1} g'(z) f(z)\right\}_{z=0} \\
&= (y!)^{-1} D^{y-1} \left\{(g(z))^y f'(z)\right\}_{z=0},
\end{aligned}$$

which is the same as (2.7). Thus, every member of the class of Lagrangian probability distributions in (2.9) becomes a member of the class of Lagrangian distributions in (2.7).

Converse. The probability mass function of the class of Lagrangian distributions in (2.7) is

$$\begin{aligned}
P(X = x) &= (x!)^{-1} D^{x-1} \left\{(g(z))^x f'(z)\right\}\Big|_{z=0} \\
&= (x!)^{-1} D^{x-1} \left\{(g(z))^x f'(z) + x\,(g(z))^{x-1} g'(z) f(z)\right\}\Big|_{z=0} \\
&\quad - (x!)^{-1} \binom{x}{1} D^{x-1} \left\{(g(z))^{x-1} g'(z) f(z)\right\}\Big|_{z=0}
\end{aligned}$$

Table 2.4. Some Lagrangian probability distributions in $L_1(f_1; g; y)$

$[0 < p = 1 - q < 1,\ 1 < m < p^{-1},\ \theta > 0,\ 0 < \lambda < 1,\ n > 0,\ 0 < p_1 = 1 - q_1 < 1]$

No	Name	$f_1(z)$	$g(z)$	$L_1(f_1;\ g;\ y)$
	Basic Distributions			*Range of* $y = 1, 2, 3, \ldots$
1.	Weighted geometric	z	$p + qz$	yp^2q^{y-1}
2.	Weighted Consul	z	$(q + pz)^m$	$\binom{my}{y-1}(1 - mp)p^{y-1}q^{my-y+1}$
3.	Ved	z	$\frac{q^m}{(1-pz)^m}$	$\binom{my+y-2}{y-1}(1 - mp/q)$ $\times p^{y-1}q^{my}$
4.	Sudha	z	$e^{\lambda(z-1)}$	$(1 - \lambda)e^{-y\lambda}(y\lambda)^{y-1}/(y - 1)!$
5.	Hari	z	$p + qz^2$ $0 < q < \frac{1}{2}$	$\frac{y!p(1-2q)}{\left(\frac{1}{2}y-\frac{1}{2}\right)!\left(\frac{1}{2}y+\frac{1}{2}\right)!}(pq)^{\frac{1}{2}y-\frac{1}{2}}$ $y = 1, 3, 5, 7, \ldots$
	Delta Distributions			*Range of* $y = n, n + 1, n + 2, \ldots$
6.	Weighted delta binomial	z^n	$(q + pz)^m$	$\binom{my}{y-n}(1 - mp)p^{y-n}q^{my-y+n}$
7.	Weighted delta Poisson	z^n	$e^{\lambda(z-1)}$	$(1 - \lambda)e^{-y\lambda}(y\lambda)^{y-n}/(y - n)!$
8.	Weighted delta negative binomial	z^n	$\frac{q^m}{(1-pz)^m}$	$(1 - mp/q)\binom{my+y-n-1}{y-n}$ $\times p^{y-n}q^{my}$
	General Distributions			
9.	Linear negative binomial	$\frac{q^n}{(1-pz)^n}$	$\frac{q^m}{(1-pz)^m}$	$\binom{n+my+y-1}{y}(q - mp)$ $\times p^y q^{n+my-1}$
10.	Linear function Poisson [See Jain (1975a)]	$e^{\theta(z-1)}$	$e^{\lambda(z-1)}$	$(1 - \lambda)(\theta + y\lambda)^y e^{-\theta-y\lambda}/y!$
11.	Linear function binomial [See Jain (1975a)]	$(q + pz)^n$	$(q + pz)^m$	$(1 - mp)\binom{n+my}{y}p^y q^{n+my-y}$
12.	Binomial-Poisson	$(q + pz)^n$	$e^{\lambda(z-1)}$	$\frac{(y\lambda)^y}{y!}(1 - \lambda)e^{-y\lambda}q^n$ $\times {}_2F_0(-n, -y;\ p/y\lambda q)$
13.	Binomial-negative binomial	$(q_1 + p_1z)^n$	$\frac{q^m}{(1-pz)^m}$	$(q - mp)p^y q^{my-1}q_1^n\binom{my+y-1}{y}$ $\times {}_2F_1(-y, -n;\ 1 - y - my;\ -p_1/pq_1)$

Table 2.4. (continued)

No	Name	$f_1(z)$	$g(z)$	$L_1(f_1;\ g;\ y)$
14.	Binomial-binomial	$(q_1+p_1z)^n$	$(q+pz)^m$	$q_1^n p^y q^{my-y}(1-mp)\binom{my}{y}$ $\times {}_2F_1(-y,-n;my-y+1;\frac{p_1q}{q_1p})$
15.	Poisson-binomial	$e^{\theta(z-1)}$	$(q+pz)^m$	$(1-mp)e^{-\theta}(\theta q^m)^y(y!)^{-1}$ $\times {}_2F_0(-y,-my;;\frac{p}{\theta q})$
16.	Poisson-negative binomial	$e^{\theta(z-1)}$	$\frac{q^m}{(1-pz)^m}$	$(q-mp)e^{-\theta}\frac{\theta^y q^{my-1}}{y!}$ $\times {}_2F_0(-y,my;;\frac{-p}{\theta})$
17.	Negative binomial-Poisson	$\frac{q^n}{(1-pz)^n}$	$e^{\lambda(z-1)}$	$(1-\lambda)q^n e^{-y\lambda}\frac{(y\lambda)^y}{y!}$ $\times {}_2F_0(-y,n;;\frac{-p}{y\lambda})$
18.	Double negative binomial	$\frac{q_1^n}{(1-p_1z)^n}$	$\frac{q^m}{(1-pz)^m}$	$(q-mp)q_1^n p_1^y q^{my-1}\frac{(n+y-2)!}{y!\Gamma(n)}$ $\times {}_2F_1(-y,my;\ 1-y-n;\ p/p_1)$
19.	Negative binomial-binomial	$\frac{q_1^n}{(1-p_1z)^n}$	$(q+pz)^m$	$(1-mp)q_1^n q^{my}(p/q)^y\binom{my}{y}$ $\times {}_2F_1(n,-y;\ my-y+1;\ \frac{p_1q}{-p})$
20.	Logarithmic-binomial	$\frac{\ln(1+pz/q)}{(-\ln q)}$	$(q+pz)^m$	$p^y q^{my-y}\frac{(1-mp)}{(-\ln q)}\sum_{k=1}^{y}\binom{my}{y-k}$ $\times\frac{(-1)^{k-1}}{k},\ y=1,2,3,\ldots$
21.	Logarithmic-Poisson	$\frac{\ln(1-pz)}{(\ln q)}$	$e^{\lambda(z-1)}$	$e^{-y\lambda}(y\lambda)^y\frac{1-\lambda}{(-\ln q)}\sum_{k=1}^{y}\frac{(p/y\lambda)^k}{(y-k)!k}$
22.	Logarithmic-negative binomial	$\frac{\ln(1-pz)}{(\ln q)}$	$\frac{q^m}{(1-pz)^m}$	$\frac{(q-mp)p^y q^{my-1}}{(-\ln q)}$ $\times\sum_{k=1}^{y}\binom{my+y-k-1}{y-k}\frac{1}{k}$
23.	Rectangular-Poisson	$\frac{1-z^n}{n(1-z)}$	$e^{\lambda(z-1)}$	$\frac{(1-\lambda)e^{-y\lambda}}{n}\sum_{i=0}^{a}\frac{(y\lambda)^i}{i!}$ $a=\min(y,n-1)$
24.	Rectangular-binomial	$\frac{1-z^n}{n(1-z)}$	$(q+pz)^m$	$\frac{1-mp}{n}q^{my}\sum_{i=0}^{a}\binom{my}{i}(p/q)^i,$ $a=\min(y,n-1)$

$$= (x!)^{-1}D^{x-1}\left\{D\left[(g(z))^x f(z)\right]\right\}\Big|_{z=0} - (x!)^{-1}D^x\left\{z\,(g(z))^{x-1}g'(z)f(z)\right\}\Big|_{z=0}$$

$$= (x!)^{-1}D^x\left[(g(z))^x\left\{1 - zg'(z)/g(z)\right\}f(z)\right]\Big|_{z=0}$$

$$= \left(1 - g'(1)\right)(x!)^{-1}D^x\left[(g(z))^x f_1(z)\right]\Big|_{z=0},$$

which is the class of Lagrangian probability distributions in (2.9). Thus, $L_1(f_1; g; y) = L(f; g; x)$ is given by (2.13) and $L(f; g; x) = L_1(f_1; g; y)$ when $f(z) = f_2(z) = \left(1 - g'(1)\right) f_1(z)/\left(1 - zg'(z)/g(z)\right)$ given by (2.10). □

However, it must also be noted that the replacement of $f_1(z)$ by $f(z)$ in $L_1(f_1; g; y)$ gives other families of that class and it is evident from the members given in Tables 2.1 to 2.4.

The following example will illustrate Theorem 2.1 more explicitly.

Example. The generalized Poisson distribution (GPD) belongs to the class of Lagrangian distributions in (2.7) and is listed as (6) in Table 2.3. Its probability mass function is

$$P(X = x) = e^{-\theta-\lambda x}\theta(\theta + x\lambda)^{x-1}/x!, \quad x = 0, 1, 2, \ldots.$$

For the class of Lagrangian distributions in (2.9), let $g(z) = e^{\lambda(z-1)}$ and $f_1(z) = (1-\lambda)^{-1}(1-\lambda z)e^{\theta(z-1)}$, so that $g(1) = f_1(1) = 1,\ g(0) = e^{-\lambda},\ f_1(0) = e^{-\theta}(1-\lambda)^{-1}$, and $g'(1) = \lambda$.

Now, the class of Lagrangian probability distribution in (2.9) is given as

$$P(Y = y) = (1-\lambda)(y!)^{-1}D^y\left[e^{y\lambda(z-1)}(1-\lambda)^{-1}(1-\lambda z)e^{\theta(z-1)}\right]_{z=0}$$

$$= e^{-\theta-y\lambda}(y!)^{-1}D^y\left[e^{(\theta+\lambda y)z}(1-\lambda z)\right]_{z=0}$$

$$= e^{-\theta-y\lambda}(y!)^{-1}\left[(\theta+y\lambda)^y - \lambda y(\theta+y\lambda)^{y-1}\right]$$

$$= e^{-\theta-y\lambda}\theta(\theta+y\lambda)^{y-1}/y!,$$

which is the same as the GPD, which belongs to the class of Lagrangian probability distributions in (2.7) and it is listed as (6) in Table 2.3.

Important Note: With the proof of Theorem 2.1, one may get the feeling that the results on Lagrangian probability distributions in (2.9) are redundant. But this is not true because the two classes in (2.7) and (2.9) enable us to have nice forms of two sets of probability distributions as given in Tables 2.1, 2.2, 2.3, and 2.4, and they are all different from each other. One inference we can draw from this theorem is, "Any property which is proved for the members of one class will also hold true for all the members of the other class."

2.2.2 Moments of Lagrangian Distributions

The Lagrange transformation $z = ug(z)$, when expanded in powers of u, provides the pgf $z = \psi(u)$ of the basic Lagrangian distribution. Accordingly, the descending factorial moments for the basic Lagrangian models become

$$\mu'_{(r)}\,(\text{basic}) = \left.\frac{d^r z}{du^r}\right|_{z=1}. \tag{2.14}$$

One can easily write down the successive derivatives of $z = ug(z)$ with respect to u as follows:

$$\frac{dz}{du} = g(z)[1 - u\,g'(z)]^{-1}, \tag{2.15}$$

$$\frac{d^2 z}{du^2} = \frac{2g(z)\,g'(z)}{(1 - u\,g'(z))^2} + \frac{z\,g(z)\,g''(z)}{(1 - u\,g'(z))^3}, \tag{2.16}$$

$$\frac{d^3 z}{du^3} = \frac{g(z)[6(g'(z))^2 + 2g(z)g''(z)]}{(1 - u\;g'(z))^3}$$
$$+ \frac{\{g(z)\,g''(z) + 8\,z\,g'(z)\,g''(z) + zg(z)g'''(z)\}\,g(z)}{(1 - u\;g'(z))^4} + \frac{3z^2\,g(z)(g''(z))^2}{(1 - u\,g'(z))^5}. \tag{2.17}$$

Thus, the three factorial moments of the basic Lagrangian distributions, on putting $z = u = 1$, become

$$\mu'_{(1)}\,(\text{basic}) = (1 - g')^{-1}, \quad \mu'_{(2)} = 2g'(1 - g')^{-2} + g''(1 - g')^{-3}, \tag{2.18}$$

and

$$\mu'_{(3)} = \frac{2g'' + 6(g')^2}{(1 - g')^3} + \frac{g'' + 8\,g'g'' + g'''}{(1 - g')^4} + \frac{3(g'')^2}{(1 - g')^5}. \tag{2.19}$$

The higher factorial moments can be similarly calculated for the basic Lagrangian distribution. Thus, the mean and the variance of the basic Lagrangian distributions are

$$\mu\,(\text{basic}) = \frac{1}{1 - g'}, \quad \sigma^2\,(\text{basic}) = \frac{g'}{(1 - g')^2} + \frac{g''}{(1 - g')^3}. \tag{2.20}$$

The pgf of the general Lagrangian distributions is a power series in u given by

$$f(z) = f(\psi(u)), \tag{2.21}$$

where $z = \psi(u)$ is defined in (2.1).

The descending factorial moments of the general Lagrangian distributions can be obtained by successively differentiating the pgf in (2.21) with respect to u, using (2.15) after each differentiation and by putting $z = u = 1$ in the result. The first two derivatives are

$$\frac{\partial\,f(\psi(u))}{\partial u} = f'(z)\,g(z)(1 - u\,g'(z))^{-1},$$

$$\frac{\partial^2\,f(\psi(u))}{\partial u^2} = \frac{[f''(z)g(z) + g'(z)f'(z)]g(z)}{(1 - ug'(z))^2} + \frac{f'(z)g(z)}{(1 - ug'(z))^2}\left[g'(z) + ug''(z)\frac{\partial z}{du}\right], \tag{2.22}$$

which provide the first two factorial moments $\mu'_{(1)}$ and $\mu'_{(2)}$ as

$$\mu'_{(1)} = \mu' = f'(1 - g')^{-1}, \tag{2.23}$$

$$\mu'_{(2)} = \frac{f'' + 2g'f'}{(1 - g')^2} + \frac{f'g''}{(1 - g')^3},$$

where g', f', g'', f'' denote the values of the differential coefficients of $g(z)$ and $f(z)$, respectively, at $z = 1$. The values of $\mu'_{(2)}$ and $\mu'_{(1)}$ provide the variance σ^2 as

$$\sigma^2 = \mu'_{(2)} + \mu'_{(1)} - \left(\mu'_{(1)}\right)^2$$

$$= \frac{f'g''}{(1-g')^3} + \frac{f'' + g'f' + f' - (f')^2}{(1-g')^2}. \tag{2.24}$$

The pgf of the class of Lagrangian distributions in (2.9) under the transformation $z = ug(z)$ is given by (2.10). The factorial moments of this class can be obtained by successively differentiating $H(u)$ in (2.10) with respect to u, by using (2.15) after each operation, and by putting $u = 1 = z$. The first two derivatives of $H(u)$ with respect to u are

$$(1-g'(1))^{-1}\frac{\partial H(u)}{\partial u} = \left\{f_1'(z)g(z) + f_1(z)g'(z)\right\}\left\{1 - ug'(z)\right\}^{-2}$$
$$+ zf_1(z)g''(z)\left\{1 - ug'(z)\right\}^{-3},$$

$$(1-g'(1)^{-1}\frac{\partial^2 H(u)}{\partial u^2} = \left\{f_1''(z)g(z) + 2f_1'(z)g'(z) + f_1(z)g''(z)\right\}(1-ug'(z))^{-2}\frac{\partial z}{\partial u}$$
$$+ 2\left\{f_1'(z)g(z) + f_1(z)g'(z)\right\}(1-ug'(z))^{-3}\left\{g'(z) + ug''(z)\frac{\partial z}{\partial u}\right\}$$
$$+ \left\{f_1(z)g''(z) + zf_1'(z)g''(z) + zf_1(z)g'''(z)\right\}(1-ug'(z))^{-3}\frac{\partial z}{\partial u}$$
$$+ 3zf_1(z)g''(z)(1-ug'(z))^{-4}\left\{g'(z) + ug''(z)\frac{\partial z}{\partial u}\right\}.$$

By putting $z = u = 1$ in the above expressions, substituting the value of $\partial z/\partial u$ at $z = u = 1$ from (2.15), and simplifying the expressions, we get

$$E[Y] = \mu = \frac{f_1'}{1-g'} + \frac{g'' + g' - (g')^2}{(1-g')^2} \tag{2.25}$$

and

$$E[Y(Y-1)] = \frac{f_1'' + g'' + 4f_1'g' + 2(g')^2}{(1-g')^2} + \frac{g''' + g'' + 3f_1'g'' + 5g'g''}{(1-g')^3} + \frac{3(g'')^2}{(1-g')^4},$$

where f_1', f_1'', g', g'', g''' denote the values of the successive derivatives of $f_1(z)$ and $g(z)$, respectively, at $z = 1$.

Thus, the variance σ^2 for the general Lagrangian distributions $L_1(f_1; g; y)$ becomes

$$\sigma^2 = E[Y(Y-1)] + E[Y] - (E(Y)]^2$$
$$= \frac{f_1'' + f_1' - (f_1')^2}{(1-g')^2} + \frac{\left(1+f_1'\right)\left(g'' + g' - (g')^2\right)}{(1-g')^3}$$
$$+ \frac{g''' + g''g' + 2g''}{(1-g')^3} + \frac{2\left(g''\right)^2}{(1-g')^4}. \tag{2.26}$$

2.2.3 Applications of the Results on Mean and Variance

Tables 2.1, 2.2, and 2.3 contain more than forty probability distributions, some of which are particular cases of the others. To determine the mean and variance of each one of these distributions would have taken a long time, especially because the probability mass functions for most of the models in Table 2.3 are in terms of hypergeometric functions. The formulas (2.23) and (2.24) are very powerful tools to calculate the mean and the variance of each distribution by simple differentiation of the two pgfs $g(z)$ and $f(z)$ which generate it and whose values are given in the three respective tables. The values of the means and variances of 28 models in Tables 2.1, 2.2, and 2.3 have been computed by using the formulas (2.23) and (2.24). These are given in Table 2.5 for convenience in future work on these probability models. The higher moments can also be calculated, if necessary, by using (2.19) and by calculating $\frac{d^4z}{du^4}$.

Table 2.4 contains a total of 24 probability models, some of which are particular cases of the others. The calculation of the mean and the variance of each model would take a long time and the results may not be free from errors. Since the values of the functions $f_1(z)$ and $g(z)$ for each probability model are given in Table 2.4 and these can be differentiated and evaluated at $z = 1$, one can use the formulas (2.25) and (2.26) as tools to get the mean and variance for each model. Their values have been evaluated for most of the models in Table 2.4 and are given in Table 2.6.

Most of the values of the means and variances given in Tables 2.5 and 2.6 need some simple restrictions on parameter values so that the denominators may not become zero or negative. These have not been given in order to shorten Tables 2.5 and 2.6.

2.2.4 Convolution Property for Lagrangian Distributions

A particular case of this property was proved by Consul and Shenton (1972) but the general result was proved by Consul and Shenton (1975) and by Good (1975).

Theorem 2.2. *Let X_1 and X_2 be two independent Lagrangian random variables with probability distributions $L(f_1;\ g;\ x_1)$ and $L(f_2;\ g;\ x_2)$, respectively. Then the probability distribution of the sum $Y = X_1 + X_2$ is given by the pmf $L(f_1\ f_2;\ g;\ y)$.*

Proof. Take the products of the pgfs of X_1 and X_2 and simplify.

The result can easily be generalized for any number of independent r.v.s. Though the r.v. $Y = X_1 + X_2$ has a Lagrangian distribution; i.e., it belongs to the class of Lagrangian distributions but it may not be exactly of the same type as the Lagrangian distribution given by $f_1(z)$ or by $f_2(z)$ even if $f_2(z) = f_1(z)$. Thus, the probability distribution may not really be closed under convolution.

Corollary 2.3. *The above theorem provides us the following interesting differentiation formula which appears to be new:*

$$\sum_{r=0}^{n} \binom{n}{r} \partial^{r-1} \left\{(g(z))^r \, \partial f_1(z)\right\} \partial^{n-r-1} \left\{(g(z))^{n-r} \, \partial f_2(z)\right\}\Big|_{z=0}$$

$$= \partial^{n-1} \left[(g(z))^n \, \partial \left\{f_1(z) f_2(z)\right\}\right]\Big|_{z=0}. \tag{2.27}$$

Table 2.5. The means and variances of some Lagrangian distributions $L(f; g; x)$

No.	Distribution	Mean μ	Variance σ^2
1.	Delta-Katz	$\frac{n(1-\beta)}{(1-\beta-b)}$	$nb(1-\beta)(1-\beta-b)^{-3}$
2.	Delta-binomial	$n(1-mp)^{-1}$	$nmpq(1-mp)^{-3}$
3.	Delta-Poisson	$n(1-\lambda)^{-1}$	$n\lambda(1-\lambda)^{-3}$
4.	Delta-Geeta	$\frac{n(1-\theta)}{(1-\beta\theta)}$	$n(\beta-1)\theta(1-\theta)(1-\beta\theta)^{-3}$
5.	Random walk	$n(1-2q)^{-1}$ $0<q<\frac{1}{2}$	$4npq(1-2q)^{-3}$
6.	Delta-Teja	$nq(1-3p)^{-1}$ $0<p<\frac{1}{3}$	$4npq(1-3p)^{-3}$
7.	Delta-Felix	$n(1-2\lambda)^{-1}$ $0<\lambda<1/2$	$4n\lambda(1-2\lambda)^{-3}$
8.	Delta-Ved	$n(1-3mq)^{-1}$ $0<q<(3m)^{-1}$	$9mnpq(1-3mq)^{-3}$
9.	Delta-Sunil	$n(1-2mq)^{-1}$ $0<q<(2m)^{-1}$	$4nmpq(1-2mq)^{-3}$
10.	Double binomial	$np'(1-mp)^{-1}$	$\frac{np'[q'+mp(q-q')]}{(1-mp)^3}$
11.	Generalized binomial	$np(1-mp)^{-1}$	$npq(1-mp)^{-3}$
12.	Generalized Poisson	$\theta(1-\lambda)^{-1}$	$\theta(1-\lambda)^{-3}$
13.	Binomial-Poisson	$np(1-\lambda)^{-1}$	$np(q+\lambda p)(1-\lambda)^{-3}$
14.	Binomial-negative binomial	$npq'(q'-kp')^{-1}$	$\frac{npq'(qq'^2+kp'-kp'q'q)}{(q'-kp')^3}$
15.	Poisson-binomial	$\theta(1-mp)^{-1}$	$\theta(1-mp^2)(1-mp)^{-3}$
16.	Poisson-negative binomial	$\theta q(q-kp)^{-1}$	$\theta q(kp^2+q^2)(q-kp)^{-3}$
17.	Negative binomial-binomial	$kp'(q'-mpq')^{-1}$	$\frac{kp'(mpqq'+1-mp)}{(1-mp)^3q'^2}$
18.	Negative binomial-Poisson	$kp[q(1-\lambda)]^{-1}$	$\frac{kp[1-\lambda p]}{q^2(1-\lambda)^3}$
19.	Generalized logarithmic series	$\frac{p}{(-\ln q)(1-mp)}$	$\frac{pq(-\ln q)-p^2(1-mp)}{(-\ln q)^2(1-mp)^3}$
20.	Logarithmic Poisson	$\frac{p}{q(-\ln q)(1-\lambda)}$	$\frac{(p-\lambda p^2)(-\ln q)-p^2(1-\lambda)}{q^2(1-\lambda)^3(-\ln q)^2}$
21.	Logarithmic-negative binomial	$\frac{p}{(-\ln q)(q-kp)}$	$\frac{pq(-\ln q)-p^2(q-kp)}{(-\ln q)^2(q-kp)^3}$
22.	Rectangular-binomial	$\frac{(n-1)}{2(1-mp)}$	$\frac{1}{2}\frac{(n-1)mpq}{(1-mp)^3}+\frac{(n^2-1)}{12(1-mp)^2}$
23.	Rectangular-Poisson	$\frac{(n-1)}{2(1-\lambda)}$	$\frac{n^2-1}{12(1-\lambda)^2}+\frac{(n-1)\lambda(2-\lambda)}{2(1-\lambda)^3}$
24.	Rectangular-Neg. binomial	$\frac{(n-1)q}{2(q-kp)}$	$\frac{(n-1)kpq}{2(q-kp)^3}+\frac{(n^2-1)q^2}{12(q-kp)^2}$
25.	Generalized Katz	$a(1-\beta-b)^{-1}$	$a(1-\beta)(1-\beta-b)^{-3}$
26.	Shenton	$2np(1-2mp)^{-1}$	$\frac{4npq}{(1-2mp)^3}$ $0<p<(2m)^{-1}$
27.	Modified Felix	$2\theta(1-2\lambda)^{-1}$	$4\theta(1-2\lambda)^{-3}$
28.	Modified Ved	$3np(1-3mp)^{-1}$	$9npq(1-3mp)^{-3}$

Table 2.6. The means and variances of some Lagrangian distributions $L_1(f_1; g; y)$

No.	Distribution	Mean μ	Variance σ^2
1.	Hari	$\frac{1+2q-4q^2}{(1-2q)^2}$	$\frac{12q}{(1-2q)^3}+\frac{4q^2(1+2q)}{(1-2q)^4}$
2.	Weighted delta-binomial	$\frac{n}{1-mp}+\frac{mpq}{(1-mp)^2}$	$\frac{(n+1)mpq}{(1-mp)^3}+\frac{2m(m-1)p^2q}{(1-mp)^4}$
3.	Weighted delta Poisson	$\frac{n}{1-\lambda}+\frac{\lambda}{(1-\lambda)^2}$	$\frac{(n+1)\lambda}{(1-\lambda)^3}+\frac{2\lambda^2}{(1-\lambda)^4}$
4.	Weighted delta-negative binomial	$\frac{nq}{q-mp}+\frac{mp}{(q-mp)^2}$	$\frac{(n+1)mpq}{(q-mp)^3}+\frac{2m(m+1)p^2q}{(q-mp)^4}$
5.	Linear binomial	$\frac{np}{1-mp}+\frac{mpq}{(1-mp)^2}$	$\frac{(n+mp)pq}{(1-mp)^3}+\frac{2m(m-1)p^2q}{(1-mp)^4}$
6.	Binomial Poisson	$\frac{np}{1-\lambda}+\frac{\lambda}{(1-\lambda)^2}$	$\frac{npq+n\lambda p^2+\lambda}{(1-\lambda)^3}+\frac{2\lambda^2}{(1-\lambda)^4}$
7.	Binomial-binomial	$\frac{np_1}{1-mp}+\frac{mpq}{(1-mp)^2}$	$\frac{np_1q_1+mpq+mnpp_1(q-q_1)}{(1-mp)^3}+\frac{2m(m-1)p^2q}{(1-mp)^4}$
8.	Linear Poisson	$\frac{\theta}{1-\lambda}+\frac{\lambda}{(1-\lambda)^2}$	$\frac{\theta+\lambda}{(1-\lambda)^3}+\frac{2\lambda^2}{(1-\lambda)^4}$
9.	Poisson-binomial	$\frac{\theta}{1-mp}+\frac{mpq}{(1-mp)^2}$	$\frac{\theta+mpq-\theta mp^2}{(1-mp)^3}+\frac{2m(m-1)p^2q}{(1-mp)^4}$
10.	Poisson-negative binomial	$\frac{\theta q}{q-mp}+\frac{mp}{(q-mp)^2}$	$\frac{\theta q^3+mpq(1+\theta p)}{(q-mp)^3}+\frac{2m(m+1)p^2q}{(q-mp)^4}$
11.	Negative binomial-Poisson	$\frac{np}{q(1-\lambda)}+\frac{\lambda}{(1-\lambda)^2}$	$\frac{\lambda q^2+np-\lambda np^2+2(1+\lambda)\lambda^2q^2}{q^2(1-\lambda)^3}+\frac{2\lambda^4}{(1-\lambda)^4}$
12.	Double negative binomial	$\frac{np_1q}{q_1(q-mp)}+\frac{mp}{(q-mp)^2}$	$\frac{np_1q^3+mpqq_1^2+mnpqp_1(q_1-q)}{q_1^2(q-mp)^3}+\frac{2m(m+1)p^2q}{(q-mp)^4}$
13.	Negative binomial-binomial	$\frac{np_1}{q_1(1-mp)}+\frac{mpq}{(1-mp)^2}$	$\frac{np_1+mpqq_1^2-mnpp_1(1-qq_1)}{(1-mp)^3q_1^2}+\frac{2m(m-1)p^2q}{(1-mp)^4}$
14.	Logarithmic-binomial	$\frac{p/(-\ln q)}{1-mp}+\frac{mpq}{(1-mp)^2}$	$\frac{-p^2-pq\ln q}{(1-mp)^2(\ln q)^2}+\frac{mpq(1-p/\ln q)}{(1-mp)^3}+\frac{2m(m-1)p^2q}{(1-mp)^4}$
15.	Logarithmic-Poisson	$\frac{-p}{(1-\lambda)q\ln q}+\frac{\lambda}{(1-\lambda)^2}$	$\frac{p(-\ln q-p)}{(1-\lambda)^2(q\ln q)^2}+\frac{[1-p/(q\ln q)]\lambda}{(1-\lambda)^3}+\frac{2\lambda^2}{(1-\lambda)^4}$
16.	Logarithmic-negative binomial	$\frac{-p}{(q-mp)\ln q}+\frac{mp}{(q-mp)^2}$	$\frac{p(-\ln q-p)}{(q-mp)^2(\ln q)^2}+\frac{(q-p/\ln q)mp}{(q-mp)^3}+\frac{2m(m+1)p^2q}{(q-mp)^4}$
17.	Rectangular-binomial	$\frac{n-1}{2(1-mp)}+\frac{mpq}{(1-mp)^2}$	$\frac{(n^2-1)(1-mp)+6(n+1)mpq}{12(1-mp)^3}+\frac{2m(m-1)p^2q}{(1-mp)^4}$

2.2.5 Probabilistic Structure of Lagrangian Distributions $L(f; g; x)$

Let $f(s)$ and $g(s)$ be two pgfs such that $g(0) \neq 0$ and let $X_{(n)}$ denote the variate whose pgf is the n-fold convolution of the probability distribution represented by the pgf $g(s)$.

By Maclaurin's theorem

$$P(X_{(n)} = n - 1) = \frac{1}{(n-1)!} D^{n-1} \left\{(g(s))^n\right\}\Big|_{s=0}. \tag{2.28}$$

By comparing (2.28) with (2.2) for the basic Lagrangian, we have the following theorem.

Theorem 2.4. *The probability $P(X = n)$ in the basic Lagrangian distribution defined by (2.2) equals $n^{-1} P(X_{(n)} = n - 1)$.*

Since the basic Lagrangian distribution given by the pgf $g(s) = q + ps$ is the geometric distribution (model (1), Table 2.1) whose pgf is $s = ug(s) = u(q + ps)$, i.e., $s = uq(1 - up)^{-1}$, its probability $P(X = n) = n^{-1} \binom{n}{n-1} qp^{n-1} = qp^{n-1}$.

Theorem 2.5. *The negative binomial distribution is a special case of the general Lagrangian distribution given by $f(s) = s^n = u^n q^n (1 - up)^{-n}$.*

Theorem 2.6. *The general Lagrangian distribution given by the pgf $f(s) = g(s)$, under the transformation $s = ug(s)$, is the basic Lagrangian distribution, defined by (2.2), displaced by one step to the left.*

Proof. By the result (2.6) for the general Lagrangian distribution for $f(s) = g(s)$,

$$\begin{aligned} P(X = x) &= \frac{1}{x!} D^{x-1} \left[(g(s))^x g'(s)\right]_{s=0} = \frac{1}{x!} D^{x-1} \left[\frac{D(g(s))^{x+1}}{x+1}\right]_{s=0} \\ &= \frac{1}{(x+1)!} D^x \left[(g(s))^{x+1}\right] \\ &= P(X = x + 1) \end{aligned}$$

for the basic Lagrangian distribution (2.2). □

Theorem 2.7. *The general Lagrangian distribution with pgf $\psi(u) = f(z)$ as a power series in $u = z/g(z)$ is obtained by randomizing the index parameter n in the Lagrangian probability distribution given by the pgf z^n (as a power series in u) according to the pgf $f(z)$ in z.*

Proof. Let $\{f_r\}$, $r = 0, 1, 2, \ldots$, represent the successive probabilities in the probability distribution of the pgf $f(z)$ in powers of z. Now,

$$\begin{aligned} f(z) &= \sum_{r=0}^{\infty} f_r z^r = f_0 + \sum_{r=1}^{\infty} f_r \left[\sum_{x=1}^{\infty} \frac{u^x}{x!} D^{x-1} \left\{r\, z^{r-1} (g(z))^x\right\}_{z=0}\right] \\ &= f_0 + \sum_{r=1}^{\infty} r\, f_r \sum_{x=r}^{\infty} \frac{u^x}{(x-r)!x} D^{x-r} (g(z))^x|_{z=0}. \end{aligned}$$

On rearranging the summations, it follows that

$$f(z) = f_0 + \sum_{x=1}^{\infty} \frac{u^x}{x!} D^{x-1} \left[(g(z))^x \left\{ \sum_{r=1}^{x} r f_r z^{r-1} \right\} \right]_{z=0}$$

$$= f_0 + \sum_{x=1}^{\infty} \frac{u^x}{x!} D^{x-1} \left\{ (g(z))^x Df(z) \right\}_{z=0},$$

which proves the result. □

Theorem 2.8. *Let $X_1, X_2, \ldots, X_N$ be a sequence of i.i.d. random variables having the basic Lagrangian distributions based on $g(z)$ and let N be an integer valued r.v. with another pgf $f(z)$ independent of $X_1, X_2, \ldots, X_N$. Then the sum $X = X_1 + X_2, + \cdots + X_N$ has the general Lagrangian distribution.*

Proof. Let $P(N = n) = f_n$ so that the pgf of N is $f(z) = \sum_{n=0}^{\infty} f_n z^n$. Also, let $H(u)$ be the pgf of the r.v. X. Since

$$P(X = x) = \sum_{n=0}^{\infty} P(X = x \mid N = n).P(N = n)$$

$$= \sum_{n=0}^{\infty} f_n \cdot \frac{n}{(x-n)!x} \left\{ D^{x-n} (g(z))^x \right\}_{z=0},$$

the pgf of the r.v. X becomes

$$H(u) = \sum_{x=n}^{\infty} \sum_{n=0}^{\infty} \frac{u^x n f_n}{(x-n)!x} \left\{ D^{x-n} (g(z))^x \right\}_{z=0} = \sum_{k=0}^{\infty} \sum_{n=0}^{\infty} \frac{n f_n u^{n+k}}{k!(n+k)} \left\{ D^k (g(z))^{n+k} \right\}_{z=0}$$

$$= \sum_{k=0}^{\infty} \frac{u^k}{k!} \left[D^{k-1} (g(z))^k \sum_{n=0}^{\infty} n f_n u^n (g(z))^{n-1} g'(z) \right]_{z=0}$$

$$= \sum_{k=0}^{\infty} \frac{u^k}{k!} \left[D^{k-1} (g(z))^k \sum_{n=0}^{\infty} D f_n u^n (g(z))^n \right]_{z=0}$$

$$= \sum_{k=0}^{\infty} \frac{u^k}{k!} \left[D^{k-1} (g(z))^k D \sum_{n=0}^{\infty} f_n z^n \right]_{z=0}$$

$$= \sum_{k=0}^{\infty} \frac{u^k}{k!} \left[D^{k-1} (g(z))^k f'(z) \right]_{z=0},$$

which is the pgf of the general Lagrangian distribution. □

Theorem 2.9. *The general Lagrangian distribution provided by the functions $f(z) = (q + pz)^n$ and $g(z) = (q + pz)^m$, $m \geq 1$, is the same as provided by the functions $f(z) = q^n(1 - pz)^{-n}$ and $g(z) = q^{m-1}(1 - pz)^{1-m}$, $m > 1\phi$.*

The proof is simple and the result can easily be checked by the models 2 and 11 in Table 2.3.

Weighted Lagrangian Distributions

Janardan (1987) has weighted a number of Lagrangian delta distributions by the variable x and thus he has obtained the weighted distributions for delta-binomial, delta-Poisson and delta-Geeta models in Table 2.2. He calls them size-biased forms, which belong to the class of Lagrangian distributions in (2.9).

Janardan (1987) has also given the weighted forms of the generalized Poisson (6), generalized binomial (2), generalized negative binomial (11), and generalized logarithmic series (12) in Table 2.3, with weights $(\theta+\lambda x)$, $(n+mx)$, $(n+mx)$, and x, respectively. These weighted forms of members of Lagrangian distributions in (2.7) belong to the class of Lagrangian distributions in (2.9). This property of getting the probability models of one class by weighing the models of another class of distributions provides a characterization for the two classes of distributions.

2.3 Modified Power Series Distributions

The family of *modified power series distributions* (MPSD) is a generalization of the family of *generalized power series distributions* (GPSD) (Noack, 1950; Patil, 1961, 1962) with the pmf given by

$$P(X=x)=a_x\theta^x/f(\theta), \quad x\in T\subset N, \tag{2.29}$$

where N is the set of nonnegative integers and T is a subset of N. The GPSD is obtained by the Maclaurin's expansion of a nonnegative analytic function $f(\theta)$ when $a_x\geq 0$ for all x in N. Jain (1975b) defined a class of power series distributions by using the Lagrange expansion in (1.78).

The MPSD were defined by Gupta (1974) by replacing θ^x in the GPSD (2.29) by $(\phi(\theta))^x$ and by assuming that the positive and analytic function $f(\theta)$ or $h(\theta)$ possesses a power series expansion in $\phi(\theta)$, where $\phi(\theta)$ is another positive and analytic function and $a_x\geq 0$ for all x in N. Thus, the probability mass function of an MPSD becomes

$$P(X=x)=a_x(\phi(\theta))^x/h(\theta), \quad x\in T\subset N. \tag{2.30}$$

The corresponding series function is

$$h(\theta)=\sum_{x\in T}a_x(\phi(\theta))^x. \tag{2.31}$$

Thus, the MPSD is an obvious generalization of the GPSD. Also, the MPSD is linear exponential, just like the GPSD, and can be written in the form

$$P(X=x)=\exp\{x\ \ln\phi(\theta)-\ln\ h(\theta)+\ln a_x\}. \tag{2.32}$$

Many Lagrangian probability distributions (Tables 2.1, 2.2, and 2.3) are MPSDs. All models in Table 2.1, except the Katz distribution, are MPSDs. Similarly, all the probability models in Table 2.2 are MPSDs. The generalized binomial model, the restricted generalized Poisson model, the generalized negative binomial model, the generalized logarithmic series model (models (12) and (14)), the Shenton, and the modified Ved are the only ones in Table 2.3 that are MPSDs.

Gupta (1974) did not provide any method or technique by which a series like (2.31) could be obtained, though he studied many properties of the MPSD and applied the results to many of the above well-known probability models.

2.3.1 Modified Power Series Based on Lagrange Expansions

Let $\phi(\theta) = \theta/\eta(\theta) = u$ so that $\theta = u\eta(\theta)$ is a Lagrange transformation. Also, let $\eta(0) \neq 0$ though $\phi(0) = 0$. By Lagrange expansion in (2.1),

$$\begin{aligned}\theta &= \sum_{k=1}^{\infty} \frac{u^k}{k!} D^{k-1}(\eta(\theta))^k|_{\theta=0} \\ &= \sum_{k=1}^{\infty} \frac{(\theta/\eta(\theta))^k}{k!} \left\{D^{k-1}(\eta(\theta))^k\right\}_{\theta=0} \\ &= \sum_{k=1}^{\infty} \frac{(\phi(\theta))^k}{k!} \left\{D^{k-1}(\eta(\theta))^k\right\}_{\theta=0}, \end{aligned} \tag{2.33}$$

which is a power series expansion in $\phi(\theta)$. Thus a parameter θ, defined over some domain, including $\theta = 0$, can always be expanded in a power series of $\phi(\theta) = \theta/\eta(\theta)$, where $\eta(0) \neq 0$.

In a similar manner any positive and analytic function $h(\theta)$ of a parameter θ can be expanded into a power series of the function $\phi(\theta) = \theta/\eta(\theta)$ by the two Lagrange expansions (1.78) and (1.80). The two expansions for the modified power series are

$$h(\theta) = h(0) + \sum_{k=1}^{\infty} \frac{(\phi(\theta))^k}{k!} \left[D^{k-1}\left\{(\eta(\theta))^k h'(\theta)\right\}\right]_{\theta=0} \tag{2.34}$$

and

$$\frac{h(\theta)}{1 - \eta'(\theta)\phi(\theta)} = \sum_{k=0}^{\infty} \frac{(\phi(\theta))^k}{k!} \left[D^k\left\{(\eta(\theta))^k h(\theta)\right\}\right]_{\theta=0}. \tag{2.35}$$

If the multiple derivatives with respect to θ are nonnegative for all integral values of k in the three expansions (2.33), (2.34), and (2.35), then all three of these expansions shall provide modified power series distributions. Thus, one can have numerous MPSDs by choosing suitable values for the functions $h(\theta)$ and $\eta(\theta)$. Since both functions $h(\theta)$ and $\eta(\theta) = \theta/\phi(\theta)$ depend upon a common parameter θ, these MPSDs will possibly be special cases of the Lagrangian probability distribution (2.7) which are based upon the functions $f(z)$ and $g(z)$ containing different parameters.

We apply the expansions (2.34) to obtain a number of MPSDs. The values of the functions $h(\theta)$, $\eta(\theta)$ and the MPSDs generated by them are given in Table 2.7.

The reader should try the expansion (2.33) to get another set of MPSDs by using the above values of $h(\theta)$ and $\eta(\theta)$ and see if the new MPSD can be reduced to one of the forms given above. The reader should also verify if the above models are special cases of the models given in Table 2.3.

2.3.2 MPSD as a Subclass of Lagrangian Distributions $L(f; g; x)$

Consul (1981) has proved that the MPSDs, as defined by Gupta (1974), belong to a subclass of the Lagrangian probability distributions. Accordingly, they possess all the properties of the Lagrangian distributions and have some other properties as well. The same proof is being given here. We shall first like to modify the probability of the Lagrangian distribution into a more suitable form.

Table 2.7. Some modified power series distributions

No.	Name	$h(\theta)$	$\eta(\theta)$	$P(X=x), x=0,1,2,3,\ldots$
1.	Poisson-negative binomial	$e^{a\theta}$ $a>0,$ $0<\theta<1$	$(1-\theta)^{-m}$ $m>0$	$e^{-a\theta}, x=0$ $\frac{(a\theta(1-\theta)^m)^x}{e^{a\theta}x!}$ $[{}_2F_0(1-x, mx;\,;-a^{-1})], x\geq 1$
2.	Poisson-negative binomial (*altered parameters*)	$e^{a\theta}$ $a>0, \theta>0$	$(1+\theta)^m$ $m\geq 1$	$e^{-a\theta}, x=0$ $e^{-a\theta}\frac{a^x\theta^x(1+\theta)^{-mx}}{x!}$ $\times[{}_2F_0(1-x, -mx;\,;a^{-1})], x\geq 1$
3.	Negative binomial-Poisson	$(1-\theta)^{-m}$ $0<\theta<1$	$e^{a\theta}$	$(1-\theta)^m, x=0$ $(1-\theta)^m\frac{\theta^x e^{-ax\theta}}{x!}m^{[x]}$ $\times[{}_1F_1(1-x; 2-m-x; ax)], x\geq 1$
4.	Negative binomial-binomial	$(1-\theta)^{-m}$ $0<\theta<1$	$(1+\theta)^n$ $n\geq 1$	$(1-\theta)^m, x=0$ $(1-\theta)^m\frac{\theta^x(1+\theta)^{-nx}}{x!}m^{[x]}$ $\times[{}_2F_1(1-x, -nx; -m-x+1; -1)],$ $x\geq 1$
5.	Binomial-Poisson	$(1+\theta)^m$ $\theta>0$	$e^{a\theta}$	$(1+\theta)^{-m}, x=0$ $(1+\theta)^{-m}\frac{\theta^x e^{-ax\theta}}{x!}m^{(x)}$ $\times[{}_2F_0(1-x; m-x+2;\,;-ax)], x\geq 1$
6.	Binomial-negative binomial	$(1+\theta)^m$	$(1-\theta)^{-n}$ $0<\theta<1$	$(1+\theta)^{-m}, x=0$ $(1+\theta)^{-m}\frac{\theta^x(1-\theta)^{nx}}{x!}m^{(x)}$ $\times[{}_2F_1(1-x, nx; m-x+1; -1)],$ $x\geq 1$
7.	Logarithmic-negative binomial (*altered parameters*)	$\ln(1+\theta)$ $\theta>0$	$(1+\theta)^m$ $m\geq 1$	$\frac{1}{mx}\binom{mx}{x}\frac{\theta^x(1+\theta)^{-mx}}{\ln(1+\theta)},$ $x\geq 0$
8.	Logarithmic-negative binomial	$-\ln(1-\theta)$ $0<\theta<1$	$(1-\theta)^{-m}$	$\frac{1}{mx+x}\binom{mx+x}{x}\frac{\theta^x(1-\theta)^{mx}}{[-\ln(1-\theta)]}, x\geq 0$
9.	Dev	$e^{m\theta}(1-\theta)^{-k}$ $m>0,\ k>1$	$e^{-\theta}$ $0<\theta<1$	$e^{-m\theta}(1-\theta)^k(\theta e^{-\theta})^x$ $\times\sum_{r=0}^{x}\binom{k+x-r-2}{x-r}$ $\times(m+x)^r/r!$
10.	Harish	$\frac{(1-\beta\theta)^{-k+1}}{(1-\theta)^n}$ $0<\theta<1,$ $n\geq 0,\ k\geq 1$	$(1-\theta)^{\beta-1}$ $1<\beta<\frac{1}{\theta}$	$(1-\theta)^n(1-\beta\theta)^{k+1}(\theta(1-\theta)^{\beta-1})^x$ $\times\sum_{r=0}^{x}\binom{k+r-1}{r}$ $\times\binom{n+\beta x-r}{x-r}\beta^r$

Let $z = b(v)$ be a one-to-one transformation such that $b(v) = 0$ for $v = a^k$ and $b(v) = 1$ for $v = a_1^k$. Also, let $f(z) = f(b(v)) = f_1(v)$, $g(z) = g(b(v)) = g_1(v)$ so that $g_1(a^k) \neq 0$, $g_1(a_1^k) = f_1(a_1^k) = 1$, and $0 < f_1(a^k) < 1$. Since

$$D \equiv \frac{\partial}{\partial z} = \frac{\partial v}{\partial z} \cdot \frac{\partial}{\partial v}$$

the probabilities (2.7) of the general Lagrangian distributions can be written as

$$P(X = x) = \left\{ \frac{1}{x!} \left(\frac{\partial v}{\partial z} \cdot \frac{\partial}{\partial v} \right)^{x-1} \left\{ (g_1(v))^x \frac{\partial f_1(v)/\partial v}{\partial z/\partial v} \right\} \right|_{v=a^k} \tag{2.36}$$

for $x \in T$ and zero elsewhere. Also, since the Lagrange transformation is $u = z/g(z) = b(v)/g_1(v)$, the general Lagrange series (2.6) can be written in the form

$$f_1(v) = \sum_{x \in T} \{b(v)/g_1(v)\}^x \ P(X = x) \tag{2.37}$$

so that $\sum_{x \in T} \ P(X = x) = 1$ when $v = a_1^k$.

Theorem 2.10. *The MPSD defined by*

$$P(X = x) = a_x(\phi(\theta^k))^x / h(\theta^k), \quad x \in T_1, \ k > 0, \tag{2.38}$$

and zero otherwise, where $a_x > 0$ and T_1 is a subset of the set of nonnegative integers with

$$h(\theta^k) = \sum_{x \in T_1} a_x(\phi(\theta^k))^x \tag{2.39}$$

belongs to the class of Lagrangian probability distributions in (2.7).

When $k = 1$, the above MPSD becomes the MPSD defined by Gupta (1974).

Proof. The proof will be complete if two functions $f(z)$ and $g(z)$, satisfying all the properties for the Lagrangian distributions, can be so defined that they provide the above MPSD.

Let b be the smallest integer in the set T_1 and let T_2 be the set of nonnegative integers obtained by subtracting b from all the elements, except b, of T_1. Now, dividing (2.39) by $(\phi(\theta^k))^b$, we get

$$h(\theta^k)/(\phi(\theta^k))^b = h_1(\theta^k) = a_b + \sum_{x \in T_2} c_x(\phi(\theta^k))^x, \tag{2.40}$$

where $c_x = a_{x+b}$ for $x \in T_2$.

The convergence of the power series (2.40) implies that the function $\phi(\theta^k)$ is bounded by zero on the left and by some quantity, say $M < 1$, on the right. Thus there must be some value of θ, say $\theta = t$, where $\phi(t^k) = 0$ and $h_1(t^k) = a_b$. Therefore,

$$\phi(\theta^k) = (\theta^k - t^k)^\alpha \cdot w(\theta^k) \quad \text{for} \quad \alpha > 0 \text{ and } w(\theta^k) > 0. \tag{2.41}$$

Now, we define the functions

$$z = \frac{(v - t^k)^\alpha}{(\theta^k - t^k)^\alpha}, \quad g_1(v) = \frac{w(\theta^k)}{w(v)}, \quad f_1(v) = \frac{h_1(v)}{h_1(\theta^k)}, \tag{2.42}$$

which are such that $z = 0$ when $\upsilon = t^k$ and $z = 1$ when $\upsilon = \theta^k$. Also, $g_1(t^k) \neq 0$ and $g_1(\theta^k) = f_1(\theta^k) = 1$, $0 < f_1(t^k) = h_1(t^k)/h_1(\theta^k) = a_b/h_1(\theta^k) < 1$, and $\frac{\partial z}{\partial \upsilon} = \frac{\alpha(\upsilon - t^k)^{\alpha-1}}{(\theta^k - t^k)^\alpha}$.

By substituting these values in (2.36)

$$P(X = x) = \frac{1}{x!}\left(\frac{(\theta^k - t^k)^\alpha}{\alpha(\upsilon - t^k)^{\alpha-1}}\frac{\partial}{\partial \upsilon}\right)^{x-1}\left\{\left(\frac{w(\theta^k)}{w(\upsilon)}\right)^x \frac{(\theta^k - t^k)^\alpha}{\alpha(\upsilon - t^k)^{\alpha-1}}\frac{\partial h_1(\upsilon)/\partial \upsilon}{h_1(\theta^k)}\right\}\Bigg|_{\upsilon=t^k}$$

$$= \frac{\{(\theta^k - t^k)^\alpha w(\theta^k)\}\ x\alpha^{-x}}{h_1(\theta^k)\cdot x!}\left[\left((\upsilon - t^k)^{1-\alpha}\frac{\partial}{\partial \upsilon}\right)^{x-1}\left\{\frac{(\upsilon - t^k)^{1-\alpha}}{(w(\upsilon))^x}\cdot\frac{\partial h_1(\upsilon)}{\partial \upsilon}\right\}\right]_{\upsilon=t^k}$$

$$= \frac{(\phi(\theta^k))^x}{h_1(\theta^k)}\cdot c_x \quad \text{for} \quad x \in T_3, \tag{2.43}$$

where $c_x = (x!)^{-1} \times$ (value of the $(x-1)$th derivative at $\upsilon = t^k$) and T_3 is a subset of N (set of integers). Also,

$$P(X = 0) = f_1(\upsilon)|_{\upsilon=t^k} = h_1(t^k)/h_1(\theta^k) = a_b/h_1(\theta^k).$$

It can easily be shown by the convergence of the series that the subsets T_3 and T_2 are identical. Thus

$$P(X = x) = \frac{(\phi(\theta^k))^x}{h_1(\theta^k)}\cdot a_x \ \text{for}\ x \in T_1$$

and zero otherwise.

When $\alpha = 1$ and $t = 0$, then $\upsilon = z\theta^k$ so that $g(z) = \frac{w(\theta)^k}{w(z\theta^k)}$ and $f(z) = \frac{h_1(z\theta^k)}{h_1(\theta^k)}$. Then, one can use the original formula (2.6) for the probability of the general Lagrangian distributions to get the same MPSD. Hence the MPSDs form a subclass of the general Lagrangian distributions. □

Since the MPSDs are based upon two unknown functions $h(\theta^k)$ and $\phi(\theta^k)$, the theorem is very general and many variations can be introduced in these two functions. Consul (1981) has given two good examples to illustrate these variations.

2.3.3 Mean and Variance of a MPSD

For every MPSD it is known that

$$h(\theta) = \sum_{x\in T} a_x \cdot (\phi(\theta))^x.$$

By differentiation with respect to θ,

$$h'(\theta) = \sum_{x\in T} a_x \cdot x(\phi(\theta))^{x-1}\phi'(\theta)$$

$$= \frac{h(\theta)\phi'(\theta)}{\phi(\theta)} E[X]$$

$$\therefore \text{ mean } \mu = E[X] = \frac{h'(\theta)\phi(\theta)}{h(\theta)\cdot\phi'(\theta)}. \tag{2.44}$$

Then, $\mu\, h(\theta) = \sum_{x\in T} x\ a_x(\phi(\theta))^x$.

By differentiating the above with respect to θ,

$$\frac{d\mu}{d\theta}h(\theta) + \mu\, h'(\theta) = \sum_{x\in T} x^2\, a_x(\phi(\theta))^x \cdot \phi'(\theta)/\phi(\theta)$$

$$= E[X^2]\cdot h(\theta)\phi'(\theta)/\phi(\theta).$$

$$\therefore\ E[X^2] = \frac{d\mu}{d\theta}\cdot\frac{\phi(\theta)}{\phi'(\theta)} + \mu\frac{h'(\theta)\phi(\theta)}{h(\theta)\cdot\phi'(\theta)} = \frac{d\mu}{d\theta}\cdot\frac{\phi(\theta)}{\phi'(\theta)} + \mu^2.$$

$$\therefore\ \text{Variance } \sigma^2 = E(X^2) - \mu^2 = \frac{\phi(\theta)}{\phi'(\theta)}\frac{d\mu}{d\theta}. \tag{2.45}$$

2.3.4 Maximum Entropy Characterization of some MPSDs

Kapur (1982) has described the Shannon–Bayesian entropy, its maximization, and application of the maximum entropy for the characterization of a number of MPSDs (a subclass of Lagrangian distributions) defined by

$$P(X = x) = a(x)\,(g(\theta))^x \,/\, h(\theta) \quad \text{for } x \in N,$$

where N is a subset of the set of nonnegative integers.

For each probability model, a prior $\alpha(x) \propto a(x)$ is chosen over N and then, assigning two unknowns A and b, the probability mass function $P(X = x) = P(x)$ is taken as

$$P(x) = \alpha(x) A\, b^x, \quad x \in N, \tag{2.46}$$

subject to the conditions

$$A\sum_{x\in N}\alpha(x)b^x = 1, \quad A\sum_{x\in N}x\alpha(x)b^x = M, \tag{2.47}$$

where M is the assigned mean of the probability model. Kapur stated that the above two conditions characterize the probability model and gave the values of A and b in terms of M and the parameters contained in $\alpha(x)$. It is not clear how the values of A and b were obtained from the two equations. Of course, they can easily be determined by taking the value of mean M and $b = g(\theta)$ from the model. Four examples, given by Kapur (1982), are given below.

(i) Generalized binomial distribution given by

$$f(t) = (q + pt)^n, \quad g(t) = (q + pt)^m, \quad p = 1 - q < m^{-1}.$$

Let

$$\alpha(x) \propto \frac{n}{n + mx}\binom{n + mx}{x}, \quad x = 0, 1, 2, \ldots \tag{2.48}$$

and let the mean be prescribed as M. Then

$$P(x) = A\frac{n}{n + mx}\binom{n + mx}{x} b^x,$$

where

$$A\sum_{x=0}^{\infty}\frac{n}{n+mx}\binom{n+mx}{x}b^x=1,\quad A\sum_{x=0}^{\infty}\frac{xn}{n+mx}\binom{n+mx}{x}b^x=M.$$

These give

$$A=\left(\frac{n+(m-1)M}{n+mM}\right)^n=q^n,\quad b=\frac{M}{n+mM}\left(\frac{n+(m-1)M}{n+mM}\right)^{m-1}=pq^{m-1}.$$

Therefore, the generalized negative binomial distribution (GNBD) can be characterized as the maximum Bayesian entropy distribution (MBED) when the prior probability distribution is proportional to $\alpha(x)$ in (2.48) and for which the mean M is also given.

(ii) Delta-binomial distribution for $f(t)=t^n,\ \ g(t)=(q+pt)^m$. Let

$$\alpha(x)\propto\frac{n}{n+x}\binom{m(n+x)}{x},\qquad x=0,1,2,\ldots \tag{2.49}$$

and let the mean be given as M. Then

$$P(x)=\frac{An}{n+x}\binom{m(n+x)}{x}b^x,\qquad x=0,1,2,\ldots,$$

where

$$A\sum_{x=0}^{\infty}\frac{n}{n+x}\binom{mn+mx}{x}b^x=1,\quad A\sum_{x=0}^{\infty}\frac{xn}{n+x}\binom{mn+mx}{x}b^x=M,$$

giving

$$A=\left(1-\frac{M-n}{Mn}\right)^{mn}=q^{mn},\quad b=\frac{M-n}{Mn}\left(1-\frac{M-n}{Mn}\right)^{m-1}=pq^{m-1}.$$

Thus,

$$P(x)=\frac{n}{n+x}\binom{mn+mx}{x}p^xq^{mn+mx-x},\qquad x=0,1,2,\ldots,$$

is the model characterized as the MBED with prior given by (2.49) and the mean M.

(iii) Delta-Poisson distribution. Kapur (1982) stated that by taking the prior $\alpha(y)=e^{-y}/y!$, $y=0,1,2,\ldots$, and by prescribing $\mathrm{E}[Y]=a,\ \ \mathrm{E}[\ln(Y+n)]=b,\ \ \mathrm{E}[(Y+n)\ln(Y+n)]=c$, the MBED is given by

$$P(y)=(A/y!)e^{-ay}(y+n)^{b+c(y+n)},\qquad y=0,1,2,\ldots, \tag{2.50}$$

which gives

$$P(x)=\frac{A}{(x-n)!}e^{-a(x-n)}x^{b+cx},\qquad x=n,n+1,n+2,\ldots.$$

When $c=1$ and $b=-n-1$, the above reduces to the delta-Poisson distribution.

(iv) Generalized Poisson distribution. Similar to (iii), the MBED given by the prior $\alpha(x) = e^{-x}/x!$, $x = 0, 1, 2, \ldots$, and by prescribing $\mathrm{E}[X] = a$, $\mathrm{E}[\ln(M + \theta X)] = b$, and $\mathrm{E}[(M + \theta X)\ln(M + \theta X)] = c$ is

$$P(x) = \frac{A}{x!} e^{-ax} (M + \theta x)^{b + c(M+\theta x)}, \quad x = 0, 1, 2, \ldots.$$

Its special case given by $a = \theta$, $c = \theta^{-1}$, and $b + M/\theta = -1$ is the generalized Poisson distribution.

2.4 Exercises

2.1 Find the mean and variance of the following Lagrangian probability distributions.
(a) Geeta and (7) to (11) in Table 2.1
(b) Haight distribution in Table 2.2
(c) Models (1) to (4) in Table 2.7.

2.2 Show that the mean and variance of the Katz distribution ((6) in Table 2.1) are

$$\mu = (1 - \beta)(1 - \beta - b)^{-1}, \quad \sigma^2 = b(1 - \beta)(1 - \beta - b)^{-3}.$$

2.3 Equation (2.35) is obtained by using the Lagrange expansion in (1.80) for the modified power series distribution. Use the result in (2.35) to obtain the MPSDs based on the Lagrange expansion in (1.80) for the following functions with $0 < \theta < 1$ and $m > 0$:
(a) $h(\theta) = e^{a\theta}$ and $\eta(\theta) = (1 - \theta)^{-m}$,
(b) $h(\theta) = (1 - \theta)^{-m}$ and $\eta(\theta) = e^{a\theta}$,
(c) $h(\theta) = (1 - \theta)^{-m}$ and $\eta(\theta) = (1 + \theta)^{-m}$,
(d) $h(\theta) = -\ln(1 - \theta)$ and $\eta(\theta) = (1 - \theta)^{-m}$.

2.4 Show that the negative binomial distribution is a special case of the general Lagrangian distribution given by $f(z) = z^n = u^n q^n (1 - up)^{-n}$ (see Theorem 2.5).

2.5 By choosing $f(z)$ and $g(z)$ appropriately, show that the binomial, the Poisson, and the logarithmic series distributions can be obtained as special cases of the general Lagrangian distribution given by (2.7).

2.6 Prove that the general Lagrangian distribution provided by the functions $f(z) = (1 - \theta + \theta z)^n$ and $g(z) = (1 - \theta + \theta z)^m$, $m \geq 1$, is the same as provided by the functions $f(z) = (1 - \theta)^m \, (1 - \theta z)^{-n}$ and $g(z) = (1 - \theta)^{m-1} \, (1 - \theta z)^{1-m}$, $m > 1$ (see Theorem 2.9).

2.7 Show that under the transformation $p = \theta/(1 + \theta)$, the generalized logarithmic series model (12) in Table 2.3 reduces to the logarithmic negative binomial model (7) in Table 2.7. Find the mean and variance of the logarithmic negative binomial model.

2.8 Obtain the probability mass function for the Dev distribution (see Table 2.7) by using the Lagrange expansion in (1.80). Is it possible to obtain it from the Lagrange expansion in (1.78), and if so, how?

2.9 Obtain the probability mass function for the Harish distribution (see Table 2.7) by using the Lagrange expansion in (1.80).

2.10 The basic Lagrangian distribution given by $g(z) = e^{\theta(z^2 - 1)}$ is the Felix distribution (see Table 2.1). Show that a generalized Felix distribution with $f(z) = z$ and $g(z) = e^{\theta(z^a - 1)}$ is the restricted generalized Poisson distribution.

2.11 The Sunil distribution in Table 2.1 is obtained from equation (2.2) by using the function $g(z) = (1 - \theta + \theta z^2)^m$, which depends on two parameters θ and m. Determine a function $g(z)$, which depends on three parameters $\theta,\ m$, and β so that equation (2.2) provides the generalized negative binomial distribution.

2.12 By using the Lagrange expansion in (1.77) on the functions $f(z) = z$ and $g(z) = (1 - \theta)^n (1 - \theta z^k)^{-n}$, show that the Lagrangian probability distribution from this expansion is the generalized negative binomial distribution. By using a suitable transformation, determine the value of k in terms of m that gives the generalized negative binomial model (11) in Table 2.3.

3

Properties of General Lagrangian Distributions

3.1 Introduction

The class of discrete Lagrangian probability distributions has three subclasses called the basic, the delta, and the general Lagrangian distributions. The class has numerous families of probability distributions. Some of these are given in Tables 2.1 to 2.4 and many more can be obtained by suitable choices of the functions $g(z)$ and $f(z)$. It has been shown that some families of the class of Lagrangian distributions in (2.9) are the weighted distributions of the corresponding families of the class of Lagrangian distributions in (2.7). Also, the equivalence between these two classes of distributions has been shown in Subsection 2.2.1.

The first two factorial moments and the mean and the variance of the two classes of distributions were obtained in the last chapter. Though one can use the same methods to find the higher factorial moments, these become quite laborious. Consul and Shenton (1972, 1975) and Good (1975) have given interesting methods to obtain the central moments and the cumulants of the class of Lagrangian distributions in (2.7). Consul and Famoye (2001) has given a similar method to compute the central moments of the class of Lagrangian distributions in (2.9). The mean, kth central moment and the kth cumulant of the Lagrangian distributions in (2.7) are denoted by ${}_1\mu$, ${}_1\mu_k$ and ${}_1L_k$, $k = 1, 2, 3, \ldots$, respectively. Similarly, the mean, kth central moment and the kth cumulant of the Lagrangian distributions in (2.9) are denoted by ${}_2\mu$, ${}_2\mu_k$, and ${}_2L_k$, $k = 1, 2, 3, \ldots$. In this notation ${}_1\mu_1$ and ${}_2\mu_1$ denote the first central moments and have a value of zero.

Though the functions $g(z)$ and $f(z)$ are not necessarily pgfs, for the sake of convenience we assume in this chapter that they are pgfs of some r.v.s and that G_k and F_k, $k = 1, 2, 3, \ldots$, denote the cumulants of the probability distributions generated by them.

3.2 Central Moments of Lagrangian Distribution $L(f; g; x)$

Since $g(z)$ and $f(z)$ are the pgfs of two independent r.v.s and G_k and F_k, $k = 1, 2, 3, \ldots$, denote the cumulants of their respective probability distributions, replacing z by e^S and taking logarithms, we get

$$\ln g\left(e^S\right) = \sum_{k=1} G_k S^k / k!, \tag{3.1}$$

$$\ln f\left(e^S\right) = \sum_{k=1} F_k\, S^k/k!\,. \tag{3.2}$$

Replacing z by e^S and u by e^β in the Lagrange transformation $z = ug(z)$, taking logarithms and using (3.1), we have

$$\beta = S - \ln g\left(e^S\right) = (1 - G_1)\, S - \sum_{i=2} G_i S^i/i!\,. \tag{3.3}$$

Let $P\,(X = x) = P_x$ denote the pmf of the general Lagrangian distribution $L(f; g; x)$ in (2.7), generated by $f(z)$ under the transformation $z = ug(z)$. Since ${}_1\mu$ is the arithmetic mean of the distribution, the central moment generating function of Lagrangian distribution becomes

$$\sum_{x=0} P_x\, e^{(x-{}_1\mu)\beta} = e^{-{}_1\mu\beta} \cdot \sum_{x=0} P_x \left(e^\beta\right)^x = e^{-{}_1\mu\beta} \sum_{x=0} P_x \left[e^S/g\left(e^S\right)\right]^x$$

$$= e^{-{}_1\mu\beta} \cdot f\left(e^S\right) = e^{-{}_1\mu\beta} \exp\left[\ln f\left(e^S\right)\right].$$

By using the expansions (3.3) and (3.2) on the right side and by expanding $e^{(x-{}_1\mu)\beta}$ as a power series in β on the left side we obtain

$$\sum_{k=0} {}_1\mu_k \cdot \frac{\beta^k}{k!} = \exp\left[\{F_1 - {}_1\mu\,(1 - G_1)\}\, S + \sum_{i=2} (F_i + {}_1\mu \cdot G_i)\, \frac{S^i}{i!}\right], \tag{3.4}$$

where β is to be replaced by the series (3.3).

Since ${}_1\mu_1 = 0$, the term of S with unit power vanishes on the left side of (3.4). Thus, the corresponding term of S must also vanish on the right-hand side of (3.4) and

$${}_1\mu = F_1/\,(1 - G_1)\,. \tag{3.5}$$

For obtaining the values of the other central moments of the Lagrangian distributions $L(f; g; x)$ in (2.7), the identity (3.4) can be written as

$$\sum_{k=2} \frac{{}_1\mu_k}{k!}\left[(1 - G_1)\, S - \sum_{i=2} G_i \frac{S^i}{i!}\right]^k \equiv \sum_{i=2} (F_i +_1 \mu G_i)\, \frac{S^i}{i!} + \frac{1}{2!}\left[\sum_{i=2} (F_i + {}_1\mu G_i)\, \frac{S^i}{i!}\right]^2$$

$$+ \frac{1}{3!}\left[\sum_{i=2} (F_i + {}_1\mu G_i)\, \frac{S^i}{i!}\right]^3 + \cdots. \tag{3.6}$$

By comparing the coefficients of S^2, S^3, S^4, S^5, and S^6 on both sides, we get the five relations

$${}_1\mu_2\,(1 - G_1)^2 = F_2 +_1 \mu G_2,$$

$${}_1\mu_3\,(1 - G_1)^3 - 3\,{}_1\mu_2\,(1 - G_1)\, G_2 = F_3 +_1 \mu G_3,$$

$${}_1\mu_4\,(1 - G_1)^4 - 6\,{}_1\mu_3\,(1 - G_1)^2\, G_2 +_1 \mu_2\left[3G_2^2 - 4G_3\,(1 - G_1)\right]$$
$$= F_4 +_1 \mu G_4 + 3\,(F_2 + {}_1\mu G_2)^2,$$

$${}_1\mu_5\,(1 - G_1)^5 - 10\,{}_1\mu_4 G_2\,(1 - G_1)^3 - 10\,{}_1\mu_3 G_3\,(1 - G_1)^2 + 15\,{}_1\mu_3 G_2^2\,(1 - G_1)$$
$$+_1 \mu_2\,(10G_2G_3 - 5G_4\,(1 - G_1)) = F_5 +_1 \mu G_5 + 10\,(F_3 + {}_1\mu G_3)\,(F_2 + {}_1\mu G_2),$$

$$
\begin{aligned}
&{}_1\mu_6 (1-G_1)^6 - 15G_2 (1-G_1)_1^4 \mu_5 + {}_1\mu_4 \left[-20G_3 (1-G_1)^3 + 45G_2^2 (1-G_1)^2\right] \\
&\quad - {}_1\mu_3 \left[10G_2^3 + 15G_4 (1-G_1)^2 - 10G_2G_3 (1-G_1)\right] \\
&\quad - {}_1\mu_2 \left[6G_5 (1-G_1) - 15G_2G_4 - 10G_3^2\right] \\
&\quad = F_6 + {}_1\mu G_6 + 10 (F_3 + {}_1\mu G_3)^2 + 15 (F_4 + {}_1\mu G_4)(F_2 + {}_1\mu G_2).
\end{aligned}
$$

On using the value of ${}_1\mu$ from (3.5) in the above five relations and on simplification, the other five central moments become

$$
\begin{aligned}
{}_1\mu_2 &= F_2 (1-G_1)^{-2} + F_1G_2 (1-G_1)^{-3}, \\
{}_1\mu_3 &= \frac{3{}_1\mu_2G_2}{(1-G_1)^2} + \frac{F_3 (1-G_1) + F_1G_3}{(1-G_1)^4}, \\
{}_1\mu_4 &= 3 ({}_1\mu_2)^2 + \frac{6{}_1\mu_3G_2}{(1-G_1)^2} + {}_1\mu_2 \left\{\frac{4G_3}{(1-G_1)^3} - \frac{3G_2^2}{(1-G_1)^4}\right\} \\
&\quad + \frac{F_4 (1-G_1) + F_1G_4}{(1-G_1)^5}, \\
{}_1\mu_5 &= 10\,{}_1\mu_2 ({}_1\mu_3) + \frac{10\,{}_1\mu_4G_2}{1-G_1} + {}_1\mu_3 \left[\frac{10G_3}{(1-G_1)^3} - \frac{15G_2^2}{(1-G_1)^5}\right] \\
&\quad + 5\,{}_1\mu_2 \left[\frac{G_4}{(1-G_1)^4} - \frac{6\,{}_1\mu_2G_2}{(1-G_1)^2} - \frac{2G_2G_3}{(1-G_1)^5}\right] + \frac{F_5 (1-G_1) + F_1G_5}{(1-G_1)^5}, \\
{}_1\mu_6 &= \frac{15\,{}_1\mu_5G_2}{1-G_1} + \frac{5{}_1\mu_4}{(1-G_1)^3} \left[4G_3 - \frac{9G_2^2}{1-G_1}\right] \\
&\quad + \frac{5\,{}_1\mu_3}{(1-G_1)^4} \left[3G_4 - \frac{2G_2G_3}{1-G_1} + \frac{3G_2^3}{(1-G_1)^2}\right] \\
&\quad + \frac{{}_1\mu_2}{(1-G_1)^5} \left[6G_5 + 15 (F_4 - F_4G_1 + F_1G_4) - \frac{15G_2G_3 + 10G_3^2}{1-G_1}\right] \\
&\quad + \frac{F_6 - F_6G_1 + F_1G_6 + 10 (F_3 + {}_1\mu G_3)^2}{(1-G_1)^6}.
\end{aligned}
\tag{3.7}
$$

Higher central moments can also be obtained from the identity (3.6) by comparing the coefficients of $S^7, S^8, S^9, \ldots$ on both sides.

Cumulants of Lagrangian Distributions $L(f; g; x)$

Though the central moments can be used to compute the cumulants of a distribution, we are providing an independent method for obtaining the cumulants of the general Lagrangian distribution. Since the basic Lagrangian distribution is generated by the expansion of z as a power series of u by the Lagrange expansion in (1.77) under the transformation $z = ug(z)$, its pgf is given by (2.1). Replacing z by e^S and u by e^β in $z = ug\,(z)$, and taking logarithms, provides the cumulant generating function of the basic Lagrangian distribution as a power series in β. Thus

$$S = \beta + \ln g\left(e^S\right) = \sum_{r=1} D_r \cdot \beta^r / r!, \tag{3.8}$$

where D_r, $r = 1, 2, 3, \ldots$, denote the cumulants of the basic Lagrangian distribution.

Since the power series expansion in β of $\ln f\left(e^S\right)$ is the cumulant generating function for the general Lagrangian distribution, given by $f\,(z)$ under the transformation $z = ug\,(z)$, from the relation (3.2), we have

$$\begin{aligned}\sum_{k=1} {}_1L_k \cdot \frac{\beta^k}{k!} &= \sum_{r=1} F_r \cdot \frac{S^r}{r!}\\ &= \sum_{r=1} \frac{F_r}{r!}\left(\sum_{i=1} \frac{D_i}{i!}\beta^i\right)^r \quad \text{[by (3.8)]},\end{aligned}$$

so that

$$\begin{aligned}{}_1L_k &= \frac{\partial^k}{\partial \beta^k}\left[\sum_{r=1} \frac{F_r}{r!}\left(\sum_{i=1} \frac{D_i}{i!}\beta^i\right)^r\right]_{\beta=0}\\ &= \sum_{r=1}^{k}\left[\frac{k!}{r!}F_r\left(\sum_{i=1}^{k} \frac{D_i}{i!}\beta^i\right)^r\right]_{\beta=0}.\end{aligned} \tag{3.9}$$

The coefficients of F_r, $r = 1, 2, 3, \ldots$, can be obtained by expanding the multinomial

$$\left(\sum_{i=1}^{k} D_i \beta^i / i!\right)^r$$

and evaluating the coefficient of β^k. This can be done precisely by the Faà de Bruno theorem, where all possible partitions of r are considered.

Thus, the cumulants ${}_1L_k$, $k = 1, 2, 3, \ldots$, of the general Lagrangian distribution are given by (3.9) in the form

$${}_1L_k = \sum_{r=1}^{k} \frac{k!}{r!} F_r \left[\sum \frac{r!}{\pi_1!\pi_2! \cdots \pi_k!} \prod_{j=1}^{k} \left(\frac{D_j}{j!}\right)^{\pi_j}\right], \tag{3.10}$$

where the second summation is taken over all partitions $\pi_1, \pi_2, \pi_3, \ldots, \pi_k$ of r such that

$$\pi_1 + \pi_2 + \cdots + \pi_k = r \quad \text{and} \quad \pi_1 + 2\pi_2 + 3\pi_3 + \cdots + k\pi_k = k. \tag{3.11}$$

The first few cumulants of the general Lagrangian distribution $L(f; g; x)$ can now be written down as particular cases of (3.10) and are

$$\begin{cases} {}_1L_1 = F_1 D_1, \\ {}_1L_2 = F_1 D_2 + F_2 D_1^2, \\ {}_1L_3 = F_1 D_3 + 3F_2 D_1 D_2 + F_3 D_1^3, \\ {}_1L_4 = F_1 D_4 + 3F_2 D_2^2 + 4F_2 D_1 D_3 + 6F_3 D_1^2 D_2 + F_4 D_1^4, \\ {}_1L_5 = F_1 D_5 + 5F_2 D_1 D_4 + 10F_2 D_2 D_3 + 15F_3 D_1 D_2^2 \\ \qquad +10F_3 D_1^2 D_3 + 10F_4 D_1^3 D_2 + F_5 D_1^5, \\ {}_1L_6 = F_1 D_6 + F_2 \left(6D_1 D_5 + 15D_2 D_4 + 10D_3^2\right) \\ \qquad +F_3 \left(15D_1^2 D_4 + 60D_1 D_2 D_3 + 15D_2^3\right) \\ \qquad +F_4 \left(20D_1^3 D_3 + 45D_1^2 D_2^2\right) + 15F_5 D_1^4 D_2 + F_6 D_1^6. \end{cases} \tag{3.12}$$

The above cumulants will be determined explicitly if the values of D_k, $k = 1, 2, \ldots$, the cumulants of the basic Lagrangian distribution are known. These can be obtained from (3.8), which can be rewritten in the form

$$\begin{aligned} S = \sum_{k=1} D_k \frac{\beta^k}{k!} &= \beta + \ln g\left(e^S\right) = \beta + \sum_{r=1} G_r \frac{S^r}{r!} \\ &= \beta + \sum_{r=1} \frac{Gr}{r!} \left(\sum_{i=1} \frac{Di}{i!} \beta^i \right)^r . \end{aligned} \tag{3.13}$$

Note that on transferring β to the left side in (3.13), it becomes very similar to (3.9) with the difference that ${}_1L_1$ is replaced by $D_1 - 1$, ${}_1L_k$ by D_k for $k = 2, 3, \ldots$, and F_r is replaced by G_r. Therefore, one would get the corresponding values of D_k, $k = 1, 2, 3, \ldots$, by making these changes in (3.12). Thus

$$\begin{cases} D_1 - 1 = G_1 D_1, \quad \text{i.e., } D_1 = (1 - G_1)^{-1}, \\ D_2 = G_1 D_2 + G_2 D_1^2, \quad \text{i.e. } D_2 = G_2 (1 - G_1)^{-3}, \\ D_3 = G_1 D_3 + 3G_2 D_1 D_2 + G_3 D_1^3, \quad \text{i.e., } D_3 = G_3 (1 - G_1)^{-4} \\ \qquad +3G_2^2 (1 - G_1)^{-5}, \\ D_4 = G_1 D_4 + 3G_2 D_2^2 + 4G_2 D_1 D_3 + 6G_3 D_1^2 D_2 + G_4 D_1^4 \\ \quad = G_4 (1 - G_1)^{-5} + 10G_3 G_2 (1 - G_1)^{-6} + 15G_2^3 (1 - G_1)^{-7}, \\ D_5 = G_1 D_5 + 5G_2 D_1 D_4 + 10G_2 D_2 D_3 + 15G_3 D_1 D_2^2 + 10G_3 D_1^2 D_3 \\ \qquad +10G_4 D_1^3 D_2 + G_5 D_1^5 \\ \quad = G_5 (1 - G_1)^{-6} + \left(15G_4 G_2 + 10G_3^2\right) (1 - G_1)^{-7} \\ \qquad +105G_3 G_2^2 (1 - G_1)^{-8} + 105G_2^4 (1 - G_1)^{-9}, \\ \text{and} \\ D_6 = G_6 (1 - G_1)^{-7} + (21G_5 G_2 + 35G_4 G_3) (1 - G_1)^{-8} \\ \qquad + \left(210G_4 G_2^2 + 280G_3^2 G_2\right) (1 - G_1)^{-9} \\ \qquad +1260G_3 G_2^3 (1 - G_1)^{-10} + 945G_2^5 (1 - G_1)^{-11}. \end{cases} \tag{3.14}$$

The cumulants $\{_1L_r\}$, $r = 1, 2, 3, \ldots$, of all families of distributions belonging to the class of Lagrangian distributions in (2.7) are completely determined by the relations (3.12) and (3.14) in terms of the cumulants $\{G_r\}$ and $\{F_r\}$, which are known from the functions $g(z)$ and $f(z)$, respectively. Substituting the values of the $D_1, D_2, \ldots, D_6$ from the relations (3.14) in the relations (3.12), the values of the first six cumulants $\{_1L_r\}$ become

$$\begin{cases} {}_1L_1 = F_1(1-G_1)^{-1}, \\ {}_1L_2 = F_2(1-G_1)^{-2} + F_1G_2(1-G_1)^{-3}, \\ {}_1L_3 = F_3(1-G_1)^{-3} + (3F_2G_2 + F_1G_3)(1-G_1)^{-4} + 3F_1G_2^2(1-G_1)^{-5}, \\ {}_1L_4 = F_4(1-G_1)^{-4} + (6F_3G_2 + 4F_2G_3 + F_1G_4)(1-G_1)^{-5} \\ \qquad + \left(15F_2G_2^2 + 10F_1G_2G_3\right)(1-G_1)^{-6} + 15F_1G_2^3(1-G_1)^{-7}, \\ {}_1L_5 = F_5(1-G_1)^{-5} + (10F_4G_2 + 10F_3G_3 + 5F_2G_4 + F_1G_5)(1-G_1)^{-6} \\ \qquad + \left(45F_3G_2^2 + 60F_2G_3G_2 + 15F_1G_4G_2 + 10F_1G_3^2\right)(1-G_1)^{-7} \\ \qquad +105\left(F_2G_2^3 + F_1G_3G_2^2\right)(1-G_1)^{-8} + 105F_1G_2^4(1-G_1)^{-9}, \\ {}_1L_6 = F_6(1-G_1)^{-6} + (15F_5G_2 + 20F_4G_3 + 15F_3G_4 + 6F_2G_5 + F_1G_6)(1-G_1)^{-7} \\ \qquad + \left(105F_4G_2^2 + 210F_3G_3G_2 + 105F_2G_4G_2 + 60F_2G_3^2\right. \\ \qquad \left. +21F_1G_5G_2 + 35F_1G_4G_3\right)(1-G_1)^{-8} \\ \qquad + \left(405F_3G_2^3 + 840F_2G_3G_2^2 + 210F_1G_4G_2^2 + 280F_1G_3^2G_2\right)(1-G_1)^{-9} \\ \qquad + \left(315F_2G_2^4 + 1260F_1G_3G_2^3\right)(1-G_1)^{-10} + 945F_1G_2^5(1-G_1)^{-11}. \end{cases} \tag{3.15}$$

Example. The Poisson-binomial family of Lagrangian distributions $L(f; g; x)$ is generated by $f(z) = e^{\theta(z-1)}$ and $g(\theta) = (q + pz)^m$, whose cumulants are known. Substituting the cumulants of $f(z)$ and $g(z)$ in the above relations (3.15), the first four cumulants become

$$\begin{cases} \kappa_1 = \theta(1-mp)^{-1}, \qquad \kappa_2 = \theta\left(1-mp^2\right)(1-mp)^{-3}, \\ \kappa_3 = \theta\left[\left(1-mp^2\right)^2 + 2mpq^2\right](1-mp)^{-5}, \\ \kappa_4 = \theta(1-mp)^{-4} + \theta mpq(5-2p)(1-mp)^{-5} \\ \qquad +\theta m^2p^2q^2(5+20q)(1-mp)^{-6} \\ \qquad +15\theta m^3p^3q^3(1-mp)^{-7}. \end{cases} \tag{3.16}$$

3.3 Central Moments of Lagrangian Distribution $L_1(f_1; g; y)$

By using the equivalence theorem in subsection 2.2.1, all the moments of this class can be derived from the moments of the class of Lagrangian distributions $L(f; g; x)$ in (2.7) but the process will become very complex because $f_1(z)$ is given by (2.10) as $f_1(z) = \left(1 - g'(1)\right)^{-1}\left(1 - zg'(z)/g(z)\right)f(z)$. Therefore, we will derive moments of Lagrangian distributions $L_1(f_1; g; y)$ by using their pgfs. Let $P(Y = y) = P_y$ be the pmf of the general Lagrangian distribution generated by the pgfs $f_1(z)$ and $g(z)$ by formula (2.9). On replacing z by e^S and u by e^β in the pgf (2.10) of the general Lagrangian distributions in (2.9), the moment generating function (mgf) becomes

$$H\left(e^{\beta}\right)=\frac{(1-G_1)\, f_1\left(e^{S}\right)}{1-(\partial/\partial S)\ln g\left(e^{S}\right)}=\frac{f_1\left(e^{S}\right)}{(1-G_1)^{-1}\left[1-(\partial/\partial S)\ln g\left(e^{S}\right)\right]}, \tag{3.17}$$

where the Lagrange transformation $z=ug(z)$ changes to

$$e^{S}=e^{\beta}g\left(e^{S}\right), \tag{3.18}$$

which gives

$$S=\beta+\ln g\left(e^{S}\right).$$

Let ${}_1F_k,\ k=1,2,3,\ldots$, denote the cumulants of the probability distributions given by the pgf of $f_1(z)$. Since (by (3.1) and (3.2))

$$\ln f_1\left(e^{S}\right)=\sum_{k=1}{}_1F_k\, S^k/k!,$$

$$\ln g\left(e^{S}\right)=\sum_{k=1} G_k\, S^k/k!,$$

the relation (3.18) provides

$$\beta=(1-G_1)\, S-\sum_{k=2} G_k S^k/k!. \tag{3.19}$$

Since the rth central moment ${}_2\mu_r$ of the general Lagrangian distributions $L_1(f_1; g; y)$ in (2.9) is ${}_2\mu_r=\sum_y P_y\,(y-{}_2\mu)^r$, the mgf for central moments is

$$\begin{aligned}\sum_{r=0}{}_2\mu_r\beta^r/r! &=\sum_{r=0}\sum_{y} P_y\,(y-{}_2\mu)^r\,\beta^r/r!\\ &=\sum_{y} P_y e^{(y-{}_2\mu)\beta}=e^{-{}_2\mu\beta}H\left(e^{\beta}\right),\quad \text{and by (3.17)}\\ &=\frac{\exp\left[-{}_2\mu\beta+\ln f_1\left(e^{S}\right)\right]}{(1-G_1)^{-1}\left[1-(\partial/\partial S)\ln g\left(e^{S}\right)\right]}=\frac{C}{B}.\end{aligned} \tag{3.20}$$

On substituting the values of β, $\ln f_1\left(e^{S}\right)$, and $\ln g\left(e^{S}\right)$ in the form of series given above, we have

$$C=\exp\left[\{{}_1F_1-{}_2\mu\,(1-G_1)\}\, S+\sum_{r=2}({}_1F_r+{}_2\mu G_r)\, S^r/r!\right] \tag{3.21}$$

and

$$B=1-\sum_{r=1}\frac{G_{r+1}}{1-G_1}\frac{S^r}{r!}. \tag{3.22}$$

Note that the mean and the central moments of the Lagrangian distributions in (2.9) can be obtained by differentiating the relation (3.20) successively with respect to S and by putting $\beta=0=S$ on both sides. Also, note from (3.19) that

$$\frac{\partial\beta}{\partial S}=1-\sum_{i=1} G_i\frac{S^{i-1}}{(i-1)!},\qquad \frac{\partial^k\beta}{\partial S^k}=-\sum_{i=k} G_i\frac{S^{i-k}}{(i-k)!}$$

so that

$$\partial\beta/\partial S|_{S=0} = 1 - G_1 \quad \text{and} \quad \partial^k\beta/\partial S^k\,|_{S=0} = -G_k, \quad k = 2, 3, \ldots. \tag{3.23}$$

Also,

$$\left.\frac{\partial C}{\partial S}\right|_{S=0} = C.D|_{S=0} = {}_1F_1 - {}_2\mu\,(1 - G_1),$$

where

$$D = {}_1F_1 - {}_2\mu\,(1 - G_1) + \sum_{r=2} ({}_1F_r + {}_2\mu G_r)\,\frac{S^{r-1}}{(r-1)!}.$$

Obviously,

$$\left.\frac{\partial^k D}{\partial S^k}\right|_{S=0} = {}_1F_{k+1} - {}_2\mu G_{k+1}, \quad k = 1, 2, 3, \ldots. \tag{3.24}$$

Also, by (3.22)

$$\left.\frac{\partial^k B}{\partial S^k}\right|_{S=0} = -\sum_{r=k} \left.\frac{G_{r+1}}{1 - G_1}\,\frac{S^{r-k}}{(r-k)!}\right|_{S=0} = -\frac{G_{k+1}}{1 - G_1}, \qquad k = 1, 2, 3, \ldots. \tag{3.25}$$

By differentiating the relation (3.20) with respect to S, we get

$$\sum_{r=1} {}_2\mu_r\,\frac{\beta^{r-1}}{(r-1)!}\,\frac{\partial\beta}{\partial S} = B^{-2}\left(-\frac{\partial B}{\partial S}\right)C + B^{-1}CD. \tag{3.26}$$

Since ${}_2\mu_1 = 0$, by putting $S = 0$ and $\beta = 0$ in the above relation, we get the mean ${}_2\mu$ from

$$0 = \frac{G_2}{1 - G_1} + {}_1F_1 - {}_2\mu\,(1 - G_1) \quad \text{or} \quad {}_2\mu = \frac{{}_1F_1}{1 - G_1} + \frac{G_2}{(1 - G_1)^2}. \tag{3.27}$$

The second derivative of (3.20) with respect to S is given by (3.26) as

$$\begin{aligned}
&\sum_{r=2} {}_2\mu_r\,\frac{\beta^{r-2}}{(r-2)!}\left(\frac{\partial\beta}{\partial S}\right)^2 + \sum_{r=1} {}_2\mu_r\,\frac{\beta^{r-1}}{(r-1)!}\,\frac{\partial^2\beta}{\partial S^2} \\
&= 2B^{-3}\left(-\frac{\partial B}{\partial S}\right)^2 C + B^{-2}\left(-\frac{\partial^2 B}{\partial S^2}\right)C + 2B^{-2}\left(-\frac{\partial B}{\partial S}\right)CD + B^{-1}CD^2 + B^{-1}C\frac{\partial D}{\partial S}.
\end{aligned} \tag{3.28}$$

On putting $S = 0$ and $\beta = 0$ in the above equation

$${}_2\mu_2\,(1 - G_1)^2 = \frac{3G_2^2}{(1 - G_1)^2} + \frac{G_3}{1 - G_1} + \frac{2G_2\,({}_1F_1 - {}_2\mu\,(1 - G_1))}{1 - G_1} + {}_1F_2 + {}_2\mu G_2,$$

which gives the variance (or the second central moment) as

$${}_2\mu_2 = \sigma^2 = \frac{{}_1F_2}{(1 - G_1)^2} + \frac{{}_1F_1G_2 + G_3}{(1 - G_1)^3} + \frac{2G_2^2}{(1 - G_1)^4}. \tag{3.29}$$

The third derivative of (3.20) with respect to S is given by (3.28) as

$$\sum_{r=3} {}_2\mu_r \frac{\beta^{r-3}}{(r-3)!}\left(\frac{\partial\beta}{\partial S}\right)^3 + \sum_{r=2} 3\,{}_2\mu_r \frac{\beta^{r-2}}{(r-2)!}\frac{\partial\beta}{\partial S}\frac{\partial^2\beta}{\partial S^2} + \sum_{r=1} {}_2\mu_r \frac{\beta^{r-1}}{(r-1)!}\frac{\partial^3\beta}{\partial S^3}$$

$$= 6B^{-3}C\left[B^{-1}\left(-\frac{\partial B}{\partial S}\right)^3 + \frac{\partial B}{\partial S}\frac{\partial^2 B}{\partial S^2} + \left(\frac{\partial B}{\partial S}\right)^2 D\right]$$

$$- B^{-2}C\left[\frac{\partial^3 B}{\partial S^3} + 3\frac{\partial^2 B}{\partial S^2}D + 3\frac{\partial B}{\partial S}D^2 + 3\frac{\partial B}{\partial S}\frac{\partial D}{\partial S}\right]$$

$$+ B^{-1}C\left[D^3 + 3D\frac{\partial D}{\partial S} + \frac{\partial^2 D}{\partial S^2}\right]. \tag{3.30}$$

On putting $\beta = 0$, $S = 0$ in the above, substituting the values of the terms, and on simplification, we have

$${}_2\mu_3 = \frac{3\,{}_2\mu_2 G_2}{(1-G_1)^2} + \frac{{}_1F_3}{(1-G_1)^3} + \frac{{}_1F_1 G_3 + G_4}{(1-G_1)^4} + \frac{4G_2G_3}{(1-G_1)^5} + \frac{2G_2^3}{(1-G_1)^6}. \tag{3.31}$$

The fourth derivative of (3.20) with respect to S can be written down from (3.30) as

$$\sum_{r=4} {}_2\mu_r \frac{\beta^{r-4}}{(r-4)!}\left(\frac{\partial\beta}{\partial S}\right)^4 + \sum_{r=3} 6\,{}_2\mu_r \frac{\beta^{r-3}}{(r-3)!}\left(\frac{\partial\beta}{\partial S}\right)^2\frac{\partial^2\beta}{\partial S^2} + \sum_{r=2} 3\,{}_2\mu_r \frac{\beta^{r-2}}{(r-2)!}\left(\frac{\partial^2\beta}{\partial S^2}\right)^2$$

$$+ \sum_{r=2} 4\,{}_2\mu_r \frac{\beta^{r-2}}{(r-2)!}\frac{\partial\beta}{\partial S}\frac{\partial^3\beta}{\partial S^3} + \sum_{r=1} {}_2\mu_r \frac{\beta^{r-1}}{(r-1)!}\frac{\partial^4\beta}{\partial S^4}$$

$$= 24B^{-5}\left(\frac{\partial B}{\partial S}\right)^4 C - 36B^{-4}\left(\frac{\partial B}{\partial S}\right)^2\frac{\partial^2 B}{\partial S^2}C - 24B^{-4}\left(\frac{\partial B}{\partial S}\right)^3 C + 6B^{-3}\left(\frac{\partial^2 B}{\partial S^2}\right)^2 C$$

$$+ 8B^{-3}\frac{\partial B}{\partial S}\frac{\partial^3 B}{\partial S^3}C + 24\beta^{-3}\frac{\partial B}{\partial S}\frac{\partial^2 B}{\partial S^2}CD + 12B^{-3}\left(\frac{\partial B}{\partial S}\right)^2 CD^2$$

$$+ 12B^{-3}\left(\frac{\partial B}{\partial S}\right)^2 C\frac{\partial D}{\partial S} - B^{-2}\frac{\partial^4 B}{\partial S^4}C - 4B^{-2}\frac{\partial^3 B}{\partial S^3}CD - 6B^{-2}\frac{\partial^2 B}{\partial S^2}CD^2$$

$$- 6B^{-2}\frac{\partial^2 B}{\partial S^2}C\frac{\partial D}{\partial S} - 4B^{-2}\frac{\partial B}{\partial S}CD^3 - 12B^{-2}\frac{\partial B}{\partial S}CD\frac{\partial D}{\partial S} - 4B^{-2}\frac{\partial B}{\partial S}C\frac{\partial^2 D}{\partial S^2}$$

$$+ B^{-1}CD^4 + 6B^{-1}CD^2\frac{\partial D}{\partial S} + 3B^{-1}C\left(\frac{\partial D}{\partial S}\right)^2 + 4B^{-1}CD\frac{\partial^2 D}{\partial S^2} + B^{-1}C\frac{\partial^3 D}{\partial S^3}. \tag{3.32}$$

On putting $\beta = 0$, $S = 0$ in the above and on substituting the values of the terms, we get

$$_2\mu_4 (1 - G_1)^4 - 6 {_2\mu_3} G_2 (1 - G_1)^2 + 3 {_2\mu_2} G_2^2 - 4 {_2\mu_2} G_3 (1 - G_1)$$

$$= {_1F_4} + \mu_1' G_4 + 3 \left({_1F_2} + \mu_1' G_2\right)^2 + \frac{6G_3 \left({_1F_2} + \mu_1' G_2\right) + G_5}{1 - G_1}$$

$$+ \frac{6G_3^2 + 6G_2^2 \left({_1F_2} + \mu_1' G_2\right) + 4G_2G_4}{(1 - G_1)^2} + \frac{18G_3G_2^2}{(1 - G_1)^3} + \frac{9G_2^4}{(1 - G_1)^4}.$$

On simplification of the above expression, the fourth central moment $_2\mu_4$ becomes

$$_2\mu_4 = 3 {_2\mu_2^2} + \frac{6 {_2\mu_3} G_2}{(1 - G_1)^2} - \frac{3 {_2\mu_2} G_2^2}{(1 - G_1)^4} + \frac{4 {_2\mu_2} G_3}{(1 - G_1)^3} + \frac{_1F_4}{(1 - G_1)^4}$$

$$+ \frac{{_1F_1} G_4 + G_5}{(1 - G_1)^5} + \frac{5G_4G_2 + 3G_3^2}{(1 - G_1)^6} + \frac{12G_3G_2^2}{(1 - G_1)^7} + \frac{6G_2^4}{(1 - G_1)^8}. \tag{3.33}$$

The higher central moments can similarly be evaluated by taking more derivatives with respect to S and by putting $\beta = 0$, $S = 0$ in them.

3.3.1 Cumulants of Lagrangian Distribution $L_1(f_1; g; y)$

The mgf of the class of Lagrangian distributions in (2.9) is given by (3.17). On simplification by the denominator in (3.20) and by (3.22) it becomes

$$H\left(e^{\beta}\right) = f_1\left(e^S\right) \left[1 - \sum_{r=1} \frac{G_{r+1}}{1 - G_1} \frac{S^r}{r!}\right]^{-1}. \tag{3.34}$$

On taking the logarithms on both sides, the cumulant generating function (cgf) of the Lagrangian distributions is

$$\ln H\left(e^{\beta}\right) = \ln f_1\left(e^S\right) - \ln\left[1 - \sum_{r=1} \frac{G_{r+1}}{1 - G_1} \frac{S^r}{r!}\right],$$

which gives

$$\sum_{k=1} {_2L_k} \frac{\beta^k}{k!} = \sum_{k=1} {_1F_k} \frac{S^k}{k!} + \sum_{k=1} \frac{1}{k} \left(\sum_{r=1} \frac{G_{r+1}}{1 - G_1} \frac{S^r}{r!}\right)^k. \tag{3.35}$$

Now, one can follow the same method of differentiating the relation (3.35) successively with respect to S and putting $S = 0$ and $\beta = 0$, as was followed for the central moments, to get the values of the successive cumulants. However, the Lagrange transformation $z = ug(z)$ changes to $e^S = e^{\beta} g\left(e^S\right)$ and provides the power series expansion (3.8) of S in terms of β as

$$S = \sum_i D_i \beta^i / i!. \tag{3.36}$$

By substituting the value of S from (3.36) in (3.35), we have the identity

$$\sum_{k=1} {_2L_k} \frac{\beta^k}{k!} = \sum_{k=1} \frac{_1F_k}{k!} \left(\sum_{i=1} \frac{D_i}{i!} \beta^i\right)^k + \sum_{k=1} \frac{1}{k} \left(\sum_{r=1} \frac{G_{r+1}}{1 - G_1} \frac{1}{r!} \left(\sum_{i=1} \frac{D_i}{i!} \beta^i\right)^r\right)^k. \tag{3.37}$$

On equating the coefficients of β, β^2, β^3, and β^4 on both sides in (3.37) we obtain the values of the first four cumulants of Lagrangian distribution $L_1(f_1; g; y)$ in the form

$$\begin{cases} {}_2L_1 = {}_1F_1 D_1 + G_2 D_1 (1-G_1)^{-1}, \\ {}_2L_2 = {}_1F_1 D_2 + {}_1F_2 D_1^2 + \left(G_2 D_2 + G_3 D_1^2\right)(1-G_1)^{-1} + G_2 D_1^2 (1-G_1)^{-2}, \\ {}_2L_3 = {}_1F_1 D_3 + 3{}_1F_2 D_1 D_2 + {}_1F_3 D_1^3 + \left(G_2 D_3 + 3G_3 D_1 D_2 + G_4 D_1^3\right)(1-G_1)^{-1} \\ \qquad + \left(3G_2 G_3 D_1^3 + 3G_2^2 D_1 D_2\right)(1-G_1)^{-2} + 2G_2^3 D_1^3 (1-G_1)^{-3}, \\ {}_2L_4 = {}_1F_1 D_4 + 4{}_1F_2 D_1 D_3 + 3{}_1F_2 D_2^2 + 6{}_1F_3 D_1^2 D_2 + {}_1F_4 D_1^4 \\ \qquad + \left(G_2 D_4 + 4G_3 D_1 D_3 + 3G_3 D_2^2 + 6G_4 D_1^2 D_2 + G_5 D_1^4\right)(1-G_1)^{-1} \\ \qquad + \left(4G_2^2 D_1 D_3 + 3G_2^2 D_2^2 + 3G_3^2 D_1^4 + 18 G_2 G_3 D_2 D_1^2 + 4G_4 G_2 D_1^4\right)(1-G_1)^{-2} \\ \qquad + 12\left(G_2^3 D_1^2 D_2 + G_2^2 G_3 D_1^4\right)(1-G_1)^{-3} + 6G_2^4 D_1^4 (1-G_1)^{-4}. \end{cases} \tag{3.38}$$

Similarly, one can write down the values of higher cumulants as well. By substituting the values of D_1, D_2, D_3, and D_4 from the relations (3.14) one would get the cumulants ${}_2L_k$, $k = 1, 2, 3, 4$, in terms of the cumulants of $g(z)$ and $f_1(z)$. However, it may be noted that the value of each cumulant consists of some terms which contain ${}_1F_k$, $k = 1, 2, 3, 4$, and some terms which contain G_k, $k = 1, 2, 3, 4$. The terms containing ${}_1F_k$, $k = 1, 2, 3, 4$, in each cumulant are the same as in (3.12) for the respective cumulants for Lagrangian distributions in (2.7) with $f_1(z)$ used in place of $f(z)$. Thus, on substitution of the values of D and on simplification, the first four cumulants of Lagrangian distributions $L_1(f_1; g; y)$ become

$$\begin{cases} {}_2L_1 = {}_1L_1 + G_2 (1-G_1)^{-2}, \\ {}_2L_2 = {}_1L_2 + G_3 (1-G_1)^{-3} + 2G_2^2 (1-G_1)^{-4}, \\ {}_2L_3 = {}_1L_3 + G_4 (1-G_1)^{-4} + 7G_3 G_2 (1-G_1)^{-5} + 8G_2^3 (1-G_1)^{-6}, \\ {}_2L_4 = {}_1L_4 + G_5 (1-G_1)^{-5} + \left(11 G_4 G_2 + 7G_3^2\right)(1-G_1)^{-6} \\ \qquad + 59 G_3 G_2^2 (1-G_1)^{-7} + 48 G_2^4 (1-G_1)^{-8}, \end{cases} \tag{3.39}$$

where the values of ${}_1L_k$, $k = 1, 2, 3, 4$, are given by the first four relations in (3.15) with $f(z)$ replaced by $f_1(z)$. The expressions in (3.39) and the relation (3.35) clearly prove that *each cumulant for a particular distribution (generated by $f_1(z)$ and $g(z)$, in the class of Lagrangian distributions (2.9)) has a larger value than the same cumulant for a similar distribution (generated by the same $f_1(z)$ and $g(z)$ in the class of Lagrangian distributions in (2.7)).*

3.3.2 Applications

Example 1. The linear Poisson family of the class of Lagrangian distributions in (2.9) is generated by $f_1(z) = e^{\theta(z-1)}$ and $g(z) = e^{\lambda(z-1)}$, for which all the cumulants are θ and λ, respectively.

Substituting the values of the cumulants of $f_1(z)$ and $g(z)$ in the results (3.15) and (3.39) and on simplification, the cumulants of the linear Poisson family become

$$\begin{cases} {}_2L_1 = \theta(1-\lambda)^{-1} + \lambda(1-\lambda)^{-2}, \\ {}_2L_2 = \theta(1-\lambda)^{-3} + \lambda(1+\lambda)(1-\lambda)^{-4}, \\ {}_2L_3 = \theta(1-2\lambda)(1-\lambda)^{-5} + \lambda\left(1+5\lambda+2\lambda^2\right)(1-\lambda)^{-6}, \\ {}_2L_4 = \theta\left(1+8\lambda+6\lambda^2\right)(1-\lambda)^{-7} + \lambda\left(1+15\lambda+26\lambda^2+6\lambda^3\right)(1-\lambda)^{-8}. \end{cases} \tag{3.40}$$

The same results can be obtained by calculating the central moments from the formulas (3.27), (3.29), (3.31), and (3.33).

Example 2. The Poisson-binomial family in the class of Lagrangian distributions in (2.9) has the pmf

$$P(Y=y) = (1-mp)\, e^{-\theta}\theta^y q^{my}\,(y!)^{-1}\ {}_2F_0\,(-y,-my;\,;\,p/\theta q)$$

for $y = 0, 1, 2, \ldots$ and zero otherwise, where $\theta > 0,\ 0 < p = 1-q < 1$, and $0 < m < p^{-1}$. It is given by the pgfs $f_1(z) = e^{\theta(z-1)}$ and $g(z) = (q+pz)^m$ under the Lagrange expansion in (1.80).

Since all the cumulants of the Poisson distribution given by $f_1(z)$ are θ and the first five cumulants of the binomial distribution, given by $g(z)$, are $mp, mpq, mpq\,(q-p)$, $mpq\,(1-6pq)$, and $mpq\,(q-p)\,(1-12pq)$, one can use the four formulas in (3.15) and (3.39) to write down the first four cumulants of the Poisson-binomial families of $L(f;g;x)$ and $L_1(f_1;g;y)$ classes:

$$\begin{cases} {}_1L_1 = \theta\,(1-mp)^{-1}, \\ {}_1L_2 = \theta\left(1-mp^2\right)(1-mp)^{-3}, \\ {}_1L_3 = \theta\left\{\left(1-mp^2\right)^2 - 2mpq^2\right\}(1-mp)^{-5}, \\ {}_1L_4 = \frac{\theta}{(1-mp)^4} + \frac{\theta mpq}{(1-mp)^5} + \frac{\theta mpq^2(2+3mp)}{(1-mp)^6} + \frac{\theta mpq^3(6+8mp+m^2p^2)}{(1-mp)^7}, \end{cases} \tag{3.41}$$

and

$$\begin{cases} {}_2L_1 = \theta\,(1-mp)^{-1} + mpq\,(1-mp)^{-2}, \\ {}_2L_2 = \theta\left(1-mp^2\right)(1-mp)^{-3} + mpq\,(1-2p+mp)\,(1-mp)^{-4}, \\ {}_2L_3 = {}_1L_3 + \frac{mpq}{(1-mp)^4}\left\{1-6pq+\frac{7mpq(q-p)}{1-mp}+\frac{8m^2pq^2}{(1-mp)^2}\right\}, \\ {}_2L_4 = {}_1L_4 + \frac{mpq(q-p)(1-12pq)}{(1-mp)^5} + \frac{m^2p^2q^2(18-94pq)}{(1-mp)^6} \\ \qquad + \frac{59m^3p^3q^3(q-p)}{(1-mp)^7} + \frac{48m^4p^4q^4}{(1-mp)^8}. \end{cases} \tag{3.42}$$

3.4 Relations between the Two Classes $L(f;g;x)$ and $L_1(f_1;g;y)$

A number of theorems establishing the relationship between the two classes of Lagrangian probability distributions, when $f_1(z)$ is replaced with $f(z)$ in $L_1(f_1;g;y)$, will be proved for some special cases.

Theorem 3.1. *Let $f(z) = g(z)$ and let Y and X be the r.v.s for the corresponding general $L_1(f;g;y)$ class and general $L(f;g;x)$ class, respectively. Then $P(Y=k) = (k+1)(1-g'(1))P(X=k)$ for all values of k.*

Proof. For the class of Lagrangian probability distribution in (2.7) with $f(z) = g(z)$,

$$\begin{aligned} P(X=k) &= \frac{1}{k!}\left\{D^{k-1}\left(g^k(z)\,g'(z)\right)\right\}_{z=0} \\ &= \frac{1}{(k+1)!}\left\{D^k\left(g^{k+1}(z)\right)\right\}_{z=0}. \end{aligned}$$

For the class of Lagrangian probability distribution in (2.9),

$$P(Y=k) = \frac{\left(1-g'(1)\right)}{k!}\left\{D^k\left(g^k(z)\cdot g(z)\right)\right\}_{z=0}$$
$$= \left(1-g'(1)\right)(k+1)P(X=k),$$

which proves the theorem. □

Theorem 3.2. *Let X have the delta Lagrangian distribution in (2.4) for $f(z)=z^n$ under the transformation $z=ug(z)$. If each probability $P(X=k)$ is weighted by the weight function $w_k=k$ and Y is the r.v. of the weighted delta Lagrangian distribution in (2.4), then Y has the corresponding delta Lagrangian distribution in (2.12).*

Proof. For the delta Lagrangian distribution in (2.4) with $f(z)=z^n$,

$$P(X=k) = \frac{n}{k!}\left\{D^{k-1}\left(g^k(z)\cdot z^{n-1}\right)\right\}_{z=0}$$
$$= \frac{n}{(k-n)!k}\left\{D^{k-n}g^k(z)\right\}_{z=0}, \; k=n, n+1, \ldots .$$

Therefore, the probability function of the weighted distribution becomes

$$P(Y=k) = \frac{kP(X=k)}{\sum_{k=n}^{\infty} kP(X=k)} = \frac{kP(X=k)}{E[X]}$$
$$= \frac{1-g'(1)}{f'(1)}\frac{n}{(k-n)!}\left\{D^{k-n}g^k(z)\right\}_{z=0} \quad \left(\text{since E}[X] = f'(1)/\left(1-G'(1)\right)\right)$$
$$= \left(1-g'(1)\right)\frac{1}{(k-n)!}\left\{D^{k-n}g^k(z)\right\}_{z=0}, \quad k=n, n+1, \ldots, \tag{3.43}$$

which is the probability function for the delta Lagrangian distribution in (2.12). □

Theorem 3.3. *Let the r.v. X have the general Lagrangian distribution in (2.7) for $f(z)=(g(z))^m$ under the transformation $z=ug(z)$, where m is a real number. If each probability function $P(X=k)$ is weighted by the weight function $w_k=m+k$ and Y is the r.v. representing the weighted Lagrangian distribution in (2.7), then Y has the corresponding Lagrangian distribution in (2.9).*

Proof. The probability function of the general Lagrangian distribution in (2.7) for $f(z)=(g(z))^m$ becomes

$$P(X=k) = \frac{m}{k!}\left\{D^{k-1}\left((g(z))^{k+m-1}g'(z)\right)\right\}_{z=0}, \qquad k=0,1,2,\ldots,$$
$$= \frac{m}{k!\,(m+k)}\left\{D^k\left((g(z))^{k+m}\right)\right\}_{z=0}. \tag{3.44}$$

Therefore, the probability function of the weighted distribution becomes

$$P(Y=k) = (m+k)\, P(X=k) \Big/ \sum_{k=0}^{\infty} (m+k)\, P(X=k)$$

$$= (m+k)\, P(X=k) \Big/ \left\{ m + f'(1) \left(1 - g'(1)\right)^{-1} \right\}.$$

Since $f'(1) = Dg^m(z)|_{z=1} = mg'(1)$, the above probability becomes

$$P(Y=k) = \left(1 - g'(1)\right) \frac{m+k}{m} P(X=k)$$

$$= \left(1 - g'(1)\right) \cdot (k!)^{-1} \left\{ D^k \left(g^k(z)\, g^m(z) \right) \right\}_{z=0} \qquad \text{(by (3.44))},$$

which is the probability function for the corresponding Lagrangian distribution in (2.9). □

Theorem 3.4. *Let the r.v. Y have a general Lagrangian distribution in (2.9) for $f(z) = (g(z))^m$ under the transformation $z = ug(z)$, m being a real number. If each probability function $P(Y = k)$ is weighted by the weight function $w_k = (m+k)^{-1}$ and X is the r.v. representing the weighted Lagrangian distribution (2.9), then X has the corresponding Lagrangian distribution (2.7) divided by m.*

Proof. The probability function of the weighted distribution of Y becomes

$$P(X=k) = \frac{w_k P(Y=k)}{\sum_{k=0}^{\infty} w_k P(Y=k)}$$

$$= \frac{(m+k)^{-1} \left(1 - g'(1)\right) \left\{ D^k g^{k+m}(z) \right\}_{z=0} / k!}{\sum_{k=0}^{\infty} (m+k)^{-1} \left(1 - g'(1)\right) \left\{ D^k g^{k+m}(z) \right\}_{z=0} / k!}$$

$$= \frac{1}{k!} \left\{ D^{k-1} \left(g^{k+m-1}(z)\, g'(z) \right) \right\}_{z=0} \Big/ \sum_{k=0}^{\infty} \frac{1}{k!} D^{k-1} \left\{ g^{k+m-1}(z)\, g'(z) \right\}_{z=0}$$

$$= \frac{1}{k!} D^{k-1} \left(g^k(z)\, g^{m-1}(z)\, g'(z) \right) \Big|_{z=0} = \frac{1}{k!m} \left\{ D^{k-1} \left(g^k(z)\, f'(z) \right) \right\}_{z=0},$$

which is the pmf for the corresponding Lagrangian distribution in (2.7) divided by m. □

Some Examples

It can easily be verified that the models (3), (4), and (5) of Table 2.4 for the basic Lagrangian distributions (2.11) are the weighted distributions of $P(X = x)$, weighted with $w_x = x$ from the basic Lagrangian distributions (2.2) as follows:

(i) Ved distribution is the weighted Geeta distribution with $g(z) = q^m(1 - pz)^{-m}$.
(ii) Sudha distribution is the weighted Borel distribution with $g(z) = e^{\lambda(z-1)}$.
(iii) Hari distribution is the weighted Otter distribution with $g(z) = p + qz^2$.

Example 4.1. Let the r.v. X have a generalized Poisson distribution GPD with the probability function

$$P(X = k) = \theta\,(\theta + k\lambda)^{k-1}\, e^{-\theta - k\lambda}/k!, \qquad k = 0, 1, 2, \ldots,$$

which is a Lagrangian distribution in (2.7) with $g(z) = e^{\lambda(z-1)}$ and $f(z) = e^{\theta(z-1)} = (g(z))^{\theta/\lambda}$.

From Theorem 3.3, the weight function is $w_k = \theta/\lambda + k$ and $E\,[w_x] = \theta/\lambda + \theta\,(1-\lambda)^{-1} = \theta\,(1-\lambda)^{-1}\,\lambda^{-1}$. Therefore, if $0 < \lambda < 1$, the weighted distribution of the GPD is

$$\begin{aligned} P\,(Y = k) &= (\theta/\lambda + k)\, P\,(X = k)\,/E\,[\theta/\lambda + X] \\ &= \frac{(\theta/\lambda + k)}{\theta/\lambda + \theta\,(1-\lambda)^{-1}}\frac{1}{k!}\theta\,(\theta + k\lambda)^{k-1}\, e^{-\theta - k\lambda},\ k = 0, 1, 2, \ldots, \\ &= (1-\lambda)\,(\theta + k\lambda)^{k}\, e^{-\theta - k\lambda}/k!, \end{aligned} \tag{3.45}$$

which is the linear Poisson distribution (model (10) of Table 2.4), defined by Jain (1975a), and is a family in the class of the Lagrangian distributions in (2.9). Thus, the linear function Poisson distribution is the weighted form of the GPD with the weight function $w_k = \theta/\lambda + k$.

Example 4.2. Let the r.v. X have a generalized negative binomial distribution (GNBD) with the probability function

$$P\,(X = k) = \frac{n}{n + mk}\binom{n + mk}{k}\theta^{k}\,(1-\theta)^{n+mk-k},\ k = 0, 1, 2, \ldots,$$

where $0 < \theta < 1$, $n > 0$, $m = 0$, or ≥ 1 and $0 < m\theta < 1$. The GNBD is a family in the class of the Lagrangian distributions in (2.7) with $g(z) = (1-\theta+\theta z)^m$ and $f(z) = (1-\theta+\theta z)^n = (g\,(z))^{n/m}$.

Since it satisfies the condition of Theorem 3.3, the weight function is $w_k = n/m + k$ and

$$E\,[w_X] = \frac{n}{m} + E\,[X] = \frac{n}{m} + \frac{n\theta}{1 - m\theta} = \frac{n}{m\,(1 - m\theta)}.$$

Thus, the weighted distribution of the GNBD is given by

$$\begin{aligned} P\,(Y = k) &= \frac{w_k}{E\,[w_x]} P\,(X = k) \\ &= \frac{(n + mk)\,(1 - m\theta)}{n}\frac{n}{n + mk}\binom{n + mk}{k}\theta^{k}\,(1-\theta)^{n+mk-k} \\ &= (1 - m\theta)\binom{n + mk}{k}\theta^{k}\,(1-\theta)^{n+mk-k}, \quad k = 0, 1, 2, \ldots, \end{aligned}$$

which is the linear binomial distribution (model (11) of Table 2.4) of the class of Lagrangian distributions in (2.9).

Example 4.3. Let the r.v. X have a generalized logarithmic series distribution (GLSD) with the probability function

$$P(X = k) = \frac{\alpha\,\Gamma(mk)\,\theta^k\,(1-\theta)^{mk-k}}{k!\,\Gamma(mk-k+1)}, \quad k = 1, 2, 3, \ldots,$$

where $\alpha = [-\ln(1-\theta)]^{-1}$, $0 < \theta < 1$, and $1 \le m < \theta^{-1}$. The GLSD is a member of the class of Lagrangian distributions in (2.7) with $g(z) = (1-\theta)^{m-1}(1-\theta z)^{1-m}$ and $f(z) = \alpha \ln(1-\theta z)$.

If the weight function is $w_k = k$, the weighted distribution of the GLSD becomes

$$P(y = k) = (1 - m\theta)\binom{mk-1}{k-1}\theta^{k-1}(1-\theta)^{mk-k}, \quad k = 1, 2, 3, \ldots, \tag{3.46}$$

which is a generalized geometric distribution and belongs to the class of Lagrangian distributions in (2.9). Its particular case for $m = 1$ is the geometric distribution.

3.5 Some Limit Theorems for Lagrangian Distributions

Consul and Shenton (1973a) have given two limit distributions for Lagrangian probability distributions. Pakes and Speed (1977) have improved these results and have modified them with a number of conditions and cases.

The Lagrangian distribution $L(f; g; x)$, defined by (2.7), has the pgf $f(z) = f(\psi(u))$ given by (2.6), where $\psi(u)$ is the pgf of the basic Lagrangian distribution whose mean and variance are $a = (1 - G_1)^{-1}$ and $D_2 = G_2 a^3$, respectively (see (3.14)). The distribution with pgf $f(z)$ has mean F_1 and variance F_2. By Theorem 2.8 the r.v. X having the Lagrangian distribution $L(f; g; x)$ can be viewed as

$$X = X_1 + X_2 + \cdots + X_N,$$

where the i.i.d. X_i, $i = 1, 2, \ldots, N$, have the basic Lagrangian distribution with pgf $\psi(u)$ and the r.v. N, independent of X_i, $i = 1, 2, 3, \ldots$, has the pgf $f(z)$. By the general results on the moments of random sums (Feller, 1968, p. 301), the r.v. X, i.e., the Lagrangian distribution, has mean ${}_1L_1$ and variance ${}_1L_2$ given by

$$\mu = {}_1L_1 = F_1 a, \qquad \sigma^2 = {}_1L_2 = d^2 F_1 + F_2 a^2 = F_1 G_2 a^3 + F_2 a^2,$$

which are the same as given in (3.15).

Rychlik and Zynal (1973) have considered the limit behavior of sums of a random number of independent r.v.s and have shown that $(X - \mu)\sigma^{-1}$ is asymptotically standard normal if the following conditions hold:

(i) $NF_1^{-1} \to 1$ (in probability) as $F_1 \to \infty$.
(ii) $(N - F_1)F_2^{-1/2} \to$ a standard normal r.v. (in distribution) as $F_1 \to \infty$.

Note that N is a r.v. having mean F_1 and variance F_2. In both conditions, the mean F_1 becomes indefinitely large and it is further implied that the r.v. N assumes a large value. Thus we have the first limit theorem for the class of Lagrangian distributions in (2.7).

Theorem 3.5. *If X is a Lagrangian variate with mean $\mu = F_1(1 - G_1)^{-1}$ and standard deviation σ, then for any specific values of G_1 (< 1) and G_2, the distribution of the standardized r.v. $Z = (X - \mu)/\sigma$ is asymptotically normal as $F_1 \to \infty$, if the conditions (i) and (ii) hold.*

It may be noted that the above theorem on limiting distribution fails when the condition $F_1 \to \infty$ reduces the distribution, given by $f(z)$, into a degenerate form. Let $f(z) = (1-p)(1-pz)^{-1}$ so that $F_1 = p(1-p)^{-1}$. Now $F_1 \to \infty$ when $p \to 1$, which changes $f(z)$ into a degenerate function.

In a similar manner, if $f(z) = \ln(1-pz)/\ln(1-p)$, then the value of $F_1 = p[-\ln(1-p)]^{-1}$, which approaches ∞ when $p \to 0$. Again, when $p \to 0$, the distribution given by the pgf $f(z)$ becomes degenerate.

Theorem 3.6. *Let X be a Lagrangian variate, defined by the pgf $f(z)$, under the transformation $z = ug(z)$, where $g(0) \neq 0$, with mean $\mu = F_1(1-G_1)^{-1}$ and variance $\sigma^2 = F_2(1-G_1)^{-2} + F_1 G_2 (1-G_1)^{-3}$. Also, let $g(z)$ be such that it does not become degenerate when $G_1 \to 1-$ and $g''(1-) > 0$. Then the limiting distribution of the variate $Y = X/\sigma$ is the inverse Gaussian density function as $F_1 \to \infty$ and $G_1 \to 1-$ such that $F_1(1-G_1) = c^2 G_2$, where c is a constant.*

For the proof of this theorem we refer the reader to Consul and Shenton (1973a) and Pakes and Speed (1977).

Minami (1999) also modified the results in Consul and Shenton (1973a) and showed that the Lagrangian distributions converged to the normal distributions under certain conditions and to the inverse Gaussian distributions under some other conditions. The conditions presented by Pakes and Speed (1977) are easier than the conditions presented by Minami (1999), which are based on higher order cumulants of the Lagrangian generating distributions.

Mutafchiev (1995) obtained an integral representation of the Lagrangian distribution (2.7) in the form

$$P[Y(\lambda,\mu) = x] = \frac{\lambda\phi'(\lambda)}{2x\pi\phi(\lambda)} \int_{-\pi}^{\pi} \left[\frac{\psi(\mu e^{i\theta})}{\psi(\mu)}\right]^m \left(\frac{\phi'(\lambda e^{i\theta})}{\phi'(\lambda)}\right) e^{-(x-1)i\theta} d\theta, \tag{3.47}$$

where $f(z) = \phi(\lambda z)/\phi(\lambda)$, $g(z) = \psi(\mu z)/\psi(\mu)$, λ and μ are parameters. In (3.47), $Y(\lambda,\mu)$ is considered as a normalized sum of $x+1$ independent r.v.s having power series distributions; x of them have the pgf $\psi(z)$ and one is defined by the derivative $\phi'(\lambda z)/\phi'(\lambda)$.

Under some assumptions on ϕ and ψ, Mutafchiev (1995) proved various local limit theorems for the class of Lagrangian distributions in (2.7) as $x \to \infty$ and the parameters λ and μ change in an appropriate way.

3.6 Exercises

3.1 Obtain the first four cumulants for the generalized logarithmic series distribution in Table 2.3.

3.2 Obtain the first four cumulants for the Poisson-negative binomial family of Lagrangian distributions in (2.7).

3.3 Obtain the first four cumulants for the double negative binomial family of the Lagrangian distributions (2.9) in Table 2.4.

3.4 Let X have the generalized logarithmic series family in the class of Lagrangian distributions in (2.7). If each probability mass $P(X = k)$ is weighed by $w_k = k$, find the weighted distribution of X.

3.5 A r.v. X has the logarithmic negative binomial distribution (model (14) in Table 2.3). Find a recurrence relation between the noncentral moments. Also, find a recurrence relation between the central moments. Hence or otherwise, obtain the first four central moments for the r.v. X.

3.6 A r.v. X has the generalized Poisson distribution given by the pgfs $g(z) = e^{\lambda(z-1)}$, $0 < \lambda < 1$, and $f(z) = e^{\theta(z-1)}$, $\theta > 1$. If each probability $P(X = k)$ is weighted by the weight function $w_k = \theta/\lambda + k$ and Y is the r.v. representing the weighted distribution, then find $P(Y = y)$ in a suitable form. Also, what will be the distribution of Y if $w_k = (\theta/\lambda + k)^{-1}$?

3.7 A random variable X has the generalized negative binomial distribution given by the pgfs $g(z) = q^m (1 - pz)^{-m}$ and $f(z) = q^n (1 - pz)^{-n}$, $0 < p = 1 - q < 1$, $m > 0$, $n > 0$, $mp < 1$. If each probability function $P(X = k)$ is weighted by the weight function $w_k = (n/m + k)^{-1}$ and Y is the r.v. representing the weighted distribution, then find $P(Y = y)$ in a suitable form.

3.8 Prove the limiting result as stated in Theorem 3.5.

3.9 Let $g(z)$ be a successively differentiable function such that $g(1) = 1$ and $g(0) \neq 0$. Let X and Y be random variables from the delta Lagrangian distribution (2.4) and delta Lagrangian distribution (2.12), respectively. Prove that

$$P(Y = k) = \frac{k}{n} \left(1 - g'(1)\right) P(X = k)$$

for all values of k. Show that this result holds when X and Y have basic Lagrangian distribution (2.2) and basic Lagrangian distribution (2.11), respectively.

4

Quasi-Probability Models

4.1 Introduction

A binomial model is based on the assumption of a repeated set of Bernoulli trials wherein the probability p of the occurrence of an event remains the same for each trial. In a laboratory experiment one can exercise controls and make sure that p is constant for each trial; however, in the real world of living beings the value of p changes according to the circumstances. These changes may be due to the inheritance of genes, psychological effects, feelings of social togetherness, previous experience, determination for success or to face a common danger, adjustments needed for changes in the environments, etc. Accordingly, the observed counts, generated in many experiments by a set of Bernoulli-type trials, do not fit a binomial model and exhibit either much greater or much lesser variation than a binomial model. This over dispersion or under dispersion is usually classified as extra-binomial variation.

Chaddha (1965) considered a binomial distribution with contagion, while Katti and Sly (1965) analyzed contagious data by developing some behavioristic models. Crow and Bardwell (1965) discussed some hyper-Poisson distributions and applied them to a number of data sets. Katti and Gurland (1961) developed the Poisson Pascal distribution for a similar type of data.

Some other scientists have dealt with such data by a generalization of the binomial model where the parameter p is taken to be a beta random variable. The beta binomial model, thus obtained, has been used by Kleinman (1973), Williams (1975), and Crowder (1978) for the analysis of overly dispersed counts.

Similar changes have been suggested to the hypergeometric probability model, but to a lesser degree. Pólya (1930) had studied a more complicated model to describe the "contagion" of different events. Friedman (1949) has described another model and Feller (1968) has shown that the conditional probabilities can be used to explain a number of physical phenomena.

A quasi-binomial model, a quasi-hypergeometric model, and a quasi-Pólya model were defined and studied by Consul (1974) to account for the variations in the observed counts. These were obtained through simple urn models dependent upon the predetermined strategy of a player. These models are generalizations of the binomial, hypergeometric, and Pólya distributions. Another set of three quasi-probability models was defined and studied by Consul and Mittal (1975). Janardan (1975, 1978) has studied some additional properties of these models. Some of these models were used by Consul (1974, 1975) for some characterization theorems. A detailed study of the quasi-binomial distribution I was made by Consul (1990d). Charalambides (1986) studied some other properties of these models under the title of Gould series distributions.

These quasi-probability models do not belong to the class of Lagrangian probability distributions. However, all these models are based upon some identities which are proved by the use of the Lagrange expansions in (1.78) and in (1.80). For the same reason these probability models are included in this chapter, as they form a separate class by themselves.

4.2 Quasi-Binomial Distribution I (QBD-I)

A discrete random variable X is said to have a quasi-binomial distribution I (QBD-I) if its probability mass function is defined by

$$P(X = x) = \binom{m}{x} p(p + x\phi)^{x-1}(1 - p - x\phi)^{m-x}, \tag{4.1}$$

for $x = 0, 1, 2, 3, \ldots, m$ and zero otherwise, where $0 \le p \le 1$ and $-p/m < \phi < (1 - p)/m$. The QBD-I reduces to the binomial model when $\phi = 0$. The r.v. X represents the number of successes in m trials such that the probability of success in any one trial is p and in all other trials is $p + x\phi$, where x is the total number of successes in the m trials. The probability of success increases or decreases as ϕ is positive or negative and is directly proportional to the number of successes. The QBD-I does not possess the convolution property but when $m \to \infty$, $p \to 0$, and $\phi \to 0$ such that $mp = \theta$ and $m\phi = \lambda$, its limiting form is the generalized Poisson distribution (Consul, 1989a) which possesses the convolution property.

Consul (1974, 1975) has proved three theorems on the characterization of the QBD-I and has shown its usefulness in the characterization of the generalized Poisson distribution. Berg and Mutafchiev (1990) have shown that the QBD-I is useful in random mappings with an attracting center.

The probabilities for the various values of x can easily be computed with the help of a simple computer program by using the recurrence relation

$$P(X = x+1) = \frac{(m - x)(p + x\phi)}{(x + 1)(1 - p - x\phi)} \left(1 + \frac{\phi}{p + x\phi}\right)^{x} \left(1 - \frac{\phi}{1 - p - x\phi}\right)^{m-x-1} P(X = x) \tag{4.2}$$

for $x = 0, 1, 2, \ldots, m - 1$ and zero otherwise, where $P(X = 0) = (1 - p)^m$.

Consul (1990d) has computed the values of the probabilities of the QBD-I for $m = 10$, $p = 0.2, 0.3, \ldots, 0.8$ and for various values of ϕ and has drawn 27 bar diagrams to show that the QBD-I is very versatile and is unimodal and that for any given set of values of m and p the distribution shifts to the right-hand side as ϕ increases in value and to the left-hand side for negative values of ϕ. Even small values of ϕ have a substantial effect on the respective probabilities and on the values of the mean, mode, and variance.

4.2.1 QBD-I as a True Probability Distribution

Under the transformation $z = u\, e^{\theta z}$, by the Lagrange expansion (1.80),

$$\frac{e^{bz}}{1 - \theta z} = \sum_{j=0}^{\infty} \frac{u^j}{j!}(b + j\theta)^j\,. \tag{4.3}$$

By considering the expansion of $e^{(a+b)z}/(1-\theta z) = e^{az} \times e^{bz}/(1-\theta z)$ in a power series of u on the left-hand side and the product of the power series in u of e^{az} by (1.78) and of $e^{bz}/(1-\theta z)$ by (4.3) on the right-hand side and by equating the coefficients of u^m on both sides, we get the identity

$$\frac{(a+b+m\theta)^m}{m!} = \sum_{k=0}^{m} \frac{a(a+k\theta)^{k-1}}{k!} \frac{(b+m\theta-k\theta)^{m-k}}{(m-k)!}. \tag{4.4}$$

On division by the expression on the left-hand side in (4.4),

$$1 = \sum_{k=0}^{m} \binom{m}{k} \frac{a}{a+b+m\theta} \left(\frac{a+k\theta}{a+b+m\theta}\right)^{k-1} \left(\frac{b+m\theta-k\theta}{a+b+m\theta}\right)^{m-k}. \tag{4.5}$$

By using the transformation $a(a+b+m\theta)^{-1} = p$ and $\theta(a+b+m\theta)^{-1} = \phi$ in the relation (4.5), the expression changes to (4.1) and it is clear that $\sum_{k=0}^{m} P(X=k) = 1$.

If $a = mn,\ b = nmp$, and $\theta = -np$ in (4.5), it gives a new and simple relation

$$1 = \sum_{k=1}^{m} \binom{m}{k} \left(1 - \frac{kp}{m}\right)^{k-1} \left(\frac{kp}{m}\right)^{m-k},$$

which gives another nice form of QBD-I with two parameters p and m as

$$P(X=k) = \binom{m}{k} \left(1 - \frac{kp}{m}\right)^{k-1} \left(\frac{kp}{m}\right)^{m-k} \tag{4.6}$$

for $k = 1, 2, 3, \ldots, m$ and zero otherwise, where $0 \le p \le 1$.

4.2.2 Mean and Variance of QBD-I

Since m is a positive integer in (4.1) and (4.6), all the moments and cumulants of the QBD-I exist. However, their expressions are not in a compact form. The mean μ and variance σ^2 of QBD-I in (4.1) can be obtained as follows:

$$\mu = E[X] = mp \sum_{x=1}^{m} \frac{(m-1)!}{(x-1)!(m-x)!} (p+x\phi)^{x-1} (1-p-x\phi)^{m-x}$$

$$= mp \sum_{x=0}^{m-1} \binom{m-1}{x} (p+\phi+x\phi)(p+\phi+x\phi)^{x-1} (1-p-\phi-x\phi)^{m-1-x}.$$

By breaking up the factor $p+\phi+x\phi$ into $p+\phi$ and $x\phi$, we get two summations such that the first sum becomes unity and the second summation is similar to the original expectation of X, but $m-1$ instead of m. Thus, after several such repetitions, it can be shown that

$$E[X] = \mu = mp\left[1 + \phi(m-1) + \phi^2(m-1)(m-2) + \cdots + \phi^{m-1}(m-1)!\right]$$

$$= p \sum_{k=0}^{m-1} (m)_{(k+1)} \phi^k. \tag{4.7}$$

When $\phi > 0$, it can easily be shown that

$$mp(1+\phi)^{m-1} < \mu < mp(1+\phi)(1-\phi(m-2))^{-1}, \tag{4.8}$$

which implies that the mean μ for QBD-I, when $\phi > 0$, is larger than mp, the mean of the corresponding binomial model. Similarly, when $\phi < 0$, the value of μ is less than mp. Also,

$$E[X(X-1)] = m(m-1)p \sum_{x=0}^{m-2} \binom{m-2}{x} (p+2\phi+x\phi)^{x+1}(1-p-2\phi-x\phi)^{m-2-x}.$$

By expressing the factor $(p+2\phi+x\phi)^{x+1}$ in the form

$$\left[(p+2\phi)^2 + x\phi(2p+5\phi) + x(x-1)\phi^2\right](p+2\phi+x\phi)^{x-1},$$

the above sum can be split into three sums such that the first sum reduces to the value $(p+2\phi)$, the second sum is similar to the evaluation of μ, and the third sum is like $E[X(X-1)]$ with different values of parameters. Thus,

$$\begin{aligned} E[X(X-1)] = \; & m(m-1)p \Bigg[(p+2\phi) + \phi(2p+5\phi)(m-2) \sum_{x=0}^{m-3} \phi^k (m-3)_{(k)} \\ & + \phi^2 \sum_{x=2}^{m-2} \frac{(m-2)!}{(x-2)!(m-2-x)!} (p+2\phi+x\phi)^{x-1}(1-p-2\phi-x\phi)^{m-2-x} \Bigg], \end{aligned}$$

and repeating the same process again and again,

$$\begin{aligned} E[X(X-1)] = \; & m_{(2)}p \Bigg[(p+2\phi) + (2p+5\phi) \sum_{k=1}^{m-2} \phi^k (m-2)_{(k)} + \phi^2(p+4\phi)(m-2)_{(2)} \\ & + (2p+9\phi) \sum_{k=3}^{m-2} \phi^k (m-2)_{(k)} + \phi^4(p+6\phi)(m-2)_{(4)} + \cdots \Bigg] \\ = \; & \sum_{k=0}^{m-2} (k+1)p^2\phi^k (m)_{(k+2)} + \frac{1}{2} p\phi \sum_{k=0}^{m-2} (k+1)(k+4)\phi^k (m)_{(k+2)}. \end{aligned} \tag{4.9}$$

Since $\sigma^2 = E[X(X-1)] + E[X] - (E[X])^2$, the expression for the variance can be written by (4.9) and (4.7). The value of σ^2 is

$$\begin{aligned} \sigma^2 = \; & m\,p(1-p) + m(m-1)p\phi[(3-4p) \\ & + \phi(6m - 10mp - 12 + 18p) + \text{ higher power of } \phi], \end{aligned} \tag{4.10}$$

which shows that σ^2 increases with the increase in the value of ϕ only if $0 < p < \frac{3}{4}$. When $p > \frac{3}{4}$ the value of σ^2 may decrease if ϕ increases. Thus the QBD-I will have positive or negative extra binomial variation depending upon the values of p and ϕ.

Mishra and Singh (1996) derived a recurrence relation among the moments of the QBD-I. By expressing the moments in terms of a "factorial power series," they obtained the first four moments about the origin for the QBD-I.

4.2.3 Negative Moments of QBD-I

The negative moments are required for the estimation of the parameters of a model and for testing the efficiency of the various types of estimators. Consul (1990d) has obtained the following 23 negative moments:

$$E\left[X(p+X\phi)^{-1}\right] = mp(p+\phi)^{-1}, \tag{4.11}$$

$$E\left[(p+X\phi)^{-1}\right] = p^{-1} - m\phi(p+\phi)^{-1}, \tag{4.12}$$

$$E\left[X(X-1)(p+X\phi)^{-1}\right] = m_{(2)}p\sum_{k=0}^{m-2}\phi^k(m-2)_{(k)}, \tag{4.13}$$

$$E\left[X(X-1)(p+X\phi)^{-2}\right] = m_{(2)}p(p+2\phi)^{-1}, \tag{4.14}$$

$$E\left[X^2(p+X\phi)^{-2}\right] = mp(p+\phi)^{-2} + m_{(2)}p^2(p+\phi)^{-1}(p+2\phi)^{-1}, \tag{4.15}$$

$$E\left[X(p+X\phi)^{-2}\right] = mp(p+\phi)^{-2} - m_{(2)}p\phi(p+\phi)^{-1}p(p+2\phi)^{-1}, \tag{4.16}$$

$$E\left[(p+X\phi)^{-2}\right] = p^{-2} - m\phi(2+\phi/p)(p+\phi)^{-2}$$
$$+ m_{(2)}\phi^2(p+\phi)^{-1}(p+2\phi)^{-1}, \tag{4.17}$$

$$E\left[\frac{X^2(X-1)}{(p+X\phi)^2}\right] = \frac{2pm_{(2)}}{p+2\phi} + p\sum_{k=0}^{m-3}\phi^k m_{(k+3)}, \tag{4.18}$$

$$E\left[X(X-1)(X-2)(p+X\phi)^{-3}\right] = m_{(3)}p(p+3\phi)^{-1}, \tag{4.19}$$

$$E\left[X(X-1)(p+X\phi)^{-3}\right] = m_{(2)}p(p+2\phi)^{-2} - m_{(3)}p\phi(p+2\phi)^{-1}(p+3\phi)^{-1}, \tag{4.20}$$

$$E\left[X^2(p+X\phi)^{-3}\right] = \frac{mp}{p+\phi}\left[\frac{1}{(p+\phi)^2} + \frac{m-1}{(p+2\phi)^2} - \frac{(m-1)\phi}{(p+\phi)(p+2\phi)}\right.$$
$$\left. - \frac{(m-1)(m-2)\phi}{(p+2\phi)(p+3\phi)}\right], \tag{4.21}$$

$$E\left[X(p+X\phi)^{-3}\right] = mp(p+\phi)^{-3} - m_{(2)}p\phi\left[(p+\phi)^{-2}(p+2\phi)^{-1}\right.$$
$$\left. + (p+\phi)^{-1}(p+2\phi)^{-1}\right]$$
$$+ m_{(3)}p\phi^2(p+\phi)^{-1}(p+2\phi)^{-1}(p+3\phi)^{-1}, \tag{4.22}$$

$$E\left[(p+X\phi)^{-3}\right] = p^{-3} - m\phi\left[p^{-2}(p+\phi)^{-1} + p^{-1}(p+\phi)^{-2}\right.$$
$$\left. +(p+\phi)^{-3}\right] + m_{(2)}\phi^2\left[p^{-1}(p+\phi)^{-1}(p+2\phi)^{-1}\right.$$
$$\left. +(p+\phi)^{-2}(p+2\phi)^{-1} + (p+2\phi)^{-2}\right]$$
$$- m_{(3)}\phi^3(p+\phi)^{-1}(p+2\phi)^{-1}(p+3\phi)^{-1}, \tag{4.23}$$

$$E\left[(1-p-X\phi)^{-1}\right] = (1-m\phi)(1-p-m\phi)^{-1}, \tag{4.24}$$

$$E\left[X(1-p+X\phi)^{-1}\right] = mp(1-p-m\phi)^{-1}, \tag{4.25}$$

$$E\left[X(1-p-X\phi)^{-2}\right] = \left[1-(m-1)\phi(1-p-m\phi+\phi)^{-1}\right]$$
$$\times\, mp(1-p-m\phi)^{-1}, \tag{4.26}$$

$$E\left[(m-X)(1-p-X\phi)^{-2}\right] = m(1+\phi-m\phi)(1-p-m\phi+\phi)^{-1}, \tag{4.27}$$

$$E\left[X^2(1-p-X\phi)^{-2}\right] = mp(1-p+\phi-m^2\phi)^{-1}(1-p-m\phi)^{-1}$$
$$\times\,(1-p-m\phi+\phi)^{-1}, \tag{4.28}$$

$$E\left[(m-X)X(1-p-X\phi)^{-2}\right] = m(m-1)p(1-p-m\phi+\phi)^{-1}, \tag{4.29}$$

$$E\left[(m-X)X^2(1-p-X\phi)^{-2}\right] = m(m-1)^2p(1-p-m\phi+\phi)^{-1} - p\sum_{k=0}^{m-3}\phi^k m_{(k+3)}, \tag{4.30}$$

$$E\left[(m-X)X(1-p-X\phi)^{-3}\right] = m_{(2)}p(1-p-m\phi+\phi)^{-2}$$
$$- m_{(3)}p\phi(1-p-m\phi+\phi)^{-1}(1-p-m\phi+2\phi)^{-1}, \tag{4.31}$$

$$E\left[(m-X)(1-p-X\phi)^{-3}\right] = m(1-m\phi+\phi)p(1-p-m\phi+\phi)^{-2}$$
$$- m_{(2)}\phi(1-m\phi+2\phi)(1-p-m\phi+\phi)^{-1}$$
$$\times\;(1-p-m\phi+2\phi)^{-1}, \tag{4.32}$$

$$E\left[(m-X)X^2(1-p-X\phi)^{-3}\right] = m(m-1)^2p(1-p-m\phi+\phi)^{-2}$$
$$- m_{(3)}p(1-p)(1-p-m\phi+\phi)^{-1}$$
$$\times\,(1-p-m\phi+2\phi)^{-1}. \tag{4.33}$$

4.2.4 QBD-I Model Based on Difference-Differential Equations

Let there be m insects, bacteria, or microbes and let θ be the initial probability of desire in each one of them to get into a particular location. On account of various factors like mutual consultation, communication, determination, prevalent conditions, and the numbers succeeding to get in that location the value of θ may increase or decrease by a small quantity ϕ. Accordingly, the probability of finding x insects, bacteria or microbes in that location will be a function of m, θ, ϕ and x and we denote it by $P_x(m;\ \theta, \phi)$. By changing each one of these two parameters we shall now provide two theorems which provide the QBD-I model.

Theorem 4.1. *If the mean μ for the distribution of the insects is increased by changing θ to $\theta + \Delta\theta$ in such a manner that*

$$\frac{\partial}{\partial\theta} P_x(m;\ \theta, \phi) = m\ P_{x-1}(m-1;\ \theta+\phi, \phi) - \frac{m-x}{1-\theta-x\theta} P_x(m;\ \theta, \phi) \tag{4.34}$$

for $x = 0, 1, 2, \ldots, m$ with the initial conditions $P_0(m;\ 0, \phi) = 1$, $P_x(m;\ 0, \phi) = 0$, and $P_x(m;\ \theta, \phi) = 0$ for $x < 0$, then the probability model is the QBD-I given by (4.1).

Proof. For $x = 0$, the differential equation (4.34) becomes

$$\frac{\partial}{\partial\theta} P_0(m;\ \theta, \phi) = -\frac{m}{1-\theta} P_0(m;\ \theta, \phi),$$

whose solution is $P_0(m;\ \theta, \phi) = (1-\theta)^m \cdot A_1(\phi)$, where $A_1(\phi)$ is an unknown function of ϕ due to integration. By the initial condition $P_0(m;\ 0, \phi) = 1$, we get $A_1(\phi) = 1$. Thus $P_0(m;\ \theta, \phi) = (1-\theta)^m$.

By putting $x = 1$ in (4.34) and by using the above value of $P_0(m;\ \theta, \phi)$, we get

$$\frac{\partial}{\partial\theta} P_1(m;\ \theta, \phi) = m(1-\theta-\phi)^{m-1} - \frac{m-1}{1-\theta-\phi} P_1(m;\ \theta, \phi),$$

which is a linear differential equation with the integrating factor of $(1-\theta-\phi)^{-m+1}$. Accordingly, the solution of the equation becomes

$$P_1(m;\ \theta, \phi) = (1-\theta-\phi)^{m-1} \cdot m\theta + A_2(\phi).$$

Since $P_1(m;\ 0, \phi) = 0$, the unknown function $A_2(\phi) = 0$. Therefore,

$$P_1(m;\ \theta, \phi) = m\theta(1-\theta-\phi)^{m-1}. \tag{4.35}$$

Then, by putting $x = 2$ in (4.34), one gets

$$\frac{\partial}{\partial\theta} P_2(m;\ \theta, \phi) + \frac{m-2}{1-\theta-2\phi} P_2(m;\ \theta, \phi) = m(m-1)(\theta+\phi)(1-\theta-2\phi)^{m-2}.$$

On integration, the solution of the above linear differential equation becomes

$$P_2(m;\ \theta, \phi) = \left[(\theta+\phi)^2\ m(m-1)/2 + A_3(\phi)\right](1-\theta-2\phi)^{m-2}.$$

By the initial condition $P_2(m;\, 0, \phi) = 0$, the value of $A_3(\phi)$ becomes $-m(m-1)\phi^2/2$ and thus

$$P_2(m;\, \theta, \phi) = \binom{m}{2} \theta(\theta + 2\phi)(1 - \theta - 2\phi)^{m-2}. \tag{4.36}$$

In a similar manner, for $x = 3$ one can show that the difference differential equation (4.34), together with (4.36) and the initial condition, gives

$$P_3(m;\, \theta, \phi) = \binom{m}{3} \theta(\theta + 3\phi)^2(1 - \theta - 3\phi)^{m-3}. \tag{4.37}$$

Now, assuming the above relation to be true for $x = k$, putting $x = k + 1$ in (4.34) and by using the initial condition $P_x(m;\, 0, \phi) = 0$, it is easy to show by the method of induction that

$$P_x(m;\, \theta, \phi) = \binom{m}{x} \theta(\theta + x\phi)^{x-1}(1 - \theta - x\phi)^{m-x}$$

for all nonnegative integral values of x from 0 to m. Hence, $P_x(m;\, \theta, \phi),\ x = 0, 1, 2, \ldots, m$, is the QBD-I in (4.1) with $p = \theta$. □

Theorem 4.2. *If the mean μ for the distribution of the insects is increased by changing ϕ to $\phi + \Delta\phi$ in such a manner that*

$$\frac{\partial}{\partial \phi} P_x(m;\, \theta, \phi) = \frac{m(x-1)\theta}{\theta + \phi} P_{x-1}(m-1;\, \theta + \phi, \phi) - \frac{x(m-x)}{1 - \theta - x\phi} P_x(m; \theta, \phi) \tag{4.38}$$

for $x = 0, 1, 2, \ldots, m$ with the initial condition

$$P_x(m;\, \theta, 0) = \binom{m}{x} \theta^x (1 - \theta)^{m-x}, \tag{4.39}$$

then the probability model given by $P_x(m;\, \theta, \phi)$ is the QBD-I.

Proof. For $x = 0$, the difference-differential equation gives the solution $P_0(m;\, \theta, \phi) = C_1(\theta)$. Since $P_0(m;\, \theta, 0) = (1-\theta)^m$ by (4.39), we get $C_1(\theta) = (1-\theta)^m$. Therefore,

$$P_0(m;\, \theta, \phi) = C_1(\theta) = (1 - \theta)^m. \tag{4.40}$$

Then for $x = 1$, the differential equation (4.38) becomes

$$\frac{\partial}{\partial \phi} P_1(m;\, \theta, \phi) + \frac{m-1}{1 - \theta - \phi} P_1(m;\, \theta, \phi) = 0,$$

whose general solution is $P_1(m;\, \theta, \phi) = (1 - \theta - \phi)^{m-1} C_2(\theta)$. By the initial condition we get $C_2(\theta) = m\theta$. Thus

$$P_1(m;\, \theta, \phi) = m\theta(1 - \theta - \phi)^{m-1}. \tag{4.41}$$

For $x = 2$, the difference-differential equation (4.38) provides

$$\begin{aligned} \frac{\partial}{\partial \phi} P_2(m;\, \theta, \phi) + \frac{2(m-2)}{1 - \theta - 2\phi} P_2(m;\, \theta, \phi) &= \frac{m\theta}{\theta + \phi} \cdot P_1(m-1;\, \theta + \phi, \phi) \\ &= m(m-1)\theta(1 - \theta - 2\phi)^{m-2} \quad \text{(by (4.41))}, \end{aligned}$$

for which the integrating factor is $(1-\theta-2\phi)^{-m+2}$. Thus the general solution of the above linear differential equation becomes

$$P_2(m;\ \theta,\phi) = [m(m-1)\theta\phi + C_2(\theta)](1-\theta-2\phi)^{m-2}.$$

By the initial condition (4.39) for $\phi = 0$, we get $C_2(\theta) = m(m-1)\theta/2$. On substitution of the value of $C_2(\theta)$ in the above and on simplification,

$$P_2(m;\ \theta,\phi) = m(m-1)\ \theta(\theta+2\phi)^{m-2}/2. \tag{4.42}$$

On putting $x = 3$ in (4.38) and on using (4.42), we have the linear equation

$$\frac{\partial}{\partial\phi} P_3(m;\ \theta,\phi) + \frac{3(m-3)}{1-\theta-3\phi} P_3(m;\ \theta,\phi) = m(m-1)(m-2)\ \theta(\theta+3\phi)(1-\theta-3\phi)^{m-3}$$

whose general solution is

$$P_3(m;\ \theta,\phi) = (1-\theta-3\phi)^{m-3}\left[\binom{m}{3}\theta(\theta+3\phi)^2 + C_3(\theta)\right].$$

By the initial condition (4.39) for $x = 3$ and $\phi = 0$, $C_3(\theta) = 0$. Hence

$$P_3(m;\ \theta,\phi) = \binom{m}{3}\theta(\theta+3\phi)^2(1-\theta-3\phi)^{m-3}. \tag{4.43}$$

In a similar manner it can be shown for successive values of $x = 4,\ldots,m$ that the unknown coefficients $C_4(\theta),\ C_5(\theta),\ldots$ are all zero and that

$$P_x(m;\ \theta,\phi) = \binom{m}{x}\theta(\theta+x\phi)^{x-1}(1-\theta-x\phi)^{m-x}, \quad x = 0,1,2,\ldots,m,$$

which is the QBD-I. □

4.2.5 Maximum Likelihood Estimation

Let a random sample of size n be taken from the QBD-I of (4.1) and let the observed values be $x_i,\ i = 1,2,\ldots,n$. The likelihood function L for the parameters p and ϕ will be proportional to

$$L \propto \prod_{i=1}^{n}\binom{m}{x_i} p(p+x_i\phi)^{x_i-1}(1-p-x_i\phi)^{m-x_i}, \tag{4.44}$$

which gives the log-likelihood function ℓ as

$$\ell \propto n\ln p + \sum_i (x_i-1)\ \ln(p+x_i\phi) + \sum_i (m-x_i)\ \ln(1-p-x_i\phi) + \sum_i \ln\binom{m}{x_i}. \tag{4.45}$$

On partial differentiation of ℓ with respect to ϕ and p, we get the two maximum likelihood (ML) equations in the form

$$S_1 = \frac{\partial\ell}{\partial\phi} = \sum_{i=1}^{n}\frac{x_i(x_i-1)}{p+x_i\phi} - \sum_{i=1}^{n}\frac{(m-x_i)x_i}{1-p-x_i\phi} = 0, \tag{4.46}$$

$$S_2 = \frac{\partial\ell}{\partial p} = \frac{n}{p} + \sum_{i=1}^{n}\frac{x_i-1}{p+x_i\phi} - \sum_{i=1}^{n}\frac{m-x_i}{1-p-x_i\phi} = 0. \tag{4.47}$$

Multiplying (4.47) by p and (4.46) by ϕ and by adding, the equation (4.47) reduces to

$$\sum_{i=1}^{n} \frac{m - x_i}{1 - p - x_i\phi} = mn. \tag{4.48}$$

The equations (4.46) and (4.48) are to be solved simultaneously for getting the ML estimates $\hat{p}$ and $\hat{\phi}$ for the two parameters. This is not an easy task, as the equations seem to imply multiple roots. Of course, many of those roots may be either complex or not admissible according to the restrictions on the parameters p and ϕ, or they may represent saddle points. To test that the roots represent the ML estimates and to evaluate them numerically we need the second-order partial derivatives. Accordingly,

$$\frac{\partial^2 \ell}{\partial \phi^2} = -\sum_{i=1}^{n} \frac{x_i^2(x_i - 1)}{(p + x_i\phi)^2} - \sum_{i=1}^{n} \frac{(m - x_i)x_i^2}{(1 - p - x_i\phi)^2}, \tag{4.49}$$

$$\frac{\partial^2 \ell}{\partial \phi \partial p} = -\sum_{i=1}^{n} \frac{x_i(x_i - 1)}{(p + x_i)^2} - \sum_{i=1}^{n} \frac{(m - x_i)x_i}{(1 - p - x_i\phi)^2}, \tag{4.50}$$

$$\frac{\partial^2 \ell}{\partial p^2} = -\frac{n}{p^2} - \sum_{i=1}^{n} \frac{x_i - 1}{(p + x_i\phi)^2} - \sum_{i=1}^{n} \frac{m - x_i}{(1 - p - x_i\phi)^2}, \tag{4.51}$$

which are all negative for all values of p and ϕ. Also, their respective expected values I_{11}, I_{12}, and I_{22} are

$$\mathrm{I}_{11} = -\frac{nm(m-1)\,p[2 + (m-3)p]}{(p + 2\phi)(1 - p - m\phi + \phi)}, \tag{4.52}$$

$$\mathrm{I}_{12} = -\frac{nm(m-1)\,p[1 - (m-1)\phi]}{(p + 2\phi)(1 - p - m\phi + \phi)} = \mathrm{I}_{21}, \tag{4.53}$$

$$\mathrm{I}_{22} = -\frac{nm}{p} - \frac{nm[p - (m-3)\phi + (m-1)(m-3)\phi^2]}{(p + 2\phi)(1 - p - m\phi + \phi)}, \tag{4.54}$$

which provide the Fisher information

$$\mathrm{I} = \mathrm{I}_{11}\mathrm{I}_{22} - \mathrm{I}_{21}\mathrm{I}_{12} \tag{4.55}$$

$$= \frac{2n^2m^2(m-1)\left[p(1-p)+4\phi-(m-1)\phi(p+2\phi)+(3m-5)p^2\phi-2(m-1)(m-2)p^2\phi^2\right]}{(p+2\phi)^2(1-p-m\phi+\phi)^2}.$$

Now the ML estimates $\hat{p}$ and $\hat{\phi}$ can be computed numerically by successive approximation using the Newton–Raphson method. The proper selection of the starting values p_0 and ϕ_0 is very important for this method. We suggest that the starting value p_0 be computed from the formula

$$p_0 = 1 - (f_0/n)^{m^{-1}}, \tag{4.56}$$

where f_0 is the observed frequency for $x = 0$. Also, the value of ϕ_0 may be computed by the formula

$$\phi_0 = [2(m-2)]^{-1}[-1 + \sqrt{\{1 + 4(m-2)(-1 + \bar{x}/mp_0)/(m-1)\}}], \tag{4.57}$$

which is obtained by taking the three terms of (4.7) and where $\bar{x}$ is the sample mean.

Another method that we found to be quite useful for getting the ML estimates $\hat{p}$ and $\hat{\phi}$ was to plot the graphs of (4.46) and (4.47) for various values of p and ϕ around both sides of p_0 and ϕ_0. For every observed sample a unique point of intersection was easily found and it provided the ML estimates $\hat{p}$ and $\hat{\phi}$.

We shall now consider three simple cases where the observed data gets reduced to a few frequency classes.

Some Particular Cases. When the sample values $x_i, i = 1, 2, \ldots, n$, are expressed in the form of a frequency distribution given by $s = 0, 1, 2, \ldots, m$ with respective frequencies as $n_0, n_1, \ldots, n_m$ and their sum $n_0 + n_1 + \cdots + n_m = n$, the two ML equations become

$$\sum_i \frac{n_i(i-1)i}{p + i\phi} - \sum_i \frac{n_i(m-i)i}{1 - p - i\phi} = 0,$$

$$\frac{n}{p} + \sum_i \frac{n_i(i-1)}{p + i\phi} - \sum_i \frac{n_i(m-i)}{1 - p - i\phi} = 0\,.$$

Case I. For $m = 1$ and $n = n_0 + n_1$, the QBD model reduces to the point binomial model with $\hat{p} = n_1(n_0 + n_1)^{-1}$.

Case II. For $m = 2$ and $n = n_0 + n_1 + n_2$, the ML equations give

$$\frac{2n_2}{p + 2\phi} - \frac{n_1}{1 - p - \phi} = 0 \quad \text{or} \quad \phi = \frac{n_2(1-p) - n_1 p/2}{n_1 + n_2},$$

and

$$\frac{n}{p} - \frac{n_0}{p} + \frac{n_2}{p + 2\phi} - \frac{2n_0}{1 - p} - \frac{n_1}{1 - p - \phi} = 0,$$

which becomes (on putting the value of ϕ)

$$\frac{n - n_0}{p} - \frac{2n_0}{1 - p} - \frac{n_1 + n_2}{2 - p} = 0.$$

The above equation reduces to $p^2 - 2p + 1 = n_0/n$, which gives $\hat{p} = 1 - \sqrt{(n_0/n)}$. Thus

$$\hat{\phi} = (n_1 + n_2)^{-1}\left[\left(n_2 + \frac{1}{2}n_1\right)\sqrt{(n_0/n)} - n_1/2\right].$$

Case III. For $m = 3$ and $n = n_0 + n_1 + n_2 + n_3$, the ML equations become

$$\frac{2n_2}{p + 2\phi} + \frac{6n_3}{p + 3\phi} - \frac{2n_1}{1 - p - \phi} - \frac{2n_2}{1 - p - 2\phi} = 0,$$

$$\frac{n}{p} - \frac{n_0}{p} + \frac{n_2}{p + 2\phi} + \frac{2n_3}{p + 3\phi} - \frac{3n_0}{1 - p} - \frac{2n_1}{1 - p - \phi} - \frac{n_2}{1 - p - 2\phi} = 0.$$

On multiplying the first equation by ϕ and the second equation by p and on adding the two equations and simplifying, we get

$$3n - \frac{3n_0}{1-p} - \frac{2n_1}{1-p-\phi} - \frac{n_2}{1-p-2\phi} = 0,$$

which gives the quadratic equation in ϕ as

$$\begin{aligned} &6\phi^2[n(1-p)-n_0] - \phi(1-p)[9n(1-p)-9n_0-4n_1-n_2] \\ &+ (1-p)^2[3n(1-p)-3n_0-2n_1-n_2] = 0. \end{aligned}$$

The above equation gives two roots for ϕ but one of them is inadmissible as it is larger than $(1-p)/m$. Accordingly, the lower value becomes the admissible ML estimate $\hat{\phi}$ given by

$$\hat{\phi} = (1-p)\frac{9n(1-p)-9n_0-4n_1-n_2-\sqrt{\{3n(1-p)-3n_0-4n_1+n_2\}^2+16n_1n_2}}{12[n(1-p)-n_0]}.$$

When this value of ϕ is substituted in any one of the two ML equations one gets a single admissible value of p in the range $0 < p < 1$.

4.3 Quasi-Hypergeometric Distribution I

Let a, b, n, and r be four positive integers. A random variable X has a *quasi-hypergeometric distribution I* (QHD-I) if its probability mass function is defined by

$$P(X=x) = \frac{\frac{a}{a+xr}\binom{a+xr}{x}\binom{b+nr-xr}{n-x}}{\binom{a+b+nr}{n}} \tag{4.58}$$

$$= \binom{n}{x}\frac{a(a+xr-1)_{(x-1)}(b+nr-xr)_{(n-x)}}{(a+b+nr)_{(n)}} \tag{4.59}$$

for $x = 0, 1, 2, 3, \ldots, n$ and zero elsewhere.

The QHD-I is a generalization of the hypergeometric distribution as it reduces to the hypergeometric model when $r = 0$. To prove that the QHD-I is a true probability distribution one has to use the Lagrange expansion of

$$\frac{(1+z)^{a+b}}{1-rz(1+z)^{-1}} = (1+z)^a \cdot \frac{(1+z)^b}{1-rz(1+z)^{-1}}, \tag{4.60}$$

under the transformation $z = u\ (1+z)^r$, as a single power series in u on the left-hand side by (1.80) and as a product of two power series in u on the right-hand side by the formulas (1.78) and (1.80). On equating the coefficients of u^n on both sides one gets the identity

$$\binom{a+b+nr}{n} = \sum_{i=0}^{n}\frac{a}{a+ir}\binom{a+ir}{i}\binom{b+nr-ir}{n-i}. \tag{4.61}$$

On division by $\binom{a+b+nr}{n}$, it is clear that $\sum_{x=0}^{n} P(X=x) = 1$.

A recurrence relation between the successive probabilities of QHD-I is

$$P(X=x+1) = \begin{cases} \frac{n-x}{x+1} \frac{(a+xr+r-1)_{(x)}(b+nr-xr-r)_{(n-x-1)}}{(a+xr-1)_{(x-1)}(b+nr-xr)_{(n-x)}} P(X=x), & x < r, \\ \frac{n-x}{x+1} \frac{(a+xr)^{[r]}(b+nr-xr-n+x)_{(r-1)}}{(a+xr-x+1)^{[r-1]}(b+nr-xr)_{(r)}} P(X=x), & r < x. \end{cases} \tag{4.62}$$

Charalambides (1986) has defined and studied the family of Gould series distributions and has shown that the QHD-I is a member of that family. He has obtained the first two factorial moments of the QHD-I and has shown that

$$\mu = E[X] = \frac{a}{(a+b+nr)_{(n)}} \sum_{k=0}^{n-1} n_{(k+1)} r^k (a+b+nr-k-1)_{(n-k-1)}$$

$$= \frac{n!\,a}{a+b+nr} \sum_{k=0}^{n-1} r^k \binom{a+b+nr-k-1}{n-k-1} \tag{4.63}$$

and

$$\sigma^2 = \frac{an}{a+b+nr} + \frac{n!\,a}{(a+b+nr)_{(n)}} \sum_{k=0}^{n-2} r^k \left\{ r(k^2+2k+2) + (k+1)\left(\frac{ak}{2} - \frac{k}{2} - 1\right) \right\}$$

$$\times \binom{a+b+nr-k-2}{n-k-2}$$

$$- \left[\frac{n!\,a}{(a+b+nr)_{(n)}} \right]^2 \sum_{k=0}^{2n-2} r^k \binom{2a+2b+2nr-k-1}{2n-2k-1}. \tag{4.64}$$

If $a,\ b$, and r become very large such that $a(a+b+nr)^{-1} = p$ and $r(a+b+nr)^{-1} = \phi$, the limiting form of the QHD-I is the QBD-I defined by (4.1).

4.4 Quasi-Pólya Distribution I

Let $a,\ b,\ c,\ r$, and n be five positive integers. A random variable X is said to have a quasi-Pólya distribution I (QPD-I) if its probability mass function is given by

$$P(X=x) = \binom{n}{x} \frac{a(a+xr)^{[x-1,c]}(b+nr-xr)^{[n-x,c]}}{(a+b+nr)^{[n,c]}} \tag{4.65}$$

for $x = 0, 1, 2, \ldots, n$ and zero elsewhere.

The above pmf can be expressed in general binomial coefficients in the following two forms:

$$P(X=x) = \frac{a}{a+xr+xc} \frac{\binom{a/c+xr/c+x}{x}\binom{(b+nr-xr)/c+n-x-1}{n-x}}{\binom{(a+b+nr)/c+n-1}{n}} \tag{4.66}$$

$$= \binom{n}{x} \frac{(-a/c)(-a/c-xr/c-1)_{(x-1)}(-b/c-nr/c-xr/c)_{(n-x)}}{(-a/c-b/c-nr/c)_{(n)}}. \tag{4.67}$$

To prove that the sum $\sum_{x=0}^{n} P(X=x)$, defined by (4.66), is unity, one has to consider the Lagrange expansions, under the transformation $z = u(1-z)^{-r/c}$, of the product

$$\frac{(1-z)^{-(a+b)/c}}{1-rz(1-z)^{-1}/c} = (1-z)^{-a/c} \cdot \frac{(1-z)^{-b/c}}{1-rz(1-z)^{-1}/c} \tag{4.68}$$

as a single power series in u by (1.80) on the left-hand side and as a product of two power series in u by (1.78) and (1.80) on the right-hand side. Then by equating the coefficients of u^n on both sides one gets the identity

$$\binom{(a+b+nr)/c+n-1}{n} = \sum_{x=0}^{n} \frac{a/c}{a/c+xr/c+x} \binom{a/c+xr/c+x}{x} \times \binom{(b+nr-xr)/c+n-x-1}{n-x}, \tag{4.69}$$

proved by Jensen (1902) and Gould (1966). The identity (4.69) proves the result (4.66).

The expression (4.67) of the QPD-I is very similar to the expression (4.59) of the QHD-I with the difference that the positive integers $a,\ b,\ r$ in (4.59) are replaced by negative rational numbers $-a/c,\ -b/c$, and $-r/c$, respectively.

In view of the above observation, the mean μ and variance σ^2 of the QPD-I can be written down from (4.63) and (4.64) by replacing $a,\ b,\ r$ by $-a/c,\ -b/c$, and $-r/c$, respectively, and by simplifying the same. Thus

$$\mu = \frac{a}{(a+b+nr)^{[n,c]}} \sum_{k=0}^{n-1} n_{(k+1)} r^k (a+b+nr+kc+c)^{[n-k-1,c]} \tag{4.70}$$

and

$$\sigma^2 = \frac{an}{a+b+nr} + \frac{n!a}{(a+b+nr)^{[n,c]}} \sum_{k=0}^{n-2} \frac{r^k}{(n-k-1)!} \left\{ r(k^2+2k+2) + \frac{(k+1)}{2}(ak+ck+2c) \right\} \times \left(\frac{a+b+nr}{c}\right)_{(n-k-1)} - \left[\frac{n!\,a}{(a+b+nr)^{[n,c]}}\right]^2 \sum_{k=0}^{2n-2} r^k \frac{(2a+2b+2nr+k+1)^{[2n-2k-1,c]}}{(2n-2k-1)!}. \tag{4.71}$$

4.5 Quasi-Binomial Distribution II

A r.v. X is said to have a *quasi-binomial distribution II (QBD-II)* if it has a pmf defined by

$$P(X=x) = \binom{n}{x} \frac{ab}{a+b} \frac{(a+x\theta)^{x-1}(b+n\theta-x\theta)^{n-x-1}}{(a+b+n\theta)^{n-1}} \tag{4.72}$$

for $x = 0, 1, 2, 3, \ldots, n$ and zero otherwise, where $a > 0,\ b > 0$, and $\theta > -a/n$.

The QBD-II is another generalization of the binomial distribution as it reduces to the binomial probability model when $\theta = 0$. If $a(a+b+n\theta)^{-1} = p$ and $\theta(a+b+n\theta)^{-1} = \alpha$, the number of parameters in (4.72) can be reduced to three and the probability distribution (4.72) can be written as

$$P(X=x) = \binom{n}{x} \frac{(1-p-n\alpha)p}{1-n\alpha}(p+x\alpha)^{x-1}(1-p-x\alpha)^{n-x-1} \tag{4.73}$$

for $0 < p < 1,\ -pn^{-1} < \alpha < (1-p)n^{-1}$, and $x = 0, 1, 2, \ldots, n$.

Special Case. If $b = aq < a$ and $\theta = -aq/n$ the expression (4.72) gives the following nice and compact form as a variant of the QBD-II with two parameters n and q:

$$P(X=x) = \binom{n}{x} \frac{q}{1+q}\left(1-\frac{xq}{n}\right)^{x-1}\left(\frac{xq}{n}\right)^{n-x-1} \tag{4.74}$$

for $x = 1, 2, 3, \ldots, n$ and zero elsewhere.

4.5.1 QBD-II as a True Probability Model

Under the transformation $z = ue^{\theta z},\ 0 \le z \le 1$, the function e^{az} can be expressed as a power series in u by Lagrange expansion (1.78) in the form

$$e^{az} = \sum_{s=0}^{\infty} a(a+s\theta)^{s-1}\, u^s/s!\,. \tag{4.75}$$

Since $e^{(a+b)z} = e^{az}.e^{bz}$, the left-hand side can be expressed as a power series in u by (4.75) and the right-hand side can be expressed as a product of two power series in u. By equating the coefficients of u^n on both sides we get the identity

$$\frac{(a+b)(a+b+n\theta)^{n-1}}{n!} = \sum_{x=0}^{n} \frac{a(a+x\theta)^{x-1}}{x!} \frac{b(b+n\theta-x\theta)^{n-x-1}}{(n-x)!}\,. \tag{4.76}$$

On division by the left-hand side, we get $\sum_{x=0}^{n} P(X=x) = 1$.

4.5.2 Mean and Variance of QBD-II

The *mean* or the *expected value* of the r.v. X, having the QBD-II defined by (4.72) can be easily determined by using the Jensen (1902) identity

$$(a+b+n\theta)^n = \sum_{s=0}^{n} \binom{n}{s} b(b+s\theta)^{s-1}(a+n\theta-s\theta)^{n-s}, \tag{4.77}$$

proved in Subsection 4.2.1. Thus,

$$\mu = E[X] = na(a+b)^{-1}. \tag{4.78}$$

Thus the mean of the QBD-II, defined by (4.72), is independent of the value of θ and is the same as for the corresponding binomial model. However, the mean μ for the QBD-II, defined by (4.73), is $np(1 - n\alpha)^{-1}$, which depends upon the additional parameter α and increases with the increase in the value of α. Its lowest value is $np(1 + p)^{-1}$ and it may approach n as $\alpha \to (1 - p)n^{-1}$. Thus, the QBD-II defined by (4.73) is far more versatile than the QBD-II defined by (4.72). Also, the mean of the QBD-II variate, defined by (4.74), is $n(1 + q)^{-1}$.

The second factorial moment for the QBD-II, defined by (4.72), is

$$E[X(X-1)] = \frac{n(n-1)ab}{(a+b)(a+b+n\theta)} \sum_{x=2}^{n} \binom{n-2}{x-2} \frac{(a+x\theta)^{x-1}(b+n\theta-x\theta)^{n-x-1}}{(a+b+n\theta)^{n-2}}.$$

By putting $n - x = s$ and by splitting the summation into two summations and writing $(a + n\theta - s\theta)/(a + b + n\theta) = 1 - (b + s\theta)/(a + b + n\theta)$, the above expression becomes

$$E[X(X-1)] = \frac{n(n-1)ab}{(a+b)(a+b+n\theta)} \left[\sum_{s=0}^{n-2} \binom{n-2}{s} \frac{(b+s\theta)^{s-1}(a+n\theta-s\theta)^{n-2-s}}{(a+b+n\theta)^{n-3}} \right.$$

$$\left. - \sum_{s=0}^{n-2} \binom{n-2}{s} \frac{(b+s\theta)^{s}(a+n\theta-s\theta)^{n-2-s}}{(a+b+n\theta)^{n-2}} \right]. \tag{4.79}$$

Gould (1972) has given the relation

$$\sum_{k=0}^{n} \binom{n}{k} (x+kz)^k (y-kz)^{n-k} = n! \sum_{k=0}^{n} \frac{(x+y)^k z^{n-k}}{k!}. \tag{4.80}$$

By using the identities (4.76) and (4.80) on (4.79) and on simplification,

$$E[X(X-1)] = \frac{n(n-1)a}{a+b} \left[1 - \sum_{s=0}^{n-2} \frac{(n-2)_{(s)}\, b\, \theta^s}{(a+b+n\theta)^{s+1}} \right],$$

which gives the variance σ^2 of the QBD-II as

$$\sigma^2 = \frac{n^2 ab}{(a+b)^2} - \frac{n(n-1)ab}{a+b} \sum_{s=0}^{n-2} \frac{(n-2)_{(s)}\, \theta^s}{(a+b+n\theta)^{s+1}}. \tag{4.81}$$

When $\theta > 0$, it can easily be proved that

$$\frac{nab}{(a+b)^2} < \sigma^2 < \frac{n^2 ab}{(a+b)^2}, \tag{4.82}$$

which implies that the QBD-II, defined by (4.72), has a variance, for $\theta > 0$, larger than the variance of the binomial model.

All the moments of QBD-I in Section 4.2 and all but the first moment of QBD-II in this section appear in term of series. Mishra, Tiwary, and Singh (1992) obtained expression for factorial moments of a family of QBD. They considered some particular cases of QBD for which the first two moments appear in simple algebraic forms. The method of moments can easily be used to estimate the parameters. One such particular case of QBD was fitted to some numerical data sets.

4.5.3 Some Other Properties of QBD-II

(i) *Limiting form of QBD-II.* When n is very large and a and θ are very small such that $na = c,\ n\theta = d$, by putting $c(b+d)^{-1} = M$ and $d(b+d)^{-1} = \phi$ and by taking the limit, it can be shown that the probability $P(X = x)$ in (4.72) becomes

$$P(X = x) = M(M + x\phi)^{x-1} e^{-M-x\phi}/x!,$$

which is the GPD studied in detail by Consul (1989a). Also, if a and b are finite, $\theta \to 0$, $n \to \infty$ such that $n\theta$ is finite, it can be shown that the QBD-II approaches the normal distribution.

(ii) *QBD-II as a conditional distribution.* Let X and Y be two independent r.v.s. having the GPDs with parameters (a, θ) and (b, θ), respectively. Then the r.v. $X + Y$ has a GPD with parameters $(a + b, \theta)$. Accordingly, the conditional distribution of X, given $X + Y = n$, becomes

$$P(X = x|X + Y = n) = \binom{n}{x} \frac{a(a + x\theta)^{x-1} b(b + n\theta - x\theta)^{n-x-1}}{(a + b)(a + b + n\theta)^{n-1}}$$

for $x = 0, 1, 2, \ldots, n$. The above is the pmf of the QBD-II defined by (4.72). Thus, the QBD-II is a conditional distribution. The converse of this property is also true. Consul (1974, 1975) has used this property of the QBD-II to characterize the generalized Poisson distribution and the QBD-II.

(iii) *Convolution property.* The QBD-II does not possess the convolution property, as each probability $P(X = x)$ is the product of two probabilities which vary differently. Charalambides (1986) has considered the QBD-II as a member of the family of Gould series distributions and has proved a general theorem showing that it does not possess the convolution property along with many other members of the Gould family.

(iv) *Maximum probability.* When n is not large one can easily determine the values of the probabilities of the QBD-II defined by (4.72) for all values of x with the help of a pocket calculator. This will enable a player to find the particular value of x for which the probability of success is the maximum. Since the guessing of the probabilities for various values of x is not easy, this model can be used to devise machines for more interesting games of chance for the players.

4.6 Quasi-Hypergeometric Distribution II

Let $a,\ b,\ n$, and θ be positive integers. A r.v. X is said to have a *quasi-hypergeometric distribution II* (QHD-II) if its probability mass function is defined by (see Consul and Mittal, 1975; Janardan, 1978)

$$P(X = x) = \binom{n}{x} \frac{ab}{a+b} \frac{(a + x\theta - 1)_{(x-1)}(b + n\theta - x\theta - 1)_{(n-x-1)}}{(a + b + n\theta - 1)_{(n-1)}} \tag{4.83}$$

for $x = 0, 1, 2, \ldots, n$ and zero elsewhere. The QHD-II is slightly different from the QHD-I defined by (4.59).

When $\theta = 0$, the QHD-II reduces to the ordinary hypergeometric model, and accordingly, it is another generalization of that model.

To prove that the QHD-II is a true probability distribution one has to use the Lagrange expansion of $(1+z)^{a+b} = (1+z)^a(1+z)^b$, under the transformation $z = u(1+z)^\theta$, as a single power series in u by (1.78) on the left-hand side and as a product of two power series in u on the right-hand side. On equating the coefficients of u^n on both sides one gets the identity

$$\frac{a+b}{a+b+n\theta}\binom{a+b+n\theta}{n} = \sum_{x=0}^{n} \frac{ab}{a+x\theta}\binom{a+x\theta}{x}\frac{b}{b+n\theta-x\theta}\binom{b+n\theta-x\theta}{n-x}. \tag{4.84}$$

On division by the left-hand side in (4.84), it follows that $\sum_{x=0}^{n} P(X=x) = 1$.

A recurrence relation between the successive probabilities of QHD-II is

$$P(X=x+1) = \begin{cases} \frac{n-x}{x+1}\frac{(a+\theta+x\theta-1)_{(x)}(b+n\theta-\theta-x\theta-1)_{(n-x-2)}}{(a+x\theta-1)_{(x-1)}(b+n\theta-x\theta-1)_{(n-x-1)}} P(X=x), & x<\theta, \\ \frac{n-x}{x+1}\frac{(a+x\theta)^{[\theta]}(b+n\theta-x\theta-n+x)_{(\theta-1)}}{(a+x\theta-x+1)^{[\theta-1]}(b+n\theta-x\theta-1)_{(\theta)}} P(X=x), & x>\theta. \end{cases} \tag{4.85}$$

The mean and variance of the QHD-II are

$$\mu = na(a+b)^{-1} \tag{4.86}$$

and

$$\sigma^2 = \frac{n^2ab}{(a+b)^2} - \frac{n(n-1)ab}{a+b}\sum_{k=0}^{n-1}\frac{(n-2)_{(k)}(\theta-1)^k}{(a+b+n\theta-n+1)^{[k+1]}}. \tag{4.87}$$

Charalambides (1986) has considered the QHD-II as a member of the family of Gould series distributions. His expression for the variance seems to be somewhat more complicated and different than (4.87).

If a, b, and θ become very large such that $a(a+b+n\theta)^{-1} = p$, $\theta(a+b+n\theta)^{-1} = \alpha$, and $(a+b+n\theta)^{-1} = 0$, then the limiting form of the QHD-II is the QBD-II defined in (4.73).

4.7 Quasi-Pólya Distribution II (QPD-II)

Let a, b, c, and θ be four positive real numbers and let n be a positive integer. A random variable X is said to have a *quasi-Pólya distribution II* (QPD-II) if the r.v. X has the probability mass function given by

$$P(X=x) = \binom{n}{x}\frac{ab(a+b+n\theta)}{(a+b)(a+x\theta)(b+n\theta-x\theta)} \times \frac{(a+x\theta)^{[x,c]}(b+n\theta-x\theta)^{[n-x,c]}}{(a+b+n\theta)^{[n,c]}} \tag{4.88}$$

for $x = 0, 1, 2, 3, \ldots, n$ and zero elsewhere (Consul and Mittal, 1975). It was called the generalized Markov–Pólya distribution by Janardan (1978). This distribution was further generalized by Sen and Jain (1996), who obtained recurrence relations between the moments.

The pmf (4.88) can also be expressed in the following two forms:

$$P(X=x)=\frac{ab(a+b+n\theta+nc)}{(a+b)(a+x\theta+xc)\{b+(n-x)(\theta+c)\}}$$
$$\times\frac{\binom{a/c+x\theta/c+x}{x}\binom{b/c+n\theta/c-x\theta/c+n-x}{n-x}}{\binom{a/c+b/c+n\theta/c+n}{n}}$$
$$=\binom{n}{x}J_x(a,c,\theta)\,J_{n-x}(b,c,\theta)/J_n(a+b,c,\theta) \tag{4.89}$$

for $x=0,1,2,\ldots,n$ and zero otherwise, where

$$J_x(a,c,\theta)=a(a+x\theta)^{-1}(a+x\theta)^{[x,c]}. \tag{4.90}$$

The QPD-II is a generalized Pólya–Eggenberger model, as its special case, given by $\theta=0$, is the Pólya-Eggenberger distribution.

By putting $a/(a+b)=p,\ b/(a+b)=q,\ c/(a+b)=r$, and $\theta/(a+b)=s$ the QPD-II can be expressed in another form with four parameters (Janardan, 1978) as

$$P(X=x)=\frac{pq(1+ns)\binom{n}{x}\prod_{i=0}^{x-1}(p+xs+ir)\prod_{i=0}^{n-x-1}(q+ns-xs+ir)}{(p+xs)(q+ns-xs)\prod_{i=0}^{n-1}(1+ns+ir)}. \tag{4.91}$$

The sum of the probabilities $P(X=x)$ in (4.88) for $x=0$ to $x=n$ becomes unity on account of the following Hagen–Rothes (1891) identity given by Gould (1972):

$$\sum_{x=0}^{n}\frac{a}{a+xz}\binom{a+xz}{x}\frac{b}{b+(n-x)z}\binom{b+nz-xz}{n-x}=\frac{a+b}{a+b+nz}\binom{a+b+nz}{n}, \tag{4.92}$$

where a, b, and z can be replaced by a/c, b/c, and $1+\theta/c$, respectively. The proof of the above identity easily follows from the Lagrange expansion in (1.78) with (1.79) for the functions $(1-z)^{-(a+b)/c}=(1-z)^{-a/c}.(1-z)^{-b/c}$, under the transformation $z=u(1-z)^{-\theta/c}$, which provide the expansions

$$\sum_{n=0}^{\infty}\frac{u^n}{n!}\frac{a+b}{c}\left(\frac{a+b+n\theta}{c}+1\right)^{[n-1]}$$
$$=\sum_{x=0}^{\infty}\frac{u^x}{x!}\frac{a}{c}\left(\frac{a+x\theta}{c}+1\right)^{[x-1]}\sum_{y=0}^{\infty}\frac{u^y}{y!}\frac{b}{c}\left(\frac{b+y\theta}{c}+1\right)^{[y-1]}. \tag{4.93}$$

Equating the coefficients of u^n on both sides with each other gives the identity. Thus $\sum_{x=0}^{n}P(X=x)=1$ and (4.88) represents a true probability distribution.

4.7.1 Special and Limiting Cases

(i) When $c=0$, the QPD-II (4.88) reduces to the QBD-II defined by (4.72).
(ii) When $c=-1$, the QPD-II (4.88) reduces to the QHD-II defined by (4.83).

(iii) If $c = -1$ and $\theta = 1$ or if $c = 1$ and $\theta = 0$, the QPD-II becomes a negative hypergeometric distribution (Patil and Joshi, 1968) given by

$$P(X = x) = \binom{a+x-1}{x}\binom{b+n-x-1}{n-x}\Big/\binom{a+b+n-1}{n}. \tag{4.94}$$

(iv) If $a/c,\ b/c$, and n/c are infinitely large quantities of the same order and if $\theta \to 0$, the QPD-II approaches the normal curve (Janardan, 1975).

(v) If $n \to \infty$ and $p \to 0,\ r \to 0,\ s \to 0$ such that $np = \theta,\ ns = \lambda$, and $nr \to 0$ in the QPD-II, given by (4.89), then its limiting form is the generalized Poisson distribution (Janardan, 1975).

(vi) *Conditional distribution.* If X and Y are two independent r.v.s having the generalized negative binomial models with parameters $(a,\ t+1,\ \alpha)$ and $(b,\ t+1,\ \alpha)$, respectively, then the conditional distribution of $X = x$, given the sum $X + Y = n$, is a QPD-II (4.88) with $c = 1$. This property characterizes the QPD-II.

4.7.2 Mean and Variance of QPD-II

By using Gould's (1966) identities, Consul and Mittal (1975) have shown that the mean of the QPD-II, defined by (4.88), is independent of θ and c and equals

$$\mu = na(a+b)^{-1}. \tag{4.95}$$

Also, they have shown that the variance σ^2 of the QPD-II is

$$\sigma^2 = \frac{n^2ab}{(a+b)^2} - \frac{n(n-1)ab}{a+b}\sum_{s=0}^{n-1}\frac{(n-1)_{(s)}(\theta+c)^s}{(a+b+n\theta+nc-c)_{(s+1,c)}}. \tag{4.96}$$

The other higher moments can also be obtained but the process of summation is rather tricky.

4.7.3 Estimation of Parameters of QPD-II

Let $f_x,\ x = 0, 1, 2, 3, \ldots, n$, denote the observed frequencies for various values of X. Also, let m_2 be the sample variance and $\bar{x}$ be the sample mean. The moment estimators of $p,\ r$, and s in QPD-II in (4.91) are

$$s^* = p^*/n, \qquad\qquad p^* = \bar{x}\Big/\sum_{i=0}^{n} f_i, \tag{4.97}$$

and

$$r^* = \frac{m_2 - np^*(1-p^*)}{n^2p^*(1-p^*) - m_2}. \tag{4.98}$$

Janardan (1975) has obtained the following ML equations by partial differentiation of the log likelihood function from (4.91):

$$\frac{\partial L}{\partial p} = \frac{q-p}{pq} + \sum_{j=1}^{k-1}\frac{1}{p+ks+jr} - \sum_{j=1}^{n-k-1}\frac{1}{q+(n-k)s+jr} = 0,$$

$$\frac{\partial L}{\partial s} = \sum_{j=1}^{k-1}\frac{k}{p+ks+jr} + \sum_{j=1}^{n-k-1}\frac{(n-k)}{q+(n-k)s+jr} - \sum_{j=1}^{N-1}\frac{n}{1+Ns+jr} = 0,$$

and

$$\frac{\partial L}{\partial r} = \sum_{j=1}^{k-1} \frac{j}{p+ks+jr} + \sum_{j=1}^{n-k-1} \frac{j}{q+(n-k)s+jr} - \sum_{j=1}^{n-1} \frac{j}{1+Ns+jr} = 0. \tag{4.99}$$

The equations will have to be solved numerically by iterative methods by using the values $p^*,\ s^*,\ r^*$ as the first set of estimates.

4.8 Gould Series Distributions

The Gould series distributions (GSD) have been defined by Charalambides (1986) by considering the expansion of some suitable functions as a series of the Gould polynomials

$$\begin{aligned} G_x(s;r) &= s(s+rx-1)_{(x-1)}, \qquad x = 1, 2, 3, \ldots, \\ &= s(s+rx-1)(s+rx-2)\cdots(s+rx-x+1), \\ G_0(s;r) &= 1. \end{aligned} \tag{4.100}$$

If $A(s;r)$ is a positive function of two parameters s and r and if

$$A(s;r) = \sum_{x \in T} a(x;r) \cdot s(s+rx-1)_{(x-1)}, \tag{4.101}$$

where $a(x;r) \geq 0,\ T$ is a subset of the set of nonnegative integers and if $a(x;r)$ are independent of the parameter s, then the series (4.101) provides the GSD whose pmf is

$$P(X=x) = [A(s;r)]^{-1}\, a(x;r)\, s(s+rx-1)_{(x-1)} \tag{4.102}$$

for $x \in T$ and zero otherwise. The domain of the parameters s and r may be positive or negative real numbers such that the terms of the expansion (4.101) are nonnegative.

Charalambides (1986) has given a method, based upon the displacement operator E, the difference operator $\Delta = E - 1$, and the Abel-difference operator ΔE^{-r}, for obtaining the function $a(x;r)$ from $A(s;r)$ as

$$a(x;r) = \frac{1}{x!} \left(\Delta E^{-r}\right)^x A(u;r)|_{u=0}. \tag{4.103}$$

The generalized negative binomial distribution (GNBD), the QHD-I, the QHD-II, the QPD-I and the QPD-II belong to this family. Charalambides has studied some properties of this family and has shown that the GSDs have applications for the busy periods in queuing processes and in the time to emptiness in dam and storage processes.

Charalambides has also given formal expressions for the pgf and the factorial mgf of the GSD; however, both expressions contain implicit functions of ΔE^{-r} and the Lagrange expansion formula is needed to compute the moments in terms of Bell polynomials. He has applied these formulas to compute the means and variances of the GNBD, QPD-I, and QPD-II.

The GSDs belong to a subclass of the Lagrangian probability models because they are based on either the Lagrange expansion or some identities that are obtained from Lagrange expansions.

4.9 Abel Series Distributions

Let $A(\theta, \lambda)$ be a positive function of two parameters θ and λ which possesses an Abel series expansion

$$A(\theta, \lambda) = \sum_{x \in T} a(x; \lambda)\, \theta(\theta + x\lambda)^{x-1}, \tag{4.104}$$

where $0 \le \theta \le \rho_1$ and $0 \le \lambda \le \rho_2$ and T is a subset of the set of nonnegative integers.

Charalambides (1990) defined the family of *Abel series distribution* (ASD) by

$$p(x; \theta, \lambda) = [A(\theta, \lambda)]^{-1} a(x, \lambda)\, \theta(\theta + x\lambda)^{x-1}, \tag{4.105}$$

for $x \in T$ and zero otherwise, if $A(\theta, \lambda)$ has the series function defined by (4.104). By using the shift operator E, derivative operator D, and the Abel operator $DE^{-\lambda} f(u) = d\, f(u - \lambda)/du$, given by Roman and Rota (1978), Charalambides gave an expression for $a(x; \lambda)$ as

$$a(x; \lambda) = [x!]^{-1} \left(DE^{-\lambda}\right)^x A(u; \lambda)|_{u=0} \tag{4.106}$$

and stated that a truncated ASD is also an ASD. He showed some applications of the ASDs in insurance, stochastic processes, length of the busy period in a queuing process, and the time of first emptiness in dam and storage processes.

The pgf of the ASD (4.105), with (4.106), is given by

$$G(z; \theta, \lambda) = [A(\theta, \lambda)]^{-1} \left[\exp\{\theta\, h^{-1}(z\, D\, E^{-\lambda})\} A(u; \lambda)\right]_{u=0}, \tag{4.107}$$

where $w = h^{-1}(v)$ is the inverse of $v = h(w) = w\, e^{-\lambda w}$. The expression (4.107) is not a closed form because the inverse function w is an infinite series based on the Lagrange transformation $v = w\, e^{-\lambda w}$.

Charalambides (1990) obtained an expression for the factorial mgf and showed that the factorial moments of the ASD can be expressed in a closed form in terms of the Bell partition polynomials.

The GPD, the QBD-I, and the QBD-II are three important examples of the ASD. Charalambides has obtained the means and variances for all three of them (but the proofs seem to be long) and has shown that the GPD is the only member of the ASD which is closed under convolution. Nandi and Das (1994) considered the Abel series distributions. They noted that the QBD-I, QBD-II, and GPD are also ASDs. The ASDs belong to a subclass of the Lagrangian probability models.

4.10 Exercises

4.1 Suppose a random variable X has the quasi-binomial distribution II given by (4.73). By using the method of differentiation (with respect to p), obtain a recurrence relation between the noncentral moments of X. Using your recurrence relation or otherwise, obtain the first three noncentral moments for X.

4.2 Prove that the QHD-I given by (4.59) tends to the QBD-I defined in (4.1) when the parameters $a, b, r \to \infty$ under the conditions that $a(a+b+nr)^{-1} = p$ and $r(a+b+nr)^{-1} = \phi$. Furthermore, show that the limiting form of the QHD-II defined by (4.83) is the QBD-II defined in (4.72), where $a, b, \theta \to \infty$ in such a way that $a(a+b+n\theta)^{-1} = p$, $b(a+b+n\theta)^{-1} = \alpha$ and $\phi(a+b+n\theta)^{-1} = \phi$.

4.3 Show that the mean and variance of QPD-I defined in (4.65) are given by the expressions in (4.70) and (4.71) respectively.
4.4 Show that the inequality in (4.8) holds for the mean of QBD-I.
4.5 Consider the QBD-II defined by (4.73). Obtain the likelihood equations for the two parameters p and α. By taking the second partial derivatives with respect to the two parameters, obtain the entries in the Fisher's information matrix.
4.6 Use the method of proportion of "zero" and the sample mean to estimate the two parameters p and α for the QBD-II given by (4.73).
 (a) Obtain the Lagrange expansion of $(1+z)^{a+b}[1-rz(1+z)^{-1}]^{-1}$ under the transformation $z = u(1+z)^r$ as a power series in u. (Hint: Use the formula given in (1.80)).
 (b) Under the transformation in (a), use the formula in (1.80) to expand $(1+z)^b[1-rz(1+z)^{-1}]^{-1}$ and the formula in (1.78) to expand $(1+z)^a$.
 (c) By taking the products of the two expansions in (b), and equating the coefficient of u^n to that of u^n in the expansion in (a), show that the QHD-I is a true probability distribution.
4.7 Show that the QHD-II represents a true probability model.
4.8 Let X_i, $i = 1, 2, \ldots, n$, be a random sample of size n, taken from the probability model (4.6), with two parameters p and m. Find the moment and the ML estimators for p and m.
4.9 Obtain some suitable applications for the GSDs.
4.10 Let X_i, $i = 1, 2, \ldots, k$, be a random sample of size k, taken from the probability model (4.74), which has two parameters $0 < q < 1$ and n (a positive integer). Obtain the moment and the ML estimators for q and n.
4.11 Indicate some suitable applications for the ASDs.
4.12 Suppose each probability $P(X = x)$ of a QBD-II in (4.72) is weighted by the weight function $\omega_k = k$ and Y is the random variable representing the weighted distribution; find the $P(Y = y)$. Obtain the mean and the variance of the r.v. Y.
4.13 In the QBD-II in (4.72), suppose n is very large and a and θ are very small such that $na = c$, $n\theta = d$. By writing $c/(b+d) = M$ and $d/(b+d) = \phi$, show that the QBD-II tends to a GPD with parameters M and ϕ. Suppose further that a and b are finite and if n is very large and $\theta \to 0$ in such a way that $n\theta$ is finite, show that the QBD-II approaches the normal distribution.

5

Some Urn Models

5.1 Introduction

Urn models are constructed by considering a number of urns which contain balls of various colors together with some sequences of experiments (trials) for drawing the balls at random from the urns under certain rules. These rules prescribe the addition of some balls to and/or the removal of some balls from certain urns at different stages of the experiment. It is presumed that all the balls in an urn are equally likely to be drawn in a draw. Thus, if an urn contains n balls, the probability that a specified ball is chosen in a draw is n^{-1}. The complete process for an urn model can be broken up into simple steps or trials which enable the scientist to calculate the probabilities for the various stages of the model. Such urn models are generally used to compute the distributions of drawing a number of balls of various types in the urns or to compute the waiting time distributions until a particular condition is satisfied.

In their excellent book on urn models, Johnson and Kotz (1977) have given numerous references showing that urn models have been in use since the seventeenth century and that they have been used by various researchers to analyze a number of complex problems concerning physical phenomena like contagious events, random walks, ballot problems, occupancy problems, and games of chance. Some other prominent writers on urn models in the present century are Pólya (1930), Friedman (1949), and Feller (1968). In all these models, an individual's strategy or decision plays no significant role.

One cannot deny that most living beings and/or their leaders do make some decisions in specific situations and that they have to face the consequences as determined by the laws of nature. On account of such decisions, which become a part of their behavior, some tribes become dormant and gradually become extinct while others succeed in migrating to new places and adapt themselves nicely to the new circumstances. The cells in the human body develop immunity against antibiotics on successive use. Similarly, insects develop immunity against insecticides with the passage of time. Thus, the introduction of some factor, based on strategy or decisions, into the probability models seems desirable for explaining the observed patterns, especially those that deal with the behavior of living beings. Consul (1974), Consul and Mittal (1975, 1977), Famoye and Consul (1989b), and Consul (1994c) have described a number of urn models that depend upon the strategy of the individual. All these urn models provide probability distributions that either are particular families of Lagrangian probability models or are associated with them.

5.2 A Generalized Stochastic Urn Model

An urn contains w white balls and b black balls. A ball is randomly drawn from the urn by a player and is given to an umpire without the player seeing it. The umpire returns the same ball to the urn together with s (≥ -1) balls of the same color and mixes them up. The value $s = -1$ implies that the umpire does not return the ball to the urn. This operation (experiment) is repeated again and again by the player and the umpire under the following rules of the game.

At any particular time let X denote the number of draws which gave black balls and Y be the number of draws which gave white balls.

(i) The player chooses his strategy by selecting two integers n (≥ 1) and β (≥ -1) and declares them to the umpire.
(ii) The player will continue the process of drawing the balls so long as the number of white balls drawn from the urn exceeds β times the number of black balls drawn from the urn until that time, and the player will be declared as a winner of the game if he stops as soon as the number of black balls and white balls drawn from the urn are exactly x and $y = n + \beta x$, respectively.

Before deciding the bets for the game, the house wants to determine the probabilities for a win by the player for the various values of $X = 0, 1, 2, 3, \ldots$. Let $P(X = x)$ denote the probability that the player wins the game after drawing x black balls. The player will keep playing the game of drawing the balls if the number of white balls drawn from the urn always exceeds βx. Thus, if he draws the xth black ball, he must have drawn at least $(\beta x + 1)$ white balls in order to remain in the game.

Since the player wins the game when exactly x black balls and exactly $n + \beta x = y$ white balls are drawn, let $f(x, y)$ be the number of sequences in which the number y of white balls drawn from the urn always exceeds βx. The determination of $f(x, y)$ gives rise to three cases: (i) $\beta = -1$, (ii) $\beta = 0$, and (iii) $\beta \geq 1$.

Case I. The player selects $\beta = -1$. Since the value of β is -1, for any number of x draws of black balls, the number of white balls will always exceed $\beta x = -x$. Therefore, the player will win with x draws of black balls if he draws a total of n balls containing $y = (n - x)$ draws of white balls in any order. Thus,

$$f(x, y) = \binom{y + x}{x} = \binom{n}{x} \tag{5.1}$$

and

$$P(X = x) = \binom{n}{x} \frac{b^{[x,s]} w^{[n-x,s]}}{(b + w)^{[n,s]}}, \quad x = 0, 1, 2, \ldots, n, \tag{5.2}$$

and zero elsewhere. Since the player is sure to win for some value of x, $P(X = x)$ in (5.2) represents a true probability distribution. It is called the Pólya–Eggenberger distribution (Eggenberger and Pólya, 1923). This probability model reduces to a number of well-known distributions for different values of the parameters:

(i) For $s = 0$, (5.2) gives the binomial model

$$P(X = x) = \binom{n}{x} \frac{b^x w^{n-x}}{(b + w)^n} = \binom{n}{x} \theta^x (1 - \theta)^{n-x}, \quad x = 0, 1, 2, \ldots, n,$$

where $0 < b(b + w)^{-1} = \theta < 1$.

(ii) For $s = -1$, (5.2) becomes the hypergeometric distribution

$$P(X = x) = \binom{b}{x}\binom{w}{n-x}\bigg/\binom{b+w}{n}, \quad x = 0, 1, 2, \ldots, \min(b, n).$$

(iii) For $s = +1$, the model (5.2) becomes the beta-binomial distribution (Kemp and Kemp, 1956)

$$P(X = x) = \binom{b+x-1}{x}\binom{w+n-x-1}{n-x}\bigg/\binom{b+w+n-1}{n}, \quad x = 0, 1, 2, \ldots, n.$$

(iv) When s is a positive integer ≥ 2, by putting $b/s = c$ and $w/s = d$, the model (5.2) can be written in the form

$$P(X = x) = \binom{c+x-1}{x}\binom{d+n-x-1}{n-x}\bigg/\binom{c+d+n-1}{n}, \quad x = 0, 1, 2, \ldots, n,$$

which is the beta-binomial distribution (Kemp and Kemp, 1956) in (iii) above.

(v) For $w = b = s$, the model (5.2) reduces to the discrete uniform distribution

$$P(X = x) = (n+1)^{-1}, \quad x = 0, 1, 2, \ldots, n.$$

(vi) For $w = s$ and $b/s = c > 0$, the model (5.2) gives the Ascending Factorial distribution (Berg, 1974)

$$P(X = x) = \binom{c+x-1}{x}\bigg/\binom{c+n}{n}, \quad x = 0, 1, 2, \ldots, n.$$

Case II. The player selects $\beta = 0$. Since the number of white balls drawn from the urn must always exceed $\beta x = 0$, to continue in the game the first draw must be a white ball and then the player can draw the black balls and white balls in any order. The player will win if he draws x black balls and $y = n + \beta x = n$ white balls in all, i.e., $(n-1)$ white balls after the first white ball. Therefore,

$$f(x, y) = \binom{y+x-1}{x} = \binom{n+x-1}{x} \tag{5.3}$$

and

$$P(X = x) = \binom{n+x-1}{x}\frac{b^{[x,s]}w^{[n,s]}}{(b+w)^{[n+x,s]}}, x = 0, 1, 2, \ldots, \tag{5.4}$$

and zero elsewhere and subject to the values of the other parameters. Since the player is sure to win for some value of x, the model (5.4) is a true probability distribution. It is called the inverse Pólya distribution (Sarkadi, 1957). It gives a number of particular cases:

(i) For $s = 0$, the model (5.4) reduces to the negative binomial distribution

$$P(X = x) = \binom{n+x-1}{x}\theta^x(1-\theta)^n, \quad x = 0, 1, 2, \ldots, \infty,$$

where $0 < b(b+w)^{-1} = \theta < 1$. When $n = a/\theta$, the above is called the Katz distribution (Katz, 1965).

(ii) For $s = -1$, the model (5.4) becomes the negative hypergeometric distribution (Kemp and Kemp, 1956)

$$P(X = x) = \frac{n}{n+x} \binom{b}{x} \binom{w}{n} \Big/ \binom{b+w}{n+x}, \quad x = 0, 1, 2, \ldots, b,$$

where $n \le w$.

(iii) For $s = +1$, the model (5.4) gives the beta-Pascal distribution (Ord, 1972)

$$P(X = x) = \frac{b}{b+w} \binom{b+x-1}{x} \binom{w+n-1}{n-1} \Big/ \binom{b+w+n+x-1}{n+x-1},$$

$$x = 0, 1, 2, \ldots, \infty.$$

(iv) When s is a positive integer ≥ 2, by putting $b/s = c$ and $w/s = d$, the model (5.4) becomes

$$P(X = x) = \frac{d}{c+d} \binom{c+x-1}{x} \binom{d+n-1}{n-1} \Big/ \binom{c+d+n+x-1}{n+x-1},$$

$$x = 0, 1, 2, \ldots, \infty,$$

which is also the beta-Pascal distribution.

(v) For $w = b = s$, the model (5.4) reduces to the Waring distribution (Irwin, 1965)

$$P(X = x) = \frac{n}{(n+x)(n+x+1)}, \quad x = 0, 1, 2, \ldots, \infty.$$

(vi) For $w = s$ and $b/s = c > 0$, the model (5.4) gives the inverse factorial distribution (Berg, 1974)

$$P(X = x) = nc(1+x)^{[n-1]}/(c+x)^{[n+1]}, \quad x = 0, 1, 2, \ldots, \infty.$$

Case III. The player selects the integer $\beta \ge 1$. To be in the game the player must draw at least $\beta x + 1$ white balls before he draws x black balls for $x = 0, 1, 2, \ldots$; i.e., at least $\beta + 1$ white balls must be drawn before the first black ball is drawn, at least $2\beta + 1$ white balls must be drawn before the second black ball is drawn, and so on. Thus, at least $(\beta x + 1)$ white balls must be drawn before the xth black ball is drawn, and this must hold for each value of $x = 0, 1, 2, 3, \ldots$. The function $f(x, y)$ denotes the number of sequences in which the y draws of white balls is always $\ge \beta x + 1$, where x denotes the number of black balls. Obviously, if $y = \beta x + 1$, then the last draw must be a black ball but when $y \ge \beta x + 2$, the last draw can be either a black ball or a white ball. Accordingly,

$$f(x, y) = 0, \text{ for } y < \beta x + 1, \tag{5.5}$$

$$= f(x-1, y), \text{ for } y = \beta x + 1, \tag{5.6}$$

$$= f(x-1, y) + f(x, y-1), \text{ for } y = \beta x + k, \quad k \ge 2. \tag{5.7}$$

We also know that

$$f(1, 0) = 0, \qquad f(0, y) = 1. \tag{5.8}$$

Also, by equations (5.5) and (5.6), we observe that

$$f(x, \beta x) = 0, \quad f(1, \beta + 1) = 1. \tag{5.9}$$

By successive applications of (5.7) and the boundary conditions (5.8) and (5.9), for all $y = \beta + k$, $k \geq 1$,

$$\begin{aligned} f(1, y) &= f(0,\ y) + f(1,\ y-1) = f(0,\ y) + f(0,\ y-1) + f(1,\ y-2) \\ &= 1 + 1 + \cdots + f(0,\ y-k+1) + f(1,\ y-k) \\ &= k + f(1,\ \beta) = y - \beta. \end{aligned}$$

Similarly, for all $y = 2\beta + k$, $k \geq 1$, and $k = y - 2\beta$, (5.7) gives

$$\begin{aligned} f(2,\ y) &= f(1,\ y) + f(2,\ y-1) = f(1,\ y) + f(1,\ y-1) + f(2,\ y-2) \\ &= \sum_{i=1}^{k} f(1,\ y+1-i) + f(2,\ y-k) \\ &= \sum_{i=1}^{k} (y+1-i-\beta) + f(2,\ 2\beta) = (y+1-\beta)k - \frac{1}{2}k(k+1) \\ &= \frac{1}{2}(y-2\beta)(y+1) = \frac{y-2\beta}{y+2}\binom{y+2}{2}. \end{aligned} \tag{5.10}$$

Also, for all $y = 3\beta + k$, $k \geq 1$, and $k = y - 3\beta$, repeated use of (5.7) gives

$$\begin{aligned} f(3,\ y) &= f(2,\ y) + f(3,\ y-1) = f(2,\ y) + f(2,\ y-1) + f(3,\ y-2) \\ &= \sum_{i=1}^{k} f(2,\ y+1-i) + f(3,\ y-k) \\ &= \sum_{i=1}^{k} \frac{1}{2}(y+1-i-2\beta)(y+1-i+1) \quad \text{(by (5.10))} \\ &= \frac{1}{2}(y+1-2\beta)(y+2)k - \frac{1}{4}(2y+3-2\beta)k(k+1) + \frac{1}{12}k(k+1)(2k+1). \end{aligned}$$

By putting $k = y - 3\beta$ and on simplifying, the above gives

$$f(3,\ y) = \frac{1}{6}(y-3\beta)(y+2)(y+1) = \frac{y-3\beta}{y+3}\binom{y+3}{3}. \tag{5.11}$$

Since the results for $f(1,\ y)$, $f(2,\ y)$, and $f(3,\ y)$ are similar and have a pattern, to obtain the general solution for the difference equations (5.7) and (5.6), we use the method of mathematical induction and assume for some given value of x and for $y = \beta x + k$, $k \geq 1$, and $k = y - \beta x$,

$$f(x,\ y) = \frac{y-\beta x}{y+x}\binom{y+x}{x}. \tag{5.12}$$

By successive applications of (5.7), we obtain for $y = \beta(x+1)+k$, $k \geq 1$,

$$f(x+1,\ y) = \sum_{i=1}^{k} f(x,\ y+1-i), \quad \text{where} \quad k = y - \beta(x+1),$$

$$= \sum_{i=1}^{k} \frac{y+1-i-\beta x}{y+1-i+x} \binom{y+1-i+x}{x}$$

$$= \sum_{i=1}^{k} \binom{y+1-i+x}{x} - (\beta+1) \sum_{i=1}^{k} \binom{y+x-i}{x-1}.$$

Now,

$$\sum_{i=1}^{k} \binom{y+1+x-i}{x} = \text{coefficient of } t^x \text{ in } \sum_{i=1}^{k} (1+t)^{y+x+1-i}$$

$$= \text{coefficient of } t^x \text{ in } (1+t)^{y+x} \left[\frac{1-(1+t)^{-k}}{1-(1+t)^{-1}} \right]$$

$$= \text{coefficient of } t^x \text{ in } (1+t)^{y+x} \left[\frac{(1+t)-(1+t)^{-k+1}}{t} \right]$$

$$= \text{coefficient of } t^{x+1} \text{ in } \left[(1+t)^{y+x+1} - (1+t)^{y+x+1-k} \right]$$

$$= \binom{y+x+1}{x+1} - \binom{y+x+1-k}{x+1}$$

$$= \binom{y+x+1}{x+1} - \binom{(\beta+1)(x+1)}{x+1}.$$

Similarly,

$$\sum_{i=1}^{k} \binom{y+x-i}{x-1} = \binom{y+x}{x} - \binom{y+x-k}{x}, \quad k = y - \beta(x+1),$$

$$= \binom{y+x}{x} - \binom{\beta(x+1)+x}{x}.$$

Therefore,

$$f(x+1,\ y) = \binom{y+x+1}{x+1} - \binom{(\beta+1)(x+1)}{x+1} - (\beta+1) \left[\binom{y+x}{x} - \binom{(\beta+1)x+\beta}{x} \right],$$

which can be easily simplified to

$$f(x+1,\ y) = \frac{y-\beta(x+1)}{y+x+1} \binom{y+x+1}{x+1}. \tag{5.13}$$

The above (5.13) is precisely the same as (5.12) with x replaced by $x + 1$. Hence, the relation (5.12) is true for all values of x.

Since the player wins the game when exactly x black balls are drawn and exactly $y = n+\beta x$ white balls are drawn, such that $y \geq \beta x + 1$, for values of $x = 0, 1, 2, 3, \ldots$, the probabilities of a win of the game by the player become

$$P(X = x) = \frac{y - \beta x}{y + x}\binom{y + x}{x}\frac{b^{[x,\, s]}w^{[y,\, s]}}{(b + w)^{[y+x,\, s]}}, \quad \text{where } y = n + \beta x,$$

$$= \frac{n}{n + \beta x + x}\binom{n + \beta x + x}{x}\frac{b^{[x,\, s]}w^{[n+\beta x,\, s]}}{(b + w)^{[n+\beta x+x,\, s]}}, \tag{5.14}$$

for $x = 0, 1, 2, 3, \ldots$ and subject to the other restrictions based on the values of the parameters $\beta,\ s,\ n,\ b$, and w.

We name (5.14) for $\beta \geq 1$ together with (5.2) for $\beta = -1$ and (5.4) for $\beta = 0$ as the *Prem distribution* with the five parameters $n,\ \beta,\ s,\ b$, and w. It is a genuine probability model when $s = -2, -1$ and $s = 0$. When s is a positive integer ≥ 1, the model (5.14) still represents the probabilities of a win of the player for $x = 0, 1, 2, 3, \ldots$, but the sum of the probabilities will be less than 1 because as $x \to \infty$, the total number of added balls $xs + (n+\beta x)s = (n+\beta x+x)s$ becomes infinitely large, which is an impossibility.

As an example, let $\beta = 1$ and $b = w = s$ in (5.14). Then,

$$P(X = x) = \frac{n}{n + 2x}\binom{n + 2x}{x}\frac{x!(n + x)!}{(n + 2x + 1)!} = \frac{n}{(n + 2x)(n + 2x + 1)},\ x = 0, 1, 2, 3, \ldots.$$

Therefore,

$$\sum_{x=0}^{\infty} P(X = x) = \sum_{x=0}^{\infty} \frac{n}{(n + 2x)(n + 2x + 1)}$$

$$< \sum_{x=0}^{\infty} \frac{n}{(n + x)(n + x + 1)} = \sum_{x=0}^{\infty}\left[\frac{n}{n + x} - \frac{n}{n + x + 1}\right] = 1.$$

Particular Cases. The Prem distribution in (5.14) gives the following well-known probability models as particular cases.

(i) For $n = \beta = 1,\ s = 0,\ b(b + w)^{-1} = \theta$, and $x = y - 1$, we get the Haight distribution

$$P(Y = y) = \frac{1}{2y - 1}\binom{2y - 1}{y}\theta^{y-1}(1 - \theta)^y, \quad y = 1, 2, 3, \ldots, \infty.$$

(ii) For $n = m,\ s = 0,\ \beta = m - 1, x = y - 1$, and $b(b + w)^{-1} = \theta$, it gives the Consul distribution (Consul and Shenton, 1975)

$$P(Y = y) = \frac{1}{y}\binom{my}{y - 1}\theta^{y-1}(1 - \theta)^{my-y+1}, \quad y = 1, 2, 3, \ldots, \infty.$$

(iii) For $n = \beta = m - 1,\ s = 0,\ x = y - 1$, and $b(b + w)^{-1} = \theta$, it reduces to the Geeta distribution (Consul, 1990b)

$$P(Y = y) = \frac{1}{my - 1}\binom{my - 1}{y}\theta^{y-1}(1 - \theta)^{my-y}, \quad y = 1, 2, 3, \ldots, \infty.$$

(iv) For $n = mk,\ \beta = m - 1,\ s = 0,\ x = y - k$, and $b(b + w)^{-1} = \theta$, it becomes the delta-binomial distribution (Consul and Shenton, 1972)

$$P(Y = y) = \frac{k}{y}\binom{my}{y-k}\theta^{y-k}(1-\theta)^{k+my-y}, \quad y = k, k+1, k+2, \ldots, \infty.$$

(v) For $\beta = 1,\ s = 0$, and $b(b + w)^{-1} = \theta$, it gives the displaced lost games distribution

$$P(X = x) = \frac{n(n + 2x - 1)!}{x!(n + x)!}\theta^x(1-\theta)^{n+x}, \quad x = 0, 1, 2, \ldots, \infty.$$

(vi) For $s = 0,\ \beta = m - 1$, and $b(b + w)^{-1} = \theta$, it reduces to the generalized negative binomial distribution (Jain and Consul, 1971)

$$P(X = x) = \frac{n}{n + mx}\binom{n + mx}{x}\theta^x(1-\theta)^{n+mx-x}, \quad x = 0, 1, 2, \ldots, \infty.$$

(vii) For $\beta = 1,\ s = -1,\ w > n + x$, we get the negative hypergeometric distribution

$$P(X = x) = \frac{n}{n + 2x}\binom{b}{x}\binom{w}{n+x}\Big/\binom{b+w}{n+2x}, \quad x = 0, 1, 2, \ldots, b.$$

(viii) For $\beta = m,\ s = -1,\ b > x$, and $w > n + mx$, we obtain the inverse hypergeometric distribution as

$$P(X = x) = \frac{n}{n + mx + x}\binom{b}{x}\binom{w}{n+mx}\Big/\binom{b+w}{n+mx+x}, \quad x = 0, 1, 2, \ldots, b.$$

(ix) For $s = -2,\ b > 2x$, and $w > 2(n + \beta x)$, we get another inverse hypergeometric distribution

$$P(X = x) = \frac{n}{n + \beta x + x}\binom{b/2}{x}\binom{w/2}{n+\beta x}\Big/\binom{b/2 + w/2}{n+\beta x+x}, \quad x = 0, 1, 2, \ldots, b/2.$$

Possibly, many other models can be obtained from the Prem distribution by assigning other values to the parameters or as limiting forms of the models given in (i) to (viii).

The Prem probability distribution can also be obtained as a generalization of the generalized negative binomial distribution in (vi) by assuming θ to be a beta variable with parameters $(\xi,\ \eta)$, which gives

$$P(X = x) = \frac{n}{n + mx + x}\binom{n + mx + x}{x}\frac{\Gamma(\xi+\eta)\Gamma(\xi+x)\Gamma(\eta+n+mx)}{\Gamma(\xi)\Gamma(\eta)\Gamma(\xi+\eta+n+mx+x)}, \tag{5.15}$$

and then replacing m with β, ξ, with b/s and η with w/s, which are rational numbers. It may also be noted that all the probability models, given by $s = 0$ in (i) to (vi) above, belong to the class of Lagrangian probability distributions.

5.2.1 Some Interrelations among Probabilities

If the probabilities $P\ (X = x)$ of the Prem model (5.14) are represented by the symbol $f\ (x; n, b, w)$, then the successive probabilities can easily be determined by the recurrence relation

$$f\,(x+1;\; n,b,w) = \frac{n\,(n+\beta+\beta x+x+1)\,(b+sx)}{(n+\beta)\,(x+1)\,(b+w+s\,(n+\beta+\beta x+x))} f\,(x;\; n+\beta,b,w), \tag{5.16}$$

where $f\,(0;\; n,b,w) = \frac{w^{[n,s]}}{(b+w)^{[n,s]}}$.

Two other relations between these probabilities are

$$f\,(x;\; n,b+s,w-s) = \frac{(w-s)\,(b+sx)}{b\,(w+ns+\beta xs-s)}\, f\,(x;\; n,b,w), \tag{5.17}$$

$$f\,(x;\; n,b-s,w+s) = \frac{(b-s)\,(w+ns+bxs)}{w\,(b+xs-s)}\, f\,(x;\; n,b,w). \tag{5.18}$$

5.2.2 Recurrence Relation for Moments

Denoting the kth moment about the origin by $M'_k\,(n,b,w)$, of the Prem distribution, we have

$$\begin{aligned}
M'_k\,(n,b,w) &= \sum_{x=0}^{\infty} x^k \frac{n(n+\beta x+x-1)!}{x!(n+\beta x)!} \frac{b^{[x,s]}w^{[n+\beta x,s]}}{(b+w)^{[n+\beta x+x,s]}} \\
&= n\sum_{x=0}^{\infty}(1+x)^{k-1}\frac{(n+\beta x+x+\beta)!}{x!(n+\beta x+\beta)!}\frac{b^{[x+1,s]}w^{[n+\beta x+\beta,s]}}{(b+w)^{[n+\beta x+x+\beta+1,s]}} \\
&= n\sum_{j=0}^{k-1}\binom{k-1}{j}\sum_{x=0}^{\infty}x^j\frac{(n+\beta x+x+\beta-1)!}{x!(n+\beta x+\beta)!}\frac{(b+s)^{[x,s]}w^{[n+\beta x+\beta,s]}}{(b+s+w)^{[n+\beta x+x+\beta,s]}} \\
&\quad\times\frac{b}{b+w}\,[n+(\beta+1)x+\beta] \\
&= \frac{nb}{b+w}\sum_{j=0}^{k-1}\binom{k-1}{j} \\
&\quad\times\left[M'_j(n+\beta,\; b+s,\; w)+\frac{\beta+1}{n+\beta}M'_{j+1}(n+\beta,\; b+s,\; w)\right]
\end{aligned} \tag{5.19}$$

for $k = 1, 2, 3, \ldots$.

The above is a recurrence relation between the moments about the origin. Obviously,

$$M'_0(n,b,w) = 1. \tag{5.20}$$

The mean for Prem distribution. By putting $k = 1$ in the above recurrence relation,

$$M'_1(n,b,w) = \frac{nb}{b+w}\left[1+\frac{\beta+1}{n+\beta}M'_1(n+\beta,b+s,w)\right]. \tag{5.21}$$

By using the relation (5.21) repeatedly, we obtain the first moment about the origin (or mean μ) in the form

$$\mu = \frac{nb}{b+w}+\frac{nb^2(\beta+1)}{(b+w)(b+w+s)}+\frac{nb^3(\beta+1)^2}{(b+w)(b+w+s)(b+w+2s)}+\cdots. \tag{5.22}$$

The Prem distribution provides the probability models for $s = 0$ and $s = -1$ only. Therefore, when $s = 0$ and $b(b+w)^{-1} = \theta$, we have from (5.22)

$$\mu = n\theta\left[1 + (\beta+1)\theta + (\beta+1)^2\theta^2 + (\beta+1)^3\theta^3 + \cdots\right]$$

$$= n\theta\,[1 - (\beta+1)\theta]^{-1} \quad \text{for} \quad (\beta+1)\theta < 1. \tag{5.23}$$

When $s = -1$, the maximum value of x is b. Accordingly, the value of the mean μ is given by (5.22) as

$$\mu = \sum_{i=0}^{b} \frac{nb^{i+1}(\beta+1)^i}{(b+w)(b+w-1)\cdots(b+w-i)}. \tag{5.24}$$

Variance of Prem distribution. Now, by putting $k = 2$ in the recurrence relation (5.19), we get the recurrence relation for the second moment about zero as

$$M_2'(n,b,w) = \frac{nb}{b+w}\left[1 + \frac{\beta+1}{n+\beta}M_1'(n+\beta, b+s, w) + M_1'(n+\beta, b+s, w)\right.$$

$$\left. + \frac{\beta+1}{n+\beta}M_2'(n+\beta, b+s, w)\right].$$

On simplification by (5.21), it gives

$$M_2'(n,b,w) = M_1'(n,b,w) + \frac{nb}{b+w}M_1'(n+\beta, b+s, w) + \frac{nb(\beta+1)}{(b+w)(n+\beta)}M_2'(n+\beta, b+s, w). \tag{5.25}$$

When $s = 0$ and $b(b+w)^{-1} = \theta$, by using (5.23) the formula (5.25) gives

$$M_2'(n,b,w) = \frac{n\theta}{1-(\beta+1)\theta} + \frac{n\theta(n+\beta)\theta}{1-(\beta+1)\theta} + \frac{n\theta(\beta+1)}{n+\beta}M_2'(n+\beta, b, w). \tag{5.26}$$

By using the formula (5.26) repeatedly, the second moment can be expressed in the form of three infinite series as

$$M_2'(n,b,w) = [1-(\beta+1)\theta]^{-1}\left[n\theta\sum_{i=0}^{\infty}((\beta+1)\theta)^i + n^2\theta^2\sum_{i=0}^{\infty}((\beta+1)\theta)^i\right.$$

$$\left. + n\beta\theta^2\sum_{i=1}^{\infty} i\,((\beta+1)\theta)^{i-1}\right]$$

$$= n\theta\,[1-(\beta+1)\theta]^{-2} + n^2\theta^2\,[1-(\beta+1)\theta]^{-2} + n\beta\theta^2\,[1-(\beta+1)\theta]^{-3}.$$

Therefore,

$$\text{variance } \sigma^2 = M_2'(n,b,w) - \mu^2$$

$$= n\beta\theta^2\,[1-(\beta+1)\theta]^{-3} + n\theta\,[1-(\beta+1)\theta]^{-2}. \tag{5.27}$$

When $s = -1$, the use of (5.24) in (5.25) provides the formula

$$M_2'(n, b, w) = \sum_{i=0}^{b} \frac{nb^{i+1}(\beta+1)^i}{(b+w)(b+w-1)\cdots(b+w-i)}$$
$$+ \sum_{i=0}^{b-1} \frac{n(n+\beta)b(b-1)^{i+1}(\beta+1)^i}{(b+w)(b+w-1)\cdots(b+w-i-1)}$$
$$+ \frac{nb(\beta+1)}{(n+\beta)(b+w)} M_2'(n+\beta, b-1, w). \tag{5.28}$$

By repeated use of (5.28), the values of $M_1'(n, b, w)$ can be expressed as a finite number of series, which can be used to get the variance for the model when $s = -1$.

5.2.3 Some Applications of Prem Model

(i) *Infectious diseases.* We are living in a very complex world where bacteria of all kinds are always floating around us. These bacteria are multiplying with their own cycles and are attacking human beings, who have varying powers of resistance. When the effect of a particular kind of bacteria equals or exceeds the resistance level of a person, then the person becomes a victim of the disease, the bacteria multiply much faster in the body, and this person becomes infectious. As medicines are used to increase the resistance among persons and to kill the bacteria, they (bacteria) increase their own strength and their resistance to the medicines, and they attack other persons with a greater vigor. This process keeps going on with an increase of the diseased persons in the form of an epidemic and then the control of the epidemic. The black balls can represent the number of bacteria of the disease, the white balls can represent the number of disease-fighting cells in the body, s may represent the increase or decrease in their numbers after each attack, while β and n may represent the threshold numbers necessary for getting infected. Thus, $P(X = x)$ will represent the probability of x persons getting infected by the disease.

(ii) *Sales of a new product.* Whenever a new consumer product is introduced in the market, the manufacturer generates some market for it through TV and newspapers ads and by giving special incentives to the salespeople. Since consumers have been using other products and have been happy with them, there is some resistance due to these old preferences. Some people who are affected by the ads buy the new product and use it. If they like the product, they speak to their friends about it. Also, the manufacturer hires people who advertise the product by saying they have used it and it is far better than other available products. Thus, more buyers are generated. The whole process becomes like the spread of an infectious disease. In this case, the data on the number of persons using the product each day in the country will possibly be according to the Prem distribution.

(iii) *Environmental toxicology.* Hoover and Fraumeni (1975) have considered changes in the incidence of diseases due to exposure to various toxicants and pollutants. The pollutants increase due to reproduction and due to immigration or emigration. They are also checked by the use of chemicals. The problem becomes very similar to the one described in (i). Accordingly, the Prem model will be applicable, and for any observed data, the parameters of interest can be estimated by the Prem model.

(iv) *Games of pleasure.* The different special models of the Prem distribution can easily be used to develop a number of computer games for students and the public.

5.3 Urn Model with Predetermined Strategy for Quasi-Binomial Distribution I

Let there be two urns, marked A and B. The urn A contains w white balls and urn B contains w white and b black balls. Let n and s be two other known positive integers. In a game of chance, a player is allowed to choose his strategy by selecting an integer k such that $0 \leq k \leq n$ and then draws the balls one by one from the urns under the following conditions.

(i) ks black balls are added to urn A and ks white balls and $(n-k)\,s$ black balls are added to urn B before any draw is made.
(ii) The player randomly draws a ball from urn A. If the ball drawn is black, the player loses the game and gets no opportunity to draw balls from urn B.
(iii) If the ball drawn from urn A is white, the player will make n draws of one ball each time, with replacement, from urn B and will be declared to be a winner if the n draws (trials) contain exactly k white balls.

The probability of drawing a white ball from urn A is $w\,(w+ks)^{-1}$. Since the urn B contains $w+ks$ white balls and $b+(n-k)\,s$ black balls the chance of drawing a white ball in each trial is $(w+ks)\,/\,(w+b+ns)$. Therefore, the probability of drawing exactly k white balls in n trials of one ball each time, with replacement, is

$$\binom{n}{k}\left(\frac{w+ks}{w+b+ns}\right)^{k}\left(\frac{b+(n-k)\,s}{w+b+ns}\right)^{n-k}.$$

Thus, the joint probability of drawing a white ball from urn A and then drawing k white balls in n draws from urn B becomes

$$\frac{w}{w+ks}\binom{n}{k}\left(\frac{w+ks}{w+b+ns}\right)^{k}\left(\frac{b+ns-ks}{w+b+ns}\right)^{n-k}.$$

Accordingly, the probability of a win by the player is

$$P\,(X=k)=\binom{n}{k}\left(\frac{w}{w+b+ns}\right)\left(\frac{w+ks}{w+b+ns}\right)^{k-1}\left(\frac{b+ns-ks}{w+b+ns}\right)^{n-k} \tag{5.29}$$

for $k=0,1,2,\ldots,n$.

By using the transformation $w\,(w+b+ns)^{-1}=p$ and $s\,(w+b+ns)^{-1}=\phi$, the above gives the three-parameter QBD-I

$$P\,(X=k)=\binom{n}{k}p\,(p+k\phi)^{k-1}\,(1-p-k\phi)^{n-k} \tag{5.30}$$

for $k=0,1,2,\ldots,n$ and zero otherwise with $0<p+n\phi\leq 1$.

5.3.1 Sampling without Replacement from Urn B

In the above urn model, if the n draws in condition (iii) are made without replacement, then the joint probability of drawing a white ball from urn A and of drawing exactly k white balls in n draws from urn B becomes

$$P(X=k)=\binom{n}{k}\frac{w}{w+ks}\frac{(w+ks)_{(k)}\,(b+ns-ks)_{(n-k)}}{(w+b+ns)_{(n)}} \tag{5.31}$$

for $k=0,1,2,\cdots,n$ and where $(a)_{(k)}=a\,(a-1)\,(a-2)\cdots(a-k+1)$.

The urn model (5.31) represents the QHD-I, which can be expressed in the form

$$P(X=x)=\frac{w}{w+xs}\binom{w+xs}{x}\binom{b+ns-xs}{n-x}\bigg/\binom{w+b+ns}{n} \tag{5.32}$$

for $x=0,1,2,\ldots,n$.

5.3.2 Pólya-type Sampling from Urn B

In the urn model of section 5.3 the condition (iii) is modified to Pólya-type sampling in which if the player gets a white ball from urn A in the first draw, he gets the chance of making n draws of one ball each time from urn B. After each draw from the urn B the ball is returned to urn B, c balls of the same color as the ball drawn are added to urn B, and they are thoroughly mixed before performing the next draw. The player is declared a winner if he gets exactly k white balls in these n draws from urn B.

In this new Pólya-type sampling the probability of the player becoming a winner is the joint probability of drawing a white ball from urn A in one draw and of drawing exactly k white balls in n independent draws from urn B, under the new condition, and is given by

$$P(X=k)=\frac{w}{w+ks}\cdot\binom{n}{k}\frac{(w+ks)^{[k,c]}\,(b+ns-ks)^{[n-k,c]}}{(w+b+ns)^{[n,c]}} \tag{5.33}$$

for $k=0,1,2,3,\ldots,n$ and zero otherwise.

The above urn model is the QPD I. By putting $w/c=a$, $b/c=d$, and $s/c=h$, the above model can also be expressed in the form

$$P(X=x)=\frac{a}{a+xh}\frac{\binom{a+xh+x-1}{x}\binom{d+nh-xh+n-x-1}{n-x}}{\binom{a+d+nh+n-1}{n}}. \tag{5.34}$$

5.4 Urn Model with Predetermined Strategy for Quasi-Pólya Distribution II

In a four-urn model, let each one of the urns A and D contain a white balls and b black balls; let urn B contain b black balls and urn C contain a white balls. Given two other positive integers n and θ, a player decides his winning strategy by choosing a positive integer k $(0\le k\le n)$ when he has to make the draws under the following conditions:

(i) $(n-k)\,\theta$ white balls and $k\theta$ black balls are added to the urns B and C, respectively, and $k\theta$ white balls and $(n-k)\,\theta$ black balls are added into urn D and the contents of each urn are mixed thoroughly.

(ii) The player is to draw one ball from urn A. The next draw by the player will be from urn B or from urn C according to whether the first ball drawn is white or black, respectively.

(iii) If the two balls drawn by the player are of the same color, the player loses his opportunity for further draws and loses the game; but if the two balls are of different colors, the player is allowed to make n random drawings of one ball each from urn D, and after each draw the particular ball and c additional balls of the same color are added to the urn D.

The player is declared a winner of the game if he gets exactly k white balls in the n draws (trials) from urn D. This urn model was given by Consul and Mittal (1975) and in a slightly modified form by Janardan (1978) by the name of the Markov–Pólya urn model. It can be represented by the flow-diagram below:

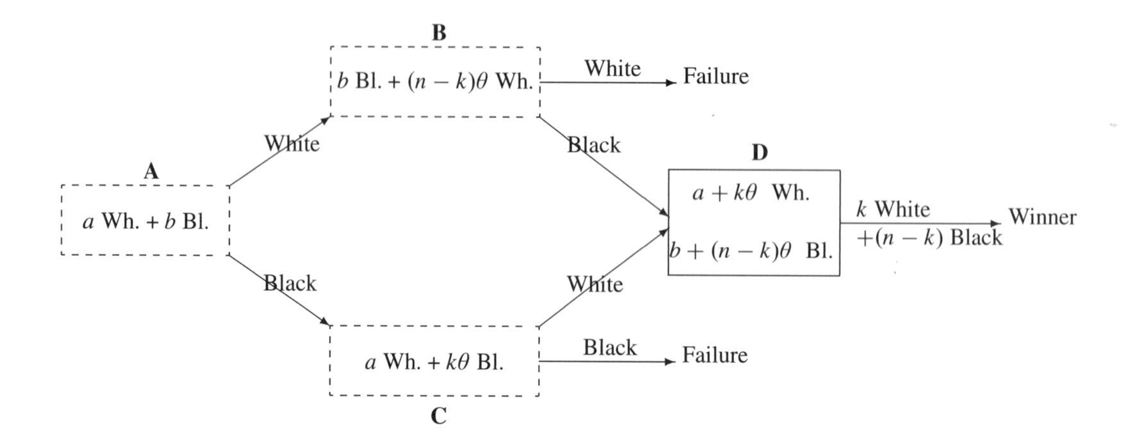

The probability that the player will get the opportunity of n drawings from urn D is

$$\frac{a}{a+b}\frac{b}{b+(n-k)\,\theta}+\frac{b}{a+b}\frac{a}{a+k\theta}$$
$$=\frac{ab(a+b+n\theta)}{(a+b)\,(a+k\theta)\,(b+n\theta-k\theta)},$$

and the conditional probability of success in the n drawings from urn D is

$$\binom{n}{k}\frac{(a+k\theta)^{[k,c]}\,(b+n\theta-k\theta)^{[n-k,c]}}{(a+b+n\theta)^{[n,c]}}.$$

Thus, the joint probability $P\,(X=k)$ of the player becoming a winner is given by

$$P\,(X=k)=\binom{n}{k}\frac{ab}{(a+b)\,(a+k\theta)}\frac{(a+k\theta)^{[k,c]}\,(b+n\theta-k\theta+c)^{[n-k-1,c]}}{(a+b+n\theta+c)^{[n-1,c]}} \tag{5.35}$$

$$=\binom{n}{k}J_k\,(a,\theta,c)\,J_{n-k}\,(b,\theta,c)\,/\,J_n\,(a+b,\theta,c)\,, \tag{5.36}$$

where $k=0,1,2,3,\ldots,n$ and zero otherwise and

$$J_k\,(a,\theta,c)=a\,(a+k\theta)^{-1}\,(a+k\theta)^{[k,c]}\,. \tag{5.37}$$

The probabilities (5.35) or (5.36) represent the QPD-II.

5.4.1 Sampling with Replacement from Urn D

In the above urn model, if the n draws in condition (iii) are made with replacement and no other balls are added, i.e., if $c=0$, then the joint probability of becoming a winner, by drawing k white balls in n draws from urn D, is given by (5.35) as

$$P(X=k)=\binom{n}{k}\frac{ab}{(a+b)(a+k\theta)}\left(\frac{a+k\theta}{a+b+n\theta}\right)^{k}\left(\frac{b+n\theta-k\theta}{a+b+n\theta}\right)^{n-k-1} \tag{5.38}$$

for $k=0,1,2,\ldots,n$ and zero otherwise. The above is the QBD-II.

5.4.2 Sampling without Replacement from Urn D

In the above urn model, if the n draws from urn D are made without replacement and no balls of the same color are added, then c becomes -1. By substituting $c=-1$ in (5.35), the joint probability for winning the urn model game becomes

$$P(X=k)=H_k(a,\theta)\,H_{n-k}(b,\theta)\,/\,H_n(a+b,\theta) \tag{5.39}$$

for $k=0,1,2,\ldots,n$ and zero otherwise. This is the QHD-II (Janardan, 1978), where

$$H_k(a,\theta)=\frac{a}{a+k\theta}\binom{a+k\theta}{k}. \tag{5.40}$$

5.4.3 Urn Model with Inverse Sampling

In the four-urn model with predetermined strategy the initial composition of the four urns remains the same but the rules of the game are changed. The player is given one positive integer θ and is required to decide the winning strategy by choosing two integers k and r ($k\geq 0$, $r\geq 1$) before making any draws from the urns under the following conditions:

(i) $r\theta$ white balls are added to urn B, $k\theta$ black balls are added to urn C, and $k\theta$ white balls plus $r\theta$ black balls are added to urn D. The contents of each urn are thoroughly mixed.
(ii) The player is to draw one ball from urn A. The next draw by the player will be from urn B or from urn C according to whether the first ball is white or black, respectively.
(iii) If the two balls drawn by the player are of the same color, the opportunity for further draws is lost and the player loses the game; but if the two balls are of different colors, the player is allowed to make $(k+r)$ random drawings of one ball each from urn D, and after each draw the particular ball and c additional balls of the same color are added to the urn before the next draw.

The player will be declared as a winner of the game if the rth black ball is obtained in the $(k+r)$th draw from urn D; otherwise the player will lose the game. The probabilities of success of the player under this urn model, for various values of k and r, are given by

$$\begin{aligned}P(X=k)&=\frac{ab}{a+b}\left(\frac{1}{b+r\theta}+\frac{1}{a+k\theta}\right)\binom{r+k-1}{k}\frac{(a+k\theta)^{[k,c]}\,(b+r\theta)^{[r,c]}}{(a+b+k\theta+r\theta)^{[r+k,c]}}\\&=\binom{r+k-1}{k}J_k(a,\theta,c)\,J_r(b,\theta,c)\,/\,J_{r+k}(a+b,\theta,c)\end{aligned} \tag{5.41}$$

for $k=0,1,2,3,\ldots$ and zero otherwise.

If Y denotes the number of draws from urn D for the success of the player, then the above probability distribution can be expressed as

$$P(Y = y) = \binom{y-1}{r-1} J_r(b, \theta, c) J_{y-r}(a, \theta, c) / J_n(a + b, \theta, c) \tag{5.42}$$

for $y = r, r + 1, r + 2, \ldots$ and zero otherwise.

Janardan (1978) has given a three-urn model for the same probability distribution. When $\theta = 0$, the probability distribution (5.41) reduces to the inverse-Pólya distribution, given by Patil and Joshi (1968), and when $\theta = 1, c = -1$ or when $\theta = 0, c = 1$, the pmf (5.41) yields the beta-Pascal distribution, derived by Ord (1972) as a model for the counts of diamondiferous stones.

5.5 Exercises

5.1 Consider a random variable X with the conditional distribution

$$P(X = x \mid \Theta = \theta) = \frac{m}{m + \beta x} \binom{m + \beta x}{x} \theta^x (1 - \theta)^{m + \beta x - x},$$

where parameter Θ is a beta random variable with parameters (ξ, η). Show that the unconditional distribution of X is given by the Prem distribution.

5.2 Verify the two relations (5.17) and (5.18) between the probabilities of Prem distribution.

5.3 Consider the Prem distribution given by (5.14). Show that if $\beta = m - 1$, $s = 0$, and $\theta = b(b + w)^{-1}$, the Prem distribution reduces to the generalized negative binomial distribution.

5.4 Show that the Prem distribution reduces to the inverse factorial distribution (model (vi) in Case II) if $\beta = 0$, $b/s = c$, and $w = s$.

5.5 Describe five possible applications for the QBD-I with suitable interpretations for the parameters and variables of the model.

5.6 Obtain the mean and variance of the three-parameter QBD-I, defined by (5.30), and of the QHD-I, defined by (5.32).

5.7 Describe two suitable applications of the QPD-II in (5.35).

5.8 Obtain the mean and variance of the four-parameter QBD-II in (5.38). Also, change the model (5.38) into the one with three parameters by suitable substitution.

5.9 Show that the Prem distribution in (5.14) reduces to the delta-binomial distribution when $n = mk$, $\beta = m - 1$, $s = 0$, $x = y - k$, and $b/(b + \omega) = \theta$. Find a recurrence relation between the central moments of the delta-binomial distribution. Using your recurrence relation or otherwise, find the mean, variance, and a measure of skewness for the delta-binomial distribution.

5.10 The Prem distribution in (5.14) reduces to the displaced lost games distribution when $\beta = 1$, $s = 0$, and $b/(b + \omega) = \theta$. Find the mean, variance, a measure of skewness, and a measure of kurtosis for the lost games distribution. What are the values of θ for which the mean of lost games distribution is greater than, equal to, or smaller than the variance of lost games distribution?

6

Development of Models and Applications

6.1 Introduction

The study of chance mechanisms enables a person to understand the principles by which the observable events are taking place in nature and provide the different models which can be used as formulas for future guidance, if they are applicable. A number of such probabilistic mechanisms for Lagrangian probability distributions are being described in this chapter.

The branching (multiplication) process contains a sequence of random variables X_0, X_1, $X_2, \ldots$ representing the number of objects (cells or individuals) in zeroth, first, second, ... generations. In section 6.2, it is shown that the general Lagrangian probability distribution is the model for the total progeny in a branching process. Such processes have many applications in the study of population growth, the spread of rumors, and nuclear chain reactions. The queuing process with a single server under the queue discipline of first-come, first-served is discussed in section 6.3. It is shown that the general Lagrangian distribution is the probability model which describes the total number of customers served in a busy period under certain conditions.

A stochastic process for epidemics is discussed in section 6.4. The total number of infected individuals is shown to have the general Lagrangian probability distribution. The enumeration of trees and a cascade process have been considered in sections 6.5 and 6.6, respectively.

6.2 Branching Process

In a branching process, each one of the initial objects can give rise to more objects of the same type or of different types. The objects produced, before dying, reproduce a certain number of new objects. If the reproduction process starts with a single individual, we are interested in the probability distribution of the total progeny at the moment of extinction or at a particular generation. Also, the probability distribution of the total progeny, when the reproduction process starts with a random number of ancestors, will be considered.

Let $X_0, X_1, X_2, \ldots, X_n, \ldots$ denote the total number of objects in the zeroth, first, second, $\ldots$, nth, $\ldots$ generations. Let the probability distribution of the number of objects produced by each object remain unaltered over successive generations and let its pgf be denoted by $g(z)$. Let the pgf for the total number of objects in the nth generation be

$$g_n(z) = \mathrm{E}(z^{X_n}).$$

Suppose $X_0 = 1$ with probability 1 (i.e., the branching process starts with a single ancestor). Then

$$g_0(z) = z \quad \text{and} \quad g_1(z) = g(z).$$

In a branching process, the conditional distribution of X_{n+1} given $X_n = j$ is that of the sum of j independent random variables, each having the same distribution as X_1 (Harris, 1947).

Now for $n = 2, 3, 4, \ldots$

$$\begin{aligned}
g_{n+1}(z) &= \sum_{k=0}^{\infty} P(X_{n+1} = k) z^k \\
&= \sum_{k=0}^{\infty} z^k \sum_{j=0}^{\infty} P(X_{n+1} = k | X_n = j) P(X_n = j) \\
&= \sum_{j=0}^{\infty} P(X_n = j)\,(g(z))^j \\
&= g_n(g(z)).
\end{aligned}$$

Also, $g_2(z) = g_1(g(z)) = g(g(z)) = g(g_1(z))$, and similarly $g_{n+1}(z) = g(g_n(z))$. The above branching process will stop as soon as $P(X_n = 0) = 1$ for some large positive integer n. The necessary and sufficient condition for this to happen is that $g'(1) < 1$.

Suppose the branching process stops after the nth generation. Let

$$Y_n = X_0 + X_1 + X_2 + \cdots + X_n,$$

where for the moment $X_0 = 1$. Let $G_n(z)$ be the pgf of

$$U_n = X_1 + X_2 + \cdots + X_n.$$

Then, $G_1(z) = g_1(z) = g(z)$, and the pgf of Y_1 is given by

$$\mathrm{E}\left(z^{Y_1}\right) = \mathrm{E}\left(z^{X_0 + X_1}\right) = zg(z) = R_1(z), \text{ say.}$$

Since each object of X_1 will start a new generation, the pgf of U_2 becomes

$$G_2(z) = g(zG_1(z)) = g(R_1(z)),$$

and similarly,

$$G_n(z) = g(R_{n-1}(z)), \quad n = 2, 3, 4, \ldots.$$

Hence, the pgf of $Y_n = 1 + U_n$ is

$$R_n(z) = zG_n(z) = zg(R_{n-1}(z)).$$

On taking the limit as n increases,

$$G(z) = \lim_{n \to \infty} R_n(z) = zg\left(\lim_{n \to \infty} R_{n-1}(z)\right),$$

and so

$$G(z) = zg\left(G(z)\right). \tag{6.1}$$

On putting $G(z) = t$ in (6.1), we obtain $t = zg(t)$.

Given that the branching process started with one object, the Lagrange expansion of t as a function of z may be used by putting $f(z) = z$ to obtain the probability distribution of the total number of objects at generation n. This leads to the basic Lagrangian probability distribution given in (2.2).

Suppose $X_0 = M$ with probability 1. Then $f(z) = z^M$, and the probability distribution of the total progeny is the delta Lagrangian probability distribution defined in (2.4) with n replaced by M. If X_0 is a random variable with a pgf $f(z)$, the probability distribution of the total progeny at the nth generation is given by the general Lagrangian probability distribution in (2.7).

Shoukri and Consul (1987) and Consul and Shoukri (1988) proved the above and had shown that the generalized Poisson model and a number of other Lagrangian models represented the probability distributions of the total number of infected individuals when the infection got started by a random number of infectives.

These branching processes are also applicable to the spread of fashions and sales, where a random number of customers are generated by television commercials and these customers (infectives) subsequently generate other customers by their use, associations, appreciation, etc. In a similar manner, the probability distributions of salespeople in dealerships (like AVON, AMWAY, Tupperware, cleaning chemicals, etc.) will also be Lagrangian because each distributor (infected) collects a group of persons (susceptibles) and enrolls some of them (infects them). The branching process continues on and on until it dies out.

The probability distribution of purchases of a product by customers will be of a similar nature because the sales potential is generated by advertising campaigns, and then others follow the trend. The distributions of burnt trees in forest fires will be of a basic Lagrangian form because the fire gets started with a single spark to one tree. Some of these have been further discussed in section 6.4.

6.3 Queuing Process

The Lagrangian probability distributions with generating functions $g(z)$ and $f(z)$ are defined in chapter 2. When $f(z) = z$, we obtain the basic Lagrangian distribution and when $f(z) = z^n$ we obtain the delta Lagrangian distribution. The pgf of the delta Lagrangian distribution is provided in (2.3). The coefficients of $u^{n+i}, i = 0, 1, 2, \ldots,$ are functions of n and also the successive derivatives of $[g(z)]^{n+i}$ at $z = 0$. These Lagrangian probabilities can be denoted by $f_{n+i}^{(n)}$ and the pgf in (2.3) can be rewritten as

$$z^n = u^n f_n^{(n)} + u^{n+1} f_{n+1}^{(n)} + u^{n+2} f_{n+2}^{(n)} + \cdots + u^{n+i} f_{n+i}^{(n)} + \cdots. \tag{6.2}$$

Consul and Shenton (1973a) used probabilistic arguments to show that the Lagrangian probability distributions play important roles in queuing theory. Suppose the number of customers served during a busy period when the queue was initiated by n customers is denoted by N_n. Let $f_r^{(n)}$ denote the probability of a busy period of r customers initiated by a queue of n customers.

Therefore,

$$f_r^{(n)} = P(N_n = r) \quad \text{with} \quad f_r^{(n)} = 0 \quad \text{if} \quad r < n.$$

Let $g_n(s),\ s = 1, 2, 3, \ldots,$ denote the probability of s arrivals during the service time of n customers. If $g(z)$ is the pgf of a univariate random variable, so also is $[g(z)]^n$, whose expansion can be written as

$$[g(z)]^n = \sum_{s=0}^{\infty} z^n g_n(s). \tag{6.3}$$

Note that $g_n(s)$ represents the probability generated by the function $[g(z)]^n$.

Let b be the service time per customer. If no new customers arrive during the total service period nb of initial n customers, we have

$$f_n^{(n)} = g_n(0), \tag{6.4}$$

where $g_n(0)$ is the probability of zero arrivals in a total service period $t = nb$.

If some new customers arrive during the service period of initial n customers, the busy period will continue and $N_n - n$ more customers will be served before the queue ends. Therefore, for $N_n = n + i,\ i = 1, 2, 3, \ldots,$ we obtain

$$f_{n+i}^{(n)} = \sum_{s=1}^{i} g_n(s) f_i^{(s)}. \tag{6.5}$$

On multiplying (6.4) by u^n and (6.5) by u^{n+i} and summing over i from 0 to ∞, we get

$$\sum_{i=0}^{\infty} u^{n+i} f_{n+i}^{(n)} = u^n g_n(0) + \sum_{i=1}^{\infty} \sum_{s=1}^{i} u^{n+i} g_n(s) f_i^s$$

$$= u^n \sum_{i=0}^{\infty} g_n(s) \left\{ \sum_{i=s}^{\infty} u^i f_i^{(s)} \right\}.$$

By using the results in (6.2) and in (6.3), we have

$$z^n = u^n \sum_{s=0}^{\infty} g_n(s) z^s = u^n [g(z)]^n = [ug(z)]^n.$$

We observe that $z = ug(z)$, the transformation for the Lagrangian probability distribution, is one of the roots of the above equation. Hence, the delta Lagrangian distribution represents the distribution of number of customers served in a busy period if the service was initiated by n customers.

Sibuya, Miyawaki, and Sumita (1994) reviewed Lagrangian distributions and related them to the busy period in queuing systems. The result of Consul and Shenton (1973a) presented above was generalized by Kumar (1981). Kumar's work is presented in subsection 6.3.1.

6.3.1 G|D|1 Queue

Consider a single server queue satisfying the following conditions: (i) arrivals during the service period of each customer are independent of each other and are identically distributed, and (ii) service time for each customer is constant.

Suppose X denotes the number of customers served in a busy period. Let the random variable Y denote the number of arrivals during one service. Let the random variable N be the number of customers initiating the queue. As shown earlier, Consul and Shenton (1973a) gave the result for the distribution of X when N is a constant. For a random variable N, Kumar (1981) stated and proved the following theorem.

Theorem 6.1. *Let X and N be two discrete random variables taking on nonnegative integer values. Let $g(z)$ and $f(z)$ be two pgfs such that $g(0) \neq 0$ and $f(z) = E(Z^N)$. Then*

$$P(X = x|N = n) = \frac{n}{x!} D_z^{x-1} \left[z^{n-1} \{g(z)\}^x \right] \Big|_{z=0} \tag{6.6}$$

if and only if

$$P(X = x) = \frac{1}{x!} D_z^{x-1} \left[f'(z) \{g(z)\}^x \right] \Big|_{z=0}, \tag{6.7}$$

where D_z denotes $\frac{\partial}{\partial z}$.

Proof. Let

$$P(X = x|N = n) = \frac{n}{x!} D_z^{x-1} \left[z^{n-1} \{g(z)\}^x \right] \Big|_{z=0}.$$

Then

$$P(X = x) = \sum_{n=1}^{x} \frac{1}{(n-1)!} \frac{1}{x!} D_z^{x-1} \left[z^{n-1} \{g(z)\}^x \right] \Big|_{z=0} D_z^n \left[f(z) \right] |_{z=0}$$

$$= \sum_{n=1}^{x} \frac{1}{x!} \binom{x-1}{n-1} D_z^{x-n} \left[\{g(z)\}^x \right] \Big|_{z=0} D_z^{n-1} \left[f'(z) \right] \Big|_{z=0}$$

$$= \frac{1}{x!} D_z^{x-1} \left[\{g(z)\}^x f'(z) \right] \Big|_{z=0}.$$

The converse of the theorem can be proved by tracing the steps backward. In the statement of the theorem we observe that (6.6) is the delta Lagrangian distribution, while (6.7) is the general Lagrangian distribution. □

Suppose the initial queue length is denoted by the random variable N. Let $f(z)$ be the pgf of N. By using the theorem, we conclude that the distribution of the number of customers served in a busy period is given by the general Lagrangian distribution provided in (2.7).

6.3.2 M|G|1 Queue

The M|G|1 queue is a single server queue with Poisson arrivals and arbitrary service-time distribution denoted by $B(x)$. Suppose the service time probability function is denoted by $b(x)$. Suppose further that the number of customers X served in a busy period has pgf $k(s)$, which is given by Kleinrock (1975, p. 184) as

$$k(s) = s B_*(\lambda - \lambda k(s)), \tag{6.8}$$

where

$$B_*(t) = \int_0^\infty e^{-tx} b(x) dx$$

is the Laplace transform of $b(x)$ and λ is the arrival rate.

On using the Lagrange expansion in (1.77) under the transformation $z = k(s)$, the numerically smallest nonzero root of (6.8) is given by

$$z = k(s) = \sum_{i=1}^{\infty} \frac{s^i}{i!} D_z^{x-1} \left[B_*^x(\lambda - \lambda z)\right]\Big|_{z=0}. \tag{6.9}$$

The above gives

$$P(X = x) = \frac{1}{x!} D_z^{x-1} \left[B_*^x(\lambda - \lambda z)\right]\Big|_{z=0} \tag{6.10}$$

for $x = i, i+1, i+2, \ldots$

The probability function in (6.10) is the basic Lagrangian distribution with $t = 1$ and $g(z) = B_*(\lambda - \lambda z)$. Suppose there are N (a constant) customers initiating the queue. The distribution of X is given by the ith convolution of (6.10) to get

$$P(X = x | N = i) = \frac{i}{x!} D_z^{x-1} \left[B_*^x(\lambda - \lambda z) z^{i-1}\right]\Big|_{z=0}$$

for $x = i, i+1, i+2, \ldots$.

If we further assume that N is a random variable with pgf $f(z)$, by using the theorem given before, we get

$$P(X = x) = \frac{1}{x!} D_z^{x-1} \left[B_*^x(\lambda - \lambda z) f'(z)\right]\Big|_{z=0} \tag{6.11}$$

for $x = 1, 2, 3, \ldots$

The above probability distribution given in (6.11) is the general Lagrangian distribution in (2.7).

Special cases of the queuing process.

(a) Let $f(z) = z,\ g(z) = e^{\theta(z-1)},\ (\theta < 1)$. These give

$$P(X = x) = \frac{(x\theta)^{x-1} e^{-\theta x}}{x!}, \quad x = 1, 2, 3, \ldots,$$

which is the Borel distribution.

(b) Let $f(z) = z^n,\ g(z) = e^{\theta(z-1)}\ (\theta < 1)$. These give

$$P(X = x) = \frac{n}{x} \frac{(x\theta)^{x-n} e^{-\theta x}}{(x-n)!}, \quad x = n, n+1, n+2, \ldots,$$

which is the Borel–Tanner distribution. This represents the distribution derived by Tanner (1953) as the distribution of the busy period for a single server queue with Poisson input.

(c) Let $f(z) = z^n,\ g(z) = (q + pz)^m\ (mp < 1)$. These generating functions give

$$P(X = x) = \frac{n}{x} \binom{mx}{x-n} q^{mx} (p/q)^{x-n}, \quad x = n, n+1, n+2, \ldots,$$

which represents the distribution of the busy period for a single server queue with binomial input.

(d) Let $f(z) = z^n$, $g(z) = q^k(1 - pz)^{-k}$ $(kp/q < 1)$. These generating functions give

$$P(X = x) = \frac{n}{x}\binom{kx + x - n - 1}{x - n} p^{x-n}(q)^{kx}, \quad x = n, n + 1, n + 2, \ldots,$$

which is a queue distribution indicating the stochastic law of the busy period for a single server queue with negative binomial input. A particular case of this probability distribution was discussed by Haight (1961).

(e) Let $f(z) = (1 - \theta)e^{\lambda(z-1)}/(1 - \theta z)$, $g(z) = e^{\theta(z-1)}$ $(\theta < 1)$. These two generating functions lead to

$$P(X = x) = \frac{(1 - \theta)(\lambda + \theta x)^x}{x!} e^{-\lambda - \theta x}, \quad x = 1, 2, 3, \ldots,$$

which is the linear Poisson distribution (model (10), Table 2.4), and is a queue distribution representing the number of customers served during the busy period for a single server queue under certain conditions.

(f) Let $f(z) = e^{\theta(z-1)}$, $g(z) = e^{\lambda(z-1)}$ $(\lambda < 1)$. These two generating functions lead to

$$P(X = x) = \frac{\theta(\theta + \lambda x)^{x-1}}{x!} e^{-\theta - \lambda x}, \quad x = 0, 1, 2, \ldots,$$

which is the GPD defined and studied in chapter 9. Thus, the GPD represents the distribution of the number of customers served in a busy period when the arriving customers have a Poisson distribution and the number of customers waiting, before the service begins, also has a Poisson distribution.

6.4 Stochastic Model of Epidemics

An important problem in the theory of epidemics is finding the probability distribution of the total number $N(u)$ of infected anywhere in the habitat, starting from those infected by a single infectious individual at location u and up to the time of extinction of the epidemic.

Neyman and Scott (1964) discussed a stochastic model for epidemics. They assumed that the number $\nu(u)$ of susceptibles infected by an infectious individual depends on the location of u of the infectious in the habitat. They also assumed that an individual infected at one point, say u, of the habitat does not usually remain at the point through his incubation period but moves away and becomes infectious himself at a different point X, where X is a random variable with its own probability distribution. With this set up, a stochastic model for epidemics appears as an extension of the branching process discussed in section 6.2.

Let $g(z|u)$ be the pgf of the number $\nu(u)$ of susceptibles who would be infected at point u if there is an infectious individual. Neyman (1965) assumed that $g(z|u) = g(z)$ is independent of u and showed that

$$G(z) = g(zG(z)) = g(\tau),$$

where $\tau = zG(z) = zg(\tau)$, and $G(z)$ is the pgf of the random variable N, the number of infected individuals before the extinction of the epidemics.

Under the transformation $\tau = zg(\tau)$, we expand $g(\tau)$ as a power series in z. This gives the basic Lagrangian distribution defined in (2.2). If the epidemics started with $X_0 = M$, a constant, the distribution of N, the number of infected individuals before the extinction of the epidemic is the delta Lagrangian distribution defined in (2.4). If the epidemic started with a r.v. X_0, which has a pgf $f(z)$, the distribution of N is the general Lagrangian distribution defined in (2.7).

Consul and Shoukri (1988) considered an epidemic example in which tourists arrive at different destinations into a country. The number of tourists is assumed to be large and the probability that a tourist carries an infectious bacteria is very small. Each infectious tourist comes in contact with a large number of persons in the country who are susceptible to infection. Upon contact with the population of susceptible, each infected tourist may infect a random number of susceptible. Consul and Shoukri assumed that the probability distribution of the number of persons who get infected among the susceptible is a Poisson random variable whose pgf is $g(z) = e^{\lambda(z-1)}$.

Suppose $X_0 = 1$ with probability 1. Then $f(z) = z$ and the probability distribution of the total number of infected individuals is

$$P(Y = y|X_0 = 1) = \frac{(\lambda y)^{y-1}e^{-\lambda y}}{y!}, \quad y = 1, 2, 3, \ldots,$$

which is known as the Borel distribution.

Suppose $X_0 = M$ with the probability 1. Then $f(z) = z^M$ and the probability distribution of the total number of infected individuals is

$$P(Y = y|X_0 = M) = \frac{M}{(y-M)!} y^{y-M-1} \lambda^{y-M} e^{-\lambda M}, \quad y = M, M+1, M+2, \ldots,$$

which is known as the Borel–Tanner distribution.

Suppose X_0 is a random variable with pgf $f(z) = e^{\theta(z-1)}$, and the probability distribution of the total number of infected individuals is given by the generalized Poisson distribution defined and studied in chapter 9.

Kumar (1981) considered an application in which $g(z) = (q + pz)^m$ and the epidemic is initiated by a random number of infected individuals having a binomial distribution, given by the pgf $f(z) = (q + pz)^m$. The distribution of the total number infected before the epidemic dies out is

$$P(Y = y) = \frac{m}{m+my} \binom{m+my}{y} p^y q^{m+my-y}, \quad y = 0, 1, 2, \ldots,$$

which is a special case of the generalized negative binomial distribution.

6.5 Enumeration of Trees

Cayley (1857) was the first to discuss the mathematical theory of trees. Henze and Blair (1931) developed recursive formulas for counting the number of rooted trees having the same number of vertices, where the number of branches at a vertex is at most four, except for a root vertex which has at most three branches. Pólya (1937) used power series as generating functions to study the number of rooted trees. Pólya's work is mainly concerned with trees and rooted trees, which are of interest to chemists.

Otter (1948, 1949) discussed the use of generating functions and the Lagrange expansion to study the enumeration of trees.

A tree is defined as any finite, connected topological graph without cycles. A vertex or a node of a tree is an end point of a line segment occurring in the tree. A rooted tree is one in which exactly one vertex, called the root, is distinguished from all other nodes in the tree. The ramification number of a vertex is the number of line segments which have that vertex in common. Otter (1948) considered trees and rooted trees where the ramification number of each vertex is not greater than a certain arbitrarily selected positive integer.

The root of a tree is generally said to be in the zeroth generation. A node is said to be in the kth generation ($k = 1, 2, \ldots$) if it is connected to one in the $(k-1)$th generation. If P is a node in the kth generation, then the set of nodes in the $(k+1)$th generation to which P is connected is called the litter of P.

Good (1965) discussed the various types of trees. He distinguished between Cayley's trees and ordered trees and noted that the Lagrange expansion is applicable only to ordered trees. An ordered tree is a rooted tree embedded in a plane. For an ordered tree, the order from left to right within each generation is important. Two trees with different orderings are not identical.

Consider ordered trees in which there is no constraint on the size of a litter. Suppose the enumeration (probability) generating function (pgf) for a litter is $g(z)$. Let the pgf for the number of nodes in the whole tree be denoted by $G(z)$. The subtrees that start at the first, instead of the zeroth, generation have the same pgf as $G(z)$. If the root has exactly n children, the pgf for the whole tree would be $z[G(z)]^n$. Good (1965) showed that the pgf for the whole tree satisfies the relation

$$G(z) = zg(G(z)) = u, \quad \text{say.}$$

Thus $u = zg(u)$, and on using the Lagrange expansion in (1.77) to expand $g(u)$ as a power series in z, we get the basic Lagrangian distribution provided in (2.2). Hence, the distribution of number of nodes in the whole tree is that of Lagrangian probability distribution.

If a forest has $X_0 = M$ trees where M is a constant, the distribution of the number of nodes in the M trees will be delta Lagrangian distribution given by (2.4). In a similar way, if the forest has X_0 trees, where X_0 is a random variable with pgf $f(z)$, the distribution of the number of nodes is the general Lagrangian distribution given in (2.7).

6.6 Cascade Process

Suppose a class of individuals gives rise seasonally to a number of new individuals, which will be referred to as children. Let the probabilities of an individual having $x (x = 0, 1, 2, \ldots)$ children be denoted by P_x $(x = 0, 1, 2, \ldots)$. Good (1949) considered the number of individuals in a cascade process by assuming that the probabilities P_x $(x = 0, 1, 2, \ldots)$ are the same for all individuals and are independent. The individuals formed in one season are called a new generation, and only this generation can reproduce in the next season. Cascade process is similar to branching process.

Let $g(z)$ be the pgf of the number of children of an individual. Good (1949) showed that the pgf of the number of individuals in the nth generation, given that there is exactly one individual in the zeroth generation, is

$$g(g(g(\cdots g(z)\cdots))) = g_n(z),$$

where $g_1(z) = g(z)$ and $g_{n+1}(z) = g(g_n(z)), n = 1, 2, 3, \ldots$. Let the total number of the individuals in each generation be denoted by X_i $(i = 0, 1, 2, \ldots, n, \ldots)$. For the moment, $X_0 = 1$. The total number of descendants of an individual in the first n generation is

$$U_n = X_1 + X_2 + \cdots + X_n.$$

Let $G_n(z)$ be the pgf of U_n. The pgf of U_n, the total number of descendants of an individual in the first n generation of a cascade process, is given by

$$G_n(z) = g(zg(zg(\cdots zg(z)\cdots))),$$

where

$$G(z) = G_1(z) = g(z), \quad \text{and}$$

$$G_n(z) = g(zG_{n-1}(z)). \tag{6.12}$$

On taking the limit in (6.12) as n increases, we obtain

$$G(z) = g[zG(z)] = g(\tau),$$

where $\tau = zG(z) = zg(\tau)$ and $G(z)$ is the pgf of U_n. Under the transformation $\tau = zg(\tau)$, we expand $g(\tau)$ as a power series in z. This gives the basic Lagrangian distribution provided in (2.2).

Suppose there are $X_0 = M$ individuals in the zeroth generation. If M is a constant, then $f(z) = z^M$ and the distribution of the total number of descendants of M individuals in the first n generations is given by the delta Lagrangian distribution. If there are X_0 individuals in the zeroth generation, and X_0 is a random variable with pgf $f(z)$, the total number of descendants is given by the general Lagrangian distribution in (2.7).

6.7 Exercises

6.1 Suppose the probability that an individual (in the zeroth generation) transmits a rumor is p. If the individual transmits the rumor, it is picked up by a random number X_1 of individuals which make up the first generation. Each of the X_1 including the originator become transmitters to the second generation X_2, and this process continues. Ultimately, the rumor becomes extinct or dies down. Suppose the individuals in the first, second, ... generations have independent probability distribution whose pgf is $g(t) = e^{\lambda(t-1)}$. Find the distribution, mean, and variance of the total number of individuals who picked up the rumor.

6.2 Suppose there are k customers waiting for service in a queue at a counter when the service is initiated. Suppose there are r additional customers, whose arrivals follow a negative binomial distribution, before the end of the first busy period. Customers are joining the queue under the condition of first-come, first-served, and service time is constant.

(a) If k is a random variable with negative binomial distribution, find the distribution, mean, and variance of $X = k + r$, the total number of customers served in the first busy period.

(b) If k is a constant, find the distribution, mean, and variance of $X = k + r$, the total number of customers served in the first busy period.

(c) Using your result in (b) or otherwise, deduce the distribution for $X = k + r$ if the service is initiated by a single customer.

6.3 Consider the relocation of five individuals who have been exposed to an infectious disease. The probability that an individual from the group of five will be infected is p. Upon relocation to a large city, those who are infected can transmit the disease to X_1 individuals in the city. The process will continue until the disease dies out. Suppose $X_1, X_2, X_3, \ldots$ are the number of infected individuals of the first, second, third, ... generations. If the distribution of $X_1, X_2, X_3, \ldots$ are independent Poisson with mean λ, find the distribution, mean, and variance of the total number of individuals infected by the disease at the nth generation.

6.4 Consider a forest with X_0 number of trees, where X_0 is a random variable with Poisson distribution. The X_0 roots of the trees belong to the zeroth generation. Suppose there are $X_1, X_2, X_3, \ldots$ nodes in the first, second, third, ... generations whose independent distribution is that of Poisson, whose mean differs from that of X_0. Show that the probability distribution of the total number of nodes in the forest at the nth generation is that of GPD.

6.5 The number of customers waiting in a single server queue, when the service begins, is a random number N whose pgf is $f(z)$. The arrivals during the service period of each customer are independent and are identically distributed with pgf $g(z)$, $g(0) \neq 0$. Let the queue discipline be first-come, first-served, the service time for each customer be constant, and X be the number of customers served in the busy period. Show that the probability distribution of X is a Lagrangian distribution.

6.6 Show that the GPD represents the probability distribution of the number of customers served in a busy period when the arrivals during the service period of each customer are independent and have the same Poisson distribution, when queue discipline is first-come, first-served, and when the number of customers waiting in the queue before the service begins has another Poisson distribution.

6.7 Suppose that n customers are waiting for service in a queue with a single server when the service begins. The arrivals for service during the service time of each customer are independent of each other and are given by a binomial model with pgf $g(z) = (q + pz)^m$, $0 < p = 1 - q < 1$, $mp < 1$. Obtain the probability distribution of the number of customers served in the busy period. What additional conditions will be necessary to obtain the solution?

6.8 Let X_0 be the total number of individuals at the zeroth generation of a cascade process. The total number of descendants of each individual in the first n generations is $U_n = X_1 + X_2 + \cdots + X_n$ and $g(t) = (1 - \theta + \theta t)^m$ is the pgf of the number of children of an individual.

(a) If $X_0 = 1$, find the distribution of the total number of descendants.

(b) If X_0 is a constant k, find the distribution of the total number of descendants.

(c) If X_0 is a random variable with pgf $f(t) = e^{\theta(t-1)}$, find the distribution of the total number of descendants.

7

Modified Power Series Distributions

7.1 Introduction

The class of modified power series distributions (MPSDs), represented by

$$P(X = x) = a_x(\phi(\theta))^x / h(\theta), \qquad x \in T \subset N, \tag{7.1}$$

and originally defined and studied by R. C. Gupta (1974), was described in section 2.3. It was shown that all modified power series distributions (MPSDs) are linear exponential and that they form a subclass of the class of Lagrangian probability distributions $L(f; g; x)$. Some important probability models belonging to the MPSD class are given in Table 2.7. A truncated MPSD is also a MPSD. The expressions for the mean μ and the variance σ^2, given in (2.44) and (2.45), are

$$\mu = E[X] = \frac{\phi(\theta)h'(\theta)}{\phi'(\theta)h(\theta)} \tag{7.2}$$

and

$$\sigma^2 = \frac{\phi(\theta)}{\phi'(\theta)} \frac{d\mu}{d\theta} . \tag{7.3}$$

Gupta (1976) studied the application of MPSDs in genetics and derived a general expression for the correlation coefficient between the number of boys and the number of girls when the family size has a MPSD. This subclass of probability models has been studied in great detail by numerous researchers. Some of the important research papers on this subject are by Consul (1981, 1990c), Famoye and Consul (1989a), R. C. Gupta (1975a, 1975b, 1976, 1977, 1984), P. L. Gupta (1982), P. L. Gupta and Singh (1981), R. C. Gupta and Singh (1982), R. C. Gupta and Tripathi (1985), Jani (1978a,b), Jani and Shah (1979a, 1979b), Kumar and Consul (1980), Patel and Jani (1977), and Tripathi et al. (1986).

Nikulin and Voinov (1994) considered a chi-square goodness-of-fit test for the MPSD. R. C. Gupta and Tripathi (1992) developed statistical inferences concerning the parameters of the MPSD based on samples from the corresponding length-biased MPSD model.

Johnson, Kotz, and Kemp (1992) have described some elementary properties of the MPSDs. A more detailed description of the properties and the estimation of the MPSDs will be given in this chapter.

7.2 Generating Functions

It has been shown in (2.33) that θ can be expressed as a function of $\phi(\theta)$, say $\psi(\phi(\theta))$, by means of the Lagrange expansion. Accordingly, the pgf of a r.v. X having a MPSD can be written in the form

$$g_X(t) = E\left[t^X\right] = \sum_{x\in T} t^x a_x (\phi(\theta))^x / h(\theta)$$

$$= \sum_x a_x (t\phi(\theta))^x / h(\theta)$$

$$= \frac{h\{\psi(t\phi(\theta))\}}{h\{\psi(\phi(\theta))\}} . \tag{7.4}$$

The mgf $M_X(t)$ of a MPSD, if it exists, is given by

$$M_X(t) = E\left[e^{tX}\right] = \sum_{x\in T} \left(e^t - 1 + 1)^x\right) a_x (\phi(\theta))^x / h(\theta)$$

$$= \sum_x \sum_{i=0}^{x} \binom{x}{i} a_x (\phi(\theta))^x \left(e^t - 1\right)^{x-i} / h(\theta)$$

$$= \sum_{i=0}^{\infty} \sum_{y=0}^{\infty} \binom{y+i}{i} y!\, a_{y+i} (\phi(\theta))^{y+i} \left(e^t - 1\right)^y / [h(\theta) y!] .$$

By expanding $(e^t - 1)^y$ as a power series in t, Gupta and Singh (1982) obtained the mgf for the MPSD as

$$M_X(t) = \sum_{s=0}^{\infty} \frac{t^s}{s!} \sum_{y=0}^{\infty} \sum_{i=0}^{\infty} \binom{y+i}{i} y!\, a_{y+i} (\phi(\theta))^{y+i} \; S(s, y)/h(\theta), \tag{7.5}$$

where $S(s, y)$ denote the Stirling numbers of the second kind, defined in subsection 1.2.4.

7.3 Moments, Cumulants, and Recurrence Relations

By differentiating the rth noncentral moment

$$\mu'_r = [h(\theta)]^{-1} \sum_{x\in T} x^r a_x (\phi(\theta))^x \tag{7.6}$$

on both sides with respect to θ, we get

$$\frac{d\mu'_r}{d\theta} = \frac{1}{h(\theta)} \sum_{x\in T} x^{r+1} a_x (\phi(\theta))^x \frac{\phi'(\theta)}{\phi(\theta)} - \frac{h'(\theta)}{h^2(\theta)} \sum_{x\in T} x^r a_x (\phi(\theta))^x$$

$$= \frac{\phi'(\theta)}{\phi(\theta)} \mu'_{r+1} - \frac{h'(\theta)}{h(\theta)} \mu'_r ,$$

which gives the recurrence relation for noncentral moments as

$$\mu'_{r+1} = \frac{\phi(\theta)}{\phi'(\theta)} \frac{d\mu'_r}{d\theta} + \mu'_1 \mu'_r, \qquad r = 0, 1, 2, \ldots, \tag{7.7}$$

where $\mu'_0 = 1$.

By using a similar technique (see Exercises 7.1 and 7.2), Gupta (1974) obtained the following recurrence relations for the central moments μ_r, $r = 1, 2, 3, \ldots$, and the ascending factorial moments $\mu^{[r]}$, $r = 1, 2, 3, \ldots$

$$\mu_{r+1} = \frac{\phi(\theta)}{\phi'(\theta)} \frac{d\mu_r}{d\theta} + r\, \mu_2 \mu_{r-1}, \quad r = 1, 2, 3, \ldots \tag{7.8}$$

with $\mu_0 = 1,\ \mu_1 = 0$, and

$$\mu^{[r+1]} = \frac{\phi(\theta)}{\phi'(\theta)} \frac{d\mu^{[r]}}{d\theta} + \mu^{[r]}(\mu - r), \quad r = 1, 2, 3, \ldots. \tag{7.9}$$

Noack (1950) proved a general relation between the noncentral moments and the cumulants K_j, $j = 1, 2, 3, \ldots$, of any discrete probability distribution as

$$\mu'_r = \sum_{j=1}^{r} \binom{r-1}{j-1} \mu'_{r-j} K_j, \qquad r = 1, 2, 3, \ldots. \tag{7.10}$$

On differentiating (7.10) with respect to θ, we obtain

$$\frac{d\mu'_r}{d\theta} = \sum_{j=1}^{r} \binom{r-1}{j-1} \left\{ \frac{d\mu'_{r-j}}{d\theta} K_j + \mu'_{r-j} \frac{dK_j}{d\theta} \right\}. \tag{7.11}$$

By combining (7.7), (7.8), and (7.11), we get

$$\sum_{j=1}^{r+1} \binom{r}{j-1} \mu'_{r+1-j} K_j$$

$$= \sum_{j=1}^{r} \binom{r-1}{j-1} \left\{ \left[\frac{\phi(\theta)}{\phi'(\theta)} \frac{d\mu'_{r-j}}{d\theta} + \mu'_1\, \mu'_{r-j} \right] K_j + \frac{\phi(\theta)}{\phi'(\theta)} \mu'_{r-j} \frac{dK_j}{d\theta} \right\}.$$

By using (7.7) and keeping only the $(r+1)$th term on the left side, Gupta (1974) obtained the recurrence relation

$$K_{r+1} = \frac{\phi(\theta)}{\phi'(\theta)} \sum_{j=1}^{r} \binom{r-1}{j-1} \mu'_{r-j} \frac{dK_j}{d\theta} - \sum_{j=2}^{r} \binom{r-1}{j-2} \mu'_{r+1-j} K_j \tag{7.12}$$

for $r = 1, 2, 3, \ldots$, between the cumulants and the noncentral moments of the MPSD.

Also, the mgf in (7.5) provides the rth moment about the origin as

$$\mu'_r = \sum_{x=0}^{r} \sum_{i=0}^{\infty} a_{x+i} \frac{(x+i)!}{i!} \frac{(\phi(\theta))^{x+i}}{h(\theta)} S(r, x), \tag{7.13}$$

and the rth ascending factorial moment as

$$\mu^{[r]} = \sum_{i=0}^{\infty} \frac{a_{r+i}(r+i)!}{i!} \frac{(\phi(\theta))^{r+i}}{h(\theta)}. \tag{7.14}$$

Negative and Incomplete Moments

Suppose a discrete r.v. X has a MPSD with $\phi(0) = 0$ and let k be a nonnegative number such that $k + x \neq 0$ for $x \in T$. We define the *rth negative moment of* X by

$$M(r, k) = E(X + k)^{-r} = \sum_{x \in T} \frac{a_x \phi^x(\theta)}{(x + k)^r h(\theta)} . \tag{7.15}$$

By differentiating (7.15) with respect to θ, we get

$$\begin{aligned} M'(r, k) &= \sum_{x \in T} \frac{a_x}{(x + k)^r} \frac{\phi^x(\theta)}{h(\theta)} \left\{ x \frac{\phi'(\theta)}{\phi(\theta)} - \frac{h'(\theta)}{h(\theta)} \right\} \\ &= \frac{\phi'(\theta)}{\phi(\theta)} M(r - 1, k) - \left\{ \frac{k\phi'(\theta)}{\phi(\theta)} + \frac{h'(\theta)}{h(\theta)} \right\} M(r, k). \end{aligned}$$

After rearrangement, the above gives the linear differential equation

$$M'(r, k) + M(r, k) \left\{ \frac{k\phi'(\theta)}{\phi(\theta)} + \frac{h'(\theta)}{h(\theta)} \right\} = \frac{\phi'(\theta)}{\phi(\theta)} M(r - 1, k). \tag{7.16}$$

On multiplying (7.16) by the factor $h(\theta)\phi^k(\theta)$ and integrating from 0 to θ, Kumar and Consul (1979) obtained the recurrence relation for successive negative moments as

$$M(r, k) = \frac{1}{h(\theta)\phi^k(\theta)} \int_0^\theta M(r - 1, k)\phi'(\theta)h(\theta)\phi^{k-1}(\theta)d\theta \tag{7.17}$$

for $r = 1, 2, 3, \ldots$, and where $M(0, k) = E(X + k)^0 = 1$.

From (7.17),

$$M(1, k) = E[(X + k)^{-1}] = \frac{\int_0^\theta \phi'(\theta)h(\theta)\phi^{k-1}(\theta)d\theta}{h(\theta)\phi^k(\theta)} . \tag{7.18}$$

Let X be a r.v. defined over a subset of the set of real numbers with distribution function $F(\cdot)$. Then

$$\mu_r'(t) = \int_{-\infty}^t x^r dF(x) \tag{7.19}$$

is called the rth *incomplete moment* of X. The rth incomplete moment for the MPSD in (7.1) is given by

$$\mu_r'(t) = \sum_{x=1}^t x^r a_x \phi^x(\theta)/h(\theta). \tag{7.20}$$

By differentiating both sides of (7.20) with respect to θ, Tripathi, Gupta, and Gupta (1986) obtained a recurrence relation between the incomplete moments as

$$\mu_{r+1}'(t) = \frac{\phi(\theta)}{\phi'(\theta)} \frac{d\mu_r'(t)}{d\theta} + \mu_1' \, \mu_r'(t). \tag{7.21}$$

Let the rth *incomplete ascending factorial moment* for MPSD be given by

$$\mu^{[r]}(t) = \sum_{x=1}^{t} x^{[r]}\, a_x \phi^x(\theta)/h(\theta), \qquad t \geq r. \tag{7.22}$$

By differentiating both sides of (7.22) with respect to θ, a recurrence relation between the incomplete ascending factorial moments for MPSD (Tripathi, Gupta, and Gupta 1986) is

$$\mu^{[r+1]}(t) = \frac{\phi(\theta)}{\phi'(\theta)} \frac{d\mu^{[r]}(t)}{d\theta} + \mu^{[r]}(t)\{\mu - r\}. \tag{7.23}$$

Jani and Shah (1979b) considered a situation where the observations corresponding to $x = 1$ are sometimes erroneously misclassified as those corresponding to $x = 0$ with probability α, $0 \leq \alpha \leq 1$. They obtained recurrence relations for moments about the origin, moments about the mean, and factorial moments of such a model with misclassification.

7.4 Other Interesting Properties

Theorem 7.1. *A necessary and sufficient condition for the variance of a random variable having MPSD to equal its mean for all values of θ is*

$$h(\theta) = c.\exp(k\phi(\theta)), \tag{7.24}$$

where $k > 0$ and c are arbitrary constants and where $h(\theta)$ generates the Poisson distribution.

We shall prove only the necessary part, as the sufficient part is straightforward.

Proof. Condition is necessary. If the mean and variance are equal, then $\mu = \mu_2$, which gives

$$\mu = \frac{\phi(\theta)}{\phi'(\theta)} \frac{d\mu}{d\theta}. \tag{7.25}$$

On using (7.2) in (7.25), the expression can be written in the form

$$\left[1 - \frac{d}{d\theta}\frac{\phi(\theta)}{\phi'(\theta)}\right] \frac{\phi(\theta)}{\phi'(\theta)} \frac{d}{d\theta} \ln h(\theta) = \left(\frac{\phi(\theta)}{\phi'(\theta)}\right)^2 \frac{d^2 \ln h(\theta)}{d\theta^2},$$

i.e.,

$$\frac{d^2 \ln h(\theta)}{d\theta^2} \Bigg/ \frac{d \ln h(\theta)}{d\theta} = \frac{\phi'(\theta)}{\phi(\theta)} \left[1 - \frac{d}{d\theta}\left(\frac{\phi(\theta)}{\phi'(\theta)}\right)\right]. \tag{7.26}$$

On integrating (7.26) with respect to θ, we have

$$\ln\left\{\frac{d}{d\theta} \ln h(\theta)\right\} = \ln \phi(\theta) - \ln\left(\frac{\phi(\theta)}{\phi'(\theta)}\right) + \ln k,$$

where $k > 0$ is an arbitrary constant. On taking the antilog of the above,

$$\frac{d}{d\theta} \ln h(\theta) = k\phi'(\theta). \tag{7.27}$$

On integrating (7.27) with respect to θ, we obtain

$$\ln h(\theta) = k\phi(\theta) + d_1,$$

where d_1 is an arbitrary constant and which gives the condition

$$h(\theta) = c \cdot \exp(k\phi(\theta)) \quad (\text{with } c = \exp(d_1)),$$

which represents a Poisson distribution. □

Theorem 7.2. *A MPSD has its variance greater than the mean if and only if*

$$h(\theta) = c \cdot \exp\left\{k \int P(\theta)\phi'(\theta)d\theta\right\}, \tag{7.28}$$

where $P(\theta) > 0$ is a monotone increasing function of θ.

Proof. Condition is necessary. By using (7.26) in the proof of Theorem 7.1, variance of a MPSD is greater than its mean if

$$\frac{d^2 \ln h(\theta)}{d\theta^2} \bigg/ \frac{d \ln h(\theta)}{d\theta} > \frac{\phi'(\theta)}{\phi(\theta)}\left[1 - \frac{d}{d\theta}\left(\frac{\phi(\theta)}{\phi'(\theta)}\right)\right],$$

which can be written in the form

$$\frac{d^2}{d\theta^2} \ln h(\theta) \bigg/ \frac{d}{d\theta} \ln h(\theta) = \frac{\phi'(\theta)}{\phi(\theta)} - \frac{\phi'(\theta)}{\phi(\theta)} \frac{d}{d\theta}\left(\frac{\phi(\theta)}{\phi'(\theta)}\right) + \frac{P'(\theta)}{P(\theta)}, \tag{7.29}$$

where $P(\theta) > 0$ is a monotonically increasing function of θ so that $P'(\theta) > 0$. On integrating (7.29) with respect to θ, we get

$$\begin{aligned}\ln\left\{\frac{d}{d\theta} \ln h(\theta)\right\} &= \ln\phi(\theta) - \ln\left(\frac{\phi(\theta)}{\phi'(\theta)}\right) + \ln P(\theta) + \ln k \\ &= \ln\{P(\theta)\phi'(\theta) \cdot k\},\end{aligned}$$

where $k > 0$ is an arbitrary constant. Thus,

$$\frac{d}{d\theta} \ln h(\theta) = k \cdot P(\theta)\phi'(\theta). \tag{7.30}$$

On integrating (7.30) again, we obtain

$$\ln h(\theta) = k \int P(\theta)\phi'(\theta)d\theta + d_1.$$

Thus

$$h(\theta) = c \cdot \exp\left\{k \int P(\theta)\phi'(\theta)d\theta\right\},$$

where c is an arbitrary constant. This completes the proof of the necessary part. □

Similarly, a MPSD has its variance smaller than the mean if and only if

$$h(\theta) = c^* \cdot \exp\left\{k^* \int Q(\theta)\phi'(\theta)d\theta\right\}, \tag{7.31}$$

where $k^* > 0$ *and* c^* *are arbitrary constants and* $Q(\theta) > 0$ *is a monotone decreasing function of* θ. (See Exercise 7.4.)

Jani and Shah (1979a) obtained integral expressions for the tail probabilities of the MPSD by proving the following theorem.

Theorem 7.3. *For the MPSD defined by (7.1), there exists a family of absolutely continuous distributions*

$$f(z; x) = \begin{cases} c(x; z)a_x\phi^x(z)/h(z), & 0 \le z < \rho, \\ 0, & \text{otherwise}, \end{cases} \tag{7.32}$$

where $x \in T$ *such that*

$$q(x; \theta) = \sum_{j=0}^{x} a_j\phi^j(\theta)/h(\theta) = \int_\theta^\rho \frac{c(x; z)a_x\phi^x(z)}{h(z)}dz, \tag{7.33}$$

if and only if $h(\rho) = \infty$. *The function* $c(x; z)$ *is given by*

$$c(x; z)a_x\phi^x(z) = \frac{\phi'(\theta)}{\phi(\theta)}\sum_{j=0}^{x}(\mu - j)a_j\phi^j(\theta), \tag{7.34}$$

where μ *is the mean of MPSD.*

Proof. Suppose $f(z; x)$ is a probability function such that (7.32) holds. Then

$$q(x; \theta) = \int_\theta^\rho f(z; x)dz = \int_0^\rho f(z; x)dz - \int_0^\theta f(z; x)dz.$$

If $\theta \to \rho$, then

$$q(x; \theta) \to q(x; \rho) = 1 - \int_0^\rho f(z; x)dz = 0, \tag{7.35}$$

which implies that $h(\rho) = \infty$.

The converse of (7.35) is straightforward. To prove (7.34), differentiate (7.33) with respect to θ and use (7.2) to obtain

$$\frac{dq(x; \theta)}{d\theta} = \frac{\phi'(\theta)}{\phi(\theta)}\sum_{j=0}^{x}(j - \mu)\, a_j\phi^j(\theta). \tag{7.36}$$

On differentiating (7.35) with respect to θ, we get

$$\frac{dq(x; \theta)}{d\theta} = -f(\theta; x). \tag{7.37}$$

By using (7.36) and (7.37), the result in (7.34) follows. □

7.5 Estimation

The MPSD as defined by (7.1) may have many other unknown parameters besides the parameter θ. However, this section deals with the estimation of the parameter θ only.

7.5.1 Maximum Likelihood Estimation of θ

Let $X_1, X_2, \ldots, X_n$ be a random sample of size n from the MPSD (7.1) and let $\bar{x}$ denote the sample mean. The likelihood function L is given by

$$L = \prod_{i=1}^{n} \left\{ a_{x_i} \phi^{x_i}(\theta) / h(\theta) \right\}. \tag{7.38}$$

On differentiating the loglikelihood function $\ln L$ with respect to θ, we obtain the likelihood equation for θ as

$$\frac{1}{L}\frac{dL}{d\theta} = n\frac{\phi'(\theta)}{\phi(\theta)}(\bar{x} - \mu(\theta)) = 0. \tag{7.39}$$

The solution of (7.39) is given by

$$\bar{x} = \mu(\theta). \tag{7.40}$$

If $\mu(\theta)$ is invertible, the MLE of θ, obtained by inverting (7.40), is given by

$$\hat{\theta} = \mu^{-1}(\bar{x}).$$

If $\mu(\theta)$ is not invertible, one may solve (7.40) iteratively using the Newton–Raphson method. Haldane and Smith (1956) showed that the amount of bias $b(\hat{\theta})$ of the MLE of any parameter θ is given by

$$b(\hat{\theta}) = -(2n)^{-1} B_1 \, A_1^{-2}, \tag{7.41}$$

where

$$A_1 = \sum_x \frac{1}{P(X=x)} \left(\frac{dP(X=x)}{d\theta} \right)^2$$

and

$$B_1 = \sum_{x \in T} \frac{1}{P(X=x)} \left(\frac{dP(X=x)}{d\theta} \right) \left(\frac{d^2 P(X=x)}{d\theta^2} \right).$$

Now,

$$\begin{aligned} A_1 &= \sum_{x \in T} \frac{1}{P(X=x)} \left(\frac{d}{d\theta} P(X=x) \right)^2 \\ &= \sum_{x \in T} P(X=x) \cdot \left\{ x \frac{\phi'(\theta)}{\phi(\theta)} - \frac{h'(\theta)}{h(\theta)} \right\}^2 \\ &= \left(\frac{\phi'(\theta)}{\phi(\theta)} \right)^2 \sum_{x \in T} (x - \mu_1')^2 P(X=x) \\ &= \left(\frac{\phi'(\theta)}{\phi(\theta)} \right)^2 \mu_2. \end{aligned} \tag{7.42}$$

Also

$$B_1 = \sum_{x\in T} \frac{1}{P(X=x)} \left(\frac{d}{d\theta} P(X=x)\right)\left(\frac{d^2}{d\theta^2} P(X=x)\right)$$

$$= \sum_{x\in T} \left(x\frac{\phi'(\theta)}{\phi(\theta)} - \frac{h'(\theta)}{h(\theta)}\right)^3 P(X=x)$$

$$+ \sum_{x\in T} \left(x\frac{\phi'(\theta)}{\phi(\theta)} - \frac{h'(\theta)}{h(\theta)}\right)\left[x\frac{\phi(\theta)\phi''(\theta) - (\phi'(\theta))^2}{(\phi(\theta))^2} - \frac{h(\theta)h''(\theta) - (h'(\theta))^2}{(h(\theta))^2}\right] P(X=x)$$

$$= \sum_{x\in T} \left(\frac{\phi'(\theta)}{\phi(\theta)}\right)^3 (x-\mu_1')^3\, P(X=x)$$

$$+ \sum_{x\in T} \frac{\phi'(\theta)}{\phi(\theta)} (x-\mu_1')\cdot x \cdot \frac{\phi(\theta)\phi''(\theta) - [\phi'(\theta)]^2}{[\phi(\theta)]^2} \cdot P(X=x)$$

$$= \left(\frac{\phi'(\theta)}{\phi(\theta)}\right)^3 \mu_3 + \left(\frac{\phi'(\theta)}{\phi(\theta)}\right)^3 \sum_x (x-\mu_1')^2\, P(X=x) \cdot \frac{\phi(\theta)\phi''(\theta) - [\phi'(\theta)]^2}{[\phi(\theta)]^2}$$

$$= \left(\frac{\phi'(\theta)}{\phi(\theta)}\right)^3 \left\{\mu_3 + \frac{\phi(\theta)\phi''(\theta) - [\phi'(\theta)]^2}{[\phi(\theta)]^2}\mu_2\right\}, \tag{7.43}$$

where $\mu_3 = \frac{\phi(\theta)}{\phi'(\theta)} \frac{d\mu_2}{d\theta}$.

By using (7.42) and (7.43) in (7.41), the bias of $\hat{\theta}$ becomes

$$b(\hat{\theta}) = \frac{-1}{n\mu_2^2} \frac{\phi(\theta)}{\phi'(\theta)} \left\{\mu_3 + \frac{\phi(\theta)\phi''(\theta) - [\phi'(\theta)]^2}{[\phi'(\theta)]^2}\mu_2\right\}. \tag{7.44}$$

The MLE $\hat{\theta}$ is unbiased when $B_1 = 0$, i.e., when

$$\mu_3 = \frac{\phi(\theta)}{\phi'(\theta)} \frac{d\mu_2}{d\theta} = -\frac{\{\phi(\theta)\phi''(\theta) - [\phi'(\theta)]^2\}}{[\phi'(\theta)]^2}\mu_2. \tag{7.45}$$

Equation (7.45) yields

$$\frac{1}{\mu_2}\frac{d\mu_2}{d\theta} = -\frac{\phi''(\theta)}{\phi'(\theta)} + \frac{\phi'(\theta)}{\phi(\theta)},$$

and on integration with respect to θ, we obtain

$$\ln \mu_2 = -\ln \phi'(\theta) + \ln \phi(\theta) + \ln k$$

$$= \ln\left(\frac{\phi(\theta)}{\phi'(\theta)} k\right).$$

Thus

$$\mu_2 = k\,\phi(\theta)/\phi'(\theta), \tag{7.46}$$

where k is a constant independent of θ. A necessary and sufficient condition for the MLE $\hat{\theta}$ of θ to be unbiased is that (7.46) hold.

The asymptotic variance of $\hat{\theta}$ is given by

$$Var(\hat{\theta}) = \frac{\phi(\theta)}{\phi'(\theta)} \bigg/ \left[n \frac{d\mu}{d\theta} \right]. \tag{7.47}$$

If $\psi = \omega(\theta)$ is a one to one function of θ, then the MLE of ψ is given by

$$\hat{\psi} = \omega(\hat{\theta}). \tag{7.48}$$

The bias and the asymptotic variance of $\hat{\psi}$ (see Exercise 7.5) are given, respectively, by

$$b(\hat{\psi}) = \frac{-1}{2n\mu_2^2} \frac{\phi(\theta)}{\phi'(\theta)} \frac{d\psi}{d\theta} \left\{ \mu_3 - \mu_2 \left[1 - \frac{\phi(\theta)\phi''(\theta)}{[\phi'(\theta)]^2} + \frac{\phi(\theta)}{\phi'(\theta)} \frac{d^2\psi}{d\theta^2} \bigg/ \frac{d\psi}{d\theta} \right] \right\} \tag{7.49}$$

and

$$Var(\hat{\psi}) = \frac{\phi(\theta)}{\phi'(\theta)} \left(\frac{d\psi}{d\theta} \right)^2 \bigg/ (n\mu_2). \tag{7.50}$$

7.5.2 Minimum Variance Unbiased Estimation

Let $I^r = \{r, r+1, r+2, \ldots\}$, where r is a nonnegative integer, and let T be such that $T \subseteq I^0$. Let

$$Y = \sum_{i=1}^{n} X_i$$

be the sample sum. Then Y is a complete and sufficient statistic for θ in (7.1) and the distribution of Y (Kumar and Consul 1980) is also a MPSD given by

$$P(Y = y) = b(n, y)\phi^y(\theta)/h^n(\theta), \quad y \in D_n, \tag{7.51}$$

where

$$D_n = \left\{ y | y = \sum x_i, \quad x_i \in T, \quad i = 1, 2, \ldots, n \right\} \tag{7.52}$$

and

$$b(n, y) = \sum a_{x_1} a_{x_2} \ldots a_{x_n}.$$

The summation extends over all ordered n-tuples $(x_1, x_2, \ldots, x_n)$ of integers $x_i \in T$ under the condition $\sum_{i=1}^n x_i = y$.

If $\ell(\theta)$ is a given function of θ and there exists some positive integer n such that

$$\ell(\theta)h^n(\theta) = \sum_{i \in E_n} c(n, i)\phi^i(\theta), \tag{7.53}$$

where $c(n, i) \neq 0$ for $i \in E_n \subseteq I^0$, then we write $\ell(\theta) \in L(n, \phi(\theta), h(\theta))$.

Kumar and Consul (1980) have given a necessary and sufficient condition for a function $\ell(\theta)$ to admit a unique minimum variance unbiased (MVU) estimator.

Theorem 7.4. *The function $\ell(\theta)$ of the parameter θ in the MPSD (7.1) is MVU estimable if and only if there exists a positive integer n such that $\ell(\theta) \in L(n, \phi(\theta), h(\theta))$ and $E_n \subseteq D_n$, where the sets D_n and E_n are defined in (7.52) and (7.53), respectively, and the MVU estimator $f(y)$ of $\ell(\theta)$, when MVU estimable, is given by*

$$f(y) = \begin{cases} c(n, y)/b(n, y), & y \in E_n, \\ 0, & \textit{otherwise}. \end{cases} \tag{7.54}$$

Proof. Necessary. Suppose $\ell(\theta)$ is estimable. Then there exists a function $f(y)$ of the sample sum Y for some $n \in I^1$ such that

$$E\{f(y)\} = \ell(\theta) \quad \text{for all} \quad \theta. \tag{7.55}$$

From (7.55) and (7.51), we have

$$\sum_{y \in D_n} f(y) b(n, y) \phi^y(\theta) = \ell(\theta) h^n(\theta), \tag{7.56}$$

which shows that $\ell(\theta) \in L(n, \phi(\theta), h(\theta))$. By using (7.53), we rewrite (7.56) as

$$\sum_{y \in D_n} f(y) b(n, y) \phi^y(\theta) = \sum_{y \in E_n} c(n, y) \phi^y(\theta), \tag{7.57}$$

where $c(n, y) \neq 0$ for $y \in E_n$. Therefore, for every $y \in E_n$, $b(n, y) \neq 0$, and so $y \in D_n$, which shows that $E_n \subseteq D_n$. On equating the coefficients of $\phi^y(\theta)$ in (7.57),

$$f(y) = \begin{cases} c(n, y)/b(n, y), & y \in E_n \\ 0, & \text{otherwise}. \end{cases}$$

Sufficiency. See Necessary. Suppose $\ell(\theta) \in L(n, \phi(\theta), h(\theta))$ and $E_n \subseteq D_n$. From (7.53), we have

$$\begin{aligned} \ell(\theta) &= \sum_{i \in E_n} c(n, i) \phi^i(\theta) / h^n(\theta) \\ &= \sum_{i \in E_n} \frac{c(n, i)}{b(n, i)} \cdot \frac{b(n, i) \phi^i(\theta)}{h^n(\theta)}. \end{aligned} \tag{7.58}$$

From the distribution of Y in (7.51) and (7.54), we write (7.58) as

$$\ell(\theta) = \sum_{i \in D_n} f(i) P(Y = i)$$

and so $f(Y)$ is an unbiased estimator of $\ell(\theta)$. Since $f(Y)$ is a function of the complete and sufficient statistic Y, it must be an MVU estimator for $\ell(\theta)$. □

MVU estimators of θ and some function $\ell(\theta)$ have been derived by Gupta (1977) and by Kumar and Consul (1980) for some members of the MPSD class. Gupta and Singh (1982) obtained the MVU estimator for the probability function in (7.1) as

$$P\left(X = x | Y = \sum^n x_i = y\right) = \begin{cases} a_x b(n-1, y-x)/b(n, y), & y \in (n-1)[T] + x, \\ 0, & \text{otherwise}. \end{cases} \tag{7.59}$$

Abu-Salih (1980) considered the resolution of a mixture of observations from two MPSDs. The method of ML was used to identify the population of origin of each observation and to estimate the population parameters.

7.5.3 Interval Estimation

Famoye and Consul (1989a) considered the problem of interval estimation for θ in the class of MPSD. Confidence intervals (CIs) $(\theta_\ell,\ \theta_u)$ for θ in small samples are obtained by solving the equations

$$\sum_{x=y}^{\infty} a_x \phi^x(\theta_\ell)/h^n(\theta_\ell) = \frac{1}{2}\alpha \tag{7.60}$$

and

$$\sum_{x=0}^{y} a_x \phi^x(\theta_u)/h^n(\theta_u) = \frac{1}{2}\alpha, \tag{7.61}$$

where $y = \sum_{i=1}^{n} x_i$.

CIs for θ in a large sample may be based on the statistic

$$W = (\bar{X} - \mu)\sqrt{n}/\sigma, \tag{7.62}$$

which converges stochastically to a normal distribution with mean zero and variance unity. The upper bound θ_u and the lower bound θ_ℓ are the solutions of the equations

$$\bar{x} - \frac{\phi(\theta)h'(\theta)}{\phi'(\theta)h(\theta)} \pm z_{\alpha/2}\left(\frac{\phi(\theta)}{n\phi'(\theta)}\frac{d\mu}{d\theta}\right)^{1/2} = 0, \tag{7.63}$$

where $z_{\alpha/2}$ is the critical value from the normal probability tables.

The result in (7.63) is based on a single statistic $\bar{X}$. Famoye and Consul (1989a) also provided a CI for θ based on two statistics, the sample mean $\bar{X}$ and the sample variance S^2. A two-sided $100(1-\alpha)\%$ CI for θ in a large sample is obtained from

$$1 - \alpha = \Pr\left\{\bar{X} - z_{\alpha/2}\frac{S}{\sqrt{n}} < \frac{\phi(\theta)h'(\theta)}{\phi'(\theta)h(\theta)} < \bar{X} + z_{\alpha/2}\frac{S}{\sqrt{n}}\right\}. \tag{7.64}$$

An advantage of (7.64) over (7.63) is that the inequality in (7.64) can often be solved algebraically for θ and the result can therefore be expressed in the form

$$1 - \alpha = \Pr\{\theta_l < \theta < \theta_u\}, \tag{7.65}$$

where (θ_l, θ_u) is the two-sided $100(1-\alpha)\%$ CI for θ.

7.6 Some Characterizations

Gupta (1975b) gave a characterization of the MPSD by using a length-biased distribution. A slightly different theorem is being given here.

Theorem 7.5. *Let X be a discrete r.v. with a MPSD and let Y be the r.v. representing its length-biased distribution. Then*

$$P(Y = y) = \frac{yP(X = y)}{\mu}. \tag{7.66}$$

If $E(Y-1) = E(X)$, then $h(\theta) = d\exp(k\phi(\theta))$, where $k > 0$ and d are arbitrary constants.

Proof. By (7.66), the probability distribution of Y becomes

$$P(Y = y) = y a_y \phi^y(\theta)/h_1(\theta), \ y = 1, 2, 3, \ldots, \tag{7.67}$$

where $h_1(\theta) = \mu(\theta) h(\theta)$, where $\mu(\theta)$ is the mean for the r.v. X. But (7.67) is itself a MPSD with

$$E(Y) = \frac{\phi(\theta) h_1'(\theta)}{\phi'(\theta) h_1(\theta)} = \mu(\theta) + \mu_2/\mu(\theta), \tag{7.68}$$

where μ_2 is the variance of X.

From $E(Y-1) = E(X)$, we obtain

$$\mu(\theta) + \mu_2/\mu(\theta) - 1 = \mu(\theta).$$

Hence, $\mu_2 = \mu(\theta)$ and by using Theorem 7.1, we get

$$h(\theta) = d \cdot \exp(k\phi(\theta)).$$

□

Theorem 7.6. *A discrete probability distribution is a MPSD if and only if the recurrence relation between its cumulants is*

$$K_{r+1} = \frac{\phi(\theta)}{\phi'(\theta)} \frac{dK_r}{d\theta}, \quad r = 1, 2, 3, \ldots \tag{7.69}$$

(Jani, 1978b).

Proof. Condition is necessary. We put $\phi(\theta) = e^u$ so that $\theta = \psi(\phi(\theta)) = \psi(e^u)$ and $h(\theta) = h\{\psi(e^u)\} = G(u)$, say. By (7.4) the mgf $M_X(t) = M(t)$ becomes

$$M(t) = E[e^{tX}] = \frac{h\{\psi(e^t \cdot e^u)\}}{G(u)} = \frac{h\{\psi(e^{t+u})\}}{G(u)} = \frac{G(t+u)}{G(u)}. \tag{7.70}$$

Then, the cgf $K(t, u)$ for the MPSD, with parameter u, is given by

$$K(t, u) = \ln M(t) = \ln G(t+u) - \ln G(u).$$

On differentiation with respect to t,

$$\begin{aligned} \frac{\partial}{\partial t} K(t, u) &= S(t, u) \\ &= \frac{G'(t+u)}{G(t+u)} = G_1(t+u), \quad \text{say.} \end{aligned} \tag{7.71}$$

Because of symmetry in $G_1(t+u)$ of t and u, one has

$$\left(\frac{\partial}{\partial t}\right)^r G_1(t+u) = \left(\frac{\partial}{\partial u}\right)^r G_1(t+u) = G_1^r(t+u), \ \text{say.}$$

Then

$$K_r = \left\{\left(\frac{\partial}{\partial t}\right)^r K(t, u)\right\}_{t=0} = G_1^{r-1}(u).$$

Hence,

$$K_{r+1} = G_1^r(u) = \frac{\partial}{\partial u} K_r, \tag{7.72}$$

which is equivalent to

$$K_{r+1} = \phi(\theta) \frac{\partial K_r}{\partial \phi(\theta)} \tag{7.73}$$

and so

$$K_{r+1} = \frac{\phi(\theta)}{\phi'(\theta)} \frac{d K_r}{d\theta}.$$

Sufficiency. Now

$$S(t, u) = \frac{\partial}{\partial t} K(t+u) = \frac{\partial}{\partial t} \sum_{r=1}^{\infty} K_r t^r / r! = \sum_{r=0}^{\infty} K_{r+1} t^r / r!.$$

From (7.72),

$$K_{r+1} = \left(\frac{\partial}{\partial u} \right)^r K_1.$$

By using the last result and the fact that $K_1 = S(0, u)$, equation (7.73) reduces to

$$S(t, u) = \sum \left(\frac{\partial}{\partial u} \right)^r \{S(0, u)\} t^r / r!. \tag{7.74}$$

Also, we have

$$S(t, u) = \sum \left\{ \left(\frac{\partial}{\partial t} \right)^r [S(t, u)] \right\}_{t=0} t^r / r!. \tag{7.75}$$

From (7.74) and (7.75), we have the identity

$$\left(\frac{\partial}{\partial u} \right)^r S(0, u) = \left\{ \left(\frac{\partial}{\partial t} \right)^r [S(t, u)] \right\}_{t=0}. \tag{7.76}$$

By substituting $u = 0$ in (7.76) and using the equivalence of two power series expansions, we have

$$S(0, u) = S(u, 0) = m(u), \text{ say.}$$

With the transformation $u = c + u'$, where c is an arbitrary constant and applying the same argument for (7.76), we obtain

$$S(0, c + u') = S(u' + c).$$

Thus, $S(t, u) = m(t + u)$. On integrating both sides of this equation with respect to t, we have

$$\int S(t, u) dt = \int m(t + u) dt$$

or

$$K(t, u) = M(t + u) - M(t), \tag{7.77}$$

where M is any functional relation. By the properties of transforms, a distribution with cgf of the form (7.77) is MPSD when the random variable is of discrete type. This completes the proof. □

Theorem 7.7. *Let $X_1, X_2, \ldots, X_n$ be n i.i.d. nonnegative integer-valued r.v.s such that $\sum_{i=1}^{n} X_i = Y$. For any x and each fixed y, the conditional probability of X_1 at $X_1 = x$ for given $Y = y$ is*

$$P(X_1 = x \mid Y = y) = a_x b(n-1, y-x)/b(n, y), \quad 0 \le x \le y,$$

and where

$$b(n, y) = \sum a_{x_1} a_{x_2} \ldots a_{x_n}$$

and the summation extends over all ordered n-tuples $(x_1, x_2, \ldots, x_n)$ of integers under the condition $\sum_{i=1}^{n} X_i = Y = y$, if and only if X_1 follows the MPSD (Jani (1985)).

Proof. Condition is necessary. The random variables $X_1, X_2, \ldots, X_n$ are i.i.d. as (7.1). By using (7.59),

$$\begin{aligned} P(X_1 = x \mid Y = y) &= \frac{P(X_1 = x,\, Y = y)}{P(Y = y)} = \frac{P(X_1 = x)\, P(Y - X_1 = y - x)}{P(Y = y)} \\ &= \frac{\left[a_x\, (\phi(\theta))^x \,/\, h(\theta)\right]\left[b(n-1, y-x)\, (\phi(\theta))^{y-x} \,/\, (h(\theta))^{n-1}\right]}{b(n, y)\, (\phi(\theta))^y \,/\, (h(\theta))^n} \\ &= a_x\, b(n-1, y-x)/b(n, y). \end{aligned}$$

Sufficiency. Let $P(X_1 = x) = f(x) > 0$ and $\sum_{x=0}^{\infty} f(x) = 1$, and $P(Y = y) = g(y) > 0$, $P(Y - X_1 = y - x) = g(y - x) > 0$; $0 \le x \le y$. The conditional probability gives

$$P(X_1 = k \mid Y = y) = \frac{f(k)\, g(y-k)}{g(y)} = \frac{a_k\, b(n-1, y-k)}{b(n, y)},$$

which holds for all values of $k \le y$ and for all values of y. For $y \ge 1$ and $0 \le k \le y$, the above yields the functional relation

$$\frac{f(k)\, g(y-k)}{f(k-1)\, g(y-k+1)} = \frac{a_k}{a_{k-1}} \frac{b(n-1, y-k)}{b(n-1, y-k+1)}. \tag{7.78}$$

By replacing k and y by $k+1$ and $y+1$, respectively, in (7.78) and by dividing it by (7.78), one gets $f(k+1)\, f(k-1)/[f(k)]^2$ on the left-hand side and a messy expression on the right-hand side. Since the left-hand side is independent of y and n, the right-hand side must also be independent of y and n. Thus,

$$\frac{f(k+1)\, f(k-1)}{[f(k)]^2} = \frac{a_{k+1}\, a_{k-1}}{(a_k)^2}. \tag{7.79}$$

Putting $k = 1, 2, 3, \ldots, x-1$ in (7.79) and multiplying them together,

$$\frac{f(x)}{f(x-1)} = \frac{f(1)}{f(0)} \frac{a_0}{a_1} \frac{a_x}{a_{x-1}}, \quad x \ge 1.$$

The above gives

$$f(x) = \frac{a_0\, f(1)}{a_1\, f(0)} \frac{a_x}{a_{x-1}} f(x-1) = \left(\frac{a_0\, f(1)}{a_1\, f(0)}\right)^x \frac{a_x}{a_0} f(0). \tag{7.80}$$

The above relation (7.80) can be written as

$$f(x) = a_x\, (\phi(\theta))^x \,/\, h(\theta),$$

where $a_0/f(0) = h(\theta)$. Since $\sum_{x=0}^{\infty} f(x) = 1$, $f(x)$ represents the MPSD. □

Theorem 7.8. *Suppose X has a MPSD given by (7.1). Then*

$$\frac{h'(\theta)}{h(\theta)} = c\frac{\phi'(\theta)}{\phi(\theta)} \exp\left\{\int \psi(\theta)\frac{\phi'(\theta)}{\phi(\theta)}d\theta\right\}, \tag{7.81}$$

where c is a constant and $\psi(\theta) = Var(X)/E(X)$ *(Gupta, 1977).*

Proof. We have

$$\psi(\theta) = \mu_2/\mu_1' = \frac{h(\theta)}{h'(\theta)}\frac{d\mu_1'}{d\theta} = 1 + \frac{\phi(\theta)}{\phi'(\theta)}\left\{\frac{h''(\theta) - h'(\theta)h'(\theta)}{h'(\theta)h(\theta)} - \frac{\phi''(\theta)}{\phi'(\theta)}\right\}.$$

On rearranging the above, we have

$$\frac{h''(\theta) - [h'(\theta)]^2}{h'(\theta)h(\theta)} = (\psi(\theta) - 1)\frac{\phi'(\theta)}{\phi(\theta)} + \frac{\phi''(\theta)}{\phi'(\theta)},$$

which is equivalent to

$$\frac{d}{d\theta}\ln\left(\frac{h'(\theta)}{h(\theta)}\right) = \psi(\theta)\frac{\phi'(\theta)}{\phi(\theta)} - \frac{d}{d\theta}\ln\phi(\theta) + \frac{d}{d\theta}\ln\phi'(\theta). \tag{7.82}$$

On integrating (7.82) with respect to θ, we obtain

$$\begin{aligned}\ln\left(\frac{h'(\theta)}{h(\theta)}\right) &= \int \psi(\theta)\frac{\phi'(\theta)}{\phi(\theta)}d\theta - \ln\phi(\theta) + \ln\phi'(\theta) + \ln c\\ &= \ln\left(\frac{c\phi'(\theta)}{\phi(\theta)}\right) + \int \psi(\theta)\frac{\phi'(\theta)}{\phi(\theta)}d\theta,\end{aligned}$$

which gives

$$\frac{h'(\theta)}{h(\theta)} = c\left(\frac{\phi'(\theta)}{\phi(\theta)}\right)\exp\left\{\int \psi(\theta)\frac{\phi'(\theta)}{\phi(\theta)}d\theta\right\}.$$

□

7.7 Related Distributions

7.7.1 Inflated MPSD

Murat and Szynal (1998) studied the class of inflated MPSD where inflation occurs at any of the support points. They gave expressions for the moments, factorial moments, and central moments. Furthermore, they considered the MLE of the parameters. They derived the distribution of the sum of i.i.d. r.v.s from inflated MPSD with inflation point s. The results of Gupta, Gupta, and Tripathi (1995) are special cases when the inflation point is at $s = 0$.

A discrete r.v. X is said to have an MPSD inflated at the point s if

$$P(X = x) = \begin{cases}\phi + (1-\phi)a(x)\,[g(\theta)]^x\,/f(\theta), & x = s,\\ (1-\phi)a(x)\,[g(\theta)]^x\,/f(\theta), & x \neq s,\end{cases} \tag{7.83}$$

where $0 < \phi \leq 1,\ x$ is a nonnegative integer, $f(\theta) = \sum_x a(x)\,[g(\theta)]^x$ and $g(\theta)$ are positive, finite and differentiable, and the coefficients $a(x)$ are nonnegative and free of θ.

Let $m'_r,\ m'_{(r)},\ \mu'_r$ denote the moments about zero, factorial moments, and central moments of MPSD, respectively, and $m_r,\ m_{(r)},\ \mu_r$ denote the moments about zero, factorial moments, and central moments of inflated MPSD, respectively. Murat and Szynal (1998) gave

$$m_r = \phi s^r + (1-\phi)m'_r,$$

$$m_{(r)} = \phi s^{(r)} + \phi m'_{(r)},$$

$$\mu_r = (1-\phi)\phi(s-m'_1)\left[(1-\phi)^{r-1} - (-\phi)^{r-1}\right]$$

$$+ (1-\phi)\sum_{j=2}^{r}\binom{r}{j}\left[\phi(m'_1-s)\right]^{r-j}\mu'_j,$$

where $r \geq 1$.

The pgf $G_X(t)$ of inflated MPSD in terms of the pgf $G_Y(t)$ of MPSD is given by

$$G_X(t) = \phi t^s + (1-\phi)G_Y(t).$$

The recurrence relation between the ordinary moments of inflated MPSD is

$$m_{r+1} = \frac{g(\theta)}{g'(\theta)}\frac{dm_r}{d\theta} + \frac{m_1 m_r}{1-\phi} - \frac{\phi s}{1-\phi}\left[m_r + s^{r-1}(m_1 - s)\right].$$

For the factorial moments, we have

$$m_{(r+1)} = \frac{g(\theta)}{g'(\theta)}\frac{dm_{(r)}}{d\theta} - \left[r - \frac{m_1}{1-\phi}\right]m_{(r)} + \frac{\phi}{1-\phi}\left[s^{(r)}(s-m_1) + sm_{(r)}\right].$$

The recurrence relation between the central moments is

$$\mu_{r+1} = \frac{g(\theta)}{g'(\theta)}\left[\frac{d\mu_r}{d\theta} + r\frac{dm_1}{d\theta}\mu_{r-1}\right] - \frac{\phi(s-m_1)}{1-\phi}\mu_r + \frac{\phi}{1-\phi}(s-m_1)^{r-1}.$$

A special case of the inflated MPSD is the zero-inflated GPD given by $s = 0$ and the MPSD is the restricted GPD in (9.77). Gupta, Gupta, and Tripathi (1996) studied the zero-inflated GPD, which is useful for a situation when the proportion of zeros in the data is higher or lower than that predicted by the GPD model (9.77). The three parameters of the model are estimated by the method of maximum likelihood. Angers and Biswas (2003) considered Bayesian analysis of zero-inflated GPD. They discussed some prior distributions and obtained the posterior distributions by using Monte-Carlo integration.

7.7.2 Left Truncated MPSD

Let X be a r.v. having a left truncated MPSD given by

$$P(X = x) = a_x\phi^x(\theta)/h(r,\theta), \tag{7.84}$$

where $x \in I^r = \{r, r+1, r+2, \ldots\}$ and $h(r,\theta) = \sum_{x=r}^{\infty} a_x\phi^x(\theta)$. If r is known, (7.84) is a MPSD as it can be reduced to (7.1). For the rest of this section, we assume that r is an unknown truncation point. If $X_1, X_2, \ldots, X_n$ is a random sample from (7.84), then $Y = \sum_{i=1}^n X_i$ and $Z = \min(X_1, X_2, \ldots, X_n)$ are jointly sufficient and complete statistics for the parameters θ and r (Fraser, 1952). Kumar and Consul (1980) gave the joint distribution of (Y, Z) as

$$P(Y = y, Z = z) = \frac{[A(y, n, z) - A(y, n, z + 1)]}{h^n(r, \theta)} \phi^y(\theta), \tag{7.85}$$

$z = r, r + 1, r + 2, \ldots; y = nz, nz + 1, \ldots,$ and

$$A(y, n, z) = \sum a_{x_1} a_{x_2} \cdots a_{x_n}, \tag{7.86}$$

where the summation extends over all ordered n-tuples $(x_1, x_2, \ldots, x_n)$ under the conditions $x_i \geq z$ for $i = 1, 2, \ldots, n$ and $\sum_{i=1}^n x_i = y$.

Kumar and Consul (1980) stated without proof the following theorem on the MVU estimation of r^m in a left truncated MPSD with unknown truncation point r.

Theorem 7.9. *For a random sample of size n taken from the distribution (7.84), the MVU estimator $w(Y, Z)$ of the parametric function r^m, $m \in I'$, is given by*

$$w(Y, Z) = Z^m - \frac{A(Y, n, Z + 1)}{A(Y, n, Z) - A(Y, n, Z + 1)} \sum_{i=1}^{m-1} \binom{m}{i} Z^i. \tag{7.87}$$

Proof. By Rao–Blackwell and Lehmann–Scheffe theorems, $w(Y, Z)$ is the unique MVU estimator of r^m. From the unbiasedness property,

$$E\{w(Y, Z)\} = r^m$$

for all θ and every $r \geq 1$. From (7.85), we obtain

$$\sum_{y=rn} \sum_{z=r} w(y, z)\{A(y, n, z) - A(y, n, z + 1)\} \phi^y(\theta)$$

$$= \sum_{y=rn} \sum_{z=r} r^m \{A(y, n, z) - A(y, n, z + 1)\} \phi^y(\theta),$$

which holds if and only if

$$\sum_{z=r}^{[y/n]} w(y, z)\{A(y, n, z) - A(y, n, z+1)\} = \sum_{z=r}^{[y/n]} r^m \{A(y, n, z) - A(y, n, z+1)\} = r^m A(y, n, r) \tag{7.88}$$

with $A(y, n, [y/n] + 1) = 0$. By replacing r in (7.88) with $r + 1$, we have

$$\sum_{z=r+1}^{[y/n]} w(y, z)\{A(y, n, z) - A(y, n, z + 1)\} = (r + 1)^m A(y, n, r + 1). \tag{7.89}$$

On subtracting (7.89) from (7.88), we obtain

$$w(y, r)\{A(y, n, z) - A(y, n, z + 1)\}$$

$$= r^m \{A(y, n, r) - A(y, n, r + 1)\} - A(y, n, r + 1) \sum_{i=0}^{m-1} \binom{m}{i} r^i,$$

which holds for every $r \geq 1$. Hence, we obtain the estimator $w(Y, Z)$ for r^m as

$$w(Y, Z) = Z^m - \frac{A(Y, n, Z+1)}{A(Y, n, Z) - A(Y, n, Z+1)} \sum_{i=0}^{m-1} \binom{m}{i} Z^i.$$

□

Theorem 7.10. *The function $l(\theta)$ of the parameter θ in a left truncated MPSD is MVU estimable if $l(\theta)$ has a power series expansion*

$$l(\theta) = \sum_{i \in E} a_i \phi^i(\theta), \quad E \subseteq I^0;$$

and for any random sample of size n, the MVU estimator $f(Y, Z)$ of the function $l(\theta)$ is given by

$$f(Y, Z) = \frac{\sum_{i=nZ}^{Y} \eta(Y-i) A(i, n, Z) - \sum_{i=n(Z+1)}^{Y} \eta(Y-i) A(i, n, Z+1)}{A(Y, n, Z) - A(Y, n, Z+1)} \tag{7.90}$$

for $Y = nZ, nZ+1, \ldots,\ Z = r, r+1, \ldots,$ where

$$\eta(i) = \begin{cases} a_i, & \text{if } i \in E, \\ 0, & \text{otherwise,} \end{cases}$$

and $A(Y, n, Z)$ is given by (7.86).

The above theorem was given by Kumar and Consul (1980) and its proof is similar to the proof of Theorem 7.9 (see Exercise 7.7).

Jani (1978b) proved a theorem on the MVU estimator of the probability of left truncated MPSD.

Theorem 7.11. *Whenever the MVU estimator $P(Y, Z)$ of the left truncated MPSD defined in (7.84) exists, it is given by*

$$P(Y, Z) = \frac{a(r)\{A(Y-r, n-1, Z) - A(Y-r, n-1, Z+1)\}}{A(Y, n, Z) - A(Y, n, Z+1)} \tag{7.91}$$

for $Y = nZ, nZ+1, \ldots,\ Z = r, r+1, \ldots$ and zero otherwise.

Proof. By the condition of unbiasedness,

$$E\{P(Y, Z)\} = P(X = x).$$

By using (7.85) and (7.86), we obtain

$$\sum_{y=nr} \sum_{z=r} P(y, z)\{A(y, n, z) - A(y, n, z+1)\}\phi^y(\theta)$$

$$= \sum_{y=nr} \sum_{z=r} a_r\{A(y, n-1, z) - A(y, n-1, z+1)\}\phi^{y+r}(\theta)$$

$$= \sum_{y=nr} \sum_{z=r} a_r\{A(y-r, n-1, z) - A(y-r, n-1, z+1)\}\phi^y(\theta).$$

On equating the coefficients of $\phi^y(\theta)$ on both sides of the above, we get

$$P(y,z)\{A(y,n,z)-A(y,n,z+1)\}=a_r\{A(y-r,n-1,z)-A(y-r,n-1,z+1)\}.$$

Hence, the MVU estimator is given by

$$P(Y,Z)=\begin{cases}\frac{a_r\{A(Y-r,n-1,Z)-A(Y-r,n-1,Z+1)\}}{A(Y,n,Z)-A(Y,n,Z+1)}, & \begin{array}{l}Y=nZ,nZ+1,\ldots,\\ Z=r,r+1,\ldots,\end{array}\\ 0, & \text{otherwise.}\end{cases}$$

□

Remarks. The MPSD is a wide class consisting of restricted generalized Poisson distribution, the generalized negative binomial distribution, and the generalized logarithmic series distribution. Some of the results derived in this chapter will be used in subsequent chapters on distributions that belong to the MPSD class.

7.8 Exercises

7.1 If X is a r.v. with a MPSD, show that a recurrence relation between the central moments is given by

$$\mu_{r+1}=\frac{\phi(\theta)}{\phi'(\theta)}\frac{d\mu_r}{d\theta}+r\mu_2\mu_{r-1},$$

where $\mu_0=1$, $\mu_1=0$, and $r=1,2,3,\ldots$.

7.2 For a MPSD, show that a recurrence relation between the factorial moments is given by

$$\mu^{[r+1]}=\frac{\phi(\theta)}{\phi'(\theta)}\frac{d\mu^{[r]}}{d\theta}+\mu^{[r]}(\mu^{[1]}-r),\quad r=1,2,3,\ldots.$$

7.3 If X has a MPSD with $h(0)\neq 0$, find $E\{(X+1)^{-1}\}$.

7.4 Prove that a MPSD has its variance smaller than the mean if and only if

$$f(\theta)=\exp\left\{d+k\int\phi(\theta)\phi'(\theta)d\theta\right\},$$

where $k>0$ and d are arbitrary constants and $\phi(\theta)>0$ is a monotone decreasing function of θ.

7.5 Suppose $\Phi=\omega(\theta)$ is a one to one function of θ in a MPSD. Verify the expressions for the bias $b(\hat{\Phi})$ in (7.49) and the asymptotic variance $Var(\hat{\Phi})$ in (7.50).

7.6 For a MPSD, find the MVU estimators of $\phi^k(\theta)$ and $h^k(\theta)$.

7.7 Prove Theorem 7.10.

7.8 For a left truncated MPSD, show that the MVU estimator of $\phi^k(\theta)$, whenever it exists, is given by

$$\varphi(Y,Z)=\frac{A(Y-k,n,Z)-A(Y-k,n,Z+1)}{A(Y,n,Z)-A(Y,n,Z+1)}$$

for $Y=nZ,nZ+1,\ldots$, $Z=r,r+1,\ldots$, and zero otherwise.

7.9 Describe the application of MPSDs in genetics considered by Gupta (1976), wherein he obtained the correlation coefficient between the number of boys and the number of girls when the family size has a MPSD.

7.10 By using the distribution in (7.83), write down the probability distribution for the zero-inflated MPSD. Derive the recurrence relations between (i) the noncentral moments, and (ii) the central moments of zero-inflated MPSD. Use your results to obtain the mean and variance of zero-inflated MPSD.

8

Some Basic Lagrangian Distributions

8.1 Introduction

Let $g(z)$ be a successively differentiable function such that $g(1) = 1$ and $g(0) \neq 0$. The function $g(z)$ may or may not be a pgf. On applying the Lagrange expansion to the variable z under the transformation $z = ug(z)$, the basic Lagrangian distributions in (2.2) of chapter 2 is obtained as

$$P(X = x) = \frac{1}{x!}\left[D^{x-1}\left(g(z)\right)^x\right]_{z=0}, \quad x = 1, 2, 3, \ldots. \tag{8.1}$$

Examples of some important basic Lagrangian distributions are provided in Table 2.1 of chapter 2. In this chapter three members of this family, the Geeta distribution, the Consul distribution, and the Borel distribution, are described in detail, as they appear to be more important than others. These three distributions are L-shaped like the generalized logarithmic series distribution described in chapter 11.

8.2 Geeta Distribution

8.2.1 Definition

Suppose X is a discrete r.v. defined over positive integers. The r.v. X is said to have a Geeta distribution with parameters θ and β if

$$P(X = x) = \begin{cases} \frac{1}{\beta x - 1}\binom{\beta x - 1}{x}\theta^{x-1}(1-\theta)^{\beta x - x}, & x = 1, 2, 3, \ldots, \\ 0, & \text{otherwise}, \end{cases} \tag{8.2}$$

where $0 < \theta < 1$ and $1 < \beta < \theta^{-1}$. The upper limit on β has been imposed for the existence of the mean. When $\beta \to 1$, the Geeta distribution degenerates and its probability mass gets concentrated at the point $x = 1$. Consul (1990a) defined and studied some of the properties of the model in (8.2).

The Geeta distribution is L-shaped (or reversed J-shaped) for all values of θ and β. Its tail may be short or long and heavy depending upon the values of θ and β. The mean and the variance of Geeta distribution are given by

$$\mu = (1-\theta)(1-\theta\beta)^{-1} \tag{8.3}$$

and

$$\sigma^2 = (\beta - 1)\theta(1 - \theta)(1 - \theta\beta)^{-3}. \tag{8.4}$$

Consul (1990b) defined the class of location-parameter discrete probability distributions (LDPDs). By expressing the parameter θ in terms of μ in (8.3), the location-parameter form of the Geeta distribution becomes

$$P(X = x) = \begin{cases} \frac{1}{\beta x-1}\binom{\beta x-1}{x}\left(\frac{\mu-1}{\beta\mu-1}\right)^{x-1}\left(\frac{\mu(\beta-1)}{\beta\mu-1}\right)^{\beta x-x}, & x = 1, 2, 3, \ldots, \\ 0, & \text{otherwise,} \end{cases} \tag{8.5}$$

where the parameters are $\mu \geq 1$ and $\beta > 1$. Consul (1990b) showed that the Geeta distribution is characterized by its variance $\sigma^2 = (\beta - 1)^{-1}\ \mu(\mu - 1)(\beta\mu - 1)$, when the probability is nonzero over all integral values of X. The Geeta distribution satisfies properties of both over-dispersion and under-dispersion. The over-dispersion property is satisfied when the variance is larger than the mean. Since

$$\sigma^2/\mu = (\beta - 1)\theta(1 - \beta\theta)^{-2}$$

and

$$\frac{\partial}{\partial\theta}\left[\frac{\sigma^2}{\mu}\right] = (\beta - 1)(1 + \beta\theta)(1 - \beta\theta)^{-3} > 0,$$

the quantity σ^2/μ is a monotonic increasing function of θ and its value varies from zero (at $\theta = 1$) to ∞ (at $\theta = \beta^{-1}$). Thus, for smaller values of θ, the Geeta distribution has under-dispersion and for larger values of $\theta < \beta^{-1}$ it has over-dispersion. The variance σ^2 equals the mean μ when

$$\theta = \frac{1}{2}\left[(3\beta - 1) + \sqrt{(3\beta - 1)^2 + 4\beta^2}\right]\beta^{-2}.$$

8.2.2 Generating Functions

The Geeta distribution can be generated by using Lagrange expansion on the parameter $\theta(0 < \theta < 1)$ under the transformation $\theta = u(1 - \theta)^{-\beta+1}$ for $\beta > 1$. This leads to

$$\theta = \sum_{x=1}^{\infty}\frac{u^x}{x!}\left[\left(\frac{\partial}{\partial\theta}\right)^{x-1}(1 - \theta)^{-\beta x+x}\right]_{\theta=0},$$

i.e.,

$$\theta = \sum_{x=1}^{\infty}\frac{1}{\beta x - 1}\binom{\beta x - 1}{x}\theta^x(1 - \theta)^{\beta x - x}. \tag{8.6}$$

On dividing (8.6) by θ, the quantity on the right-hand side is the sum of the Geeta distribution probabilities in (8.2) and this sum is 1.

The pgf of a Geeta variate X, with parameters θ and β, is given by the Lagrange expansion of

$$f(u) = z, \quad \text{where} \quad z = u(1 - \theta)^{\beta-1}(1 - \theta z)^{-\beta+1},\ 1 < \beta < \theta^{-1}. \tag{8.7}$$

By putting $z = e^s$ and $u = e^r$ in (8.7), one obtains the mgf for the Geeta distribution as

$$M_x(r) = e^s, \text{ where } e^s = e^r(1 - \theta)^{\beta-1}(1 - \theta e^s)^{-\beta+1}. \tag{8.8}$$

8.2.3 Moments and Recurrence Relations

All the moments of Geeta distribution exist if $0 < \theta < 1$ and $1 < \beta < \theta^{-1}$. Using μ_k to denote the kth central moment, Consul (1990a) obtained the following recurrence relations between the central moments of Geeta distribution in (8.2):

$$\mu_{k+1} = \theta\mu \frac{d\mu_k}{d\theta} + k\mu_2\mu_{k-1}, \quad k = 1, 2, 3, \ldots, \tag{8.9}$$

where $\mu_0 = 1$, $\mu_1 = 0$, and $\mu_2 = \sigma^2$, as defined in (8.4). The third and the fourth central moments are given by

$$\mu_3 = (\beta - 1)\theta(1 - \theta)(1 - 2\theta + 2\theta\beta - \theta^2\beta)(1 - \theta\beta)^{-5} \tag{8.10}$$

and

$$\mu_4 = 3\mu_2^2 + (\beta - 1)\theta(1 - \theta)A(1 - \theta\beta)^{-7}, \tag{8.11}$$

where $A = 1 - 6\theta + 6\theta^2 + 2\theta\beta(4 - 9\theta + 4\theta^2) + \theta^2\beta^2(6 - 6\theta - 4\theta^2 + 5\theta^3)$.

For the Geeta distribution defined in (8.5), a recurrence relation between the central moments is given by

$$\mu_{k+1} = \sigma^2 \left\{ \frac{d\mu_k}{d\mu} + k\mu_{k-1} \right\}, \quad k = 1, 2, 3, \ldots. \tag{8.12}$$

On using (8.12), the third central moment for model (8.5) is

$$\mu_3 = \sigma^2(3\beta\mu^2 - 2\beta\mu - 2\mu + 1)(\beta - 1)^{-1}, \tag{8.13}$$

which is equivalent to (8.10).

8.2.4 Other Interesting Properties

Theorem 8.1 (Convolution Property). *Suppose $X_1, X_2, \ldots, X_n$ are i.i.d. Geeta r.v.s, as defined in (8.2). The sample sum $Y = \sum X_i$ has a Geeta-n distribution given by*

$$P(Y = y) = \begin{cases} \frac{n}{y}\binom{\beta y - n - 1}{y - n}\theta^{y-n}(1 - \theta)^{\beta y - y}, & y = n, n + 1, \ldots, \\ 0, & \text{otherwise.} \end{cases} \tag{8.14}$$

The Geeta-n distribution in (8.14) reduces to the Haight distribution when $\beta = 2$.

Proof. See Exercise 8.1.

Theorem 8.2 (Unimodality Property). *The Geeta distribution is unimodal but not strongly unimodal for all values of θ and β and the mode is at the point $x = 1$.*

Proof. See Exercise 8.2.

The successive probabilities for the Geeta distribution in (8.5) can be computed by using the recurrence relation

$$P(X = k+1) = \prod_{i=1}^{k} \left(1 + \frac{\beta}{\beta k - i}\right) \frac{\mu - 1}{\mu} \left(\frac{(\beta - 1)\mu}{\beta\mu - 1}\right)^{\beta} \cdot P(X = k), \; k = 2, 3, 4, \ldots, \tag{8.15}$$

where

$$P(X = 1) = \left(\frac{(\beta - 1)\mu}{\beta\mu - 1}\right)^{\beta - 1} \tag{8.16}$$

and

$$P(X = 2) = \frac{\mu - 1}{\mu} \left(\frac{(\beta - 1)\mu}{\beta\mu - 1}\right)^{2\beta - 1}. \tag{8.17}$$

8.2.5 Physical Models Leading to Geeta Distribution

Consul (1990c) gave two stochastic models for the Geeta distribution. He showed that the Geeta distribution can be obtained as an urn model and that it is also generated as a model based on a difference-differential equation.

An urn model. Let an urn A contain some white and black balls such that the probability of drawing a black ball with replacement is θ. Thus, the probability of drawing a white ball is $1 - \theta$. As soon as a ball is drawn from urn A, it is returned to A and another ball of the same color is put into an urn B which is initially empty before the first draw. A game is played under the following conditions:

(a) The player selects a strategy by choosing an integer $\beta \geq 2$.
(b) The player is allowed to draw balls one by one from urn A as long as the number of white balls in urn B exceeds $(\beta - 1)$ times the number of black balls in urn B and loses the game as soon as this condition is violated.
(c) The player wins the game when the number of black and white balls in urn B are exactly $x - 1$ and $(\beta - 1)x$, respectively, where $x = 1, 2, 3, \ldots$.

Suppose $P(X = x)$ is the probability that a player wins the game. Now, one can consider the derivation of $P(X = x)$ for the various values of $x = 1, 2, 3, \ldots$. When $x = 1$,

$$\begin{aligned} P(X = 1) &= P(\text{drawing } \beta - 1 \text{white balls and 0 black balls}) \\ &= (1 - \theta)^{\beta - 1} \theta^0 = (1 - \theta)^{\beta - 1}. \end{aligned} \tag{8.18}$$

When $x = 2$,

$$\begin{aligned} P(X = 2) &= P(\text{drawing } 2\beta - 2 \text{ white balls and 1 black ball such} \\ &\quad \text{that the black ball is drawn in the last } \beta - 1 \text{ draws}) \\ &= \binom{\beta - 1}{1} \theta^1 (1 - \theta)^{2\beta - 2} = (\beta - 1)\theta(1 - \theta)^{2\beta - 2}. \end{aligned} \tag{8.19}$$

When $x > 2$, let the probability of drawing exactly $(x - 1)$ black balls and $y = \beta x - x$ white balls be

$$P(X = x) = f(x - 1, y)\theta^{x - 1}(1 - \theta)^{y}, \tag{8.20}$$

where $f(x - 1, y)$ denotes the number of sequences in which y always exceeds $(\beta - 1)(x - 1)$.

To win the game with any number $(x - 1)$ of black balls, by condition (c), the player has to draw a total of $(\beta - 1)x + x - 1 = \beta x - 1$ balls. Let each one of the $(\beta - 1)x$ white balls be denoted by a (-1) and each of the black balls be denoted by a $(+1)$. By condition (b), the player stays in the game until his selection is such that the number of white balls in urn B always exceeds $(\beta - 1)(x - 1)$. Thus, the player must have selected $[(\beta - 1)(x - 1) + 1]$ white balls before drawing the $(x - 1)$th black ball. Therefore, the partial sum S_x of -1 and $+1$ is $\leq x - 1 - [(\beta - 1)(x - 1) + 1] = (2 - \beta)(x - 1) - 1$. So, the player stays in the game as long as

$$S_x \leq (2 - \beta)(x - 1) - 1, \quad x = 1, 2, 3, \ldots. \tag{8.21}$$

Since y in (8.20) must always exceed $(\beta - 1)(x - 1)$, we have

$$f(x - 1, y) = 0 \quad \text{for} \quad y \leq (\beta - 1)(x - 1). \tag{8.22}$$

The last draw in the game can either be a white ball (-1) or a black ball $(+1)$. Therefore,

$$f(x - 1, y) = f(x - 2, y) + f(x - 1, y - 1), \quad y > (\beta - 1)(x - 1) + 1, \tag{8.23}$$

and

$$f(x - 1, y) = f(x - 2, y), \quad y > (\beta - 1)(x - 1) + 1. \tag{8.24}$$

We have the initial conditions

$$f(0, y) = f(1, 0) = 1. \tag{8.25}$$

From (8.22), we have $f(x - 1, \ (\beta - 1)(x - 1)) = 0$.

The solution of the system of equations in (8.23) and (8.24) with the boundary conditions in (8.25) is

$$f(x - 1, y) = \frac{y - (x - 1)(\beta - 1)}{y + x - 1} \binom{y + x - 1}{x - 1}. \tag{8.26}$$

By using $y = \beta x - x$ and (8.26) in (8.20), we obtain the probability that a player wins the game as

$$P(X = x) = \frac{1}{\beta x - 1} \binom{\beta x - 1}{x} \theta^{x-1} (1 - \theta)^{\beta x - x}, \quad x = 1, 2, 3, \ldots,$$

which is the Geeta distribution in (8.2).

Model based on difference-differential equations. Consider a regenerative process which is initiated by a single microbe, bacteria, or cell and which may grow into any number. Let the probability of x cells in a location be $P_x(\theta, \beta)$.

Theorem 8.3. *If the mean μ for the distribution of the microbes is increased by changing θ to $\theta + \Delta\theta$ in such a way that*

$$\frac{dP_x(\theta, \beta)}{d\theta} + \frac{x(\beta - 1)}{1 - \theta} P_x(\theta, \beta) = \frac{x - 1}{x} \frac{(\beta x - x)^{[x-1]} (1 - \theta)^{\beta - 1}}{(\beta x - x - \beta + 1)^{[x-2]}} P_{x-1}(\theta, \beta) \tag{8.27}$$

for all integral values of $x \geq 1$ with the initial conditions

$$P_1(0, \beta) = 1 \text{ and } P_x(0, \beta) = 0 \quad \text{for} \quad x \geq 2, \tag{8.28}$$

then the probability model $P_x(\theta, \beta)$ is the Geeta distribution, where $a^{[k]} = a(a + 1) \cdots (a + k - 1)$.

Proof. For $x = 1$, equation (8.27) becomes

$$\frac{dP_1(\theta,\beta)}{d\theta} + \frac{\beta-1}{1-\theta} P_1(\theta,\beta) = 0,$$

which is a simple differential equation with the solution

$$P_1(\theta,\beta) = (1-\theta)^{\beta-1}. \tag{8.29}$$

For $x = 2$, equation (8.27) with (8.29) gives

$$\frac{dP_2(\theta,\beta)}{d\theta} + \frac{2(\beta-1)}{1-\theta} P_2(\theta,\beta) = (\beta-1)(1-\theta)^{2\beta-2}.$$

The solution to the above equation is

$$P_2(\theta,\beta) = (2\beta-2)^{[1]}\theta(1-\theta)^{2\beta-2}/2!. \tag{8.30}$$

By putting $x = 3$ in equation (8.27) and by using (8.30), we obtain another linear differential equation whose solution is

$$P_3(\theta,\beta) = (3\beta-3)^{[2]}\theta^2(1-\theta)^{3\beta-3}/3!. \tag{8.31}$$

By using the principle of mathematical induction, one can show that the solution for $x = k$ is given by

$$\begin{aligned} P_k(\theta,\beta) &= (\beta k - k)^{[k-1]}\,\theta^{k-1}(1-\theta)^{k(\beta-1)}/k! \\ &= \frac{1}{\beta k - 1}\binom{\beta k - 1}{k}\,\theta^{k-1}(1-\theta)^{\beta k - k}, \end{aligned}$$

which is the Geeta distribution defined in (8.2). □

8.2.6 Estimation

Let a random sample of size n be taken from the Geeta distribution and let the observed frequencies be denoted by n_x, $x = 1, 2, \ldots, k$, such that $\sum_{x=1}^{k} n_x = n$. The sample mean and sample variance are given, respectively, by

$$\bar{x} = n^{-1}\sum_{x=1}^{k} x n_x$$

and

$$s^2 = (n-1)^{-1}\sum_{x=1}^{k}(x-\bar{x})^2 n_x.$$

Three methods of estimation for the parameters μ and β in model (8.5) are given below.

Moment estimation. By equating the parameters μ and σ^2 with their corresponding sample values $\bar{x}$ and s^2, one gets

$$\mu = \bar{x} \quad \text{and} \quad \sigma^2 = \mu(\mu - 1)(\beta\mu - 1)(\beta - 1)^{-1} = s^2,$$

which provide the moment estimates as

$$\tilde{\mu} = \bar{x} \quad \text{and} \quad \tilde{\beta} = \left[s^2 - \bar{x}(\bar{x} - 1)\right] \Big/ \left[s^2 - \bar{x}^2(\bar{x} - 1)\right]. \tag{8.32}$$

The moment estimate $\tilde{\beta}$ of β is greater than 1 when $s^2 > \bar{x}^2(\bar{x}-1)$. Thus, a necessary condition for the applicability of the Geeta model to an observed data set is that

$$s^2 > \bar{x}^2(\bar{x} - 1).$$

Method based on sample mean and first frequency. Let the estimates be denoted by μ^* and β^*. Equating $P(X = 1)$ with the corresponding sample value,

$$\left(1 - \frac{\mu - 1}{\beta\mu - 1}\right)^{\beta - 1} = \frac{n_1}{n}. \tag{8.33}$$

On combining equation (8.33) with the condition $\mu^* = \bar{x}$, one gets the expression

$$J(\beta) = \left(\frac{(\beta - 1)\bar{x}}{\beta\bar{x} - 1}\right)^{\beta - 1} = \frac{n_1}{n}. \tag{8.34}$$

On differentiating $J(\beta)$ in (8.34) with respect to β,

$$\begin{aligned}
\frac{dJ(\beta)}{d\beta} &= J(\beta)\left\{\frac{\bar{x} - 1}{\beta\bar{x} - 1} + \log\left(1 - \frac{\bar{x} - 1}{\beta\bar{x} - 1}\right)\right\} \\
&= J(\beta)\left\{-\frac{1}{2}\left(\frac{\bar{x} - 1}{\beta\bar{x} - 1}\right)^2 - \frac{1}{3}\left(\frac{\bar{x} - 1}{\beta\bar{x} - 1}\right)^3 - \cdots\right\} < 0.
\end{aligned}$$

Therefore, $J(\beta)$ is a decreasing function of β and so equation (8.34) has a unique root and it can be solved iteratively to obtain β^*. The initial value of β can be taken to be the moment estimate $\tilde{\beta}$ of β in (8.32).

Maximum likelihood estimation. The log likelihood function for the Geeta distribution can be written as

$$\begin{aligned}
\log L = n\bar{x}\,\{(\beta - 1)\log[\mu\,(\beta - 1)] - \beta\,\log(\beta\mu - 1) + \log(\mu - 1)\} + n\,\{\log(\beta\mu - 1) \\
- \log(\mu - 1)\} + \sum_{x=2}^{k}\sum_{i=2}^{x} n_x \log(\beta x - i) - \sum_{x=2}^{k} n_x \log(x!).
\end{aligned} \tag{8.35}$$

On taking partial derivatives of (8.35) with respect to μ and β and equating to zero, one gets the equations

$$\frac{\partial \log L}{\partial \mu} = \frac{(\beta - 1)\bar{x}}{\mu} - \frac{\beta^2\bar{x}}{\beta\mu - 1} + \frac{\bar{x} - 1}{\mu - 1} + \frac{\beta}{\beta\mu - 1} = 0 \tag{8.36}$$

and

$$\frac{\partial \log L}{\partial \beta} = n\bar{x}\left\{\log(\beta-1)+\log\bar{x}-\log(\beta\bar{x}-1)\right\}+\sum_{x=2}^{k}\sum_{i=2}^{x}\frac{xn_x}{\beta x-i}=0. \tag{8.37}$$

On simplification, equation (8.36) gives the ML estimate $\hat{\mu}$ of μ as

$$\hat{\mu}=\bar{x}. \tag{8.38}$$

Also, the relation (8.37) can be written in the form

$$G(\beta)=\frac{(\beta-1)\bar{x}}{\beta\bar{x}-1}=e^{-H(\beta)}, \tag{8.39}$$

where

$$H(\beta)=\frac{1}{n\bar{x}}\sum_{x=2}^{k}\sum_{i=2}^{x}\frac{xn_x}{\beta x-i}. \tag{8.40}$$

The function $H(\beta)$ is a monotonically decreasing function of β and so $e^{-H(\beta)}$ is a monotonically increasing function of β. Also, the function

$$G(\beta)=(\beta-1)\,\bar{x}/(\beta\bar{x}-1) \tag{8.41}$$

is a monotonically increasing function of β.

Thus the ML estimation of β will be the unique point of intersection of the graphs of (8.41) and of $e^{-H(\beta)}$, where $H(\beta)$ is given by (8.40). Alternatively, one can solve equation (8.39) iteratively for $\hat{\beta}$, the ML estimate of β. The moment estimate of β can be used as the initial guess for parameter β.

Minimum variance unbiased estimation. Consul (1990a) gave the mvu estimates for some functions of parameter θ, which are similar to the results obtained for the MPSD considered in chapter 7.

8.2.7 Some Applications

(i) Suppose a queue is initiated with one member and has traffic intensity with negative binomial arrivals, given by the generating function $g(z)=(1-\theta)^{\beta-1}(1-\theta z)^{-\beta+1}$ and constant service time. Then the Geeta distribution represents the probability that exactly x members will be served before the queue vanishes.

(ii) In the branching process, discussed in section 6.2, started by a single member, let the member before dying reproduce a certain number of new members with a probability given by the negative binomial distribution, and each member of the new generation, before dying, reproduces new members in the same manner. If the branching process continues in this manner, then the probability distribution of the total progeny at the nth generation is given by the Geeta model.

(iii) In the stochastic model of epidemics (section 6.4), let $X_0=1$ with probability 1 and let the number of new persons infected, among the susceptibles, by each infected person be distributed according to the negative binomial distribution. If the process of infection continues in this manner again and again, then the probability distribution of the total number of persons infected at any given time will be given by the Geeta model.

(iv) The Geeta model will be applicable to the sales and spread of fashions as well whenever the conditions provided in (ii) and/or (iii) above are satisfied. Similarly, the probability distributions of salespeople in dealerships like AVON or AMWAY, etc., where the process gets started by a single dealer, will be the Geeta model if each dealer succeeds in enrolling new members according to the negative binomial distribution.

8.3 Consul Distribution

8.3.1 Definition

A discrete r.v. X is said to have a Consul distribution if its probability function is given by

$$P(X=x)=\begin{cases}\frac{1}{x}\binom{mx}{x-1}\theta^{x-1}(1-\theta)^{mx-x+1}, & x=1,2,3,\ldots,\\ 0, & \text{otherwise,}\end{cases} \tag{8.42}$$

where $0<\theta<1$ and $1\le m\le\theta^{-1}$. The mean and the variance of the model exist when $m<\theta^{-1}$. The Consul distribution reduces to the geometric distribution when $m=1$. Famoye (1997a) obtained the model in (8.42) by using Lagrange expansion on the pgf of a geometric distribution and called it a generalized geometric distribution. He studied some of its properties and applications.

The probability model (8.42) belongs to the class of location-parameter discrete distributions studied by Consul (1990b). This class of discrete distributions is characterized by their variances.

The mean and variance of Consul distribution are given by

$$\mu=(1-\theta m)^{-1} \quad \text{and} \quad \sigma^2=m\theta(1-\theta)(1-\theta m)^{-3}. \tag{8.43}$$

From the mean, one obtains the value of θ as $\theta=(\mu-1)/m\mu$. On using this value of θ in (8.42), the Consul distribution can be expressed as a location-parameter discrete probability distribution in the form

$$P(X=x)=\begin{cases}\frac{1}{x}\binom{mx}{x-1}\left(\frac{\mu-1}{m\mu}\right)^{x-1}\left(1-\frac{\mu-1}{m\mu}\right)^{mx-x+1}, & x=1,2,3,\ldots,\\ 0, & \text{otherwise,}\end{cases} \tag{8.44}$$

where the mean $\mu>1$ and $m\ge 1$. In the form (8.44), the variance of the Consul distribution is

$$\sigma^2=\mu(\mu-1)(m\mu-\mu+1)/m. \tag{8.45}$$

It can easily be shown from (8.43) that $\sigma^2/\mu=m\theta(1-\theta)(1-\theta m)^{-2}$ is a monotonically increasing function of θ. Similarly, by (8.45) $\sigma^2/\mu=\mu(\mu-1)-(\mu-1)^2/m$ is also a monotonically increasing function of μ as well as of m. Accordingly, the minimum value for the ratio σ^2/μ is at the point $m=1$ and the maximum value is when $m\to\infty$. Thus, we have that

$$\mu-1\le\sigma^2/\mu<\mu(\mu-1). \tag{8.46}$$

The Consul probability distribution satisfies the dual properties of under-dispersion and over-dispersion. The model is under-dispersed for all values of $m\ge 1$ when $\mu\le(\sqrt{5}+1)/2$ and is over-dispersed for all values of $m\ge 1$ when $\mu>2$. When $(\sqrt{5}+1)/2<\mu<2$, the Consul model is under-dispersed for $1<m<(\mu-1)^2(\mu^2-\mu-1)^{-1}$ and over-dispersed for $m>(\mu-1)^2(\mu^2-\mu-1)^{-1}$. The mean and variance of the Consul distribution are equal when $\mu=2$ and $m=1$.

8.3.2 Generating Functions

Let $f(z) = (1-\theta)z(1-\theta z)^{-1}$ and $g(z) = (1-\theta)^{m-1}(1-\theta z)^{-m+1}$, $m > 1$, be two functions of z where $f(z)$ is the pgf of a geometric distribution and $g(z)$ may or may not be a pgf. It is clear that both functions $f(z)$ and $g(z)$ are analytic and are successively differentiable with respect to z any number of times. Also, $f(1) = g(1) = 1$ and $g(0) \neq 0$. Under the transformation $z = ug(z)$, the function $f(z)$ can be expressed as a power series expansion of u by the Lagrange expansion to obtain

$$f(z) = \sum_{x=1}^{\infty} \frac{u^x}{x} \binom{mx}{x-1} \theta^{x-1}(1-\theta)^{mx-x+1}, \tag{8.47}$$

which is the pgf for Consul distribution. Thus the pgf can be written as

$$h(u) = (1-\theta)z(1-\theta z)^{-1}, \quad \text{where} \quad z = u(1-\theta)^{m-1}(1-\theta z)^{-m+1}. \tag{8.48}$$

Another pgf for the Consul distribution is given by

$$h(u) = z, \quad \text{where} \quad z = u(1-\theta+\theta z)^m. \tag{8.49}$$

The Consul distribution can also be obtained by taking the Lagrange expansion of the function $\theta(1-\theta)^{-1}$ $(0 < \theta < 1)$ under the transformation $\theta = u(1-\theta)^{-m+1}$ and then dividing the series by $\theta(1-\theta)^{-1}$.

By putting $z = e^s$ and $u = e^\beta$ in (8.48), one obtains the mgf for the Consul distribution as

$$M_X(\beta) = (1-\theta)e^s(1-\theta e^s)^{-1} \quad \text{where} \quad e^s = e^\beta(1-\theta)^{m-1}(1-\theta e^s)^{-m+1}. \tag{8.50}$$

8.3.3 Moments and Recurrence Relations

All the moments of Consul distribution exist for $0 < \theta < 1$ and $1 \le m < \theta^{-1}$. Let the kth noncentral moment be denoted by μ'_k. This is given by

$$\mu'_k = E(X^k) = \sum_{x=1}^{k} x^{k-1} \binom{mx}{x-1} \theta^{x-1}(1-\theta)^{mx-x+1}. \tag{8.51}$$

By differentiating (8.51) with respect to θ and simplifying the expression, one can obtain the recurrence relation

$$\mu'_{k+1} = \theta(1-\theta)(1-\theta m)^{-1}\frac{d\mu'_k}{d\theta} + \mu'_1\mu'_k, \quad k = 0, 1, 2, \ldots, \tag{8.52}$$

where $\mu'_1 = (1-\theta m)^{-1}$ as given by (8.43).

Denoting the kth central moment as μ_k, it can be shown that the recurrence relation between the central moments for the Consul distribution is

$$\mu_{k+1} = \theta(1-\theta)(1-\theta m)^{-1}\frac{d\mu_k}{d\theta} + k\mu_2\mu_{k-1}, \quad k = 1, 2, 3, \ldots, \tag{8.53}$$

where $\mu_2 = \sigma^2 = m\theta(1-\theta)(1-\theta m)^{-3}$. The result (8.53) gives the third and the fourth central moments for the Consul distribution as

$$\mu_3 = m\theta(1-\theta)(1-2\theta+2\theta m-\theta^2 m)(1-\theta m)^{-5} \tag{8.54}$$

and

$$\mu_4 = 3\mu_2^2 + m\theta(1-\theta)A(1-\theta m)^{-7}, \tag{8.55}$$

where $A = 1-6\theta+6\theta^2+2\theta m(4-9\theta+4\theta^2)+\theta^2 m^2(6-6\theta+\theta^2)$.

8.3.4 Other Interesting Properties

Theorem 8.4. *If $X_1, X_2, \ldots, X_n$ are i.i.d. r.v.s having Consul distribution, then their sample sum $Y = \sum X_i$ is a delta-binomial distribution given by*

$$P(Y = y) = \begin{cases} \frac{n}{y}\binom{my}{y-n}\theta^{y-n}(1-\theta)^{my-y+n}, & y = n, n+1, \ldots, \\ 0, & \text{otherwise} \end{cases} \tag{8.56}$$

(Famoye, 1997a).

Proof. Each X_i has the pgf $h(u) = (1-\theta)z(1-\theta z)^{-1}$ where $z = u(1-\theta)^{m-1}(1-\theta z)^{-m+1}$. Since the X_i's are i.i.d., the pgf of $Y = \sum X_i$ is

$$\{h(u)\}^n = (1-\theta)^n z^n (1-\theta z)^{-n} \quad \text{where} \quad z = u(1-\theta)^{m-1}(1-\theta z)^{-m+1}.$$

The Lagrange expansion of $[h(u)]^n$, under the transformation $z = u(1-\theta)^{m-1}(1-\theta z)^{-m+1}$, gives

$$[h(u)]^n = (1-\theta)^n z^n (1-\theta z)^{-n}$$

$$= \sum_{y=n}^{\infty} u^y \cdot \frac{n}{y}\binom{my}{y-n}\theta^{y-n}(1-\theta)^{my-y+n}.$$

Since $z = 1$ when $u = 1$, the above leads to

$$1 = \sum_{y=n}^{\infty} \frac{n}{y}\binom{my}{y-n}\theta^{y-n}(1-\theta)^{my-y+n},$$

which gives the probability distribution of $Y = \sum X_i$ as (8.56). □

Theorem 8.5 (Unimodality Property). *The Consul distribution, defined in (8.42), is unimodal but not strongly unimodal for all values of $m \geq 1$ and $0 < \theta < 1$ and the mode is at the point $x = 1$ (Famoye, 1997a).*

Proof. Keilson and Gerber (1971) gave a necessary and sufficient condition for the sequence $\{P_x\}$ to be strongly unimodal as

$$P_x^2 / \left[P_{x-1}\, P_{x+1}\right] \geq 1 \quad \text{for all values of } x. \tag{8.57}$$

Substituting the values of the probabilities in the above expression for $x = 2$ from the Consul model (8.42), one gets

$$P_2^2\, P_1^{-1}\, P_3^{-1} = 2m(3m-1)^{-1} < 1$$

for all values of m and θ. Therefore, the Consul distribution in (8.42) is not strongly unimodal since it does not satisfy (8.57) even for $x = 2$.

When $m = 1$, the Consul distribution reduces to the geometric distribution which is unimodal. We shall now consider the unimodality when $m > 1$. For all values of $x = 1, 2, 3, \ldots,$ (8.42) gives

$$\frac{P_{x+1}}{P_x} = \frac{1}{x+1} \frac{\Gamma(mx+m+1)\,\Gamma(mx-x+2)}{\Gamma(mx+m-x+1)\,\Gamma(mx+1)}\,\theta(1-\theta)^{m-1}. \tag{8.58}$$

But the function $\theta(1-\theta)^{m-1}$ is an increasing function of θ and it achieves its maximum as $\theta \to m^{-1}$. So,

$$\max_{\theta} \theta(1-\theta)^{m-1} = m^{-1}\left(1-\frac{1}{m}\right)^{m-1}.$$

By using this value in (8.58),

$$\begin{aligned}
\frac{P_{x+1}}{P_x} &< \frac{1}{x+1}\cdot\frac{1}{m}\left(1-\frac{1}{m}\right)^{m-1} \frac{\Gamma(mx+m+1)\,\Gamma(mx-x+2)}{\Gamma(mx+m-x+1)\,\Gamma(mx+1)} \\
&= \left(1-\frac{1}{m}\right)^{m-1} \frac{mx+1}{mx+m} \prod_{i=2}^{m}\left(1+\frac{x}{mx-x+i}\right) \\
&< \left(1-\frac{1}{m}\right)^{m-1} \frac{mx+1}{mx+m} \left(1+\frac{x}{mx-x+2}\right)^{m-1} \\
&= \frac{mx+1}{mx+m}\left[\left(1-\frac{1}{m}\right)\left(1+\frac{x}{mx-x+2}\right)\right]^{m-1} \\
&= \frac{mx+1}{mx+m}\left(1-\frac{2}{m(mx-x+2)}\right)^{m-1} < 1.
\end{aligned}$$

Therefore, the Consul distribution in (8.42) is unimodal with its mode at the point $x = 1$. Thus, the model has a maximum at $x = 1$ and is L-shaped for all values of m and θ such that $m \geq 1$ and $0 < \theta < 1$. □

Theorem 8.6 (Limit of Zero-Truncated Model). *The Consul distribution defined in (8.42) is the limit of zero-truncated GNBD (Famoye, 1997a).*

Proof. The zero-truncated GNBD is given by (10.121) in chapter 10. On taking the limit of (10.121) as the parameter $\beta \to 1$, we get

$$\lim_{\beta\to 1} f_x(\theta, \beta, m) = \frac{1}{x}\binom{mx}{x-1}\theta^{x-1}(1-\theta)^{mx-x+1},$$

which is the Consul distribution in (8.42). □

8.3.5 Estimation

Famoye (1997a) obtained the moment estimates, the estimates based upon sample mean and first frequency, and the ML estimates. Suppose n_x, $x = 1, 2, 3, \ldots, k$, are the observed frequencies in a random sample of size n and let

$$n = \sum_{x=1}^{k} n_x.$$

The sample mean and sample variance are given by

$$\bar{x} = n^{-1} \sum_{x=1}^{k} x n_x \tag{8.59}$$

and

$$s^2 = (n-1)^{-1} \sum_{x=1}^{k} (x - \bar{x})^2 n_x. \tag{8.60}$$

Moment estimation. By equating the sample mean in (8.59) and sample variance in (8.60) with the corresponding population values in (8.45), the moment estimates are given by

$$\tilde{\mu} = \bar{x} \tag{8.61}$$

and

$$s^2 = \sigma^2 = \tilde{\mu}(\tilde{\mu} - 1)(\tilde{m}\tilde{\mu} - \tilde{\mu} + 1)/\tilde{m}, \tag{8.62}$$

which provides

$$\tilde{m} = \bar{x}(\bar{x} - 1)^2 \left[\bar{x}^2(\bar{x} - 1) - s^2\right]^{-1}. \tag{8.63}$$

Since the moment estimate $\tilde{m}$ must be greater than or equal to 1, we have

$$\bar{x}^2(\bar{x} - 1) \geq s^2 \geq \bar{x}(\bar{x} - 1). \tag{8.64}$$

In subsection 8.2.6, the corresponding estimate for parameter β in Geeta distribution is greater than 1 when

$$s^2 > \bar{x}^2(\bar{x} - 1). \tag{8.65}$$

In applying either distribution to an observed data set, one has to compute $\bar{x}$ and s^2 from the sample. If $s^2 \geq \bar{x}^2(\bar{x} - 1)$, one should apply the Geeta model to the observed data set. If

$$\bar{x}(\bar{x} - 1) \leq s^2 < \bar{x}^2(\bar{x} - 1),$$

the Consul distribution is more suitable. If $s^2 < \bar{x}(\bar{x} - 1)$, then none of these models may be applicable.

Method based on sample mean and first frequency. By equating the probability of $x = 1$ with the corresponding sample proportion, one obtains

$$\frac{n_1}{n} = P_1 = (1 - \theta)^m = \left(1 - \frac{\mu - 1}{m\mu}\right)^m. \tag{8.66}$$

On combining (8.66) with (8.61) and solving the two equations, we get

$$\mu^* = \bar{x}$$

and

$$H(m) = m \log\left(1 - \frac{\bar{x} - 1}{m\bar{x}}\right) - \log\left(\frac{n_1}{n}\right) = 0. \tag{8.67}$$

By expanding the log term into a series and by differentiation of (8.67),

$$\frac{\partial H(m)}{\partial m} = \left\{ \frac{1}{2m^2}\left(\frac{\bar{x}-1}{\bar{x}}\right)^2 + \frac{2}{3m^3}\left(\frac{\bar{x}-1}{\bar{x}}\right)^3 + \frac{3}{4m^4}\left(\frac{\bar{x}-1}{\bar{x}}\right)^4 + \cdots \right\} > 0.$$

Hence, the root of equation (8.67) is unique. This equation can be solved iteratively to obtain m^*, the estimate based on the sample mean and the first frequency. The initial estimate of m for the iterative process can be taken as the moment estimate $\tilde{m}$ in (8.63).

Maximum likelihood estimation. The log likelihood function for the Consul distribution can be written as

$$\log L = n(\bar{x}-1)\log(\mu-1) + n(m\bar{x}-\bar{x}+1)\log(m\mu-\mu+1)$$
$$- nm\bar{x}\log(m\mu) - \sum_{x=1}^{k} n_x \log(x!) + \sum_{x=2}^{k}\sum_{i=0}^{x-2} n_x \log(mx-i). \tag{8.68}$$

The partial derivatives of (8.68) with respect to μ and m are given by

$$\frac{\partial \log L}{\partial \mu} = \frac{\bar{x}-1}{\mu-1} + \frac{(m-1)(m\bar{x}-\bar{x}+1)}{m\mu-\mu+1} - \frac{m\bar{x}}{\mu} \tag{8.69}$$

and

$$\frac{\partial \log L}{\partial m} = \bar{x}\log\left[(m\mu-\mu+1)/m\mu\right] + \frac{\mu(m\bar{x}-\bar{x}+1)}{m\mu-\mu+1} - \bar{x} + \frac{1}{n}\sum_{x=2}^{k}\sum_{i=0}^{x-2}\frac{xn_x}{mx-i}. \tag{8.70}$$

On equating (8.69) to zero, it gives

$$\hat{\mu} = \bar{x} \tag{8.71}$$

as the ML estimate for parameter μ.

On equating (8.70) to zero, using (8.71) in (8.70) and on simplifying, we obtain

$$0 = \log\left(1 - \frac{\bar{x}-1}{m\bar{x}}\right) + \frac{1}{n\bar{x}}\sum_{x=2}^{k}\sum_{i=0}^{x-2}\frac{xn_x}{mx-i}. \tag{8.72}$$

Equation (8.72) can be rewritten as

$$\frac{m\bar{x}-\bar{x}+1}{m\bar{x}} = e^{-H(m)} = G(m),$$

where

$$H(m) = \frac{1}{n\bar{x}}\sum_{x=2}^{k}\sum_{i=0}^{x-2}\frac{xn_x}{mx-i}. \tag{8.73}$$

But the function $G(m) = (m\bar{x}-\bar{x}+1)/m\bar{x}$ is a monotonically increasing function of m as m increases from $m = 1$ to $m \to \infty$. Also, the function $H(m)$ in (8.73) is a monotonically decreasing function of m as m increases from $m = 1$ to $m \to \infty$. Therefore, the function $G(m) = e^{-H(m)}$ represents a monotonically increasing function over $m = 1$ to $m \to \infty$. Since $e^{-H(m)}$ and the function $G(m) = (m\bar{x}-\bar{x}+1)/m\bar{x}$ are both monotonically increasing over the same values of m, the two functions can have at most a single point of intersection. Hence, the ML estimate $\hat{m}$ of m from (8.72) is unique.

By starting with the moment estimate $\tilde{m}$ as the initial value, the equation (8.72) is solved iteratively to obtain the ML estimate $\hat{m}$ of m.

8.3.6 Some Applications

(i) Suppose a queue is initiated with one member and has traffic intensity with binomial arrivals, given by the generating function $g(z) = (1-\theta+\theta z)^m$ and constant service time. Then the Consul distribution represents the probability that exactly x members will be served before the queue vanishes.

(ii) In the branching process, discussed in section 6.2, started by a single member, let the member before dying reproduce a certain number of new members with a probability given by the binomial distribution and each member of the new generation, before dying, reproduces new members in the same manner. If the branching process continues in this manner, then the probability distribution of the total progeny at the nth generation is given by the Consul model.

(iii) In the stochastic model of epidemics (section 6.4), let $X_0 = 1$ with probability 1 and let the number of new persons infected, among the susceptibles, by each infected person be distributed according to the binomial distribution. The number of susceptibles is a finite number m as used by Kumar (1981). If the process of infection continues in this manner again and again, then the probability distribution of the total number of persons infected at any given time will be given by the Consul model.

(iv) The Consul model will be applicable to the sales and spread of fashions as well whenever the conditions provided in (ii) and/or (iii) above are satisfied. Similarly, the probability distributions of salespeople in dealerships like AVON or AMWAY, etc., where the process gets started by a single dealer, will be the Consul model if each dealer succeeds in enrolling new members according to the binomial distribution.

(v) *Molecular size distribution in polymers.* Yan (1978) derived a molecular size distribution in linear and nonlinear polymers. For the derivation, the following three assumptions were made: (a) all functional groups A's are equally reactive; (b) no intramolecular reactions occur; (c) the weight of materials lost during condensation is negligible.
Let $P(X = x)$ be the distribution of the number of an x-mer. Yan (1978) considered the size distribution to be of the form of a generalized power series distribution given by

$$P(X = x) = a_x\,[\phi(\theta)]^x \big/ h(\theta), \quad x = 1, 2, 3, \ldots,$$

for $a_x \geq 0$, $\theta > 0$, and $h(\theta) = \sum_{x=1}^{\infty} a_x\,[\phi(\theta)]^x$. The parameter θ is the fraction of reacted functional groups.
For the condensation of an RA_β monomer, consider the generating function $\phi(\theta) = \theta(1-\theta)^{\beta-2}$ and $h(\theta) = \theta(1-\beta\theta/2)(1-\theta)^{-2}$. On using the Lagrange expansion in (2.34) on $h(\theta)$ under the transformation $\phi(\theta) = \theta/\eta(\theta)$, one obtains a Lagrange power series which provides the pgf of the molecular size distribution.
Yan (1979) considered the condensation of $ARB_{\beta-1}$ monomers, where β is the number of functional groups. In this condensation, A may react with B, but reactions between like functional groups are excluded. Suppose the generating function $\phi(\theta) = \theta(1-\theta)^{\beta-2}$ and $h(\theta) = \theta/(1-\theta)$. By using the Lagrange expansion in (2.34) on $h(\theta)$ under the transformation $\phi(\theta) = \theta/\eta(\theta)$, we obtain

$$P(X = x) = \frac{1}{x}\binom{(\beta-1)x}{x-1}\theta^{x-1}(1-\theta)^{(\beta-1)x-x-1},$$

which is the Consul distribution in (8.42).

8.4 Borel Distribution

8.4.1 Definition

A discrete random variable X is said to have the Borel distribution if its probability mass function is given by

$$P(X=x)=\begin{cases}\frac{(x\lambda)^{x-1}}{x!}e^{-\lambda x}, & x=1,2,3,\ldots,\\ 0, & \text{otherwise,}\end{cases} \tag{8.74}$$

where $0<\lambda<1$. The probability distribution in (8.74) was first obtained by Borel (1942).

Suppose a queue is initiated with one member and has traffic intensity under the Poisson arrivals and constant service time. Haight and Breuer (1960) pointed out that the Borel distribution in (8.74) represents the probability that exactly x members of the queue will be served before the queue vanishes.

The mean and variance of the distribution in (8.74) are given by

$$\mu=(1-\lambda)^{-1} \quad \text{and} \quad \sigma^2=\lambda(1-\lambda)^{-3}. \tag{8.75}$$

The Borel model satisfies the properties of under-dispersion and over-dispersion. There is over-dispersion when λ satisfies the inequality $3/2-\sqrt{5}/2<\lambda<1$. The model is under-dispersed when $\lambda<3/2-\sqrt{5}/2$. The mean and the variance are both equal when $\lambda=3/2-\sqrt{5}/2$.

8.4.2 Generating Functions

The Borel distribution can be generated by using Lagrange expansion on parameter λ ($0<\lambda<1$) under the transformation $\lambda=ue^{\lambda}$. This leads to

$$\lambda=\sum_{x=1}^{\infty}\frac{u^x}{x!}\left[\left(\frac{\partial}{\partial\lambda}\right)^{x-1}e^{\lambda x}\right]_{\lambda=0},$$

which gives

$$\lambda=\sum_{x=1}^{\infty}\frac{(x\lambda)^{x-1}}{x!}e^{-\lambda x}, \tag{8.76}$$

which shows that the sum of the Borel distribution in (8.74) is 1.

The pgf of the Borel distribution is given by the Lagrange expansion of

$$f(u)=z, \quad \text{where} \quad z=ue^{\lambda(z-1)}, \quad 0<\lambda<1. \tag{8.77}$$

By putting $z=e^s$ and $u=e^{\beta}$ in (8.77), the mgf for Borel distribution becomes

$$M_X(\beta)=e^s, \quad \text{where} \quad s=\beta+\lambda(z-1). \tag{8.78}$$

8.4.3 Moments and Recurrence Relations

All the moments of the Borel distribution exist for $0 < \lambda < 1$. Let μ_k' denote the kth noncentral moment for the Borel distribution. Thus,

$$\mu_k' = E\left(X^k\right) = \sum_{x=1}^{\infty} x^k \frac{(x\lambda)^{x-1}}{x!} e^{-x\lambda}. \tag{8.79}$$

On differentiating (8.79) with respect to λ and on simplifying the result, it gives the recurrence relation

$$\mu_{k+1}' = \lambda(1-\lambda)^{-1}\frac{d\mu_k'}{d\lambda} + \mu_1'\mu_k', \quad k = 0, 1, 2, \ldots, \tag{8.80}$$

where $\mu_1' = (1-\lambda)^{-1}$.

By using μ_k to denote the central moments, a recurrence relation between the central moments is given by

$$\mu_{k+1} = \lambda(1-\lambda)^{-1}\frac{d\mu_k}{d\lambda} + k\mu_2\mu_{k-1}, \quad k = 1, 2, 3, \ldots, \tag{8.81}$$

where $\mu_2 = \sigma^2$ is given by (8.75). The third and the fourth central moments can be obtained by using (8.81) and these are given by

$$\mu_3 = \lambda(1+2\lambda)(1-\lambda)^{-5} \tag{8.82}$$

and

$$\mu_4 = \lambda(1+8\lambda+6\lambda^2)(1-\lambda)^{-7} + 3\mu_2^2. \tag{8.83}$$

By using the values of μ_2, μ_3, and μ_4, the expressions for the coefficient of skewness (β_1) and kurtosis (β_2) become

$$\beta_1 = \frac{1+2\lambda}{\sqrt{\lambda(1-\lambda)}} \tag{8.84}$$

and

$$\beta_2 = 3 + \frac{1+8\lambda+6\lambda^2}{\lambda(1-\lambda)}. \tag{8.85}$$

These values in (8.84) and (8.85) are similar to those obtained for the generalized Poisson distribution in chapter 9. It is clear from (8.84) that the Borel distribution is always positively skewed. From (8.85), it is clear that β_2 is always greater than 3 and so the Borel distribution is leptokurtic.

8.4.4 Other Interesting Properties

Theorem 8.7. *Suppose $X_1, X_2, \ldots, X_n$ are i.i.d. r.v.s with Borel distribution in (8.74), and the sample sum $Y = \sum X_i$ has a Borel–Tanner distribution given by*

$$P(Y=y) = \begin{cases} \frac{n}{y}\frac{(\lambda y)^{y-n}}{(y-n)!}e^{-\lambda y}, & y = n, n+1, \ldots, \\ 0, & \text{otherwise.} \end{cases} \tag{8.86}$$

Proof. By using Lagrange expansion on $f(u) = z^n$ under the transformation $z = e^{\lambda(z-1)}$, one can show that the distribution of $Y = \sum X_i$ is the Borel–Tanner distribution.

The Borel–Tanner distribution in (8.86) was given by Haight and Breuer (1960) as a probability model to describe the number of customers served in the first busy period when the queue is initiated by n customers.

Theorem 8.8 (Unimodality Property). *The Borel distribution is unimodal but not strongly unimodal for all values of λ in $0 < \lambda < 1$, and the mode is at the point $x = 1$.*

Proof. See Exercise 8.9.

The successive probabilities for the Borel distribution can be computed by using the recurrence relation

$$P_{x+1} = \left(1 + \frac{1}{x}\right)^{x-1} \lambda\, e^{-\lambda}\, P_x, \quad x = 1, 2, 3, \ldots, \tag{8.87}$$

where $P_1 = e^{-\lambda}$.

8.4.5 Estimation

Suppose a random sample of size n is taken from the Borel distribution and let the observed frequencies be denoted by n_x, $x = 1, 2, \ldots, k$, such that $\sum_{x=1}^{k} n_x = n$. The sample mean is given by $\bar{x} = n^{-1} \sum_{x=1}^{k} x n_x$. We now provide three methods for estimating the parameter λ in model (8.74).

Moment estimation. On equating the sample mean with the population mean, the moment estimate $\tilde{\lambda}$ of λ is given by

$$\tilde{\lambda} = 1 - \frac{1}{\bar{x}}. \tag{8.88}$$

Method based on first frequency. On equating $P(X = 1)$ to the corresponding sample proportion,

$$\frac{n_1}{n} = P_1 = e^{-\lambda},$$

which gives the first frequency estimate of λ as

$$\lambda^* = \log\left(\frac{n}{n_1}\right). \tag{8.89}$$

Maximum likelihood estimate. The log likelihood function for the Borel distribution can be written as

$$\log L = n(\bar{x} - 1) \log \lambda - n\bar{x}\lambda + \sum_{x=1}^{k} n_x \left[(x - 1) \log x - \log(x!)\right]. \tag{8.90}$$

On differentiating (8.90) with respect to λ and equating to zero, it can easily be shown that the ML estimate $\hat{\lambda}$ of λ is

$$\hat{\lambda} = 1 - \frac{1}{\bar{x}}, \tag{8.91}$$

which is the same as the moment estimate in (8.88).

8.5 Weighted Basic Lagrangian Distributions

The class $L_1(f_1; g; x)$ of Lagrangian probability distributions were defined in chapter 2 by using the Lagrange expansion given in (1.80). On using the pgf for the basic Lagrangian distributions under the Lagrange expansion in (1.80), we obtain the basic Lagrangian distributions provided in (2.11) as

$$P_1(X = x) = \frac{1 - g'(1)}{(x-1)!}\left[\left(\frac{\partial}{\partial z}\right)^{x-1} (g(z))^x\right]_{z=0}, \quad x = 1, 2, 3, \ldots, \tag{8.92}$$

where $\left[1 - g'(1)\right]^{-1}$ is the mean of corresponding basic Lagrangian distribution.

Suppose X, the number of individuals in a group, is a random variable with probability function $P(X = x)$. Suppose that a group gets recorded only when at least one of the members in the group is sighted. If each individual has an independent probability q of being sighted, Patil and Rao (1978) showed that the probability that an observed group has x individuals is

$$P_w\left(X = x\right) = w(x)\; P\left(X = x\right) / E\left[w(x)\right], \tag{8.93}$$

where $w(x) = 1 - (1 - q)^x$. The limit of (8.93) as $q \to 0$ is given by

$$P_0(X = x) = \frac{x P(X = x)}{E(X)}, \tag{8.94}$$

which provides a size biased (or weighted) distribution. It is interesting to note that all the basic L_1 distributions are the corresponding size biased (or weighted) Lagrangian distributions given by (2.2). From (8.92), one can rewrite the basic L_1 distributions as

$$\begin{aligned} P_1(X = x) &= \frac{x\left[\left(\frac{\partial}{\partial z}\right)^{x-1} (g(z))^x\right]_{z=0}}{x!E(X)} \\ &= \frac{x P_0(X = x)}{E(X)}, \end{aligned} \tag{8.95}$$

where $P_0(X = x)$ is the probability mass at $X = x$ for the basic Lagrangian distribution. Thus, $P_1(X = x)$ represents the size biased (or weighted) distribution. From (8.95), the moments of weighted distribution given by $P_1(X = x)$ can be obtained from the moments of the basic Lagrangian distribution given by $P_0(X = x)$.

Weighted Geeta distribution. A r.v. X is said to have a weighted Geeta distribution if its probability function is given by

$$P(X = x) = \begin{cases} (1 - \theta\beta)\binom{\beta x - 2}{x - 1}\theta^{x-1}(1 - \theta)^{\beta x - x - 1}, & x = 1, 2, 3, \ldots, \\ 0, & \text{otherwise}, \end{cases} \tag{8.96}$$

where $0 < \theta < 1$ and $1 < \beta < \theta^{-1}$. The condition $1 < \beta < \theta^{-1}$ is imposed for the existence of all moments. The mean and variance of weighted Geeta distribution in (8.96) are given by

$$\mu = (1 - \theta)(1 - \theta\beta)^{-2} - \theta(1 - \theta\beta)^{-1} = (1 - 2\theta + \theta^2\beta)(1 - \theta\beta)^{-2} \tag{8.97}$$

and

$$\sigma^2 = 2(\beta - 1)\,\theta(1-\theta)\,(1-\beta\theta)^{-4}. \tag{8.98}$$

Weighted Consul distribution. A r.v. X is said to follow a weighted Consul distribution if its pmf is given by

$$P(X=x) = \begin{cases} (1-\theta m)\binom{mx}{x-1}\theta^{x-1}(1-\theta)^{mx-x-1}, & x = 1,2,3,\ldots, \\ 0, & \text{otherwise}, \end{cases} \tag{8.99}$$

where $0 < \theta < 1$ and $1 \le m < \theta^{-1}$. For all the moments to exist, we impose the condition $1 \le m < \theta^{-1}$. The mean and variance of the model in (8.99) are given by

$$\mu = (1 - m\theta^2)(1-\theta m)^{-2}$$

and

$$\sigma^2 = 2m\theta(1-\theta)^2(1-\theta m)^{-4}.$$

Weighted Borel distribution. A discrete r.v. X is said to have a weighted Borel distribution if its pmf is

$$P(X=x) = \begin{cases} (1-\lambda)\,\frac{(\lambda x)^{x-1}}{(x-1)!}\,e^{-\lambda x}, & x = 1,2,3,\ldots, \\ 0, & \text{otherwise}, \end{cases} \tag{8.100}$$

where $0 < \lambda < 1$. The mean and variance of the model in (8.100) are given by

$$\mu = (1-\lambda)^{-2}$$

and

$$\sigma^2 = 2\lambda(1-\lambda)^{-4}.$$

8.6 Exercises

8.1 Suppose that $X_1, X_2, \ldots, X_n$ is a random sample from Geeta distribution with parameters θ and β. Show that the distribution of $Y = \sum X_i$ has a Geeta-n distribution

$$P(Y=y) = \frac{n}{y}\binom{\beta y - n - 1}{y-n}\theta^{y-n}(1-\theta)^{\beta y - y}, \quad y = n, n+1, \ldots.$$

8.2 Prove that the Geeta distribution with parameters θ and β is unimodal but not strongly unimodal and that the mode is at the point $x = 1$.

8.3 Suppose that $X_1, X_2, \ldots, X_n$ is a random sample from Borel distribution with parameter λ. Show that the distribution of $Y = \sum X_i$ has the Borel–Tanner distribution

$$P(Y=y) = \frac{n}{y}\left(\frac{(\lambda y)^{y-n}}{(y-n)!}\right)e^{-\lambda y}, \quad y = n, n+1, \ldots.$$

8.4 Find the recurrence relations between the noncentral moments of the following distributions:

(a) weighted Borel distribution,
(b) weighted Consul distribution,
(c) weighted Geeta distribution.

8.5 (a) Suppose X is a Geeta r.v. with known value of β. Find the MVU estimators for the mean and variance of X.
(b) Suppose a r.v. X has the Consul distribution with known value of m. Find the MVU estimators for the mean and variance of X.

8.6 Consider a weighted Borel distribution. Obtain the moment estimate for λ and derive an equation for finding the ML estimate of λ. Also, obtain the ML estimate of λ.

8.7 Consider a zero-truncated GNBD with parameters θ, β, and m given as

$$P(X=x)=\frac{\beta}{\beta+mx}\binom{\beta+mx}{x}\frac{\theta^x(1-\theta)^{\beta+mx-x}}{1-(1-\theta)^\beta}, \qquad x=1,2,\ldots.$$

(a) Show that the Consul distribution is a particular case of the zero-truncated GNBD when $\beta=1$.
(b) What is the limiting form of the zero-truncated GNBD when $\beta\to 0$?

8.8 Consider a regenerative process which is initiated by a single cell that may grow into any number of cells. Let the probability of x cells be $P_x(\theta, m)$. Suppose the mean μ for the distribution of cells is increased by changing θ to $\theta+\Delta\theta$ in such a manner that

$$\frac{dP_x(\theta,m)}{d\theta}+\frac{mx-x+1}{1-\theta}P_x(\theta,m)=\frac{x-1}{x}\frac{(mx-x+2)^{[x-1]}(1-\theta)^{m-1}}{(mx-m-x+3)^{[x-2]}}P_{x-1}(\theta,m)$$

for all $x\geq 1$, $a^{[k]}=a(a+1)(a+2)\cdots(a+k-1)$, with the initial conditions

$$P_1(0,m)=1 \quad \text{and} \quad P_x(0,m)=0 \quad \text{for all} \quad x\geq 2.$$

Show that $P_x(\theta,m)$ is the Consul distribution with parameters θ and m.

8.9 Prove that the Borel distribution in (8.74) is unimodal but not strongly unimodal and that the mode is at the point $x=1$.

8.10 Show that the Geeta probability distribution is L-shaped for all values of θ and β and that its tail may be either thin or heavy depending upon the values of θ and β.

8.11 Develop an urn model, based upon two or three urns, which may lead to the Consul probability distribution.

8.12 (a) When β is very large and θ is very small such that $\beta\theta=\lambda$, show that the Geeta distribution in (8.2) approaches the Borel distribution.
(b) When m is very large and θ is very small such that $m\theta=\lambda$, show that the Consul distribution in (8.42) approaches the Borel distribution.

9

Generalized Poisson Distribution

9.1 Introduction and Definition

Let X be a discrete r.v. defined over nonnegative integers and let $P_x(\theta, \lambda)$ denote the probability that the r.v. X takes a value x. The r.v. X is said to have a GPD with parameters θ and λ if

$$P_x(\theta, \lambda) = \begin{cases} \theta(\theta + \lambda x)^{x-1} e^{-\theta-\lambda x}/x!, & x = 0, 1, 2, \ldots, \\ 0, & \text{for } x > m \quad \text{if} \quad \lambda < 0, \end{cases} \tag{9.1}$$

and zero otherwise, where $\theta > 0$, $\max(-1, \ -\theta/m) \le \lambda \le 1$, and m (≥ 4) is the largest positive integer for which $\theta + m\lambda > 0$ when $\lambda < 0$. The parameters θ and λ are independent, but the lower limits on λ and $m \ge 4$ are imposed to ensure that there are at least five classes with nonzero probability when λ is negative. The GPD model reduces to the Poisson probability model when $\lambda = 0$. Consul and Jain (1973a, 1973b) defined, studied, and discussed some applications of the GPD in (9.1).

The GPD belongs to the class of Lagrangian distributions $L(f; g; x)$, where $f(z) = e^{\theta(z-1)}$, $\theta > 0$, and $g(z) = e^{\lambda(z-1)}$, $0 < \lambda < 1$, and is listed as (6) in Table 2.3. It belongs to the subclass of MPSD. Naturally, it possesses all the properties of these two classes of distributions.

When λ is negative, the model includes a truncation due to $P_x(\theta, \lambda) = 0$ for all $x > m$ and the sum $\sum_{x=0}^{m} P_x(\theta, \lambda)$ is usually a little less than unity. However, this truncation error is less than 0.5% when $m \ge 4$ and so the truncation error does not make any difference in practical applications.

The multiplication of each $P_x(\theta, \lambda)$ by $[F_m(\theta, \lambda)]^{-1}$, where

$$F_m(\theta, \lambda) = \sum_{x=0}^{m} P_x(\theta, \lambda), \tag{9.2}$$

has been suggested for the elimination of this truncation error. (See Consul and Shoukri (1985), Consul and Famoye (1989b).)

Lerner, Lone, and Rao (1997) used analytic functions to prove that the GPD will sum to 1. Tuenter (2000) gave a shorter proof based upon an application of Euler's difference lemma.

The properties and applications of the GPD have been discussed in full detail in the book *Generalized Poisson Distribution: Properties and Applications*, by Consul (1989a). Accordingly, some important results only are being given in this chapter. The GPD and some of its properties have also been described in *Univariate Discrete Distributions* by Johnson, Kotz, and Kemp (1992).

9.2 Generating Functions

The pgf of the GPD with parameters (θ, λ) is given by the Lagrange expansion in (1.78) as

$$G(u) = e^{\theta(z-1)}, \text{ where } z = u\, e^{\lambda(z-1)}. \tag{9.3}$$

The above pgf can also be stated in the form

$$G(u) = e^{\theta(w(u)-1)}, \tag{9.4}$$

where the function $w(u)$ is defined by the relation

$$w(u) = u\ \exp\{\lambda(w(u)-1)\}. \tag{9.5}$$

The function $w(u)$ is 0 at $u = 0$ and 1 at $u = 1$, and its derivative is

$$w'(u) = \left[e^{-\lambda(w(u)-1)} - u\lambda\right]^{-1}. \tag{9.6}$$

By putting $z = e^s$ and $u = e^\beta$ in (9.3), one obtains the mgf for the GPD model as

$$M_x(\beta) = e^{\theta(e^s-1)}, \text{ where } s = \beta + \lambda(e^s - 1). \tag{9.7}$$

Thus, the cgf of the GPD becomes

$$\psi(\beta) = \ln M_x(\beta) = \theta(e^s - 1), \text{ where } s = \beta + \lambda(e^s - 1). \tag{9.8}$$

It has been shown in chapter 2 that the GPD is a particular family of the class of Lagrangian distributions $L(f; g; x)$ and that the mean μ and variance σ^2 are

$$\mu = \theta(1-\lambda)^{-1}, \qquad \sigma^2 = \theta(1-\lambda)^{-3}. \tag{9.9}$$

The variance σ^2 of the GPD is greater than, equal to, or less than the mean μ according to whether $\lambda > 0$, $\lambda = 0$, or $\lambda < 0$, respectively.

Ambagaspitiya and Balakrishnan (1994) have expressed the pgf $P_x(z)$ and the mgf $M_x(z)$ of the GPD (9.1) in terms of Lambert's W function as

$$M_x(z) = \exp\left\{-(\lambda/\theta)\left[W(-\theta\exp(-\theta - z)) + \theta\right]\right\}$$

and

$$P_x(z) = \exp\left\{-(\lambda/\theta)\left[W(-\theta z\exp(-\theta)) + \theta\right]\right\},$$

where W is the Lambert's function defined as

$$W(x)\exp(W(x)) = x.$$

They have derived the first four moments from them, which are the same as those given in (9.9) and (9.13).

9.3 Moments, Cumulants, and Recurrence Relations

All the cumulants and moments of the GPD exist for $\lambda < 1$. Consul and Shenton (1975) and Consul (1989a) have given the following recurrence relations between the noncentral moments μ'_k and the cumulants K_k:

$$(1-\lambda)\,\mu'_{k+1} = \theta\,\mu'_k + \theta\frac{\partial \mu'_k}{\partial \theta} + \lambda\frac{\partial \mu'_k}{\partial \lambda}, \qquad k = 0, 1, 2, \ldots, \tag{9.10}$$

$$(1-\lambda)K_{k+1} = \lambda\frac{\partial K_k}{\partial \lambda} + \theta\frac{\partial K_k}{\partial \theta}, \qquad k = 1, 2, 3, \ldots. \tag{9.11}$$

A recurrence relation between the central moments of the GPD is

$$\mu_{k+1} = \frac{k\theta}{(1-\lambda)^3}\,\mu_{k-1} + \frac{1}{1-\lambda}\left\{\frac{d\,\mu_k(t)}{dt}\right\}_{t=1}, \qquad k = 1, 2, 3, \ldots, \tag{9.12}$$

where $\mu_k(t)$ is the central moment μ_k with θ and λ replaced by θt and λt, respectively. The mean and variance of GPD are given in (9.9). Some other central moments of the model are

$$\left.\begin{aligned}
\mu_3 &= \theta(1+2\lambda)(1-\lambda)^{-5},\\
\mu_4 &= 3\theta^2(1-\lambda)^{-6} + \theta(1+8\lambda+6\lambda^2)(1-\lambda)^{-7},\\
\mu_5 &= 10\theta^2(1+2\lambda)(1-\lambda)^{-8} + \theta(1+22\lambda+58\lambda^2+24\lambda^3)(1-\lambda)^{-9},\\
&\text{and}\\
\mu_6 &= 15\theta^3(1-\lambda)^{-9} + 5\theta^2(5+32\lambda+26\lambda^2)(1-\lambda)^{-10}\\
&\quad +\theta(1+52\lambda+328\lambda^2+444\lambda^3+120\lambda^4)(1-\lambda)^{-11}.
\end{aligned}\right\} \tag{9.13}$$

By using the values of $\mu_2 = \sigma^2$ in (9.9), μ_3 and μ_4 in (9.13), the expressions for the coefficients of skewness (β_1) and kurtosis (β_2) are given by

$$\beta_1 = \frac{1+2\lambda}{\sqrt{\theta(1-\lambda)}} \quad \text{and} \quad \beta_2 = 3 + \frac{1+8\lambda+6\lambda^2}{\theta(1-\lambda)}. \tag{9.14}$$

For any given value of λ, the skewness of the GPD decreases as the value of θ increases and becomes zero when θ is infinitely large. Also, for any given value of θ, the skewness is infinitely large when λ is close to unity. The skewness is negative for $\lambda < -\frac{1}{2}$.

For all values of θ and for all values of λ in $0 < \lambda < 1$, the GPD is leptokurtic as $\beta_2 > 3$. When

$$-\frac{1}{6}\sqrt{10} - \frac{2}{3} < \lambda < \frac{1}{6}\sqrt{10} - \frac{2}{3},$$

the GPD becomes platykurtic since β_2 becomes less than 3.

The expressions for the mean deviation, the negative integer moments, and the incomplete moments are all given in the book by Consul (1989a).

9.4 Physical Models Leading to GPD

The GPD does relate to a number of scientific problems and can therefore be used to describe many real world phenomena.

Limit of Generalized Negative Binomial Distribution

The discrete probability model of GNBD, discussed in chapter 10, is given by

$$P(X=x)=\frac{m}{m+\beta x}\binom{m+\beta x}{x}p^x(1-p)^{m+\beta x-x}, \qquad x=0,1,2,\ldots, \tag{9.15}$$

and zero otherwise, where $0<p<1$ and $1\le\beta<p^{-1}$ for $m>0$.

Taking m and β to be large and p to be small such that $mp=\theta$ and $\beta p=\lambda$, the GNBD approaches the GPD.

Limit of Quasi-Binomial Distribution

While developing urn models dependent upon predetermined strategy, Consul (1974) obtained a three-parameter QBD-I defined in (4.1). If $p\to 0$, $\phi\to 0$, and n increases without limit such that $n\,p=\theta$ and $n\,\phi=\lambda$, the QBD-I approaches the GPD.

Consul and Mittal (1975) gave another urn model which provided a three-parameter QBD-II defined in (4.73). When $p\to 0$, $\alpha\to 0$ and n increases without limit such that $n\,p=\theta$ and $n\alpha=\lambda$, the QBD-II approaches the GPD.

Limit of Generalized Markov–Pólya Distribution

Janardan (1978) considered a four-urn model with predetermined strategy and obtained the generalized Markov–Pólya distribution given by

$$P(X=k)=\frac{pq(1+Ns)\binom{N}{k}\prod_{j=0}^{k-1}(\theta+ks+jr)\prod_{j=0}^{N-k-1}(q+Ns-ks+jr)}{(p+ks)(q+Ns-ks)\prod_{j=0}^{N-1}(q+Ns+jr)} \tag{9.16}$$

for $k=0,1,2,\ldots,N$ and zero otherwise, where $0<p<1$, $q=1-p$, $r>0$, $s>0$, and N is a positive integer.

When N increases without limit and $p\to 0$, $r\to 0$, $s\to 0$ such that $Np=\theta$, $Ns=\lambda$, and $Nr\to 0$, the Markov–Pólya distribution in (9.16) approaches the GPD.

Models Based on Difference-Differential Equations

Consul (1988) provided two models, based on difference-differential equations, that generate the GPD model. Let there be an infinite but countable number of available spaces for bacteria or viruses or other micro-organisms. Let the probability of finding x micro-organisms in a given space be $P_x(\theta,\lambda)$.

Model I. Suppose the mean $\mu(\theta,\lambda)$ for the probability distribution of the micro-organisms is increased by changing the parameter θ to $\theta+\Delta\theta$ in such a way that

$$\frac{dP_0(\theta,\lambda)}{d\theta}=-P_0(\theta,\lambda), \tag{9.17}$$

and

$$\frac{dP_x(\theta,\lambda)}{d\theta}=-P_x(\theta,\lambda)+P_{x-1}(\theta+\lambda,\lambda), \tag{9.18}$$

for all integral values of $x>0$ with the initial conditions $P_0(0,\lambda)=1$ and $P_x(0,\lambda)=0$ for $x>0$, then the probability model $P_x(\theta,\lambda)$, $x=0,1,2,\ldots$, is the GPD.

Proof. See Consul (1989a).

Model II. Suppose the mean $\mu(\theta, \lambda)$ for the distribution of the micro-organisms is increased by changing the parameter λ to $\lambda + \Delta\lambda$ in such a way that

$$\frac{dP_0(\theta, \lambda)}{d\lambda} = 0 \tag{9.19}$$

and

$$\frac{dP_x(\theta, \lambda)}{d\lambda} = -xP_x(\theta, \lambda) + \frac{(x-1)\theta}{\theta + \lambda} P_{x-1}(\theta + \lambda, \lambda) \tag{9.20}$$

for all integral values of $x > 0$ with the initial conditions $P_x(\theta, 0) = e^{-\theta}\,\theta^x/x!$ for all values of x. Then the probability given by $P_x(\theta, \lambda), x = 0, 1, 2, \ldots$ is the GPD.

Proof. See Consul (1989a).

Queuing Process

Let $g(z)$, the pgf of a Poisson distribution with mean λ, denote the pgf of the number of customers arriving for some kind of service at a counter and let X be a random variable which denotes the number of customers already waiting for service at the counter before the service begins. Also, let $f(z)$, the pgf of another Poisson distribution with mean θ, denote the pgf of X. Consul and Shenton (1973a) showed that the number of customers Y served in a busy period of the counter is a GPD. (See more on queuing process in chapter 6.)

Branching Process

Suppose

(a) the total number of units in a group is large,
(b) the probability of acquiring a particular characteristic by a unit in the group is small,
(c) each of the units having the particular characteristic becomes a spreader of the characteristic for a short time, and
(d) the number of members in the group where each spreader having the particular characteristic is likely to spread it is also large.

Consul and Shoukri (1988) proved that the total number of individuals having the particular characteristic (i.e., the total progeny in the branching process) is the GPD. (See more on the branching process in chapter 6.)

Thermodynamic Process

Consul (1989a) described generating GPD from a thermodynamic process with forward and backward rates. Let the forward and backward rates be given, respectively, by

$$a_k = (\theta + k\lambda)^{1-k} \quad \text{and} \quad b_k = ke^{\lambda}(\theta + k\lambda)^{1-k}, \tag{9.21}$$

which become smaller and smaller in value as k increases. Under the above forward and backward rates, the steady state probability distribution of a first-order kinetic energy process is that of the GPD model.

Proof. See Consul (1989a).

9.5 Other Interesting Properties

Theorem 9.1 (Convolution Property). *The sum $X+Y$ of two independent GP variates X and Y, with parameters (θ_1, λ) and (θ_2, λ), respectively, is a GP variate with parameters $(\theta_1 + \theta_2, \lambda)$ (Consul, 1989a).*

Theorem 9.2 (Unimodality). *The GPD models are unimodal for all values of θ and λ and the mode is at $x = 0$ if $\theta e^{-\lambda} < 1$ and at the dual points $x = 0$ and $x = 1$ when $\theta e^{-\lambda} = 1$, and for $\theta e^{-\lambda} > 1$ the mode is at some point $x = M$ such that*

$$\left(\theta - e^{-\lambda}\right)\left(e^{\lambda} - 2\lambda\right)^{-1} < M < a, \tag{9.22}$$

where a is the smallest value of M satisfying the inequality

$$\lambda^2 M^2 + M\left[2\lambda\theta - (\theta + 2\lambda)\, e^{\lambda}\right] + \theta^2 > 0 \tag{9.23}$$

(Consul and Famoye (1986a)).

A number of useful relations on the derivatives, integrals, and partial sums of the GPD probabilities are given in the book by Consul (1989a).

9.6 Estimation

Let a random sample of n items be taken from the GPD and let $x_1, x_2, \ldots, x_n$ be their corresponding values. If the sample values are classified into class frequencies and n_i denotes the frequency of the ith class, the sample sum y can be written as

$$y = \sum_{j=1}^{n} x_j = \sum_{i=0}^{k} i n_i, \tag{9.24}$$

where k is the largest of the observations, $\sum n_i = n$, and $\bar{x} = y/n$ is the sample mean. The sample variance is given by

$$s^2 = (n-1)^{-1} \sum_{i=0}^{k} n_i (i - \bar{x})^2 = (n-1)^{-1} \sum_{j=1}^{n} (x_j - \bar{x})^2. \tag{9.25}$$

9.6.1 Point Estimation

Moment Estimation

Consul and Jain (1973a) gave the moment estimators in the form

$$\tilde{\theta} = \sqrt{\frac{\bar{x}^3}{s^2}} \quad \text{and} \quad \tilde{\lambda} = 1 - \sqrt{\frac{\bar{x}}{s^2}}. \tag{9.26}$$

Shoukri (1980) computed the asymptotic biases and the asymptotic variances of the moment estimators correct up to the second order of approximation. They are

$$b(\tilde{\theta}) \simeq \frac{1}{4n}\left[5\theta + \frac{3\lambda\,(2+3\lambda)}{1-\lambda}\right], \tag{9.27}$$

$$b(\tilde{\lambda}) \simeq -\frac{1}{4n\theta}\left[5\theta(1-\lambda) + \lambda\left(10 + 9\lambda^2\right)\right], \tag{9.28}$$

$$V(\tilde{\theta}) \simeq \frac{\theta}{2n}\left[\theta + \frac{2 - 2\lambda + 3\lambda^2}{1-\lambda}\right], \tag{9.29}$$

$$V(\tilde{\lambda}) \simeq \frac{1-\lambda}{2n\theta}\left[\theta - \theta\lambda + 2\lambda + 3\theta^2\right], \tag{9.30}$$

and

$$Cov(\tilde{\theta}, \tilde{\lambda}) \simeq -\frac{1}{2n}\left[\theta(1-\lambda) + 3\lambda^2\right]. \tag{9.31}$$

Bowman and Shenton (1985) stated that the ratio of the sample variance to the sample mean estimates a simple function of the GPD dispersion parameter λ. They provided moment series to order n^{-24} for related estimators and obtained exact integral formulations for the first two moments of the estimator.

Estimation Based on Sample Mean and First (Zero-Class) Frequency

When the frequency for the zero-class in the sample is larger than most other class frequencies or when the graph of the sample distribution is approximately L-shaped (or reversed J-shaped), estimates based upon the mean and zero-class frequency may be appropriate. These estimates are given by

$$\theta^* = -\ln(n_0/n) \qquad \text{and} \qquad \lambda^* = 1 - \theta^*/\bar{x}. \tag{9.32}$$

Their variances and covariance up to the first order of approximation are

$$V(\theta^*) \simeq \frac{1}{n}\left(e^{\theta} - 1\right), \tag{9.33}$$

$$V(\lambda^*) \simeq \frac{1-\lambda}{n\theta^2}\left[(1-\lambda)\left(e^{\theta}-1\right) + \theta(2\lambda - 1)\right], \tag{9.34}$$

and

$$Cov(\theta^*, \lambda^*) \simeq -\frac{1-\lambda}{n\theta}\left(e^{\theta} - \theta - 1\right). \tag{9.35}$$

Maximum Likelihood Estimation

The ML estimate $\hat{\lambda}$ of λ is obtained by solving

$$H(\lambda) = \sum_{x=0}^{k} \frac{x(x-1)n_x}{\bar{x} + (x - \bar{x})\lambda} - n\bar{x}. \tag{9.36}$$

The ML estimate $\hat{\theta}$ of θ is obtained from

$$\hat{\theta} = \bar{x}(1 - \hat{\lambda}).$$

Consul and Shoukri (1984) proved that the ML estimates $\hat{\theta} > 0$ and $\hat{\lambda} > 0$ are unique when the sample variance is larger than the sample mean. Consul and Famoye (1988) showed that if the sample variance is less than the sample mean, the ML estimates $\hat{\theta} > 0$ and $\hat{\lambda} < 0$ are also unique.

The GPD satisfies the regularity conditions given by Shenton and Bowman (1977). The variances and covariance of the ML estimators up to the first order of approximation are

$$V(\hat{\theta}) \simeq \frac{\theta(\theta+2)}{2n}, \tag{9.37}$$

$$V(\hat{\lambda}) \simeq \frac{(\theta+2\lambda-\theta\lambda)(1-\lambda)}{2n\theta}, \tag{9.38}$$

and

$$Cov(\hat{\theta}, \hat{\lambda}) \simeq -\frac{\theta(1-\lambda)}{2n}. \tag{9.39}$$

Consul and Shoukri (1984) gave the asymptotic biases as

$$b(\hat{\theta}) \simeq \frac{-\theta(5\theta^2+28\theta\lambda-6\theta\lambda^2+24\lambda^2)}{2n(1-\lambda)(\theta+2\lambda)^2(\theta+3\lambda)} \tag{9.40}$$

and

$$b(\hat{\lambda}) \simeq \frac{5\theta^3(1-\lambda)-2\theta^2\lambda\left(2\lambda^2+9\lambda-13\right)+4\theta\lambda^2(11-6\lambda)+24\lambda^2}{2n(1-\lambda)\theta(\theta+2\lambda)^2(\theta+3\lambda)}. \tag{9.41}$$

In comparing other estimators with the ML estimators, Consul (1989a) gave the asymptotic relative efficiency (ARE). The ARE for the moment estimators $\tilde{\theta}$ and $\tilde{\lambda}$ is given by

$$\text{ARE}(\tilde{\theta}, \tilde{\lambda}) = 1 - \frac{3\lambda^2}{\theta(1-\lambda)+\lambda(2+\lambda)}. \tag{9.42}$$

The ARE in (9.42) decreases monotonically as the value of λ increases, while it increases monotonically with θ. It was suggested that the moment estimators were reliable when $-0.5 < \lambda < 0.5$ and $\theta > 2$.

The ARE for the estimators based on mean and zero-class frequency is given by

$$\text{ARE}(\theta^*, \lambda^*) = \frac{\lambda+\theta/2}{\lambda+\left(e^{\theta}-1-\theta\right)\theta^{-1}} < 1. \tag{9.43}$$

For small values of θ, the estimators based on the mean and zero-class frequency will be better than the moment estimators.

Empirical Weighted Rates of Change Estimation

Famoye and Lee (1992) obtained point estimates for parameters θ and λ by using the empirical weighted rates of change (EWRC) method. Let $f_x = n_x/n$ be the observed frequency proportion for class x. The GPD likelihood equations can be written as

$$\sum_x f_x \frac{\partial}{\partial\theta_i} \ln P_x(\theta, \lambda) = 0, \quad i = 1, 2,$$

where $\theta_1 = \theta$ and $\theta_2 = \lambda$. From the fact that $\sum_x P_x(\theta, \lambda) = 1$, we obtain

$$\sum_x P_x \frac{\partial}{\partial \theta_i} \ln P_x(\theta, \lambda) = 0.$$

On combining the above with the likelihood equations, Famoye and Lee (1992) obtained the weighted discrepancies estimating equations as

$$\sum_x [f_x - P_x(\theta, \lambda)] \left[\frac{x(\theta + \lambda)}{\theta(\theta + \lambda x)} - 1 \right] = 0$$

and

$$\sum_x [f_x - P_x(\theta, \lambda)] \left[\frac{x(x-1)}{(\theta + \lambda x)} - x \right] = 0 \, .$$

The score function

$$\frac{\partial}{\partial \theta_i} \ln P_x(\theta, \lambda)$$

is viewed as the relative rates of change in the probabilities as the parameters θ and λ change. This score function is being weighted by the relative frequency in the case of the MLE method and is weighted by the discrepancy between the relative frequency and the estimated probability in the case of the weighted discrepancies estimation method. Famoye and Lee (1992) considered the combination of these two methods to define the EWRC estimators. The EWRC estimating equations are

$$\sum_x f_x [f_x - P_x(\theta, \lambda)] \left[\frac{x(\theta + \lambda)}{\theta(\theta + \lambda x)} - 1 \right] = 0 \tag{9.44}$$

and

$$\sum_x f_x [f_x - P_x(\theta, \lambda)] \left[\frac{x(x-1)}{(\theta + \lambda x)} - x \right] = 0 \, . \tag{9.45}$$

The bias under the EWRC estimation is as small or smaller than the bias from ML and moment estimation methods.

Lee and Famoye (1996) applied several methods to estimate the GPD parameters for fitting the number of chromosome aberrations under different dosages of radiations. They compared the methods of moments, ML, minimum chi-square, weighted discrepancy, and EWRC. They found that the EWRC method provided the smallest mean square error and mean absolute error for most dosages of radiation.

9.6.2 Interval Estimation

When the parameter λ is fixed at λ_0 in a small sample, Famoye and Consul (1990) showed that a $100(1-\alpha)\%$ CI (θ_ℓ, θ_u) for θ can be obtained by solving for θ_u and θ_ℓ in equations

$$\sum_{j=0}^{y} n\theta_u (n\theta_u + j\lambda_o)^{j-1} \, e^{-n\theta_u - j\lambda_o}/j! = \frac{\alpha}{2} \tag{9.46}$$

and

$$\sum_{j=y}^{\infty} n\theta_\ell (n\theta_\ell + j\lambda_o)^{j-1} e^{-n\theta_\ell - j\lambda_o}/j! = \frac{\alpha}{2}, \tag{9.47}$$

where y is the sample sum.

When the ML point estimate of θ is more than 10, a sharper $100(1-\alpha)\%$ CI for θ may be obtained by using the property of normal approximation. Thus, a $100(1-\alpha)\%$ CI for θ is given by

$$\frac{\bar{x}(1-\lambda_0)^3}{(1-\lambda_0)^2 + z_{\alpha/2}} < \theta < \frac{\bar{x}(1-\lambda_0)^3}{(1-\lambda_0)^2 - z_{\alpha/2}}. \tag{9.48}$$

For large sample size, a $100(1-\alpha)\%$ CI for θ is given by

$$\frac{\left(\bar{x} - z_{\alpha/2}s\right)(1-\lambda_0)}{\sqrt{n}} < \theta < \frac{\left(\bar{x} + z_{\alpha/2}s\right)(1-\lambda_0)}{\sqrt{n}}. \tag{9.49}$$

The statistic s in (9.49) may be dropped for $\sigma^2 = \theta(1-\lambda_0)^{-3}$. By using only the sample mean $\bar{x}$, a $100(1-\alpha)\%$ CI for θ becomes

$$\frac{\bar{x}(1-\lambda_0)^3\sqrt{n}}{(1-\lambda_0)^2\sqrt{n} + z_{\alpha/2}} < \theta < \frac{\bar{x}(1-\lambda_0)^3\sqrt{n}}{(1-\lambda_0)^2\sqrt{n} - z_{\alpha/2}}. \tag{9.50}$$

When the parameter θ is fixed at θ_0 in a small sample, a $100(1-\alpha)\%$ CI for λ can be obtained from equations (9.46) and (9.47) by replacing θ_u and θ_ℓ with θ_0 and by replacing λ_0 in (9.46) with λ_u and λ_0 in (9.47) with λ_ℓ. For large samples, a $100(1-\alpha)\%$ CI for λ, when statistics $\bar{x}$ and s are used, is given by

$$1 - \frac{\theta_0}{\bar{x} - z_{\alpha/2}s/\sqrt{n}} < \lambda < 1 - \frac{\theta_0}{\bar{x} + z_{\alpha/2}s/\sqrt{n}}. \tag{9.51}$$

If only the sample mean $\bar{x}$ is used, a $100(1-\alpha)\%$ CI for λ is given by finding the smallest value of λ that satisfies the inequality

$$[(1-\lambda)\bar{x} - \theta_0]\sqrt{n(1-\lambda)} + \sqrt{\theta_0}\, z_{\alpha/2} > 0, \tag{9.52}$$

and the largest value of λ that satisfies the inequality

$$[(1-\lambda)\bar{x} - \theta_0]\sqrt{n(1-\lambda)} - \sqrt{\theta_0}\, z_{\alpha/2} < 0. \tag{9.53}$$

The smallest and largest values of λ satisfying (9.52) and (9.53), respectively, may be determined with the help of a computer program, as given in the book by Consul (1989a).

Suppose the two parameters θ and λ are unknown and we wish to obtain CIs for one of the parameters. The parameter, which is not of interest, becomes a nuisance parameter and has to be eliminated before any inference can be made. The method of "maximization of likelihood" for eliminating the nuisance parameter can be applied. Let $\hat{\theta}$ and $\hat{\lambda}$ be the ML estimates of θ and λ, respectively. Famoye and Consul (1990) applied the method of "maximization of likelihood" and derived a U-shaped likelihood ratio statistic for determining a $100(1-\alpha)\%$ CI for θ when λ is a nuisance parameter. The statistic is given by

$$T_m(\theta) = -2n\left[\ln(\theta/\hat{\theta}) - \theta + \hat{\theta} - \bar{x}(\tilde{\lambda}(\theta) - \hat{\lambda})\right] - \sum_{i=1}^{n} 2(x_i - 1)\ln\left(\frac{\theta + \tilde{\lambda}(\theta)x_i}{\hat{\theta} + \hat{\lambda}x_i}\right). \tag{9.54}$$

A $100(1-\alpha)\%$ limits for θ are the values θ_ℓ and θ_u of θ at which the straight line

$$T_m(\theta) = \chi^2_{\alpha,1}$$

intersects the graph of the function $T_m(\theta)$ against θ. The value $\chi^2_{\alpha,1}$ is the upper percentage point of the chi-square distribution with 1 degree of freedom. The corresponding statistic for constructing a $100(1-\alpha)\%$ CI for parameter λ when θ is a nuisance parameter is given by

$$T_m(\lambda) = -2n\left[\ln(\tilde{\theta}(\lambda)/\hat{\theta}) + \hat{\theta} - \tilde{\theta}(\lambda) + \bar{x}(\hat{\lambda} - \lambda)\right] - \sum_{i=1}^{n} 2(x_i - 1)\ln\left(\frac{\tilde{\theta}(\lambda) + \lambda x_i}{\hat{\theta} + \hat{\lambda} x_i}\right). \tag{9.55}$$

9.6.3 Confidence Regions

The partial derivatives of the log likelihood function can be used to obtain an approximate chi-square expression for constructing confidence regions when the sample size is large. Famoye and Consul (1990) derived the bivariate log likelihood function

$$T(\theta, \lambda) = \frac{\theta(\theta + 2\lambda)}{n(\theta - \theta\lambda + 2\lambda)}\left\{\left[\frac{n(1-\theta)}{\theta} + \sum_{i=1}^{n}\frac{x_i - 1}{\theta + \lambda x_i}\right]^2 + \frac{1-\lambda}{2(\theta + 2\lambda)}\right.$$
$$\left.\times\left[n\theta - n\bar{x}(2 - \lambda + 2\lambda/\theta) + (1 + 2\lambda/\theta)\sum_{i=1}^{n}\frac{x_i(x_i - 1)}{\theta + \lambda x_i}\right]^2\right\}, \tag{9.56}$$

which has an asymptotic chi-square distribution with two degrees of freedom. An approximate $100(1-\alpha)\%$ confidence region for (θ, λ) is the set of values of θ and λ for which

$$T(\theta, \lambda) \le \chi^2_{\alpha,2}\,.$$

9.7 Statistical Testing

9.7.1 Test about Parameters

Consul and Shenton (1973a) showed that if the r.v. X has a GPD and if $\lambda < 0.5$, the standardized variate

$$z = \frac{X - \mu}{\sigma}$$

tends to a standard normal form as θ increases without limit. Accordingly, a test for

$$H_0 : \lambda = \lambda_0 \le 0.5 \quad \text{against} \quad H_1 : \lambda > \lambda_0$$

can be based on normal approximation. Famoye and Consul (1990) based the test on $\bar{X} = \frac{1}{n}\sum X_i$, and the critical region at a significance level α is $\bar{X} > C$, where

$$C = \theta_0(1 - \lambda_0)^{-1} + z_\alpha\sqrt{\frac{\theta_0(1-\lambda_0)^{-3}}{n}}\,. \tag{9.57}$$

The power of the test is given by

$$\pi = 1 - \beta = P(\bar{X} > c \mid H_1) .$$

In large samples, the test for θ or λ can be based on the likelihood ratio test. For a test about θ, the likelihood ratio test statistic is

$$T = -2\left[n\hat{\theta} - n\theta_0 + y\hat{\lambda} - y\tilde{\lambda}(\theta_0) + n\,\ln(\theta_0/\hat{\theta}) + \sum_{i=1}^{n}(x_i - 1)\ln\left(\frac{\theta_0 + \tilde{\lambda}(\theta_0)x_i}{\hat{\theta} + \hat{\lambda}x_i}\right)\right]. \tag{9.58}$$

The null hypothesis $H_0 : \theta = \theta_0$ (against the alternative $H_1 : \theta \neq \theta_0$) is rejected if

$$T > \chi^2_{\alpha,1} .$$

To test the hypothesis $H_0 : \lambda = \lambda_0$ against an alternative composite hypothesis $H_1 : \lambda \neq \lambda_0$, a similar likelihood ratio test statistic as in (9.58) can be used.

Famoye and Consul (1990) obtained the power of the likelihood ratio test of $H_0 : \theta = \theta_0$ against $H_1 : \theta \neq \theta_0$. The power is approximated by

$$\pi = 1 - \beta \simeq \int_a^{\infty} d\,\chi^2\left(1 + \frac{\gamma_1}{1 + 2\gamma_1}\right), \tag{9.59}$$

where $\chi^2(r)$ is a central chi-square variate with r degrees of freedom,

$$a = (1 + \gamma_1)(1 + 2\gamma_1)^{-1}\,\chi^2_{\alpha,1} \tag{9.60}$$

and

$$\gamma_1 = (\theta_1 - \theta_0)\,\frac{2n\left[\theta_1 - \theta_1\tilde{\lambda}(\theta_1) + 2\tilde{\lambda}(\theta_1)\right]}{\theta_1\left[\theta_1 + 2\tilde{\lambda}(\theta_1)\right]} . \tag{9.61}$$

In (9.61), θ_1 is the specified value of θ under the alternative hypothesis.

Fazal (1977) has considered an asymptotic test to decide whether there is an inequality between the mean and the variance in Poisson-like data. If the test indicates inequality between them, then the GPD is the appropriate model for the observed data.

9.7.2 Chi-Square Test

The goodness-of-fit test of the GPD can be based on the chi-square statistic

$$\chi^2 = \sum_{x=0}^{k}(O_x - E_x)^2/E_x, \tag{9.62}$$

where O_x and E_x are the observed and the expected frequencies for class x. The parameters θ and λ are estimated by the ML technique. The expected value E_x is computed by

$$E_x = nP_x(\theta, \lambda), \tag{9.63}$$

where n is the sample size.

The random variable χ^2 in (9.62) has an asymptotic chi-square distribution with $k - 1 - r$ degrees of freedom where r is the number of estimated parameters in the GPD.

9.7.3 Empirical Distribution Function Test

Let a random sample of size n be taken from the GPD model (9.1) and let n_x, $x = 0, 1, 2, \ldots, k$, be the observed frequencies for the different classes, where k is the largest of the observations.

An empirical distribution function (EDF) for the sample is defined as

$$F_n(x) = \frac{1}{n}\sum_{i=0}^{x} n_i, \quad x = 0, 1, 2, \ldots, k. \tag{9.64}$$

Let the GPD cdf be

$$F(x;\theta,\lambda) = \sum_{i=0}^{x} P_i(\theta,\lambda), \quad x \geq 0, \tag{9.65}$$

where $P_i(\theta,\lambda)$ is given by (9.1). The EDF statistics are those that measure the discrepancy between $F_n(x)$ in (9.64) and $F(x;\theta,\lambda)$ in (9.65). Let $\tilde{\theta}$ and $\tilde{\lambda}$ be the moment estimators of the GPD, based on the observed sample. To test the goodness-of-fit of the GPD, Famoye (1999) defined some EDF statistics analogous to the statistics defined for the continuous distributions.
(a) The Kolmogorov–Smirnov statistic

$$K_d = \sup_x \left| F_n(x) - F(x;\tilde{\theta},\tilde{\lambda}) \right|. \tag{9.66}$$

(b) The modified Cramer–von Mises statistic

$$W_d^* = n\sum_{x=0}^{k}\left[F_n(x) - F(x;\tilde{\theta},\tilde{\lambda})\right]^2 P_x(\tilde{\theta},\tilde{\lambda}).$$

(c) The modified Anderson–Darling statistic

$$A_d^* = n\sum_{x=0}^{k}\frac{\left[F_n(x) - F(x;\tilde{\theta},\tilde{\lambda})\right]^2 P_x(\tilde{\theta},\tilde{\lambda})}{F(x;\tilde{\theta},\tilde{\lambda})\left[1 - F(x;\tilde{\theta},\tilde{\lambda})\right]}.$$

The EDF test statistics use more information in the data than the chi-square goodness-of-fit test. By using the parametric bootstrap method, Famoye (1999) has carried out Monte Carlo simulations to estimate the critical values of the above three EDF test statistics and has shown that, in general, the Anderson–Darlin statistic A_d^* is the most powerful of all the EDF test statistics for testing the goodness-of-fit of the GPD model.

9.8 Characterizations

A large number of characteristic properties of the GPD have been provided in section 9.1 through section 9.5. Ahsanullah (1991a) used the property of infinite divisibility to characterize the GPD. We next consider some general probabilistic and statistical properties which lead to different characterizations of the GPD. For the proofs of these characterization theorems, the reader is referred to the book by Consul (1989a).

The following characterizations are based on the conditional probability.

Theorem 9.3. *Let X_1 and X_2 be two independent discrete r.v.s whose sum Z is a GP variate with parameters θ and λ as defined in (9.1). Then X_1 and X_2 must each be a GP variate defined over all nonnegative integers (Consul, 1974).*

Theorem 9.4. *If X_1 and X_2 are two independent GP variates with parameters (θ_1, λ) and (θ_2, λ), respectively, then the conditional probability distribution of X_1 given $X_1 + X_2 = n$ is a QBD-II (Consul, 1975).*

Theorem 9.5. *If a nonnegative GP variate N is subdivided into two components X and Y in such a way that the conditional distribution $P(X = k, \ Y = n - k \mid N = n)$ is QBD-II with parameters (n, p, θ), then the random variables X and Y are independent and have GP distributions (Consul, 1974).*

Theorem 9.6. *If X and Y are two independent r.v.s defined on the set of all nonnegative integers such that*

$$P(X = k \mid X + Y = n) = \frac{\binom{n}{k} p_n \pi_n (p_n + k\lambda)^{k-1} [\pi_n + (n-k)\lambda]^{n-k-1}}{(1 + n\lambda)^{n-1}} \tag{9.67}$$

for $k = 0, 1, 2, \ldots, n$, and zero otherwise, where $p_n + \pi_n = 1$, then

(a) *p_n is independent of n and equals a constant p for all values of n, and*
(b) *X and Y must have GP distributions with parameters $(p\alpha, \lambda\alpha)$ and $(\pi\alpha, \lambda\alpha)$, respectively, where $\alpha (> 0)$ is an arbitrary number (Consul, 1974).*

Theorem 9.7. *If X and Y are two independent nonnegative integer-valued r.v.s such that*

(a)

$$P(Y = 0 \mid X + Y = n) = \frac{\theta_1(\theta_1 + n\alpha)^{n-1}}{(\theta_1 + \theta_2)(\theta_1 + \theta_2 + n\alpha)^{n-1}}, \tag{9.68}$$

(b)

$$P(Y = 1 \mid X + Y = n) = \frac{n\theta_1\theta_2(\theta_1 + n\alpha - \alpha)^{n-2}}{(\theta_1 + \theta_2)(\theta_1 + \theta_2 + n\alpha)^{n-1}}, \tag{9.69}$$

where $\theta_1 > 0, \ \theta_2 > 0, \ 0 \le \alpha \le 1$. Then X and Y are GP variates with parameters $(\theta_1 p, \alpha p)$ and $(\theta_2 p, \alpha p)$, respectively, where p is an arbitrary number $0 < p < 1$ (Consul, 1975).

Situations often arise where the original observations produced by nature undergo a destructive process and what is recorded is only the damaged portion of the actual happenings. Consul (1975) stated and proved the following three theorems on characterizations by damage process.

Theorem 9.8. *If N is a GP variate with parameters $(\theta, \alpha\theta)$ and if the destructive process is QBD-II, given by*

$$S(k \mid n) = \binom{n}{k} \frac{p\pi}{1 + n\alpha} \left(\frac{p + k\alpha}{1 + n\alpha}\right)^{k-1} \left(\frac{\pi + (n-k)\alpha}{1 + n\alpha}\right)^{n-k-1}, \quad k = 0, 1, 2, \ldots, n, \tag{9.70}$$

where $0 < p < 1, \ \ p + \pi = 1, \ \alpha > 0$, and Y is the undamaged part of N, then

(a) *Y is a GP variate with parameters $(p\theta, \alpha\theta)$,*

(b) $P(Y = k) = P(Y = k \mid N \text{ damaged}) = P(Y = k \mid N \text{ undamaged})$, *and*
(c) $S_k = 0$ *for all k if* $S_k = P(Y = k) - P(Y = k \mid N \text{ undamaged})$ *does not change its sign for any integral value of k.*

Theorem 9.9. *Suppose that* $S(k \mid n)$ *denotes the QBD-II given by (9.70). Then* $P(Y = k) = P(Y = k \mid X \text{ undamaged})$ *if and only if* $\{P_x\}$ *is a GP variate.*

Theorem 9.10. *If a GP variate N, with parameters* $(\theta, \theta\alpha)$*, gets damaged by a destructive process* $S(k \mid n)$ *and is reduced to a variate Y such that*

$$P(Y = k) = P(Y = k \mid N \text{ undamaged}),$$

the destructive process $S(k \mid n)$ *must be QBD-II.*

Theorem 9.11. *Let* $X_1, X_2, \ldots, X_N$ *be a random sample taken from a discrete population possessing the first three moments. Let*

$$\Lambda = X_1 + X_2 + \cdots + X_N\,,$$

and let a statistic T be defined in terms of the eight subscripts $g, h, \ldots, n$*, by*

$$\begin{aligned} T &= 120 \sum X_g X_h \cdots X_m \left[28 X_n + (N-7)(14 - 3X_m)\right] \\ &- 20(N-6)^{(2)} \sum X_g \cdots X_k X_\ell^2 \left[X_\ell - 3X_k + 6\right] \\ &+ 6(N-5)^{(3)} \sum X_g X_h X_i X_j^2 X_k^2 \left[2X_k + 2 - 3X_i\right] \\ &- (N-4)^{(4)} \sum X_g X_h \cdots X_i^2 X_j^2 \left[X_n X_j - \frac{1}{3} X_i X_j + 2X_n - 18 X_g X_n\right], \end{aligned} \tag{9.71}$$

where $(N)^{(j)} = N!/(N-j)!$ *and the summations are taken over all subscripts* $g, h, i, \ldots, n$ *which are different from each other and vary from 1 to N. Then the population must be a GPD if and only if the statistic T has a zero regression on the statistic* Λ *(Consul and Gupta, 1975).*

9.9 Applications

Chromosomes are damaged one or more times in the production process, and zero or more damages are repaired in the restitution process. The undamaged chromosomes form a queue in the production process and the damaged ones form a queue in the restitution process. Janardan and Schaeffer (1977) have shown that if X is the net number of aberrations (damaged chromosomes) awaiting restitution in the queue, then the probability distribution of the r.v. X is the GPD given by (9.1). It was suggested that the parameter λ in the GPD represented an equilibrium constant which is the limit of the ratio of the rate of induction to the rate of restitution, and thus the GPD could be used to estimate the net free energy for the production of induced chromosome aberrations.

Consul (1989a) described the application of GPD to the number of chromosome aberrations induced by chemical and physical agents in human and animal cells. Janardan, Schaeffer, and DuFrain (1981) observed that a three-parameter infinite mixture of Poisson distributions is slightly better than the GPD when the parameters are estimated by the moment method. Consul (1989a) used the ML estimates and found that the fit by the GPD model was extremely good.

Schaeffer et al. (1983) provided a formal link between the GPD and the thermodynamic free energy by using a Markov-chain model and estimated that the free energy required to produce isochromatid breaks or dicentrics is about 3.67 KJ/mole/aberration and 18.4 KJ/mole, which is in good agreement with free energy estimates on the formation of DNA. A detailed description of this model can be studied either in their paper or in Consul (1989a), where many other models and applications are also given.

Janardan, Kerster, and Schaeffer (1979) considered sets of data on spiders and sow bugs, weevil eggs per bean, and data on sea urchin eggs. They showed that the observed patterns can be easily explained and described by the GPD models. Interpretative meanings were given to the parameters θ and λ in GPD.

Consul (1989a) also described the use of GPD to study shunting accidents, home injuries, and strikes in industries. Meanings were given to both parameters θ and λ in the applications.

The number of units of different commodities purchased by consumers over a period of time follows the GPD model. Consul (1989a) suggested the following interpretations for the parameter values: θ reflects the basic sales potential for the product and λ represents the average rates of liking generated by the product among consumers.

Other important applications discussed by Consul (1989a) are references of authors, spatial patterns, diffusion of information, and traffic and vehicle occupancy.

Tripathi, Gupta, and Gupta (1986) have given a very interesting use of the GPD in the textile manufacturing industry. Since the Poisson distribution, a particular case of the GPD, is generally used in the industry, they compared it with the GPD for increasing the profits. Let X be a random variable which represents the characteristic of a certain product and let x be its observed value. The profit $P(x)$ equals the amount received for good product plus the amount received for scrap (unusable) product minus the manufacturing cost of the total product and minus the fixed cost. They took $\mathrm{E}[P(X)]$ and obtained the condition for its maximization. By considering different values for the various parameters in the problem they found that in each case the profits were larger when the GPD was used instead of the Poisson distribution.

Itoh, Inagaki, and Saslaw (1993) showed that when clusters are from Poisson initial conditions, the evolved Eulerian distribution is generalized Poisson. The GPD provides a good fit to the distribution of particle counts in randomly placed cells, provided the particle distributions evolved as a result of gravitational clustering from an initial Poisson distribution. In an application in astrophysics, Sheth (1998) presented a derivation of the GPD based on the barrier crossing statistics of random walks associated with Poisson distribution.

9.10 Truncated Generalized Poisson Distribution

Consul and Famoye (1989b) defined the truncated GPD by

$$\Pr(X = x) = P_x(\theta, \lambda)/F_m(\theta, \lambda), \qquad x = 0, 1, 2, \ldots, m, \tag{9.72}$$

and zero otherwise, where $\theta > 0$, $-\infty < \lambda < \infty$ and $F_m(\theta, \lambda)$ is given by (9.2). In (9.72), m is any positive integer less than or equal to the largest possible value of x such that $\theta + \lambda x > 0$.

When $\lambda < -\theta/2$, the truncated GPD reduces to the point binomial model with $m = 1$ and probabilities

$$P(X = 0) = \left(1 + \theta e^{-\lambda}\right)^{-1} \tag{9.73}$$

and

$$P(X=1)=\theta e^{-\lambda}\left(1+\theta e^{-\lambda}\right)^{-1}. \tag{9.74}$$

When $\lambda > -\theta/2$, the value of m can be any positive integer $\leq -\theta/\lambda$. When $0 < \lambda < 1$, the largest value of m is $+\infty$ and the truncated GPD reduces to the GPD model (9.1) since $F_m(\theta,\lambda)=1$ for $m=\infty$. When $\lambda > 1$, the quantity $P_x(\theta,\lambda)$ is positive for all integral values of x and the largest value of m is $+\infty$. However, $F_\infty(\theta,\lambda)$ is not unity.

The mean μ_m and variance σ_m^2 of the truncated GPD can be written in the form

$$\mu_m = E(X) = [F_m(\theta,\lambda)]^{-1}\sum_{x=1}^{m} xP_x(\theta,\lambda) \tag{9.75}$$

and

$$\sigma_m^2 = [F_m(\theta,\lambda)]^{-1}\sum_{x=1}^{m} x^2P_x(\theta,\lambda)-\mu_m^2. \tag{9.76}$$

Consul and Famoye (1989b) considered the ML estimation of the parameters of truncated GPD. They also obtained estimates based upon the mean and ratio of the first two frequencies.

Shanmugam (1984) took a random sample $X_i,\ i=1,2,\ldots,n$, from a positive (zero truncated or decapitated) GPD given by

$$P(X=x)=\lambda(1+\alpha x)^{x-1}\left(\theta e^{-\alpha\theta}\right)^x\Big/x!,\quad x=1,2,3,\ldots,$$

where $\lambda=\left(e^\theta-1\right)^{-1}$, and obtained a statistic, based on the sample sum $\sum_{i=1}^n X_i=k$ to test the homogeneity of the random sample.

The probability distribution of the sample sum is given by

$$P\left(\sum_{i=1}^n X_i=k\right)=\lambda^{-n}n!\,t(k,n,\alpha)\left(\theta e^{-\alpha\theta}\right)^k\Big/k!$$

for $k=n,n+1,n+2,\ldots$ and where

$$t(k,n,\alpha)=\sum_{i=0}^{k-1}\binom{k-1}{i}(\alpha k)^{k-i-1}S(n,i+1),$$

$S(n,i+1)$ being the Stirling numbers of the second kind. The conditional distribution of X_1, given $\sum_{i=1}^n X_i=k$, is then

$$P\left(X_1=x\,\middle|\,\sum_{i=1}^n X_i=k\right)=\frac{\binom{k}{x}(1+\alpha x)^{x-1}t(k-x,n-1,\alpha)}{n\,t(k,n,\alpha)},\quad x=1,2,\ldots,k-n+1.$$

Since the positive GPD is a modified power series distribution and the above expression is independent of θ, the sum $\sum_{i=1}^n X_i$ becomes a complete sufficient statistic for θ and the above expression provides an MVU estimate for the probability of the positive GPD. The above conditional distribution provides a characterization also that the mutually independent positive integer-valued r.v.s $X_1,X_2,\ldots,X_n,\ n\geq 2$, have the same positive GPD.

Shanmugam (1984) defines a statistic

$$V = \sum_{i=1}^{n} X_i^2 \quad \text{for fixed} \quad \sum_{i=1}^{n} X_i = k,$$

and shows that though the r.v.s $X_1, X_2, \ldots, X_n$, are mutually dependent on account of the fixed sum, yet they are asymptotically mutually independent and V is asymptotically normally distributed. Under the null hypothesis

$$H_0 : (X_1, X_2, \ldots, X_n) \quad \text{is a homogeneous random sample from positive GPD},$$

Shanmugam obtains complex expressions for

$$\mu_0 = \mathrm{E}\left[V \,\middle|\, \sum_{i=1}^{n} X_i = k \right], \quad \sigma_0^2 = \mathrm{Var}\left[V \,\middle|\, \sum_{i=1}^{n} X_i = k \right].$$

The null hypothesis H_0 is rejected if

$$\sigma_0^{-1} \mid V - \mu_0 \mid \geq z_{\epsilon/2},$$

where $z_{\epsilon/2}$ is the $(1 - \epsilon/2)$th percentile of the unit normal distribution.

9.11 Restricted Generalized Poisson Distribution

9.11.1 Introduction and Definition

In many applied problems, it is known in advance that the second parameter λ in the GPD model (9.1) is linearly proportional to the parameter θ (see an example in Consul, 1989a, section 2.6). Putting $\lambda = \alpha\theta$ in the model (9.1), we get the probabilities for the restricted GPD model in the form

$$P_x(\theta, \alpha\theta) = (1 + x\alpha)^{x-1}\, \theta^x\, e^{-\theta - x\alpha\theta}/x!\,, \quad x = 0, 1, 2, \ldots, \tag{9.77}$$

and zero otherwise. For the probability model in (9.77), the domain of α is $\max(-\theta^{-1}, -m^{-1}) < \alpha < \theta^{-1}$, and accordingly, the parameter α is restricted above by θ^{-1}. In (9.77), $P_x(\theta, \alpha\theta) = 0$ for $x > m$ when $\alpha < 0$. The model reduces to the Poisson distribution when $\alpha = 0$.

The mean and variance of the restricted GPD are given by

$$\mu = \theta(1 - \alpha\theta)^{-1} \quad \text{and} \quad \sigma^2 = \theta(1 - \alpha\theta)^{-3}. \tag{9.78}$$

Other higher moments can be obtained from (9.13) by replacing λ with $\alpha\theta$.

9.11.2 Estimation

The restricted GPD is a MPSD defined and discussed in chapter 7 with

$$\phi(\theta) = \theta\, e^{-\alpha\theta} \quad \text{and} \quad h(\theta) = e^{\theta}.$$

When α is known, all the results derived for the MPSD in chapter 7 hold for the restricted GPD. Thus, the ML estimation of θ, MVU estimation of θ, and its function $\ell(\theta)$ for both restricted GPD and truncated restricted GPD are similar to the results obtained for the MPSD in chapter 7.

When both parameters θ and α are unknown in (9.77), the moment estimators are

$$\tilde{\theta} = \sqrt{\frac{\bar{x}^3}{s^2}} \quad \text{and} \quad \tilde{\alpha} = \sqrt{\frac{s^2}{\bar{x}^3}} - \frac{1}{\bar{x}}. \tag{9.79}$$

Kumar and Consul (1980) obtained the asymptotic biases, variances, and covariance of the moment estimators in (9.79) as

$$b(\tilde{\theta}) \simeq \frac{\theta}{4n}\left[5 + \frac{3\alpha(2+3\alpha\theta)}{1-\alpha\theta}\right], \tag{9.80}$$

$$b(\tilde{\alpha}) \simeq \frac{-3}{4n\theta}\left[1 + \frac{2\alpha+\alpha^2\theta}{1-\alpha\theta}\right], \tag{9.81}$$

$$V(\tilde{\theta}) \simeq \frac{\theta}{2n}\left[\theta + \frac{2-2\alpha\theta+3\alpha^2\theta^2}{1-\alpha\theta}\right], \tag{9.82}$$

$$V(\tilde{\alpha}) \simeq \frac{1}{2n\theta^2}\left[1 + \frac{2\alpha+\alpha^2\theta}{1-\alpha\theta}\right], \tag{9.83}$$

and

$$Cov(\tilde{\theta}, \tilde{\alpha}) \simeq \frac{-1}{2n}\left[1 + \frac{2\alpha+\alpha^2\theta}{1-\alpha\theta}\right]. \tag{9.84}$$

The generalized variance of $\tilde{\theta}$ and $\tilde{\alpha}$ is given by

$$\left|\sum\right| \simeq \frac{n-1}{2n^3\theta}\left[1 - \alpha\theta + \frac{n-1}{n}\left(2\alpha + \alpha^2\theta\right)\right]. \tag{9.85}$$

For estimators based on sample mean and zero-class frequency, Consul (1989a) gave the following:

$$\theta^* = -\ln\left(\frac{n_0}{n}\right) \quad \text{and} \quad a^* = (\theta^*)^{-1} - \frac{1}{\bar{x}}. \tag{9.86}$$

The variance of θ^* is the same as given in (9.33). However, the variance of α^* and the covariance of θ^* and α^* are given by

$$V(\alpha^*) \simeq \frac{1}{n\theta^4}\left[e^{\theta} - 1 - \theta(1-\alpha\theta)\right] \tag{9.87}$$

and

$$Cov(\theta^*, \alpha^*) \simeq \frac{-1}{n\theta^2}\left[e^{\theta} - 1 - \theta(1-\alpha\theta)\right]. \tag{9.88}$$

The ML estimate $\hat{\theta}$ of θ in restricted GPD is found by solving

$$\sum_{i=1}^{n} \frac{x_i(x_i-1)}{\theta(\bar{x}-x_i)+x_i\bar{x}} - n = 0 \tag{9.89}$$

iteratively. The corresponding ML estimate $\hat{\alpha}$ of α is obtained from

$$\hat{\alpha} = \hat{\theta}^{-1} - \frac{1}{\bar{x}}.$$

The uniqueness of the ML estimates for the restricted GPD has not been shown. Consul (1989a) conjectured that the estimate will be unique and that $\hat{\theta}$ will be greater than $\bar{x}$ or less than $\bar{x}$ according to whether the sample variance is greater or less than the, sample mean.

The asymptotic biases, variances, and covariance are

$$b(\tilde{\theta}) \simeq \frac{\theta(5+12\alpha)}{4n(1+3\alpha)}, \tag{9.90}$$

$$b(\hat{\alpha}) \simeq \frac{-3}{4n\theta}\,\frac{1-2\alpha}{1+3\alpha}, \tag{9.91}$$

$$V(\hat{\theta}) \simeq \frac{\theta(2+\theta)}{2n}, \tag{9.92}$$

$$V(\hat{\alpha}) \simeq \frac{1+2\alpha}{2n\theta^2}, \tag{9.93}$$

and

$$Cov(\hat{\theta}, \hat{\alpha}) \simeq \frac{-(1+2\alpha)}{2n}. \tag{9.94}$$

Let $\Phi_1 = \ln\theta$ and $\Phi_2 = \ln(1-\alpha\theta)$, so that $\Phi' = (\Phi_1, \Phi_2)$ denote the new parameters for estimation and

$$\left.\begin{aligned} \eta_1 &= \ln\mu = \ln\theta - \ln(1-\alpha\theta) = \Phi_1 - \Phi_2 \\ \eta_2 &= \ln\mu_2 = \ln\theta - 3\ln(1-\alpha\theta) = \Phi_1 - 3\Phi_2 \\ \eta_3 &= \ln(-\ln P_0) = \ln\theta = \Phi_1 . \end{aligned}\right\} \tag{9.95}$$

We note that $\ln(\bar{x})$, $\ln(s^2)$, and $\ln[-\ln(n_0/n)]$ are the sample estimates for η_1, η_2, and η_3, respectively.

By using

$$h' = \left(\ln(\bar{x}),\ \ln(s^2),\ \ln[-\ln(n_0/n)]\right)$$

and

$$\eta' = \left(\eta_1, \eta_2, \eta_3\right) \quad \text{with} \quad \eta = W\,\Phi$$

where

$$W = \begin{bmatrix} 1 & -1 \\ 1 & -3 \\ 1 & 0 \end{bmatrix},$$

Consul (1989a) derived a generalized minimum chi-square estimators $\tilde{\theta}$ and $\tilde{\alpha}$ for θ or α. The estimators are given by

$$\tilde{\theta} = \exp(\Phi_1) \quad \text{and} \quad \tilde{\alpha} = \left(1 - e^{\Phi_2}\right)/\tilde{\theta}. \tag{9.96}$$

When the sample is small and parameter α in restricted GPD is known, a $100(1-\alpha)\%$ CI for θ can be based on the statistic Y which is complete and sufficient for θ. The result is similar to the interval estimation in section 9.6 for the GPD model in (9.1).

When the two parameters θ and α are unknown in the model (9.77), the method of maximization of likelihood can be used to eliminate the nuisance parameter. In addition to this method, Famoye and Consul (1990) proposed the method of conditioning for eliminating parameter θ in the restricted GPD. This approach led to the statistic

$$T_c(\alpha) = -2\left[(n\bar{x}-1)\ln\left(\frac{1-\hat{\alpha}}{1+\alpha\bar{x}}\right) + \sum_{l=1}^{n}(x_i-1)\ln\left(\frac{1+\alpha x_i}{1+2x_i}\right)\right]. \tag{9.97}$$

The function $T_c(\alpha)$ is U-shaped and a $100(1-\alpha)\%$ CI for α can be obtained by finding the two values α_ℓ and α_u at which the straight line

$$T_c(\alpha) = \chi^2_{\alpha,1}$$

intersects the graph of $T_c(\alpha)$ in (9.97).

9.11.3 Hypothesis Testing

All the tests described in section 9.7 of this chapter are applicable to the restricted GPD model. In addition, a uniformly most powerful test for θ when α is known can be constructed. Famoye and Consul (1990) described a uniformly most powerful test for testing

$$H_0 : \theta \le \theta_0 \quad \text{against} \quad H_1 : \theta > \theta_0.$$

Consider the null hypothesis

$$H_0 : \theta = \theta_0 \quad \text{against} \quad H_1 : \theta = \theta_1 \ \ (\theta_1 > \theta_0). \tag{9.98}$$

If $X_1, X_2, \ldots$ is a sequence of independent r.v.s from the restricted GPD in (9.77) and the value of α is known, Consul (1989a) proposed a sequential probability ratio test for the hypotheses in (9.98). The test is to observe $\{X_i\}$, $i = 1, 2, \ldots, N$, successively and at any state $N \ge 1$,

(i) reject H_0 if $L(\underline{x}) \ge A$,
(ii) accept H_0 if $L(\underline{x}) \le B$,
(iii) continue observing X_{N+1} if $B < L(\underline{x}) < A$, where

$$L(\underline{x}) = \frac{\theta_1}{\theta_0}\sum_{i=1}^{N} x_i \, e^{\left(N+\alpha\sum_{i=1}^{N} x_i\right)(\theta_0-\theta_1)}. \tag{9.99}$$

The constants A and B are approximated by

$$A \simeq \frac{1-\beta_1}{\alpha_1} \quad \text{and} \quad B \simeq \frac{\beta_1}{1-\alpha_1}, \tag{9.100}$$

where α_1 and β_1 are the probabilities of type I and type II errors, respectively.

Let Z_i, $i = 1, 2, 3, \ldots, r, \ r+1$ be independent restricted GP variates with parameters (θ_i, α_i). Famoye (1993) developed test statistics to test the homogeneity hypothesis

$$H_0 : \theta_1 = \theta_2 = \cdots = \theta_{r+1} = \theta \tag{9.101}$$

against a general class of alternatives. When θ is known, the test statistic is

$$T = (1-\alpha\theta)^2 \sum_{i=1}^{r+1} \left(Z_i - \frac{\theta}{1-\alpha\theta} \right)^2 - \sum_{i=1}^{r+1} Z_i, \tag{9.102}$$

which can be approximated by a normal distribution. The mean and variance of T are

$$\mathrm{E}(T) = 0 \quad \text{and} \quad \mathrm{Var}(T) = \frac{2\theta^2\,(r+1)(1-\alpha\theta+2\alpha+3\alpha^2\theta)}{(1-\alpha\theta)^3}. \tag{9.103}$$

When θ is unknown, a test of homogeneity for the restricted GPD against a general class of alternatives is based on a large value of $\sum_{i=1}^{r+1} Z_i^2$ conditional on the sample sum $\sum_{i=1}^{r+1} Z_i = m$.

9.12 Other Related Distributions

9.12.1 Compound and Weighted GPD

Goovaerts and Kaas (1991) defined a random variable S by

$$S = X_1 + X_2 + \cdots + X_N,$$

where $X_i,\ i = 1, 2, \ldots, N$, denote the amounts of the claims under the different insurance policies and N is the number of claims produced by a portfolio of policies in a given time period. Assuming that the random variables $N, X_1, X_2, \ldots$ are mutually independent, that $X_1, X_2, \ldots, X_N$ are identically distributed r.v.s with the distribution function $F(x)$, and that N has the GPD, they obtained the distribution function of S as

$$F_S(x) = \sum_{n=0}^{\infty} F^{*n}(x)\lambda(\lambda+n\theta)^{n-1}e^{-\lambda-n\theta}/(n!)$$

and called it the compound generalized Poisson distribution (CGPD). They used a recursive method, involving Panjer's recursion, to compute the total claims distribution of S.

Ambagaspitiya and Balakrishnan (1994) have obtained the pgf $P_S(z)$ and the mgf $M_S(z)$ of the CGPD, the total claim amount distribution in terms of the Lambert's W function as

$$P_S(z) = \exp\left\{-(\lambda/\theta)\left[W(-\theta\exp(-\theta)P_X(z)) + \theta\right]\right\}$$

and

$$M_S(z) = \exp\left\{-(\lambda/\theta)\left[W(-\theta\exp(-\theta)M_X(z)) + \theta\right]\right\},$$

where $P_X(z),\ M_X(z)$, and W are defined in section 9.2. They have also derived an integral equation for the probability distribution function of CGPD, when the insurance claim severities are absolutely continuous and have given a recursive formula for the probability function of CGPD.

By using Rao's (1965) definition of a weighted distribution

$$f_x(\theta,\lambda) = \frac{\omega(\theta+\lambda x,\lambda)P_x(\theta,\lambda)}{W(\theta,\lambda) = \sum_x \omega(\theta+\lambda x,\lambda)P_x(\theta,\lambda)},$$

Ambagaspitiya (1995) has defined a weighted GPD as

$$P_x(\theta, \lambda) = \frac{W(\theta+\lambda, \lambda)}{W(\theta, \lambda)} \frac{\theta}{\theta+\lambda} \left(\lambda + \frac{\theta}{x}\right) P_{x-1}(\theta+\lambda, \lambda)$$

for $x = 1, 2, 3, \ldots$. When $\omega(\theta, \lambda) = \theta$, the above reduces to the linear function Poisson distribution (Jain, 1975a). He obtained the pgf of the weighted GPD above and showed that it satisfies the convolution property.

Also, Ambagaspitiya (1995) considered a discrete distribution family with the property

$$P_x(a, b) = \left[h_1(a, b) + x^{-1} h_2(a, b)\right] P_{x-1}(a+b, b), \quad x = 1, 2, 3, \ldots,$$

and showed that the weighted GPD, with weight of the form $\omega(\theta + \lambda x, \lambda)$, forms a subclass of this family. A recursive formula for computing the distribution function has been provided.

9.12.2 Differences of Two GP Variates

Suppose that X has the distribution $P_x(\theta_1, \lambda)$ and Y has the distribution $P_y(\theta_2, \lambda)$ and X and Y are independent. Consul (1986) showed that the probability distribution of $D = X - Y$ is

$$P(D = X - Y = d) = e^{-\theta_1 - \theta_2 - d\lambda} \sum_{y=0}^{\infty} (\theta_1, \lambda)_{y+d} (\theta_2, \lambda)_y \, e^{-2y\lambda}, \tag{9.104}$$

where

$$(\theta, \lambda)_x = \frac{\theta(\theta + x\lambda)^{x-1}}{x!} \tag{9.105}$$

and d takes all integral values from $-\infty$ to $+\infty$.

The pgf of the r.v. $D = X - Y$ is

$$G(u) = \exp\left[\theta_1(z_1 - 1) + \theta_2(z_2 - 1)\right], \tag{9.106}$$

where $z_1 = u\, e^{\lambda(z_1 - 1)}$ and $z_2 = u^{-1} e^{\lambda(z_2 - 1)}$.

From (9.106), the cgf of D is

$$\psi(\beta) = \frac{\theta_1(Z_1 - \beta)}{\lambda} + \frac{\theta_2(Z_2 - \beta)}{\lambda}, \tag{9.107}$$

where $Z_1 = \beta + \lambda\left(e^{Z_1} - 1\right)$ and $Z_2 = -\beta + \lambda\left(e^{Z_2} - 1\right)$.

Consul (1989a) denoted the cumulants by L_k, $k = 1, 2, 3, \ldots$, and obtained the relation

$$(1 - \lambda)\, L_{k+1} = \left(1 + \lambda \frac{\partial}{\partial \lambda}\right)\left(2\theta_1 \frac{\partial}{\partial \theta_1} - 1\right) L_k, \qquad k = 1, 2, 3, \ldots. \tag{9.108}$$

The first four cumulants are given by

$$L_1 = \frac{\theta_1 - \theta_2}{1 - \lambda}, \tag{9.109}$$

$$L_2 = \frac{\theta_1 + \theta_2}{(1 - \lambda)^3}, \tag{9.110}$$

$$L_3 = \frac{(\theta_1 - \theta_2)(1 + 2\lambda)}{(1 - \lambda)^5}, \tag{9.111}$$

and

$$L_4 = \frac{(\theta_1 + \theta_2)(1 + 8\lambda + 6\lambda^2)}{(1-\lambda)^7}. \tag{9.112}$$

The coefficients of skewness and kurtosis for the r.v. D are

$$\beta_1 = \frac{(\theta_1 - \theta_2)^2}{(\theta_1 + \theta_2)^3} \frac{(1+2\lambda)^2}{1-\lambda} \quad \text{and} \quad \beta_2 = 3 + \frac{1 + 8\lambda + 6\lambda^2}{(\theta_1 + \theta_2)(1-\lambda)}. \tag{9.113}$$

9.12.3 Absolute Difference of Two GP Variates

Let X_1 and X_2 be two independent GP variates and let $Y = |X_1 - X_2|$ be the absolute difference. Let the probability distributions of the two variates be given by

$$\begin{aligned} P_i^{(j)} &= \frac{(1+i\lambda)^{i-1}}{i!} \left(\theta_j \, e^{-\lambda\theta_j}\right)^i e^{-\theta_j}, \qquad i = 0, 1, 2, \ldots, \qquad j = 1, 2, \\ &= C(i, \lambda)(\varphi_j)^i \, e^{-\theta_j} \end{aligned} \tag{9.114}$$

and zero otherwise, where $C(i, \lambda) = (1 + i\lambda)^{i-1}/i!$ and $\varphi_j = \theta_j \, e^{-\lambda\theta_j}$. By using

$$\varphi_j' = \frac{d\varphi_j}{d\theta_j} = (1 - \lambda\theta_j)\, e^{-\lambda\theta_j},$$

Consul (1986) showed that the probability distribution of the r.v. Y is given by

$$P(Y = k) = \begin{cases} \sum_{i=0}^{\infty} [C(i,\lambda)]^2 \, (\varphi_1.\varphi_2)^i \, e^{-\theta_1 - \theta_2}, & k = 0, \\ \sum_{i=0}^{\infty} C(i,\lambda) C(i+k, \lambda)\, e^{-\theta_1-\theta_2} \left[\varphi_1^{i+k} \varphi_2^i + \varphi_1^i \varphi_2^{i+k}\right], & k = 1, 2, 3, \ldots . \end{cases} \tag{9.115}$$

9.12.4 Distribution of Order Statistics when Sample Size Is a GP Variate

Suppose X_i, $i = 1, 2, \ldots, N$, is a random sample of size N from a population with pdf $f(x)$ and cdf $F(x)$. Suppose $Y_1, Y_2, \ldots, Y_N$ denote the corresponding order statistics and the sample size N is a restricted GP variate with probability distribution $P_x(\theta, \alpha\theta)$ in (9.77).

Let $g_j(y \mid n)$ denote the conditional pdf of the jth order statistics Y_j for a given $N = n$, let h_j be the unconditional pdf of Y_j, and let h_{ij} be the joint unconditional pdf of Y_i and Y_j. Consul (1984) showed that the unconditional pdf of the jth order statistics Y_j is given by

$$h_j = \frac{\left(\theta\, e^{-\alpha\theta}\right)^j F_j^{j-1} f_j}{(j-1)!\, e^{\theta}\, Q_j(\theta, \alpha\theta)} \sum_{r=0}^{\infty} \frac{(1 + \alpha j + \alpha r)^{r+j-1}}{r!} \left[\theta(1 - F_j)\, e^{-\alpha\theta}\right]^r, \tag{9.116}$$

where

$$Q_j(\theta, \alpha\theta) = \sum_{i=j+1}^{\infty} P_i(\theta, \alpha\theta). \tag{9.117}$$

Also, the joint pdf of Y_i and Y_j, $i < j$, is given by

$$h_{ij} = \frac{F_j^{i-1} \left(F_j - F_i\right)^{j-i-1} f_i \, f_j \left(\theta\, e^{-\alpha\theta}\right)^j}{(i-1)!(j-i-1)!\, e^{\theta}\, Q_j(\theta, \alpha\theta)} \sum_{r=0}^{\infty} \frac{(1 + \alpha j + \alpha r)^{r+j-1}}{r!} \left(1 - F_j\right)^r \theta^r e^{-r\alpha\theta}, \tag{9.118}$$

where F_i and F_j are the distribution functions of Y_i and Y_j, respectively.

9.12.5 The Normal and Inverse Gaussian Distributions

Let X be a GP variate with parameters θ and λ. Consul and Shenton (1973a) showed that for all values of λ, the random variable

$$Z = \frac{X - \mu}{\sigma} \tag{9.119}$$

approaches the standard normal curve as θ becomes infinitely large. When $-0.5 < \lambda < 0.2$, a value of θ such as 15 makes the GPD model approximately normal.

If X is a GP variate with mean μ and variance σ^2, Consul and Shenton (1973a) showed that the distribution of

$$Y = X/\sigma$$

approaches the inverse Gaussian density function with mean c and variance 1 when $\theta \to \infty$ and $\lambda \to 1$ such that the product $\theta(1 - \lambda) = c^2$.

9.13 Exercises

9.1 Let X be a discrete random variable which has a generalized Poisson distribution with parameters $(\theta t, \lambda t)$. If $E(X - \mu_1')^k$ is denoted by μ_k, show that

$$\mu_{k+1} = k\theta(1 - \lambda)^{-3}\, \mu_{k-1} + (1 - \lambda)^{-1} \left\{ \frac{d}{dt}\, \mu_k(t) \right\}_{t=1} .$$

By using the above relation, verify the first six central moments given by equations (9.9) and (9.13).

9.2 If the kth negative integer moment of the GPD is denoted by

$$\Phi_k(\theta, r) = E\left[(X + r)^{-k} \right],$$

show that

$$\Phi_2(\theta,\, \theta/\lambda) = \frac{\lambda^2}{\theta^2} - \frac{\lambda^3}{\theta(\theta + \lambda)} - \frac{\lambda^3}{(\theta + \lambda)^2} + \frac{\lambda^4}{(\theta + \lambda)(\theta + 2\lambda)} .$$

Find a corresponding expression for the restricted GPD.

9.3 Suppose the probability distribution of finding X bacteria in a given space is denoted by $P_x(\theta, \lambda)$. Suppose further that the mean $\mu(\theta, \lambda)$ of X is increased by changing the parameter λ to $\lambda + \Delta\lambda$ in such a way that

$$\frac{dP_0(\theta, \lambda)}{d\lambda} = 0$$

and

$$\frac{dP_x(\theta, \lambda)}{d\lambda} = -x\, P_x(\theta, \lambda) + \frac{(x - 1)\theta}{\theta + \lambda}\, P_{x-1}(\theta + \lambda, \lambda)$$

for all integral values of $x > 0$ with the initial conditions $P_x(\theta, 0) = e^{-\theta}\, \theta^x / x!$ for all values of x. Show that $P_x(\theta, \lambda)$ is a GPD.

9.4 Suppose the initial number k of customers is a Poisson random variable with mean θ per unit service interval and the subsequent arrivals are also Poissonian with mean λ per unit service interval. Prove that the probability distribution of the number of customers served in the first busy period of a single server is the GPD model with parameters (θ, λ).

9.5 Verify the asymptotic biases, variances, and covariance of the moment estimators as given in the results (9.27)–(9.31).

9.6 If a nonnegative GP variate Z is subdivided into two components X and Y in such a way that the conditional distribution $P(X = k, Y = z - k|Z = z)$ is QBD-II with parameters (z, p, θ), show that the random variables X and Y are independent and that they have GP distributions.

9.7 Suppose X follows the restricted GPD with parameters θ and α. Show that a recurrence relation between the noncentral moments is given by

$$\mu'_{k+1} = \mu'_1 \left\{ \mu'_k + \frac{d\mu'_k}{d\theta} \right\}, \qquad k = 0, 1, 2, \ldots .$$

Also, obtain a corresponding recurrence relation between the central moments

$$\mu_k, \quad k = 2, 3, 4, \ldots .$$

9.8 Suppose that the probability of buying a product by a person is small and the number of persons is very large. If each buyer of the product becomes its advertiser for a short time in his or her town, which has a large population, show by using the principle of branching process that the total number of persons who will become advertisers will be given by the GPD.

9.9 Draw the graphs of the generalized Poisson distribution for the following sets of parameter values:
(a) $\theta = 8$, $\lambda = -0.1$; (b) $\theta = 8$, $\lambda = 0.2$; (c) $\theta = 8$, $\lambda = 0.8$;
(d) $\theta = 8$, $\lambda = -2.0$; (e) $\theta = 16$, $\lambda = -2.0$; (f) $\theta = 16$, $\lambda = -3.0$.

9.10 A textile mill produces bolts of cloth of length L. Let X be the actual length of each bolt. If $X \geq 1$, the bolt is sold for \$$A$ and if $X < L$, the bolt is sold as scrap at a price sx, where s is fixed and x is the observed value of X. If the production cost is $c_0 + cx$ dollars, where c_0 and c are the cost constants. Find the expectation $\mathrm{E}(P(X))$ where $P(X)$ is the profit function of the bolt of cloth when
(a) X is a Poisson random variable with mean θ, and
(b) X is a GP random variable with parameters θ and λ.
Find the maximum value of $\mathrm{E}(P(X))$ as θ increases in the two cases. (Hint: See Tripathi, Gupta, and Gupta, 1986.)

9.11 Use the recurrence relation in Exercise 9.7 to determine the mean, the second, third, and fourth central moments of the restricted GPD. Obtain a measure of skewness and a measure of kurtosis. Determine the parameter values for which the restricted GPD is negatively skewed, positively skewed, leptokurtic, and platykurtic.

9.12 Suppose $s = 0$, $\beta = m - 1$, and n are very large, and $b/(b + \omega) = p$ is very small such that $np = \theta$ and $mp = \lambda$ in Prem distribution in (5.14). Show that the Prem distribution approaches the generalized Poisson distribution in (9.1).

10

Generalized Negative Binomial Distribution

10.1 Introduction and Definition

A discrete r.v. X is said to have a *generalized negative binomial distribution (GNBD) with parameters* $\theta,\ \beta$, and m if its pmf is given by

$$P_x(\theta,\ \beta,\ m) = \frac{m}{m+\beta x}\binom{m+\beta x}{x}\theta^x(1-\theta)^{m+\beta x-x} \tag{10.1}$$

for $x = 0, 1, 2, 3, \ldots$ and zero otherwise, where $0 < \theta < 1,\ \ \beta = 0$ or $1 \le \beta \le \theta^{-1}$, and $m > 0$. Also, when $\beta = 0$, the parameter m is a positive integer. The probability model (10.1) reduces to the binomial distribution when $\beta = 0$ and to the negative binomial distribution when $\beta = 1$. Johnson, Kotz, and Kemp (1992) have given the model under the title of Lagrangian "generalized negative binomial distribution."

The GNBD model was defined, studied, and applied by Jain and Consul (1971), however in their definition the domain of the parameter β included negative values and values in (0, 1) as well. Consul and Gupta (1980) have shown that the parameter cannot take values in (0, 1). For negative values of β the GNBD gets truncated and the probabilities do not sum to unity. In an unpublished study Consul and Famoye (1985) have shown that for small negative values of β such that the values of the probability mass function (10.1) are positive for at least $x = 0, 1, 2, 3, 4$, the truncation error is less than 5%. This error is much smaller than the truncation error due to the sampling process unless very large samples (in thousands) are taken.

Famoye and Consul (1993) have defined and studied the truncated GNBD (given later in this chapter) where β can take all values in $(-\infty,\ \infty)$ and the model represents a true probability distribution.

The GNBD is a member of the class of Lagrangian distributions $L(f; g; x)$ in (2.7) and is also a member of its subclass, the MPSD. It is listed at (2) and (11) in Table 2.3. Accordingly, it possesses all the properties of the MPSD and of the Lagrangian distributions $L(f; g; x)$ discussed in the earlier chapters. According to the results in subsection 2.3.3, the mean and the variance of the GNBD are

$$\mu = m\theta(1-\theta\beta)^{-1} \quad \text{and} \quad \sigma^2 = m\theta(1-\theta)(1-\theta\beta)^{-3}, \tag{10.2}$$

which exist for $1 \le \beta < \theta^{-1}$. Also, the model exists for $\theta\beta = 1$ but its mean and variance do not exist.

10.2 Generating Functions

According to the results obtained for the Lagrangian distributions $L(f; g; x)$ in chapter 2, the pgf of the GNBD in (10.1) is

$$G(u) = f(z) = (1 - \theta + \theta z)^m, \text{ where } z = u\,(1 - \theta + \theta z)^\beta, \tag{10.3}$$

for all values of the parameters.

Another pgf for the GNBD is

$$G(u) = f(z) = (1 - \theta)^m (1 - \theta z)^{-m}, \tag{10.4}$$

where

$$z = u\,(1 - \theta)^{\beta - 1}(1 - \theta z)^{-\beta + 1}. \tag{10.5}$$

Thus, the GNBD is one of the few probability models which gets generated through two different sets of functions.

Consul and Famoye (1995) gave two other methods by which the GNBD is generated. The GNBD model can be generated through the Lagrange expansion of $(1 - \theta)^{-m}$ in powers of $u = \theta(1 - \theta)^{\beta - 1}$ under the transformation $\theta = u(1 - \theta)^{1 - \beta}$. Also, the GNBD model can be generated as a particular case of the Lagrangian Katz family discussed in chapter 12.

10.3 Moments, Cumulants, and Recurrence Relations

All the moments of the GNBD model (10.1) exist for $1 \le \beta < \theta^{-1}$. Jain and Consul (1971) obtained the first four noncentral moments by using the recurrence relation

$$M_k(m) = m\theta \sum_{j=0}^{k-1} \binom{k-1}{j} \left\{ M_j(m + \beta - 1) + \frac{\beta}{m + \beta - 1} M_{j+1}(m + \beta - 1) \right\} \tag{10.6}$$

for $k = 1, 2, 3, \ldots$, and where $M_k(m) = E(X^k)$ is a function of the parameters m, β, and θ.

The rth noncentral moment of the GNBD is

$$\mu_r' = \sum_{x=0}^{\infty} x^r \frac{m}{m + \beta x} \binom{m + \beta x}{x} \theta^x (1 - \theta)^{m + \beta x - x}. \tag{10.7}$$

On differentiating (10.7) with respect to θ, we obtain

$$\frac{d\mu_r'}{d\theta} = \frac{1 - \theta\beta}{\theta(1 - \theta)} \sum_{x=0}^{\infty} x^r P_x(\theta, \beta, m)\{x - \mu_1'\}.$$

Hence

$$\frac{\theta(1 - \theta)}{1 - \theta\beta} \frac{d\mu_r'}{d\theta} = \mu_{r+1}' - \mu_1'\mu_r',$$

and so a recurrence relation between the noncentral moments is given by

$$\mu_{r+1}' = \frac{\theta(1 - \theta)}{1 - \theta\beta} \frac{d\mu_r'}{d\theta} + \mu_1'\mu_r', \qquad r = 0, 1, 2, 3, \ldots. \tag{10.8}$$

Ali-Amidi (1978) showed that a recurrence relation between the central moments of GNBD is given by

$$\mu_k = \mu_2 \left\{ \frac{d\mu_{k-1}}{d\theta} \cdot \frac{1}{d\mu/d\theta} + (k-1)\mu_{k-2}, \right\} \tag{10.9}$$

for $k = 2, 3, 4, \ldots$.

The recurrence relation (10.9) can also be written in the form

$$\mu_{k+1} = \frac{\theta(1-\theta)}{1-\theta\beta} \frac{d\mu_k}{d\theta} + k\, \mu_2\, \mu_{k-1} \tag{10.10}$$

for $k = 1, 2, 3, \ldots$.

By using the method of differentiation one can obtain a recurrence relation (see Exercise 10.1) between the descending factorial moments of the GNBD as

$$\mu_{(r+1)} = \frac{\theta(1-\theta)}{1-\theta\beta} \frac{d\mu_r}{d\theta} + \left(r - \mu_{(1)}\right) \mu_{(r)} \tag{10.11}$$

for $r = 1, 2, 3, \ldots$.

Denoting the rth cumulant by L_r, Consul and Shenton (1975) gave the following recurrence relation between the cumulants of the GNBD model as

$$(1-\theta\beta)\, L_{r+1} = \theta(1-\theta) \frac{dL_r}{d\theta} \tag{10.12}$$

for $r = 1, 2, 3, \ldots$, where $L_1 = m\theta(1-\theta\beta)^{-1}$.

Shoukri (1980) obtained the third, fourth, fifth, and sixth central moments of the GNBD as

$$\mu_3 = m\theta(1-\theta)(1-2\theta+2\theta\beta-\theta^2\beta)(1-\theta\beta)^{-5}, \tag{10.13}$$

$$\mu_4 = 3\mu_2^2 + m\theta(1-\theta)A(1-\theta\beta)^{-7}, \tag{10.14}$$

$$\mu_5 = 10\mu_2\mu_3 + m\theta(1-\theta)B(1-\theta\beta)^{-9}, \tag{10.15}$$

and

$$\begin{aligned} \mu_6 = {} & 15\mu_2\mu_4 + 10\mu_3^2 - 30\mu_2^3 + m^2\theta^2(1-\theta)^2 C(1-\theta\beta)^{-10} \\ & + m\theta(1-\theta)(1-2\theta+8\theta\beta-7\theta^2\beta)B(1-\theta\beta)^{-11}, \end{aligned} \tag{10.16}$$

where

$$A = 1 - 6\theta + 6\theta^2 + 2\theta\beta(4 - 9\theta + 4\theta^2) + \theta^2\beta^2(6 - 6\theta + \theta^2), \tag{10.17}$$

$$\begin{aligned} B = {} & 1 - 14\theta + 36\theta^2 + 24\theta^3 + 2\theta\beta(11 - 42\theta + 28\theta^2) \\ & - \theta^2\beta(29 - 96\theta + 58\theta^2) + \theta^2\beta^2(58 - 96\theta + 29\theta^2) \\ & - 2\theta^3\beta^2(28 - 42\theta + 11\theta^2) + 2\theta^3\beta^3(12 - 9\theta + \theta^2) \\ & - \theta^4\beta^3(18 - 12\theta + \theta^2) \end{aligned} \tag{10.18}$$

and $C = \frac{dB}{d\theta}$.

By using the above moments, the measures of skewness and kurtosis become

$$\sqrt{\beta_1} = \frac{1 - 2\theta + \theta\beta(2-\theta)}{\sqrt{m\theta(1-\theta)(1-\theta\beta)}} \tag{10.19}$$

and

$$\beta_2 = 3 + \frac{A}{m\theta(1-\theta)(1-\theta\beta)}, \tag{10.20}$$

where A is given by (10.17). For any given value of θ and β, the skewness of the GNBD model decreases as the value of m increases and becomes zero when m is infinitely large. Also, for small values of β and m, the skewness is infinitely large when θ is close to unity, zero, or $1/\beta$. The skewness becomes negative when

$$\beta < (2\theta - 1)[\theta(2-\theta)]^{-1}.$$

Since

$$A = (1-\theta)^2(1 + 8\beta\theta + \beta^2\theta^2) + (1-2\theta)^2 + \theta^2(1-\beta)^2 + 4\beta^2\theta^2(1-\theta) > 0,$$

the kurtosis β_2 is always greater than 3 and so the GNBD model is leptokurtic when $\beta > 0$. The recursive relation between the negative moments of GNBD is a special case of the recursive relation between the negative moments of the MPSD discussed in chapter 7. The first three incomplete moments and the incomplete factorial moments for the GNBD are given by Tripathi, Gupta, and Gupta (1986). The recurrence relation between the incomplete moments and the incomplete factorial moments follow that of the MPSD discussed in chapter 7. Gupta and Singh (1981) provided formulas for computing the moments and factorial moments for the GNBD.

10.4 Physical Models Leading to GNBD

The GNBD is a very versatile model and has been derived by different researchers under very diverse conditions. Takács (1962) and Consul and Shenton (1975) have obtained the GNBD as queuing models while Mohanty (1966) derived it as a random walk model. Hill and Gulati (1981) obtained the GNBD as the probability of a gambler being ruined at the nth step. Some of these models are described below.

Random Walk Model

Suppose a is an integer, for example, $a = -1$ or $+1$. Consider a random walk where the particle at any stage moves either -1 with probability θ or $+1$ with probability $1-\theta$. We consider the probability of the first passage time (FPT) through $m = 1, 2, \ldots$. Let $f(x;\, \theta, a, m)$ denote the probability of FPT through m with x a's. Mohanty (1979) showed that if $a = 0$, $f(x;\, \theta, 0, m)$ is the negative binomial distribution

$$f(x;\, \theta, a, m) = \binom{m+x-1}{x}\theta^x(1-\theta)^m, \qquad x = 0, 1, 2, \ldots, \tag{10.21}$$

where $\binom{m+x-1}{x}$ is the number of paths from (0, 1) to (m, x) that do not touch the line $x = m$ except at the end. Mohanty also proved that for $a = \beta - 1$, the number of paths from (0, 0) to $(m + ax,\, x)$ that do not touch the line $x = y + m$ except at the end is given by

$$\frac{m}{m+\beta x}\binom{m+\beta x}{x}. \tag{10.22}$$

By using (10.22), the probability distribution of the FPT is given by

$$f(x;\,\theta,\beta-1,m) = \frac{m}{m+\beta x}\binom{m+\beta x}{x}\theta^x(1-\theta)^{m+\beta x-x},$$

which is the GNBD model (10.1). This is a special case of the random walk model discussed in chapter 13.

Queuing Model

Suppose that there are k customers waiting for service at a single server queue when the service is initially started. Suppose that the customers arriving during the service period of the k customers have a binomial distribution with parameters θ and β and that they are joining the queue under the condition of "first-come, first-served." The first busy period (FBP) will end when the server becomes idle. Also, suppose that the number k is a binomial random variable with parameters p and m. Consul and Shenton (1975) have shown that the probability distribution of X, the number of customers served in the FBP, is the double binomial distribution

$$P(Y=0\text{, i.e., when FBP is nonexistent}) = (1-p)^m,$$

and

$$P(Y=y) = \frac{m}{y}(1-p)^m\left[\theta(1-\theta)^{\beta-1}\right]^y \sum_{k=1}^{b}\binom{m-1}{k-1}\binom{\beta y}{y-k}\left(\frac{p(1-\theta)}{\theta(1-p)}\right)^k \tag{10.23}$$

for $y=1,2,3,\ldots,$ where $b=\min(m,y)$. A special case of the double binomial distribution in (10.23) is obtained by setting $p=\theta$. This yields the GNBD in (10.1) as a queue model.

Another special case of the queue model is the result by Takács (1962). Suppose that the customers arrive at a counter in batches of size $\beta-1$ according to a Poisson process with traffic intensity λ. The customers are served individually by a single server with i.i.d. service times distribution

$$h(w) = \begin{cases} e^{-bw}, & w>0, \\ 0, & \text{otherwise.} \end{cases} \tag{10.24}$$

The service times are independent of the arrival times. The probabilities of arrival and departure are, respectively, $\theta=\lambda(\lambda+b)^{-1}$ and $1-\theta=b(\lambda+b)^{-1}$. Takács (1962) showed that the probability that a busy period consists of $(\beta-1)x$ services is

$$P(X=x) = \frac{\beta-1}{\beta-1+\beta x}\binom{\beta-1+\beta x}{x}\theta^x(1-\theta)^{\beta-1+\beta x-x}, \tag{10.25}$$

which is a special case of the GNBD with parameters $\theta,\ \beta$, and $m=\beta-1$.

Mixture Distribution

The Lagrangian delta-binomial distribution is given in Table 2.2 as a member of delta Lagrangian distributions. Its probability function can be written as

$$P(X = x \mid N) = \frac{N}{x} \binom{\beta x}{x - N} \theta^{x-N} (1-\theta)^{N+\beta x - x}$$

for $x = N,\ N+1, \ldots$ and zero otherwise.

Suppose N is a binomial random variable with parameters θ and m $(m > 0$ and $0 < \theta < 1)$; then a binomial mixture of the delta binomial distributions is given by

$$\begin{aligned} P(X = x) &= \sum_{n=0}^{m} P(X = x \mid N = n) \cdot P(N = n) \\ &= \sum_{n=0}^{m} \frac{n}{x} \binom{\beta x}{x-n} \theta^{x-n} (1-\theta)^{n+\beta x - x} \cdot \binom{m}{n} \theta^n (1-\theta)^{m-n} \\ &= \frac{m}{m+\beta x} \binom{m+\beta x}{x} \theta^x (1-\theta)^{m+\beta x - x}, \qquad x = 0, 1, 2, \ldots, \end{aligned}$$

which is the GNBD in (10.1).

Urn Model

Let there be two urns, marked A and B. Urn A contains a fixed number of white balls and a fixed number of black balls so that the probability of drawing a black ball with replacement is θ and of drawing a white ball with replacement is $1 - \theta$. Urn B is initially empty. When a ball is drawn from urn A without being seen by the player, another ball of the same color is put in urn B. A player observes the following three conditions:

(i) The player chooses a strategy by selecting a positive integer "m" and a nonnegative integer "β."

(ii) The player is allowed to draw balls from urn A one by one, give them without seeing them to an umpire, who returns them each time to urn A and puts a ball of the same color each time in urn B. The player continues with this process of drawing balls until the number of white balls in urn B exceeds $(\beta - 1)$ times the number of black balls in urn B and will lose the game as soon as this condition is violated.

(iii) The player will be declared a winner of the game if he stops when the number of black balls and white balls in urn B are exactly x and $m + (\beta - 1)x$, respectively.
The probability of realizing condition (iii) is

$$P(X = x) = f(x, y) \cdot \theta^x (1-\theta)^y,$$

where $f(x, y)$ is the number of sequences in which the number of white balls $y = m + (\beta - 1)x$ in urn B always exceeds $(\beta - 1)x$. By using the conditions of the game, Famoye and Consul (1989b) obtained the difference equations

$$f(x, y) = \begin{cases} 0, & (\beta - 1)x \ge y, \\ f(x-1, y), & (\beta - 1)x = y - 1, \\ f(x-1, y) + f(x, y-1), & (\beta - 1)x < y - 1, \end{cases} \tag{10.26}$$

with the boundary conditions

$$f(1, 0) = f(0, y) = 1. \tag{10.27}$$

The solution of the system of equations in (10.26) is

$$f(x, y) = \frac{y - x(\beta - 1)}{y + x} \binom{y + x}{x}, \qquad y > x(\beta - 1). \tag{10.28}$$

By using $y = m + \beta x - x$ and (10.28) in (10.26), the probability that a player wins the game is given by

$$P(X = x) = \frac{m}{m + \beta x} \binom{m + \beta x}{x} \theta^x (1 - \theta)^{m + \beta x - x},$$

which is the GNBD in (10.1).

This is a special case of urn models defined and studied in chapter 5.

10.5 Other Interesting Properties

The GNBD possesses numerous interesting properties. Some of these are given below.

Convolution Property

Theorem 10.1. *The sum of two independent generalized negative binomial variates X_1 and X_2 with the parameters (θ, β, m_1) and (θ, β, m_2) is a GNB variate with the parameters $(\theta, \beta, m_1 + m_2)$ (Jain and Consul, 1971).*

Proof.

$$\begin{aligned} P(X_1 + X_2 = x) &= \sum_{j=0}^{x} P_j(\theta, \beta, m_1) \cdot P_{x-j}(\theta, \beta, m_2) \\ &= \theta^x (1 - \theta)^{m_1 + m_2 + \beta x - x} \sum_{j=0}^{x} \frac{m_1}{m_1 + \beta j} \binom{m_1 + \beta j}{j} \\ &\quad \times \frac{m_2}{m_2 + \beta(x - j)} \binom{m_2 + \beta(x - j)}{x - j}. \end{aligned} \tag{10.29}$$

By using the identity (1.85), the above gives

$$P(X_1 + X_2 = x) = \frac{m_1 + m_2}{m_1 + m_2 + \beta x} \binom{m_1 + m_2 + \beta x}{x} \theta^x (1 - \theta)^{m_1 + m_2 + \beta x - x},$$

which is a GNBD with parameters $(\theta, \beta, m_1 + m_2)$. □

Thus the GNBD possesses the convolution property and is closed under convolution. Charalambides (1986) has shown that among all Gould series distributions, the GNBD is the only distribution which is closed under convolution.

Unimodality

Lemma 10.2 (Steutel and van Harn, 1979). *Let $\{P_x\}_0^\infty$ be a probability distribution on the nonnegative integers with pgf $G(z)$ satisfying*

$$\frac{d}{dz}\log\{G(z)\} = R(z) = \sum_{k=0}^{\infty} \gamma_k z^k, \tag{10.30}$$

where the γ_k, $k = 0, 1, 2, \ldots$, are all nonnegative. Then $\{P_x\}_0^\infty$ is unimodal if $\{\gamma_x\}_0^\infty$ is nonincreasing, and $\{P_x\}_0^\infty$ is nonincreasing if and only if in addition $\gamma_0 \le 1$.

Theorem 10.3. *The GNBD is unimodal for all values of θ, β, and m (Consul and Famoye, 1986b).*

Proof. For $\beta = 0$ and $\beta = 1$, the GNBD reduces to the binomial and the negative binomial distributions whose unimodality has been well known. In this proof, we consider $\beta > 1$. The pgf of GNBD is given by (10.3). Since self-decomposability is preserved under positive powers, we consider

$$G_1(u) = (1-\theta+\theta z)^1, \quad \text{where} \quad z = u(1-\theta+\theta z)^\beta.$$

By using $f(z) = \log(1-\theta+\theta z)$ in the Lagrange expansion in (1.78), differentiating with respect to u, and changing k to $k+1$, we get

$$\frac{d}{du}\log G_1(u) = \theta(1-\theta)^{\beta-1} + \theta\sum_{k=0}^{\infty} u^k \binom{\beta k+\beta-1}{k}\theta^k(1-\theta)^{(\beta-1)(k+1)}$$

$$= \sum_{k=0}^{\infty} \gamma_k z^k.$$

Thus

$$\frac{\gamma_k}{\gamma_{k-1}} = \frac{\theta(1-\theta)^{\beta-1}}{k}\cdot\frac{(\beta k+\beta-1)!(\beta k-k)!}{(\beta k-1)!(\beta k+\beta-k-1)!} \tag{10.31}$$

for $k = 1, 2, 3, \ldots$

By using the inequalities given by Feller (1968, p. 54) on (10.31), and after much simplification, we obtain

$$\frac{\gamma_k}{\gamma_{k-1}} < \left\{\frac{(\beta-1)(\beta k+\beta-1)}{\beta(\beta k-k+\beta-1)}\right\}^\beta \cdot \frac{1}{e}\cdot\left(1+\frac{1}{\beta k-1}\right)^{\beta k}$$

$$\le \left(\frac{(\beta k+\beta-1)}{\beta(k+1)}\right)^\beta < 1.$$

So $\frac{\gamma_k}{\gamma_{k-1}} < 1$ and it follows that the sequence $\{\gamma_x\}_0^\infty$ is nonincreasing. Hence the GNBD is unimodal.

For $G_m(u) = (1-\theta+\theta z)^m$, $\gamma_0 = m\theta(1-\theta)^{\beta-1}$. If $\gamma_0 = m\theta(1-\theta)^{\beta-1} < 1$, the GNBD is nonincreasing and so the mode is at the point $x = 0$. If $\gamma_0 = m\theta(1-\theta)^{\beta-1} = 1$, the mode is at the dual points $x = 0$ and $x = 1$, as both have the same probability mass. However, if

$$\gamma_0 = m\theta(1-\theta)^{\beta-1} = m\varphi > 1,$$

the mode is at some point $x = N$ such that

$$\frac{m\varphi}{1-\varphi(2\beta-1)} < N < h, \tag{10.32}$$

where h is the value of N satisfying the inequality $\beta(\beta-1)\varphi N^2 + \{\varphi(2m\beta-m+1)-(m+2\beta-1)\}N + \varphi(m^2-1) > 0$. □

Relations between Probabilities

For computation of probabilities, the following recurrence relation between the probabilities is useful:

$$P_{x+1}(\theta,\beta,m)=\frac{m+(\beta-1)x+\beta}{x+1}\theta(1-\theta)^{\beta-1}\prod_{j=1}^{x-1}\left(1+\frac{\beta}{m+\beta x-j}\right)P_x(\theta,\beta,m) \tag{10.33}$$

for $x=1,2,3,4,\ldots$, with

$$P_0(\theta,\beta,m)=(1-\theta)^m \quad \text{and} \quad P_1(\theta,\beta,m)=m\theta(1-\theta)^{m+\beta-1}.$$

Suppose

$$F_k(\theta,\beta,m)=\sum_{x=0}^{k}P_x(\theta,\beta,m) \tag{10.34}$$

and

$$Q_k(\theta,\beta,m)=\sum_{x=k}^{\infty}P_x(\theta,\beta,m). \tag{10.35}$$

Then we obtain the following relationships between the GNBD probabilities:

$$\frac{m}{m-\beta}P_k(\theta,\beta,m-\beta)=\frac{\theta}{1-\theta}\frac{m+(\beta-1)(k-1)}{k}P_{k-1}(\theta,\beta,m), \tag{10.36}$$

$$(k+1)P_{k+1}(\theta,\beta,m)=\frac{m\theta[m+\beta+(\beta-1)k]}{(1-\theta)(m+\beta)}P_k(\theta,\beta,m+\beta), \tag{10.37}$$

$$P_k(\theta,\beta,m)=F_k(\theta,\beta,m)-F_{k-1}(\theta,\beta,m), \tag{10.38}$$

$$P_k(\theta,\beta,m)=Q_{k-1}(\theta,\beta,m)-Q_k(\theta,\beta,m), \tag{10.39}$$

$$\int_0^1 P_x(\theta,\beta,m)\,d\theta=\frac{m}{(m+\beta k)(m+\beta k+1)}, \tag{10.40}$$

$$\sum_{j=x}^{\infty}(j-x)\,P_j(\theta,\beta,m)=\sum_{j=0}^{\infty}j\,P_{j+x}(\theta,\beta,m), \tag{10.41}$$

$$P_k(\theta,\beta,m+k)=\frac{m+k}{m}P_k(\theta,\beta+1,m), \tag{10.42}$$

and

$$\theta(1-\theta)\frac{dP_k(\theta,\beta,m)}{d\theta}=(1-\theta\beta)(k-\mu)P_k(\theta,\beta,m), \tag{10.43}$$

where μ is the mean of the GNBD given in (10.2).

Completeness and Sufficiency Property

A family $\wp = \{P_\theta(X = x),\ \theta \in \Omega\}$ of probability distributions of a r.v. X is complete if for any function $\phi(x)$ satisfying

$$E_\theta\left[\phi(X)\right] = 0 \quad \text{for all } \theta \in \Omega,$$

the function $\phi(x) = 0$ for all x (except possibly on a set of probability zero), where E_θ denotes the expectation.

For the GNBD family defined by (10.1),

$$E_\theta\left[\phi(X)\right] = \sum_{x=0}^{\infty} \phi(x) \frac{m}{m+\beta x} \binom{m+\beta x}{x} \theta^x (1-\theta)^{m+\beta x - x}$$

if $E_\theta\left[\phi(X)\right] = 0$ for all θ in (0, 1). Then, assuming $\theta(1-\theta)^{\beta-1} = \psi$, the above expression implies that

$$\sum_{x=0}^{\infty} \phi(x) \frac{m}{m+\beta x} \binom{m+\beta x}{x} \psi^x = 0$$

for all ψ in $\left(0, (\beta-1)^{\beta-1}\beta^{-\beta}\right)$. Since the above sum of a power series is zero for all values of ψ, all the terms must be identically zero, i.e., $\phi(x) = 0$ for all integral values of x. Hence the family of the GNBDs defined in (10.1) is complete.

Also, if $X_i,\ i = 1, 2, \ldots, n$, represents a random sample of size n taken from the GNBD model in (10.1), the probability distribution of the sample sum $Y = \sum_{i=1}^{n} X_i$ is given by

$$\begin{aligned} P(Y = y) &= \frac{nm}{nm+\beta y} \binom{nm+\beta y}{y} \theta^y (1-\theta)^{nm+\beta y - y} \\ &= h(y)\psi_\theta(y), \end{aligned}$$

which is a product of two functions $\psi_\theta(y)$, a function of the statistic Y and θ, and of $h(y)$ which is independent of θ. Hence, by the factorization theorem, the sample sum Y is a sufficient statistic for the parameter θ.

10.6 Estimation

Let a random sample of size n be taken from the GNBD model (10.1) and let the observed values be $x_1, x_2, \ldots, x_n$. If the sample values are classified into class frequencies and n_i denotes the frequency of the ith class, the sample sum y can be written in the form

$$y = \sum_{j=1}^{n} x_j = \sum_{i=0}^{k} i n_i, \tag{10.44}$$

where k is the largest of the observations, $\sum_{i=0}^{k} n_i = n$, and $\bar{x} = y/n$ is the sample mean. The sample variance is

$$S_2 = (n-1)^{-1} \sum_{i=0}^{k} n_i (i - \bar{x})^2 = (n-1)^{-1} \sum_{j=1}^{n} (x_j - \bar{x})^2. \tag{10.45}$$

The third central moment for the sample is given by

$$S_3 = (n-1)^{-1} \sum_{i=0}^{k} n_i (i - \bar{x})^3 = (n-1)^{-1} \sum_{j=1}^{n} (x_j - \bar{x})^2. \tag{10.46}$$

10.6.1 Point Estimation

Moment Estimation

Jain and Consul (1971) gave the moment estimators of the GNBD in the form

$$\tilde{\theta} = 1 - \frac{1}{2}A + (A^2/4 - 1)^{\frac{1}{2}}, \tag{10.47}$$

where $A = -2 + (\bar{x}S_3 - 3S_2^2)^2/(\bar{x}S_2^3)$,

$$\tilde{\beta} = \left\{1 - \left(\frac{\bar{x}(1-\tilde{\theta})}{S_2}\right)^{\frac{1}{2}}\right\} \Big/ \tilde{\theta}, \tag{10.48}$$

and

$$\tilde{m} = \bar{x}(1 - \tilde{\theta}\tilde{\beta})/\tilde{\theta}. \tag{10.49}$$

Estimation Based on Moments and Zero-Class Frequency

Consul and Famoye (1995) considered the estimation method based upon zero-class frequency and the first two moments. When the frequency for the zero class in the sample is larger than most other class frequencies, estimates based upon the first two moments and the zero-class frequency may be appropriate. These estimates are obtained by solving the following equations:

$$\mu_1' = m\theta(1 - \theta\beta)^{-1} = \bar{x}, \tag{10.50}$$

$$\mu_2 = m\theta(1-\theta)(1 - \theta\beta)^{-3} = S_2, \tag{10.51}$$

and

$$P_0 = (1-\theta)^m = \frac{n_0}{n} = f_0. \tag{10.52}$$

On simplification, equations (10.50)–(10.52) lead to

$$f_1(\theta) = S_2(\ln f_0)^2/\bar{x}^3 - (1-\theta)(\ln(1-\theta))^2/\theta^2 = 0. \tag{10.53}$$

Equation (10.53) is solved iteratively to obtain θ^*, the estimate of θ based on the first two moments and the zero frequency. The initial estimate of θ can be taken to be the moment estimate of θ in (10.47). On getting θ^* from equation (10.53), one obtains the estimates for m and β as

$$m^* = \left\{(1-\theta^*)\bar{x}^3/S_2\right\}^{\frac{1}{2}} \Big/ \theta^* \tag{10.54}$$

and

$$\beta^* = \frac{1}{\theta^*} - m^*/\bar{x}. \tag{10.55}$$

Estimation Based on Moments and Ratio of Frequencies

Suppose P_1 and P_0 denote the probabilities of the "one" and "zero" classes, respectively. The ratio of the one class to the zero class is given by

$$P_1/P_0 = m\theta(1-\theta)^{\beta-1}.$$

On equating the above ratio to the corresponding sample ratio, one obtains

$$P_1/P_0 = m\theta(1-\theta)^{\beta-1} = \frac{n_1}{n_0} = f_{10}. \tag{10.56}$$

By combining equation (10.56) with the first two moment equations (10.50) and (10.51), we obtain estimates based on the first two moments and the ratio of the first two frequencies. On simplification, equations (10.50), (10.51), and (10.56) lead to

$$f_2(\theta) = \left\{\frac{2}{\theta} - \frac{2}{\theta}\left(\frac{\bar{x}(1-\theta)}{S_2}\right)^{\frac{1}{2}} - 1\right\}\ln(1-\theta) - \ln\left(S_2 f_{10}^2/\bar{x}^3\right) = 0. \tag{10.57}$$

After solving equation (10.57) iteratively to obtain $\bar{\theta}$, the estimate of θ based on the first two moments and the ratio of the first two frequencies, estimate $\bar{\beta}$ and $\bar{m}$ are given by

$$\bar{m} = \left\{(1-\bar{\theta})\bar{x}^3/S_2\right\}^{\frac{1}{2}}/\bar{\theta} \tag{10.58}$$

and

$$\bar{\beta} = \frac{1}{\bar{\theta}} - \bar{m}/\bar{x}. \tag{10.59}$$

Maximum Likelihood Estimation

Consul and Famoye (1995) considered the ML estimation method for the GNBD model. The log likelihood function of the GNBD model (10.1) is given by

$$\begin{aligned}\ell = \log L(\theta,\beta,m) &= \log\left\{\prod_{x=0}^{k}[P_x(\theta,\beta,m)]^{n_x}\right\} \\ &= (n-n_0)\log m + n\bar{x}\log\theta + n[m+(\beta-1)\bar{x}]\log(1-\theta) \\ &\quad + \sum_{x=2}^{k} n_x\left\{\sum_{i=1}^{x-1}\log(m+\beta x-i) - \log(x!)\right\}.\end{aligned} \tag{10.60}$$

On differentiating (10.60), we obtain the likelihood equations

$$\frac{\partial\ell}{\partial\theta} = \frac{n[\bar{x}-\theta(m+\beta\bar{x})]}{\theta(1-\theta)} = 0, \tag{10.61}$$

$$\frac{\partial\ell}{\partial\beta} = n\bar{x}\ \ln(1-\theta) + \sum_{x=2}^{k}\sum_{i=1}^{x-1}\frac{xn_x}{m+\beta x-i} = 0, \tag{10.62}$$

and

$$\frac{\partial \ell}{\partial m} = \frac{n - n_0}{m} + n \ln(1-\theta) + \sum_{x=2}^{k} \sum_{i=1}^{x-1} \frac{n_x}{m + \beta x - i} = 0, \tag{10.63}$$

where $\ell = \ln L(\theta, \beta, m)$. Equation (10.61) gives

$$\hat{\theta} = \bar{x}(m + \beta \bar{x})^{-1}, \tag{10.64}$$

and thus equations (10.62) and (10.63) can be written in the form

$$\frac{(n - n_0)\bar{x}}{m} - \sum_{x=2}^{k} \sum_{i=1}^{x-1} \frac{(x - \bar{x})n_x}{m + \beta x - i} = 0 \tag{10.65}$$

and

$$n\bar{x} \ln[1 - \bar{x}(m + \beta \bar{x})^{-1}] + \sum_{x=2}^{k} \sum_{i=1}^{x-1} \frac{x n_x}{m + \beta x - i} = 0. \tag{10.66}$$

The ML estimates $\hat{m}$ and $\hat{\beta}$ are obtained by solving (10.65) and (10.66) iteratively, starting with their moment estimates as the initial values. The Newton–Raphson iterative technique or some other technique can be used. Then, the value of $\hat{\theta}$ is given by (10.64).

MVU Estimation when β and m Are Known

When the parameters β and m are known in GNBD, the sample sum Y is a complete and sufficient statistic for the parameter θ. Since the GNBD is a MPSD defined in chapter 2 and discussed in chapter 7, the MVU estimators of θ and functions of θ are special cases of the MVU estimators considered in chapter 7 for the class of MPSD. Consul and Famoye (1989a) obtained the MVU estimators for μ, μ^2, and σ^2 (see Exercise 10.2) as

$$\tilde{\mu} = \frac{y}{n}, \quad y = 0, 1, 2, \ldots, \tag{10.67}$$

$$\tilde{\mu}^2 = \begin{cases} \frac{m}{n} \frac{y!}{(nm+\beta y-1)!} \sum_{i=1}^{y-1} \frac{(nm+\beta y-i-1)!}{(y-i-1)!}, & y = 2, 3, \ldots, \\ 0, & y = 0, 1, \end{cases} \tag{10.68}$$

and

$$\tilde{\sigma}^2 = \begin{cases} \frac{nm+\beta y-y}{(nm+\beta y-1)!} \frac{y!}{n} \sum_{i=0}^{y-1} \frac{\beta^i (i+1)(nm+\beta y-i-2)!}{(y-i-1)!}, & y = 1, 2, \ldots, \\ 0, & y = 0. \end{cases} \tag{10.69}$$

Gupta (1977) and Kumar and Consul (1980) obtained MVU estimators for some parametric functions of θ for the GNBD model (10.1). Let $\{\{\ell(\theta)\}\}$ denote the MVU estimator for $\ell(\theta)$. Kumar and Consul (1980) provided the following MVU estimators (see Exercise 10.3):

$$\{\{\theta^k\}\} = \frac{y!(nm + \beta k - k)(\beta y + nm - k - 1)!}{(y-k)!\; mn\; (\beta y + mn - 1)!}, \quad y \geq k \text{ and } k \in I', \tag{10.70}$$

$$\{\{(\theta(1-\theta)^{\beta-1})^k\}\} = \begin{cases} \frac{y!}{mn(mn+\beta y-1)^{(y-1)}}, & y = k, \\ \frac{y!(mn+\beta(y-k)-1)^{(y-k-1)}}{(y-k)!(mn+\beta y-1)^{(y-1)}}, & y > k, \\ 0, & \end{cases} \tag{10.71}$$

where $k \in I'$.

$$\{\{(1-\theta)^{mt}\}\} = \begin{cases} 1; & y = 0 \\ \left(\frac{n-t}{n}\right) \frac{(m(n-t)+\beta y-1)^{(y-1)}}{(mn+\beta y-1)^{(y-1)}}; & y \geq 1 \end{cases} \tag{10.72}$$

where $t(\leq n)$ is an integer, and

$$\{\{P(X=k)\}\} = \begin{cases} \frac{y!}{mn(mn+\beta y-1)^{(y-1)}}, & y = k, \\ \left(\frac{n-1}{n}\right) \frac{y^{(k)}(m(n-1)+\beta(y-k)-1)^{(y-k-1)}}{(mn+\beta y-1)^{(y-1)}}, & y > k, \end{cases} \tag{10.73}$$

where $k \in I^0$ and if $\beta = 0,\ m \geq k$.

When parameters m and β are known, the moment and the ML estimates of θ are both equal to

$$\tilde{\theta} = \hat{\theta} = \bar{x}/(m + \beta\bar{x}). \tag{10.74}$$

The bias of $\hat{\theta}$ up to order n^{-1} is

$$b(\hat{\theta}) = -\theta\beta(1-\theta)/mn, \tag{10.75}$$

and the variance of $\hat{\theta}$ up to the first order of approximation is

$$V(\hat{\theta}) = \theta(1-\theta)(1-\theta\beta)/mn. \tag{10.76}$$

The Bayes estimator of θ under a prior distribution which is beta with parameters a and b (see Exercise 10.4) is given by

$$\theta^{**} = (a+y)/(nm+\beta y+a+b). \tag{10.77}$$

Famoye (1997c) considered the estimation methods based on moments and zero-class frequency, moments and ratio of frequencies, ML, and that of minimum chi-square. Famoye (1997c) derived and compared the asymptotic relative efficiencies for these estimation methods. The minimum chi-square method is more efficient than the method based on moments and zero-class frequency and the method based on moments and ratio of frequencies. From simulation results, the method based on moments and zero-class frequency is the best when both bias and variance properties of the estimators are considered.

10.6.2 Interval Estimation

When the parameters β and m are fixed at β_0 and m_0, respectively, in a small sample, Famoye and Consul (1989a) obtained a $100(1-\alpha)\%$ CI for θ by solving for θ_ℓ and θ_u in equations

$$\sum_{x=y}^{\infty} \frac{nm_0}{nm_0+\beta_0 x} \binom{nm_0+\beta_0 x}{x} \theta_\ell^x (1-\theta_\ell)^{nm_0+\beta_0 x-x} = \frac{\alpha}{2} \tag{10.78}$$

and

$$\sum_{x=0}^{y} \frac{nm_0}{nm_0+\beta_0 x} \binom{nm_0+\beta_0 x}{x} \theta_u^x (1-\theta_u)^{nm_0+\beta_0 x-x} = \frac{\alpha}{2}. \tag{10.79}$$

When $\beta_0 = 0$, the above equations correspond to those in Johnson, Kotz, and Kemp (1992) for the binomial distribution. Similarly, $\beta_0 = 1$ provides the corresponding result for the negative binomial distribution.

When the parameters β and m are fixed and the sample is large, a $100(1-\alpha)\%$ CI can be obtained by using the method considered for the MPSD in chapter 7. By applying the method based on two statistics $\bar{X}$ and S_2, a $100(1-\alpha)\%$ CI was obtained by Famoye and Consul (1989a) as

$$\left(\frac{\bar{x} - z_{\alpha/2}S_2/\sqrt{n}}{m_0 + \beta_0(\bar{x} - z_{\alpha/2}S_2/\sqrt{n})}, \quad \frac{\bar{x} + z_{\alpha/2}S_2/\sqrt{n}}{m_0 + \beta_0(\bar{x} - z_{\alpha/2}S_2/\sqrt{n})}\right). \tag{10.80}$$

By using the method based on a single statistic $\bar{X}$, a $100(1-\alpha)\%$ CI for θ can be obtained (see Exercise 10.5).

10.7 Statistical Testing

The goodness-of-fit test of the GNBD can be based on the chi-square statistic

$$\chi^2 = \sum_{x=0}^{k} (O_x - E_x)^2 / E_x, \tag{10.81}$$

where O_x and E_x are the observed and the expected frequencies for class x and k is the largest observed value of x. The parameters θ, β, and m are estimated by the ML technique. The expected value E_x is computed by

$$E_x = nP_x(\theta, \beta, m), \tag{10.82}$$

where n is the sample size.

The r.v. χ^2 in (10.81) has a chi-square distribution with $k-1-r$ degrees of freedom where r is the number of estimated parameters in the GNBD.

Famoye (1998a) developed goodness-of-fit test statistics based on the EDF for the GNBD model. For small sample sizes, the tests are compared with respect to their simulated power of detecting some alternative hypotheses against a null hypothesis of a GNBD. The discrete Anderson–Darling-type test (defined in subsection 9.7.3) is the most powerful among the EDF tests.

When both parameters m and β are known constants and X_i, $i = 1, 2, 3, \ldots, n$, is a random sample of size n, a uniformly most powerful test for testing

$$H_0 : \theta \le \theta_0 \quad \text{against} \quad H_a : \theta > \theta_0 \tag{10.83}$$

can be constructed by using the following result in from Lehmann (1997, p. 80).

Corollary. *Let θ be a real parameter, and let X have probability density (with respect to some measure μ)*

$$P_\theta(x) = C(\theta)e^{Q(\theta)T(x)}h(x), \tag{10.84}$$

where $Q(\theta)$ is strictly monotone. Then there exists a uniformly most powerful test $\phi(x)$ for testing (10.83). If $Q(\theta)$ is increasing, the test is given by

$$\phi(x) = \begin{cases} 1, & T(x) > k, \\ \gamma, & T(x) = 0, \\ 0, & T(x) < k, \end{cases} \tag{10.85}$$

where the constant k and the quantity γ are determined by $E(\phi(X)\,|\theta_0) = \alpha$. If $Q(\theta)$ is decreasing, the inequalities are reversed.

For the GNBD model in (10.1), the random variable X is discrete and μ is a counting measure, therefore by using (10.84) for a random sample of size n, we get

$$P_\theta(\underline{x}) = e^{nm\ln(1-\theta)} e^{[\ln\theta+(\beta-1)\ln(1-\theta)]\sum x_i} \prod_{i=1}^{n} \frac{m}{m+\beta x_i}\binom{m+\beta x_i}{x_i}, \tag{10.86}$$

where $\underline{x} = (x_1, x_2, \ldots, x_n)$. By comparing (10.86) with (10.84), we note that

$$Q(\theta) = \ln\theta + (\beta - 1)\ln(1-\theta)$$

is a strictly increasing function of θ and the uniformly most powerful test is given by (10.85) with $T(x) = \sum x_i$. The quantities k and γ are determined from

$$\alpha = P\left(\sum X_i > k\,|H_0\right) + \gamma\, P\left(\sum X_i = k\,|H_0\right),$$

where the last term is only of interest if one desires randomization to obtain an exact significance level α. Quite often, statisticians ignore this last term as pointed out by Famoye and Consul (1990).

A sequential probability ratio test may be used to test the hypotheses

$$H_0 : \theta = \theta_0 \quad \text{against} \quad H_a : \theta = \theta_1. \tag{10.87}$$

Suppose $B < A$ are two constants. A sequential probability ratio test $S(B, A)$ for the test in (10.87) is defined as follows:

Observe $\{X_i\}$, $i = 1, 2, 3, \ldots, N$, sequentially, and at stage $N \geq 1$,

(i) reject H_0 if $L(\underline{x}) \geq A$,
(ii) accept H_0 if $L(\underline{x}) \leq B$,
(iii) continue by observing X_{N+1} if $B < L(\underline{x}) < A$ where

$$L(\underline{x}) = \frac{L(\theta_1, \underline{x})}{L(\theta_0, \underline{x})} = \frac{L(\theta_1;\ x_1, x_2, \ldots, x_N)}{L(\theta_0;\ x_1, x_2, \ldots, x_N)}$$

and

$$L(\theta_1, \underline{x}) = \prod_{i=1}^{N} P_{x_i}(\theta_1, \beta, m).$$

The measures of effectiveness of this test are obtained through α_0, the probability of type I error; α_1, the probability of type II error; and N, the expected sample size when H_0 is true. By using the approximations

$$B \simeq \frac{\alpha_1}{1-\alpha_0} \quad \text{and} \quad A \simeq \frac{1-\alpha_1}{\alpha_0}, \tag{10.88}$$

a sequential probability ratio test is of the form

$$K_\ell < \sum_{i=1}^{N} X_i < K_u, \tag{10.89}$$

where

$$K_\ell = \frac{\log A - \log\left[\left(\frac{1-\theta_1}{1-\theta_0}\right)^{mN}\right]}{\log\left[\frac{\theta_1}{\theta_0}\left(\frac{1-\theta_1}{1-\theta_0}\right)^{\beta-1}\right]} \tag{10.90}$$

and

$$K_u = \frac{\log B - \log\left[\left(\frac{1-\theta_1}{1-\theta_0}\right)^{mN}\right]}{\log\left[\frac{\theta_1}{\theta_0}\left(\frac{1-\theta_1}{1-\theta_0}\right)^{\beta-1}\right]}. \tag{10.91}$$

We continue to observe when (10.89) holds and the process is terminated when

$$\sum_{i=1}^{N} X_i \le K_\ell \quad \text{or} \quad \sum_{i=1}^{N} X_i \ge K_u.$$

10.8 Characterizations

A number of characteristic properties of the GNBD model are provided in sections 10.1 through 10.4. We now consider some general probabilistic and statistical properties which further characterize the GNBD model in (10.1).

Theorem 10.4. *Let X_1 and X_2 be two independent r.v.s whose sum Y is a GNB variate with parameters θ, β, and m. Then X_1 and X_2 must each be a GNB variate defined over all nonnegative integers (Famoye, 1994).*

Proof. The r.v. Y has a lattice distribution defined over all nonnegative integers. By using the arguments of Raikov (1937a, 1937b) we conclude that the r.v.s X_1 and X_2 must have lattice distributions defined over all nonnegative integers.

Denote the pgf of the random variable X_i, $i = 1, 2$, by

$$g_i(u) = \sum_{x=0}^{\infty} P_i(x) u^x, \quad |u| < 1, \tag{10.92}$$

where the probability distribution of X_i, $i = 1, 2$, is given by

$$P(X_i = x) = P_i(x).$$

Since Y is a GNB r.v. and from (10.3), its pgf is given by

$$g(u) = \Phi(z) = (1-\theta)^m (1-\theta z)^{-m}, \quad \text{where} \quad z = u(1-\theta)^{\beta-1}(1-\theta z)^{-\beta+1}.$$

Since $X + X_2 = Y$, then

$$g_1(u) \cdot g_2(u) = g(u) = \Phi(z) = (1-\theta)^m (1-\theta z)^{-m} \quad \text{with} \quad z = u(1-\theta)^{\beta-1}(1-\theta z)^{-\beta+1}. \tag{10.93}$$

By using an argument similar to that of Raikov (1937a, 1937b), the pgf of a negative binomial distribution in (10.93) can be factored into pgfs of negative binomial distributions. Thus, the factors $\Phi_1(z)$ and $\Phi_2(z)$ of $\Phi(z) = (1-\theta)^m(1-\theta z)^{-m}$ must be given by

$$\Phi_1(z) = (1-\theta)^{m_1}(1-\theta z)^{-m_1} \quad \text{and} \quad \Phi_2(z) = (1-\theta)^{m-m_1}(1-\theta z)^{-m+m_1},$$

where $m > m_1 > 0$ is an arbitrary number. Hence the pgfs of X_1 and X_2 become

$$g_1(u) = (1-\theta)^{m_1}(1-\theta z)^{-m_1} \quad \text{and} \quad g_2(u) = (1-\theta)^{m-m_1}(1-\theta z)^{-m+m_1},$$

where $z = u(1-\theta)^{\beta-1}(1-\theta z)^{-\beta+1}$. Because of the uniqueness property, the pgfs $g_1(u)$ and $g_2(u)$ must represent GNBD models. Therefore, X_1 and X_2 must be GNB variates, defined over all nonnegative integers, with respective parameters (θ, β, m_1) and $(\theta, \beta, m-m_1)$. □

Theorem 10.5. *If a nonnegative GNB variate Z is subdivided into two components X and Y such that the conditional distribution*

$$P(X=k,\ Y=z-k \mid Z=z)$$

is a generalized negative hypergeometric distribution

$$P(X=x) = \frac{\frac{m_1}{m_1+\beta x-x}\binom{m_1+\beta x-1}{x}\frac{m-m_1}{m-m_1+\beta(z-x)-(z-x)}\binom{m-m_1+\beta(z-x)-1}{z-x}}{\frac{m}{m+\beta z-z}\binom{m+\beta z-1}{z}} \tag{10.94}$$

with parameters (m, m_1, z, β), $m > m_1 > 0$, then the random variables X and Y are independent and have the GNBD (Jain and Consul, 1971).

Proof. Let Z be a GNB variate with parameters (θ, β, m). Its probability distribution is

$$P(Z=z) = \frac{m}{m+\beta z}\binom{m+\beta z}{z}\theta^z(1-\theta)^{m+\beta z-z}, \quad z = 0, 1, 2, \ldots,$$

and zero otherwise. The joint probability distribution of the r.v.s X and Y is given by the conditional probability distribution as the product

$$P(X=k,\ Y=z-k) = P(X=k,\ Y=z-k \mid Z=z)\cdot P(Z=z)$$

$$= \frac{\frac{m_1}{m_1+\beta k-k}\binom{m_1+\beta k-1}{k}\frac{m-m_1}{m-m_1+\beta(z-k)-(z-k)}\binom{m-m_1+\beta(z-k)-1}{z-k}}{\frac{m}{m+\beta z-z}\binom{m+\beta z-1}{z}}$$

$$\times \frac{m}{m+\beta z}\binom{m+\beta z}{z}\theta^z(1-\theta)^{m+\beta z-z}. \tag{10.95}$$

The result in (10.95) can be rewritten in the form

$$P(X=k,\ Y=z-k) = \frac{m_1}{m_1+\beta k}\binom{m_1+\beta k}{k}\theta^k\cdot(1-\theta)^{m_1+\beta k-k}$$

$$\times \frac{m-m_1}{m-m_1+\beta(z-k)}\binom{m-m_1+\beta(z-k)}{z-k}\theta^{z-k}(1-\theta)^{m-m_1+(\beta-1)(z-k)}$$

$$= P_k(\theta, \beta, m_1)\cdot P_{z-k}(\theta, \beta, m-m_1),$$

which is a product of two GNB probabilities corresponding to the r.v.s X and Y. Thus, the random variables X and Y are independent and have GNBD models with respective parameters (θ, β, m_1) and $(\theta, \beta, m - m_1)$. □

Theorem 10.6. *If X_1 and X_2 are two independent GNB variates with parameters (θ, β, m_1) and $(\theta, \beta, m - m_1)$, $m > m_1 > 0$, respectively, the conditional probability distribution of X_1, given $X_1 + X_2 = Z$, provides a generalized negative hypergeometric distribution with parameters (m, m_1, z, β) (Jain and Consul, 1971).*

Proof.

$$P(X_1 = x \mid X_1 + X_2 = z) = \frac{P_x(\theta, \beta, m_1) \cdot P_{z-x}(\theta, \beta, m - m_1)}{\sum_{j=0}^{z} P_j(\theta, \beta, m_1) P_{z-j}(\theta, \beta, m - m_1)}$$

$$= \frac{\frac{m_1}{m_1+\beta x-x}\binom{m_1+\beta x-1}{x}\frac{m-m_1}{m-m_1+\beta(z-x)-(z-x)}}{\frac{m}{m+\beta z-z}\binom{m+\beta z-1}{z}} \times \binom{m - m_1 + \beta(z-x) - 1}{z-x},$$

which is a generalized negative hypergeometric distribution. □

Theorem 10.7. *If X and Y are two independent nonnegative integer-valued r.v.s such that*

$$P(X = 0 \mid X + Y = z) = \frac{\frac{m-m_1}{m-m_1+\beta z-z}\binom{m-m_1+\beta z-1}{z}}{\frac{m}{m+\beta z-z}\binom{m+\beta z-1}{z}} \tag{10.96}$$

and

$$P(X=1 \mid X + Y = z) = \frac{\frac{m_1}{m_1+\beta-1}\binom{m_1+\beta-1}{1}\frac{m-m_1}{m-m_1+(\beta-1)(z-1)}\binom{m-m_1+\beta(z-1)-1}{z-1}}{\frac{m}{m+\beta z-z}\binom{m+\beta z-1}{z}}, \tag{10.97}$$

where $m > m_1 > 0$, β and z are real numbers, then X and Y are GNB variates with parameters (θ, β, m_1) and $(\theta, \beta, m - m_1)$, respectively (Famoye, 1994).

Proof. Let $P(X = x) = f(x)$ and $P(Y = y) = g(y)$. By the condition in (10.96),

$$\frac{f(0)g(z)}{\sum_{i=0}^{z} f(i)g(z-i)} = \frac{\frac{m-m_1}{m-m_1+\beta z-z}\binom{m-m_1+\beta z-1}{z}}{\frac{m}{m+\beta z-z}\binom{m+\beta z-1}{z}} \tag{10.98}$$

and by condition (10.97),

$$\frac{f(1)g(z-1)}{\sum_{i=0}^{z} f(i)g(z-i)} = \frac{\frac{m_1}{m_1+\beta-1}\binom{m_1+\beta-1}{1}\frac{m-m_1}{m-m_1+(\beta-1)(z-1)}\binom{m-m_1+\beta(z-1)-1}{z-1}}{\frac{m}{m+\beta z-z}\binom{m+\beta z-1}{z}}. \tag{10.99}$$

On dividing (10.98) by (10.99), we obtain

$$\frac{g(z)f(0)}{g(z-1)f(1)} = \frac{m - m_1 + \beta(z-1)}{m - m_1 + \beta z} \frac{\binom{m-m_1+\beta z}{z}}{\binom{m-m_1+\beta(z-1)}{z-1}} \cdot \frac{1}{m_1}.$$

Thus,

$$\frac{g(z)}{g(z-1)} = \frac{m-m_1+\beta(z-1)}{m-m_1+\beta z} \frac{\binom{m-m_1+\beta z}{z}}{\binom{m-m_1+\beta(z-1)}{z-1}} \cdot \frac{f(1)}{m_1 f(0)}$$

$$= \frac{m-m_1+\beta(z-1)}{m-m_1+\beta z} \frac{\binom{m-m_1+\beta z}{z}}{\binom{m-m_1+\beta(z-1)}{z-1}} \cdot \theta(1-\theta)^{\beta-1},$$

where

$$f(1)/[m_1 f(0)] = \theta(1-\theta)^{\beta-1}.$$

Hence

$$g(z) = \frac{m-m_1+\beta(z-1)}{m-m_1+\beta z} \frac{\binom{m-m_1+\beta z}{z}}{\binom{m-m_1+\beta(z-1)}{z-1}} \theta(1-\theta)^{\beta-1} g(z-1). \tag{10.100}$$

By using the relation (10.100) recursively, we get

$$g(y) = \frac{m-m_1}{m-m_1+\beta y} \binom{m-m_1+\beta y}{y} \left(\theta(1-\theta)^{\beta-1}\right)^y g(0).$$

By using the fact that $\sum_y g(y) = 1$ and the Lagrange expansion in (1.78), it is clear that $g(0) = (1-\theta)^{m-m_1}$. Hence,

$$P(Y=y)=g(y)=\frac{m-m_1}{m-m_1+\beta y} \binom{m-m_1+\beta y}{y} \theta^y (1-\theta)^{m-m_1+\beta y-y}, \quad y=0,1,2,\ldots.$$

Thus Y is a GNB variate with parameters θ, β, and $m - m_1$. In a similar way, it can be shown that the r.v. X is also a GNB variate with parameters θ, β, and m_1. □

Theorem 10.8. *If X and Y are two independent discrete r.v.s defined on the set of all nonnegative integers such that*

$$P(X=x|X+Y=z) = \frac{\frac{m_1}{m_1+\beta x-x}\binom{m_1+\beta x-1}{x}\frac{m_2}{m_2+(\beta-1)(z-x)}\binom{m_2+\beta(z-x)-1}{z-x}}{\frac{m_1+m_2}{m_1+m_2+\beta z-1}\binom{m_1+m_2+\beta z-1}{z}} \tag{10.101}$$

for $\beta > 1$ and $x = 0, 1, 2, \ldots, z$ and zero otherwise, then X and Y must be GNB variates with parameters (θ, β, m_1) and (θ, β, m_2), respectively (Famoye, 1994; incomplete proof given in Jain and Consul, 1971).

Proof. Let $P(X = x) = f(x) > 0$ with $\sum f(x) = 1$ and let $P(Y = y) = g(y) > 0$ with $\sum g(y) = 1$. Since X and Y are independent random variables,

$$P(X=x|X+Y=z) = \frac{f(x)g(z-x)}{\sum_{x=0}^{z} f(x)g(z-x)},$$

which is given by (10.101) for all values of $x \le z$ and for all integral values of z. For $z \ge 1$ and $0 \le x \le z$, this yields the functional relation

$$\frac{f(x)g(z-x)}{f(x-1)g(z-x-1)} = \frac{\frac{m_1}{m_1+\beta x-x}\binom{m_1+\beta x-1}{x}}{\frac{m_1}{m_1+(\beta-1)(x-1)}\binom{m_1+\beta(x-1)-1}{x-1}}$$

$$\times \frac{\frac{m_2}{m_2+(\beta-1)(z-x)}\binom{m_2+\beta(z-x)-1}{z-x}}{\frac{m_2}{m_2+(\beta-1)(z-x+1)}\binom{m_2+\beta(z-x+1)-1}{z-x+1}}. \tag{10.102}$$

Writing $z - x = 0$ in the above relation and simplifying, we get

$$\frac{f(x)}{f(x-1)} = \frac{g(1)}{m_2 g(0)} \cdot \frac{\frac{m_1}{m_1+\beta x}\binom{m_1+\beta x}{x}}{\frac{m_1}{m_1+\beta(x-1)}\binom{m_1+\beta(x-1)}{x-1}}, \tag{10.103}$$

where $g(1)/[m_2 g(0)]$ is independent of x and can be replaced by A (an arbitrary quantity). Replacing x by $x-1, x-2, \ldots, 3, 2, 1$ in (10.103) and multiplying them columnwise, we have

$$f(x) = \frac{m_1}{m_1+\beta x}\binom{m_1+\beta x}{x} A^x . f(0), \quad x = 0, 1, 2, \ldots. \tag{10.104}$$

Since $\sum f(x) = 1$, the above gives

$$\sum_{x=0}^{\infty} \frac{m_1}{m_1+\beta x}\binom{m_1+\beta x}{x} A^x = [f(0)]^{-1}. \tag{10.105}$$

The above expansion on the left side becomes the same as in (1.91) if we replace the arbitrary quantity A by another arbitrary quantity $\theta(1-\theta)^{\beta-1}$, where $0 < \theta < 1$. Thus,

$$[f(0)]^{-1} = (1-\theta)^{-m_1} \quad \text{or} \quad f(0) = (1-\theta)^{m_1}.$$

Hence, the relation (10.104) becomes

$$f(x) = \frac{m_1}{m_1+\beta x}\binom{m_1+\beta x}{x} \theta^x (1-\theta)^{m_1+\beta x-x}, \quad x = 0, 1, 2, \ldots.$$

Thus, X is a GNB variate with parameters (θ, β, m_1). Similarly, one can show that the random variable Y is also a GNB variate with parameters (θ, β, m_2). □

Let Z be a r.v. defined on nonnegative integers with the probability distribution $\{P_z\}$ and let Y be a r.v. denoting the undamaged part of the r.v. Z when it is subjected to a destructive process such that

$$P(Y = k \mid Z = z) = S(k \mid z), \qquad k = 0, 1, 2, \ldots, z.$$

Theorem 10.9. *Suppose Z is a GNB variate with parameters (θ, β, m). If the destructive process is the generalized negative hypergeometric distribution given by*

$$S(k \mid z) = \frac{\frac{m_1}{m_1+\beta k-k}\binom{m_1+\beta k-1}{k}\frac{m-m_1}{m-m_1+(\beta-1)(z-k)}\binom{m-m_1+\beta(z-k)-1}{z-k}}{\frac{m}{m+\beta z-z}\binom{m+\beta z-1}{z}} \tag{10.106}$$

for $k = 0, 1, 2, \ldots, z$ *and* $m > m_1 > 0$, *then*

(*a*) *Y, the undamaged part of Z, is a GNB variate with parameters* (θ, β, m_1),
(*b*) $P(Y = k) = P(Y = k \mid Z \text{ damaged}) = P(Y = k \mid Z \text{ undamaged})$,
(*c*) $S_k = 0$ *for all k if* $S_k = P(Y = k) - P(Y = k \mid Z \text{ undamaged})$ *does not change its sign for any integral value of k.*

(See Famoye, 1994.)

Proof. (a)

$$
\begin{aligned}
P(Y = k) &= \sum_{z=k}^{\infty} S(k \mid z) P_z(\theta, \beta, m) \\
&= \sum_{z=k}^{\infty} \frac{\frac{m_1}{m_1+\beta k-k}\binom{m_1+\beta k-1}{k}\frac{m-m_1}{m-m_1+(\beta-1)(z-k)}\binom{m-m_1+\beta(z-k)-1}{z-k}}{\frac{m}{m+\beta z-z}\binom{m+\beta z-1}{z}} \\
&\qquad \times \frac{m}{m+\beta z}\binom{m+\beta z}{z}\theta^z(1-\theta)^{m+\beta z-z} \\
&= \frac{m_1}{m_1+\beta k}\binom{m_1+\beta k}{k}\theta^k(1-\theta)^{m_1+\beta k-k}\sum_{z=k}^{\infty}\frac{m-m_1}{m-m_1+\beta(z-k)} \\
&\qquad \times \binom{m-m_1+\beta(z-k)}{z-k}\theta^{z-k}(1-\theta)^{m-m_1+(\beta-1)(z-k)} \\
&= \frac{m_1}{m_1+\beta k}\binom{m_1+\beta k}{k}\theta^k(1-\theta)^{m_1+\beta k-k} \\
&= P_k(\theta, \beta, m_1).
\end{aligned}
$$

Thus, the random variable Y is a GNB variate with parameters (θ, β, m_1).

(b)

$$
\begin{aligned}
P(Y = k \mid Z \text{ damaged}) &= \frac{\sum_{z=k}^{\infty} S(k \mid z) P_z(\theta, \beta, m)}{\sum_{k=0}^{\infty}\sum_{z=k}^{\infty} S(k \mid z) P_z(\theta, \beta, m)} \\
&= \frac{\frac{m}{m+\beta k}\binom{m+\beta k}{k}\theta^k(1-\theta)^{m+\beta k-k}}{\sum_{k=0}^{\infty}\frac{m}{m+\beta k}\binom{m+\beta k}{k}\theta^k(1-\theta)^{m+\beta k-k}} \\
&= P(Y = k),
\end{aligned}
$$

since the denominator is unity. Similarly, it can be shown that

$$
\begin{aligned}
P(Y = k \mid Z \text{ undamaged}) &= \frac{S(k \mid k) P_k(\theta, \beta, m_1)}{\sum_{k=0}^{\infty} S(k \mid k) P_k(\theta, \beta, m_1)} \\
&= P(Y = k), \quad \text{since} \quad S(k \mid k) = 1.
\end{aligned}
$$

(c) Since $\sum_{j=0}^{z} S(j \mid z) = 1$, we have

$$\frac{m}{m+\beta z - z}\binom{m+\beta z-1}{z} = \sum_{j=0}^{z} \frac{m_1}{m_1+\beta j - j}$$

$$\times \binom{m_1+\beta j-1}{j}\frac{m-m_1}{m-m_1+(\beta-1)(z-j)} \times \binom{m-m_1+\beta(z-j)-1}{z-j}.$$

Now,

$$\begin{aligned} S_k &= P(Y=k) - P(Y=k \mid z \text{ undamaged}) \\ &= \sum_{z=k}^{\infty} P_z(\theta,\beta,m)S(k \mid z) - \frac{P_k(\theta,\beta,m_1)S(k \mid k)}{\sum_{k=0}^{\infty} P_k(\theta,\beta,m_1)S(k \mid k)}. \end{aligned}$$

Summing over k from 0 to ∞, we have

$$\sum_{k=0}^{\infty} S_k = \sum_{k=0}^{\infty}\sum_{z=k}^{\infty} P_z(\theta,\beta,m)S(k \mid z) - \sum_{k=0}^{\infty}\left\{\frac{P_k(\theta,\beta,m_1)S(k \mid k)}{\sum_{k=0}^{\infty} P_k(\theta,\beta,m_1)S(k \mid k)}\right\}. \quad (10.107)$$

From the proof of (b), the second term on the right-hand side of (10.107) reduces to

$$\sum_{z=0}^{\infty} P_z(\theta,\beta,m) = 1.$$

Hence, equation (10.107) becomes

$$\begin{aligned} \sum_{k=0}^{\infty} S_k &= \sum_{k=0}^{\infty}\sum_{z=k}^{\infty} P_z(\theta,\beta,m)S(k \mid z) - 1 = \sum_{z=0}^{\infty} P_z(\theta,\beta,m)\sum_{k=0}^{\infty} S(k \mid z) - 1 \\ &= \sum_{z=0}^{\infty} P_z(\theta,\beta,m) - 1, \text{ since } \sum_{k=0}^{\infty} S(k \mid z) = 1 = 1 - 1 = 0. \end{aligned}$$

Since all $S_k,\ k = 0, 1, 2, \ldots,$ are of the same sign and their sum is zero, each of the S_k must be zero. □

Theorem 10.10. *Let X be a discrete r.v. indexed by three parameters (θ, β, m) where parameter θ is a r.v. that has beta distribution with parameters 1 and $k-1$. If a beta mixture of the probability distribution of the r.v. X is a generalized factorial distribution (generalized binomial distribution)*

$$P(X = x_*) = \frac{(k-1)(\lambda+\beta x_* - x_* - 1)^{(k-1)}}{(\lambda+\beta x_*)^{(k)}}\frac{\lambda-k+1}{\lambda-k+1+\beta x_* - x_*}, \quad (10.108)$$

then X has a GNBD with parameters $\theta,\ \beta,$ and $m = \lambda - k + 1$ (Jain and Consul, 1971).

For the proof, see the paper of Jain and Consul (1971).

Ahsanullah (1991b) used an infinite divisibility condition and a relation between the mean and variance to characterize the GNBD. A necessary and sufficient set of conditions (Katti, 1967) for a r.v. X to be infinitely divisible is that

$$\pi_i = \frac{iP_i}{P_0} - \sum_{j=1}^{i-1} \pi_{i-j} \frac{P_i}{P_0} \geq 0 \quad \text{for} \quad i = 1, 2, 3, \ldots. \tag{10.109}$$

Theorem 10.11. *A r.v. X has the GNBD with parameters θ, β, and m if and only if*

$$\pi_k = m \binom{\beta k - 1}{k - 1} \gamma^k, \quad k = 1, 2, \ldots \tag{10.110}$$

with $\gamma = \theta(1-\theta)^{\beta-1}$ and $P_0 = (1-\theta)^m$ (Ahsanullah, 1991b).

Proof. Let X be a GNBD with pmf as in (10.1), then it can be shown that (10.110) is true.

Assume (10.110) is true. Then

$$\begin{aligned}
P_1 &= \pi_1 P_0 = (1-\theta)^m \left[\theta(1-\theta)^{\beta-1}\right], \\
P_2 &= \frac{1}{2}\left[\pi_2 P_0 + \pi_1 P_1\right] \\
&= \frac{1}{2}\left[m\binom{2\beta-1}{1}\gamma^2 + m^2\gamma^2\right] P_0 \\
&= \frac{m}{m+2\beta}\binom{m+2\beta}{2}(1-\theta)^m \left[\theta(1-\theta)^{\beta-1}\right]^2, \\
P_3 &= \frac{1}{3}\left[\pi_3 P_0 + \pi_2 P_1 + \pi_1 P_2\right] \\
&= \frac{\gamma^3}{3}\left[m\binom{3\beta-1}{2} + m^2\binom{2\beta-1}{1} + m\binom{m-2\beta}{2}\frac{m}{m+2\beta}\right] P_0 \\
&= \binom{m+3\beta}{3}\frac{m}{m+3\beta}\gamma^3 P_0.
\end{aligned}$$

In general,

$$\begin{aligned}
kP_k &= \sum_{j=0}^{k-1} \pi_{k-j} P_j \\
&= m\gamma^k \sum_{j=1}^{k} \binom{m\beta - (j-1)\beta - 1}{k-j}\binom{m+(j-1)\beta}{j-1}\frac{m}{m+(j-1)\beta} P_0 \\
&= mk\gamma^k \binom{m+\beta k}{k}\frac{1}{m+\beta k} P_0.
\end{aligned}$$

Thus,

$$P_k = \binom{m+\beta k}{k}\frac{m}{m+\beta k}\gamma^k P_0,$$

which is the GNBD. □

Suppose $X_1, X_2, \ldots, X_N$ denote a random sample from a population possessing the first four moments. Also, let

$$\Lambda = \sum_{i=1}^{N} X_i$$

and a statistic T be defined in terms of the eight subscripts $g, h, \ldots, n$ by

$$\begin{aligned} T = 7! \sum X_g \ldots X_n(8 - X_n) - 240(N-7) \sum X_g \ldots X_\ell X_n^2 (9 - 5X_1 + 6X_m) \\ + 6(N-6)(N-7) \sum X_g \ldots X_k X_\ell (24X_k - 45X_j X_k + 20X_k X_\ell - 10X_\ell^2 \\ - (N-5)(N-6)(N-7) \sum X_g X_h X_i X_j^2 X_k^2 \\ \times (12X_i - 72X_h X_i + 8X_i X_k + 6X_j X_k - 3X_k^2), \end{aligned} \tag{10.111}$$

where the summations go over all subscripts $g, h, i, \ldots, n$ which are all different and vary from 1 to N. Consul and Gupta (1980) showed that the population must be a GNBD if and only if the statistic T has zero regression on the statistic Λ.

Consul (1990c) defined a new class of location-parameter discrete probability distributions (LDPDs). Furthermore, he proved that a random variable X with LDPDs satisfies

$$\sigma^2 = n(1 + an)(1 + bn), \quad a > b > 0,$$

if and only if the random variable X has a GNBD with parameters $\theta = (a-b)^n/(1+an)$, $\beta = nb + 1$, and $m = n$.

A discrete probability distribution is a GNBD if and only if the recurrence relation between its cumulants is given by

$$K_{r+1} = \frac{\theta(1-\theta)}{1-\theta\beta} \frac{\partial K_r}{\partial \theta}, \qquad r = 1, 2, 3, \ldots, \tag{10.112}$$

where $K_1 = m\theta(1 - \theta\beta)^{-1}$. This characterization is a special case of the one given by Jani (1978b) for the class of MPSD.

10.9 Applications

Queuing Process

Univariate distributions associated with Lagrange expansions have been shown by Consul and Shenton (1975) to represent the number of customers served in a busy period of a single server queuing system. The G|D|1 queue is a single server queue with arbitrary interarrival time distribution and constant service time. Let X denote the number of customers served in a busy period and let Y denote the number of arrivals during the service period of each customer. Suppose the number of customers, N, initiating the service has a binomial distribution with pgf $f(z) = (1-\theta+\theta z)^m$. Let the distribution of the r.v. Y be binomial with pgf $g(z) = (1-\theta+\theta z)^\beta$. By using Theorem 6.1, the probability distribution of X is the GNBD in (10.1).

Branching Process and Epidemics

Good (1960, 1965) showed that the class of Lagrangian probability distributions that contains the GNBD is important in the analysis of biological data and other areas where branching mechanism is involved. In a branching process, let $X_0, X_1, X_2, \ldots, X_n, \ldots$ denote the total number of objects in the zeroth, first, second, ..., nth, ... generations. Suppose $g(z) = (1 - \theta + \theta z)^\beta$ is the pgf of the probability distribution of offspring in the branching process. If X_0, is a random variable with pgf $f(z) = (1 - \theta + \theta z)^m$, then the probability distribution of the total progeny at the nth generation is given by the GNBD in (10.1).

Let X denote the total number of infected anywhere in a habitat, starting from those initially infected and up to the time of extinction of an epidemic. In section 6.4, it was shown that the probability distribution of X is a Lagrangian probability distribution. Suppose an epidemic is started with a random number X_0, which has a pgf $f(z) = (1 - \theta + \theta z)^m$. Let $g(z) = (1 - \theta + \theta z)^\beta$ be the pgf of the number ν of susceptibles who would be infected at a certain point if there is an infectious individual. The distribution of X, the total number of infected individuals before the extinction of the epidemic, is the GNBD in (10.1).

Molecular Weight Distribution in Polymers

The GNBD has an important use in chemistry in polymerization reaction where the substances formed are generally classified into unbranched linear chains and branched chains. The molecular weights distribution can be suitably represented by the GNBD. Gordon (1962) derived the distribution of branched polymers by using the cascade theory of stochastic processes. Yan (1979) also used the theory of cascade processes to obtain a univariate Stockmayer distribution for the condensation of polymer chains with an initial size distribution. The primary polymers are called "chains" and the aggregates formed after cross-linking are called "molecules."

Theories of branched polymers are generally based on the assumptions that all functional groups are equally reactive and that intramolecular reaction is excluded. Both the number and weight distributions of the chains are assumed to exist. Yan (1979) retained these assumptions to obtain the weight distribution of an x-mer molecule with x degree of polymerization.

To derive the distribution, the x-mer is considered to be a "tree" in the cascade process. The "zeroth generation" or the "root" of the tree is a randomly chosen monomeric unit of the type RA_k, where k is the number of functional groups. Let the pgf of the number of functional groups in the zeroth generation be $g_0(z) = (1 - \theta + \theta z)^k$, where θ is the fraction of reacted functional groups (or the branching probability in cascade theory) and z is related to the generating variable u by

$$z = ug_1(z), \quad 0 \leq u \leq 1, \quad 0 \leq z \leq 1,$$

where $g_1(z)$ is the pgf of the first, second, third, ..., generations. In the chemical situation, we have

$$g_0(z) \neq g_1(z) = g_2(z) = g_3(z) = \cdots$$

because the zeroth generation can produce one more offspring than each of the other generations. Thus,

$$g_1(z) = g_2(z) = \cdots = (1 - \theta + \theta z)^{k-1}.$$

The weight distribution $P(X = x)$ for a chemical condensation is given by the pgf

$$f(u) = \sum_x P(X = x)u^x = ug_0(z) = (z/g_1(z))\, g_0(z) = z\,(1 - \theta + \theta z).$$

By using the Lagrange expansion in (1.78) on $f(u)$ under the transformation $z = u\,(1-\theta+\theta z)^{k-1}$, we obtain

$$P(X=x) = \frac{k}{(k-1)x+1}\binom{(k-1)x+1}{x-1}\theta^{x-1}(1-\theta)^{(k-2)x+2}, \quad x \geq 1.$$

On replacing x by $y+1$, we get

$$P(Y=y) = \frac{k}{k+(k-1)y}\binom{k+(k-1)y}{y}\theta^{y}(1-\theta)^{k+(k-1)y-y}, \quad y \geq 0,$$

which is the GNBD in (10.1) with $m=k$ and $\beta = k-1$.

10.10 Truncated Generalized Negative Binomial Distribution

Famoye and Consul (1993) defined the truncated GNBD as

$$f_x(\theta,\beta,m) = P_x(\theta,\beta,m)/K_t, \qquad x = 0,1,2,\ldots,t, \tag{10.113}$$

and zero otherwise, where $0<\theta<1,\ m>0,\ -m<\beta<\infty$,

$$K_t = \sum_{x=0}^{t} P_x(\theta,\beta,m), \tag{10.114}$$

and $P_x(\theta,\beta,m)$ is the GNBD model in (10.1).

When $-m < \beta < \frac{1}{2}(1-m)$, the maximum value of t is 1 and so the truncated GNBD in (10.113) reduces to the point binomial probability model with two probabilities

$$P_0(\theta,\beta,m) = \left[1+m\theta(1-\theta)^{\beta-1}\right]^{-1} \tag{10.115}$$

and

$$P_1(\theta,\beta,m) = m\theta(1-\theta)^{\beta-1}\left[1+m\theta(1-\theta)^{\beta-1}\right]^{-1}. \tag{10.116}$$

When $\frac{1}{2}(1-m) < \beta < 1$, the value of t is any positive integer $\leq (m+1)(1-\beta)^{-1}$. The largest observed value of x in a random sample is usually taken as an estimate of t. When $1<\beta<\theta^{-1}$, $P_x(\theta,\beta,m)$ is positive for all integral values of x and so the largest value of t may be $+\infty$. By the Lagrange theorem, $K_t = 1$ for $t=\infty$ and so $P_x(\theta,\beta,m)$, for $x=0,1,2,\ldots$, provides a true probability distribution.

When $\beta > \theta^{-1}$, $P_x(\theta,\beta,m)$ for $x = 0,1,2,\ldots$ does not provide a true probability distribution. Consul and Gupta (1980) showed that a quantity q_0, say, can be determined such that $P_x(\theta q_0, \beta q_0, m q_0)$, $x=0,1,2,\ldots$, is a true probability distribution.

The ratio of the first two probabilities in (10.113) is

$$\frac{f_1(\theta,\beta,m)}{f_0(\theta,\beta,m)} = m\theta(1-\theta)^{\beta-1}.$$

The population mean μ_t and the variance σ_t^2 for the truncated GNBD (10.113) can be written as

$$\mu_t = E(X) = K_t^{-1}\sum_{x=0}^{t} x P_x(\theta,\beta,m) \tag{10.117}$$

and

$$\sigma_t^2 = E(X - \mu_t)^2 = K_t^{-1} \sum_{x=0}^{t} x^2 P_x(\theta, \beta, m) - \mu_t^2. \qquad (10.118)$$

Famoye and Consul (1993) proposed the following estimation methods for the truncated GNBD: (i) method based on mean and ratio of frequencies; (ii) method based on ratio of first two frequencies, mean, and second factorial moment; and (iii) method of ML. The ML method has the smallest bias while the moment method is the most efficient.

The truncated GNBD given in (10.113) is a right truncated MPSD. In a similar manner a left truncated GNBD which is also a left truncated MPSD can be defined. The MVU estimators of θ and its functions follow the general procedure described for the MPSD in chapter 7. Kumar and Consul (1980), Jani (1977), and Gupta (1977) gave MVU estimators of θ and some of its functions for the left truncated GNBD model.

In a study conducted by Consul and Famoye (1995), they found that while the fitted frequencies of GNBD are close to the observed frequencies, the parameter estimates are not close to the actual parameter values. By using the bias property, they found that the truncated GNBD model provided better parameter estimates than the corresponding nontruncated GNBD model when fitted to data sets from the nontruncated GNBD model. Consul and Famoye (1995) suggested that the error caused by truncation can be eliminated to a great extent in estimation by using a truncated model for the estimation of parameters even though the data is from a nontruncated GNBD model.

10.11 Other Related Distributions

10.11.1 Poisson-Type Approximation

The GNBD in (10.1) can be written in the form

$$P_x(\theta, \beta, m) = (m\theta)^x \prod_{i=1}^{x-1} \left(1 + \frac{\beta x - i}{m}\right) (1-\theta)^{m+\beta x - x}/x!, \quad x = 0, 1, 2, \ldots. \qquad (10.119)$$

By substituting $\theta = \alpha/m$, $\beta/m = \lambda/\alpha$ and taking the limit of (10.119) as $m \to \infty$, Jain and Consul (1971) obtained the limit

$$P_x(\lambda, \alpha) = \alpha(\alpha + \lambda x)^{x-1}\, e^{-\alpha - \lambda x}/x! \qquad (10.120)$$

for $x = 0, 1, 2, 3, \ldots$, which is the generalized Poisson distribution discussed in chapter 9.

10.11.2 Generalized Logarithmic Series Distribution-Type Limit

The zero-truncated GNBD can be obtained from (10.113) as

$$f_x(\theta, \beta, m) = \frac{m}{m + \beta x} \binom{m + \beta x}{x} \theta^x (1-\theta)^{m+\beta x - x}/[1 - (1-\theta)^m] \quad \text{for} \quad x = 1, 2, 3, \ldots. \qquad (10.121)$$

As $m \to 0$, the zero-truncated GNBD in (10.113) tends to the generalized logarithmic series distribution with parameters θ and β. The GLSD is discussed in chapter 11 and its probability function is given by

$$P_x(\theta, \beta) = [-\log(1-\theta)]^{-1} \frac{1}{\beta x} \binom{\beta x}{x} \theta^x (1-\theta)^{\beta x - x}. \qquad (10.122)$$

10.11.3 Differences of Two GNB Variates

Let X and Y be two independent r.v. having GNBDs $P_x(\theta, \beta, m)$ and $P_y(\theta, \beta, n)$, respectively. Consul (1989b) defined the distribution $P_r(D = d)$ of the difference $D = Y - X$ as

$$P_r(D = d) = \begin{cases} (1-\theta)^{m+n} \sum_x (m, \beta)_x (n, \beta)_{d+x} [\theta(1-\theta)^{\beta-1}]^{2x+d}, & d \geq 0, \\ (1-\theta)^{m+n} \sum_y (n, \beta)_y (m, \beta)_{y-d} [\theta(1-\theta)^{\beta-1}]^{2y-d}, & d < 0, \end{cases} \quad (10.123)$$

for $x = 0, 1, 2, \ldots$, and where the summations on x and y are from 0 to ∞ and $(m, \beta)_x = \frac{m}{m+\beta x}\binom{m+\beta x}{x}$, and d takes all integral values from $-\infty$ to ∞.

The pgf of the random variable D is given by

$$G_1(u, v) = (1 - \theta + \theta t_1)^n (1 - \theta + \theta t_2)^m, \quad (10.124)$$

where $t_1 = u(1-\theta+\theta t_1)^\beta$ and $t_2 = v(1-\theta+\theta t_2)^\beta$. The pgf in (10.124) is somewhat restricted as the parameters m, n, and β are positive integers. A more general pgf of the distribution of the random variable $D = Y - X$ is

$$G_2(u, v) = f(t_1, t_2) = (1-\theta)^{m+n}(1-\theta t_1)^{-n}(1-\theta t_2)^{-m}, \quad (10.125)$$

where $t_1 = u(1-\theta)^{\beta-1}(1-\theta t_1)^{-\beta+1}$ and $t_2 = v(1-\theta)^{\beta-1}(1-\theta t_2)^{-\beta+1}$.

The cgf is given by

$$\psi(U) = (\beta - 1)^{-1}[n(T_1 - U) + m(T_2 + V)], \quad (10.126)$$

where $T_1 = U + (\beta - 1)\{\log(1-\theta) - \log(1 - \theta e^{T_1})\}$ and $T_2 = U + (\beta - 1)\{\log(1-\theta) - \log(1 - \theta e^{T_2})\}$.

A recurrence relation between the cumulants is

$$\psi(U) = m\frac{\partial \psi(U)}{\partial m} + n\frac{\partial \psi(U)}{\partial n}, \quad (10.127)$$

which corresponds to

$$L_k = m\frac{\partial L_k}{\partial m} + n\frac{\partial L_k}{\partial n}, \qquad k = 1, 2, 3, \ldots, \quad (10.128)$$

where L_k is the kth cumulant and $L_1 = (m-n)\theta(1-\theta\beta)^{-1}$. Consul (1990c) obtained the next three cumulants as

$$L_2 = (m+n)\theta(1-\theta)(1-\theta\beta)^{-3}, \quad (10.129)$$

$$L_3 = (m-n)\theta(1-\theta)[1 - 2\theta + 2\theta\beta - \theta^2\beta](1-\theta\beta)^{-5}, \quad (10.130)$$

and

$$L_4 = (m+n)\theta(1-\theta)A(1-\theta\beta)^{-7}, \quad (10.131)$$

where

$$A = 1 - 6\theta + 6\theta^2 + \theta\beta(8 - 18\theta + 8\theta^2) + \theta^2\beta^2(6 - 6\theta + \theta^2). \quad (10.132)$$

The coefficients of skewness $\sqrt{\beta_1}$ and kurtosis β_2 are given by

$$\sqrt{\beta_1} = \frac{(m-n)}{m+n}\frac{(1 - 2\theta + 2\theta\beta - \theta^2\beta)}{\sqrt{(m+n)\theta(1-\theta)(1-\theta\beta)}} \quad (10.133)$$

and

$$\beta_2 = 3 + A/[(m+n)\theta(1-\theta)(1-\theta\beta)], \tag{10.134}$$

where A is defined in (10.132).

The first three moments for the absolute difference $|X - Y|$ of two GNB variates are also given in Consul (1990c).

10.11.4 Weighted Generalized Negative Binomial Distribution

Ambagaspitiya (1995) defined an $\omega(m+\beta x, \beta)$-weighted GNBD as

$$P_x(m,\beta) = \frac{\omega(m+\beta x,\beta)}{\mathrm{E}\,[\omega(m+\beta X,\beta)]}\frac{m}{m+\beta x}\binom{m+\beta x}{x}\theta^x(1-\theta)^{m+\beta x-x},$$

or

$$P_x(m,\beta) = \frac{W(m+\beta,\beta)}{W(m,\beta)}\frac{\theta^2}{(m+\beta)(1-\theta)}\left(\beta-1+\frac{m+1}{x}\right)P_{x-1}(m+\beta,\beta), \quad x=1,2,3,\ldots,$$

where

$$W(m,\beta) = \sum_x \omega(m+\beta x,\beta)\frac{m}{m+\beta x}\binom{m+\beta x}{x}\theta^x(1-\theta)^{m+\beta x-x},$$

and obtained the pgf of the distribution. When $W(m,\beta) = m$, this weighted distribution reduces to the probability distribution given by Janardan and Rao (1983). The $\omega(m+\beta x,\beta)$-weighted distribution satisfies the convolution property. Ambagaspitiya (1995) also derived a recursive formula for computing the distribution function.

The GNBD in (10.1) satisfies the relation

$$P_x(m,\beta) = \frac{m\theta}{(m+\beta)(1-\theta)}\left(\beta-1+\frac{m+1}{x}\right)P_{x-1}(m+\beta,\beta), \quad x=1,2,3,\ldots,$$

which has been used in the second definition of the weighted GNBD.

Lingappaiah (1987) discussed the relationship of the GNBD with its weighted distribution, given by $\omega(m+\beta x,\beta) = x$, and obtained a number of recurrence relations between their moments. Also, he obtained some inverse moments and other relations.

10.12 Exercises

10.1 Use the method of differentiation to show that a recurrence relation between the central moments of GNBD is given by (10.10). Also, show that a recurrence relation between the factorial moments is (10.11).

10.2 Suppose $X_1, X_2, \ldots, X_n$ is a random sample from a GNBD with unknown θ (parameters β and m are assumed known). Show that the MVU estimators of μ_1', $(\mu_1')^2$, and μ_2 are given, respectively, by (10.67), (10.68), and (10.69).

10.3 Consider Exercise 10.2. Show that the MVU estimators of θ^k and $\theta^k(1-\theta)^{\beta k-k}$ are given, respectively, by (10.70) and (10.71).

10.4 Suppose $X_1, X_2, \ldots, X_n$ is a random sample from a GNBD with parameter θ as the only unknown. Under a prior distribution which is beta with parameters a and b, show that the Bayes estimator of θ is given by (10.77).

10.5 Let $X_1, X_2, \ldots, X_n$, where n is large, be a random sample from a GNBD with parameters β and m known. By using the sample mean $\bar{X}$ and normal approximation, obtain an expression for computing a $100(1-\alpha)\%$ CI for θ.

10.6 Suppose X is a discrete r.v. indexed by three parameters θ, β, and m, where θ is a r.v. that has beta distribution with parameters $a = 1$ and $b = k - 1$. If a beta mixture of the probability distribution of the r.v. X is generalized factorial distribution in (10.108), show that X has a GNBD with parameters θ, β, and $m = \lambda - k + 1$.

10.7 If X is a generalized negative binomial random variable with mean $\mu = m\theta(1-\theta\beta)^{-1}$ and variance $\sigma^2 = m\theta(1-\theta)(1-\theta\beta)^{-3}$, show that the limiting distribution of $Z = (X-\mu)/\sigma$, as $m \to \infty$, is normal.

10.8 Let $X_1, X_2, \ldots, X_N$ be a random sample of size N from the Consul probability distribution given by

$$P(X=x) = \frac{1}{x}\binom{mx}{x-1}\theta^{x-1}(1-\theta)^{mx-x+1}, \quad x = 1, 2, 3, \ldots,$$

and zero otherwise, where $0 < \theta < 1$ and $1 \le m \le \theta^{-1}$. Show that the probability distribution of $Y = X_1 + X_2 + \cdots + X_N$, given N, is

$$P(Y=y|N) = \frac{N}{y}\binom{my}{y-N}\theta^{y-N}(1-\theta)^{N+my-y}$$

for $y = N, N+1, \ldots$, and zero otherwise. If N is a binomial r.v. with parameters θ and k $(k > 0)$, show that the probability distribution of Y is a GNBD.

10.9 Suppose a queue is initiated by N members and has traffic intensity with negative binomial arrivals, given by the generating function $g(z) = (1-\theta)^{\beta-1}(1-\theta z)^{-\beta+1}$, $0 < \theta < 1$, $1 < \beta < \theta^{-1}$, and constant service time. Find the probability that exactly x members will be served before the queue vanishes, where the queue discipline is "first-come, first-served." Also, obtain the probability distribution of the customers served in the first busy period if N is also a negative binomial r.v. with parameters θ and m, $1 < m < \theta^{-1}$.

10.10 Let Y be a r.v. with zero-truncated GNBD. Obtain a recurrence relation between the noncentral moments of Y. By using the recurrence relation, find the mean and variance of zero-truncated GNBD. Assuming that parameter m is known, find the estimates of θ and β by using the first moment and the proportion of ones.

11

Generalized Logarithmic Series Distribution

11.1 Introduction and Definition

A discrete r.v. X is said to have a generalized logarithmic series distribution (GLSD) if its probability function is given by

$$P_x(\theta, \beta) = \begin{cases} \frac{1}{\beta x}\binom{\beta x}{x}\frac{\theta^x(1-\theta)^{\beta x - x}}{\{-\ln(1-\theta)\}}, & x = 1, 2, 3, \ldots, \\ 0, & \text{otherwise,} \end{cases} \tag{11.1}$$

where $0 < \theta < 1$ and $1 \leq \beta < \theta^{-1}$. When $\beta = 1$, the probability function in (11.1) reduces to that of the logarithmic series distribution given in Johnson et al. (1992). The GLSD belongs to the class of Lagrangian distributions in (2.7) and is listed at (12) in Table 2.3. It belongs to the subclass MPSD. Thus, the GLSD possesses all the properties of the MPSDs and of the Lagrangian distributions as discussed in earlier chapters.

The GLSD in (11.1) was obtained (Jain and Gupta, 1973; Patel, 1981) as a limiting form of zero-truncated GNBD, which is defined as

$$P(X = x) = \frac{m}{m + \beta x}\binom{m + \beta x}{x}\frac{\theta^x(1-\theta)^{m+\beta x - x}}{1 - (1-\theta)^m}, x = 1, 2, 3, \ldots, \tag{11.2}$$

where $0 < \theta < 1, \; 1 \leq \beta \leq \theta^{-1}$, and $m > 0$.

On rewriting (11.2) as

$$x!P(X = x) = (m+\beta x - 1)\,(m+\beta x - 2)\cdots(m+\beta x - x + 1)\cdot\theta^x(1-\theta)^{\beta x - x}\frac{m(1-\theta)^m}{1-(1-\theta)^m},$$

and taking the limit of the above as $m \to 0$, we obtain the probability function in (11.1).

Tripathi and Gupta (1985) defined a discrete probability distribution, which they called a GLSD with two parameters. Their GLSD was developed by taking the limit of a shifted version of a GNBD given by Tripathi and Gurland (1977). Another generalization of logarithmic series distribution was obtained by Tripathi and Gupta (1988). This GLSD was based on a GNBD obtained from a generalized Poisson distribution compounded with a truncated gamma distribution. We remark here that both forms of GLSD defined by Tripathi and Gupta (1985, 1988) differ from the probability model in (11.1) and both of them have complicated probability mass functions.

Since

$$\sum_{x=1}^{\infty} P_x(\theta, \beta) = 1,$$

the mean μ_1' can easily be obtained by differentiating

$$[-\ln(1-\theta)] = \sum_{x=1}^{\infty} \frac{1}{\beta x} \binom{\beta x}{x} \left[\theta(1-\theta)^{\beta-1}\right]^x$$

with respect to θ and simplifying the result. Thus,

$$\mu_1' = \alpha\theta(1-\theta\beta)^{-1}, \quad \text{where} \quad \alpha = [-\ln(1-\theta)]^{-1}. \tag{11.3}$$

Mishra and Tiwary (1985) defined the GLSD for negative values of the parameters. By putting $\theta = -a$ and $\beta = -b$ in (11.1), where a and b are positive quantities, we obtain

$$P_x(-a, -b) = \frac{1}{x\ln(1+a)} \binom{bx+x-1}{x-1} \left(\frac{a}{1+a}\right)^x \left(\frac{1}{1+a}\right)^{bx}.$$

On substituting $\theta_1 = a/(1+a)$ and $\beta_1 = 1+b$, the above reduces to

$$P_x(-a, -b) = \frac{1}{\beta_1 x} \binom{\beta_1 x}{x} \frac{\theta_1^x (1-\theta_1)^{\beta_1 x}}{-\ln(1-\theta_1)},$$

which is of the same form as (11.1). The assumption of θ and β being positive may be dropped, and hence the GLSD is defined for negative values of θ and β.

11.2 Generating Functions

The pgf of the GLSD can be obtained by using the results obtained for the class of Lagrangian probability distributions $L(f; g; x)$ in (2.7). The pgf is given by

$$G_x(u) = \ln(1-\theta+\theta z), \tag{11.4}$$

where $z = u(1-\theta+\theta z)^\beta$.

On using $z = e^s$ and $u = e^v$ in (11.4), we obtain the mgf for the GLSD as

$$M_x(v) = \ln(1-\theta+\theta e^s), \text{ where } s = v + \beta \ln(1-\theta+\theta e^s).$$

Another pgf for the GLSD is given by

$$G_x(u) = f(z) = \ln\left(\frac{1-\theta}{1-\theta z}\right), \quad \text{where} \quad z = u\left(\frac{1-\theta}{1-\theta z}\right)^{\beta-1}. \tag{11.5}$$

Correspondingly, we obtain the mgf from (11.5) as

$$M_x(v) = \ln\left(\frac{1-\theta}{1-\theta e^s}\right), \quad \text{where} \quad s = v + (\beta-1)\ln\left(\frac{1-\theta}{1-\theta e^s}\right).$$

Since the $G_x(u)$ in (11.4) and the $G_x(u)$ in (11.5) cannot be transformed into each other, the GLSD is one of those few probability models which has two independent sets of the pgfs for its generation and has two sets of mgfs, which can be used to obtain the moments.

11.3 Moments, Cumulants, and Recurrence Relations

All the moments of GLSD exist for $1 \le \beta < \theta^{-1}$. Though one of the two sets of the above mgfs can be used to determine the moments of the GLSD about the origin, there is a recurrence relation which gives the moments more easily. The kth moment about the origin can be written as

$$\mu'_k = E\left[X^k\right] = \sum_{x=1}^{\infty} x^k \frac{1}{\beta x}\binom{\beta x}{x}\frac{\theta^x(1-\theta)^{\beta x - x}}{[-\ln(1-\theta)]}.$$

On differentiating the above with respect to θ, we get

$$\frac{d\mu'_k}{d\theta} = \sum_{x=1}^{\infty} x^k P_x(\theta,\beta)\left[\frac{x(1-\theta\beta)}{\theta(1-\theta)} - \frac{1}{(1-\theta)\{-\ln(1-\theta)\}}\right]$$

$$= \frac{1-\theta\beta}{\theta(1-\theta)}\left[\mu'_{k+1} - \mu'_1\mu'_k\right].$$

Hence,

$$\mu'_{k+1} = \frac{\theta(1-\theta)}{1-\theta\beta}\frac{d\mu'_k}{d\theta} + \mu'_1\mu'_k, \qquad k = 1, 2, 3, \ldots, \tag{11.6}$$

which is a recurrence relation between the noncentral moments. Famoye (1995) used the relation (11.6) to obtain the first six noncentral moments as

$$\mu'_1 = \alpha\theta(1-\theta\beta)^{-1} \quad \text{(given by (11.3))},$$

$$\mu'_2 = \alpha\theta(1-\theta)(1-\theta\beta)^{-3}, \tag{11.7}$$

$$\mu'_3 = \alpha\theta(1-\theta)\left[1-2\theta+\theta\beta(2-\theta)\right](1-\theta\beta)^{-5}, \tag{11.8}$$

$$\mu'_4 = \alpha\theta(1-\theta)A(1-\theta\beta)^{-7}, \tag{11.9}$$

$$\mu'_5 = \alpha\theta(1-\theta)B(1-\theta\beta)^{-9}, \tag{11.10}$$

and

$$\mu'_6 = \alpha\theta(1-\theta)\left[(1-2\theta+8\theta\beta-7\theta^2\beta)B + \theta(1-\theta)(1-\theta\beta)C\right](1-\theta\beta)^{-11}, \tag{11.11}$$

where

$$\alpha = [-\ln(1-\theta)]^{-1}, \tag{11.12}$$

$$A = 1 - 6\theta + 6\theta^2 + 2\theta\beta(4-9\theta+4\theta^2) + \theta^2\beta^2(6-6\theta+\theta^2), \tag{11.13}$$

$$B = 1 - 14\theta + 36\theta^2 - 24\theta^3 + \theta\beta(22 - 113\theta + 152\theta^2 - 58\theta^3)$$
$$+ \theta^2\beta^2(58 - 152\theta + 113\theta^2 - 22\theta^3) + \theta^3\beta^3(24 - 36\theta + 14\theta^2 - \theta^3), \tag{11.14}$$

$$C = \frac{dB}{d\theta}. \tag{11.15}$$

The kth central moment of GLSD is given by

$$\mu_k = \sum_{x=1}^{\infty} (x - \mu_1')^k \frac{1}{\beta x} \binom{\beta x}{x} \frac{\theta^x (1-\theta)^{\beta x - x}}{[-\ln(1-\theta)]} .$$

On differentiating the above with respect to θ, we obtain

$$\frac{d\mu_k}{d\theta} = \sum_{x=1}^{\infty} (x - \mu_1')^k \cdot \frac{1-\theta\beta}{\theta(1-\theta)} (x - \mu_1') P_x(\theta, \beta) - \sum_{x=1}^{\infty} k(x - \mu_1')^{k-1} \frac{d\mu_1'}{d\theta} P_x(\theta, \beta)$$

$$= \sum_{x=1}^{\infty} (x - \mu_1')^{k+1} \frac{1-\theta\beta}{\theta(1-\theta)} P_x(\theta, \beta) - k\, \mu_{k-1} \frac{d\mu_1'}{d\theta} .$$

On further simplification, a recurrence relation between the central moments becomes

$$\mu_{k+1} = \frac{\theta(1-\theta)}{1-\theta\beta} \frac{d\mu_k}{d\theta} + k\mu_2\, \mu_{k-1}, \qquad k = 2, 3, \ldots . \tag{11.16}$$

By using the mean μ_1' in (11.3) and the second noncentral moment in (11.7), the variance of GLSD becomes

$$\sigma^2 = \mu_2 = \alpha\theta(1-\theta)(1-\theta\beta)^{-3} - (\mu_1')^2. \tag{11.17}$$

The GLSD is a member of the class of modified power series distributions considered in chapter 7 with $\phi(\theta) = \theta(1-\theta)^{\beta-1}$ and $h(\theta) = \{-\ln(1-\theta)\}$. Tripathi, Gupta, and Gupta (1986) obtained the first three incomplete moments and incomplete factorial moments for the GLSD. Recurrence relations between these moments for the MPSD are given in chapter 7. The incomplete moments are given by (7.21), while the incomplete factorial moments are given by (7.23). By using these results Tripathi, Gupta, and Gupta (1986) have derived the expressions for the first three incomplete moments and the first three incomplete factorial moments for the GLSD. However, these expressions are too long.

11.4 Other Interesting Properties

Gupta (1976) obtained the distribution of $Z = \sum_{i=1}^{n} X_i$, where $X_1, X_2, \ldots, X_n$ is a random sample from the GLSD model (11.1). Since each X_i is a MPSD with $\phi(\theta) = \theta(1-\theta)^{\beta-1}$ and $h(\theta) = \{-\ln(1-\theta)\}$, the distribution of Z is of the form

$$P(Z = z) = b(z, n)\, \phi^z(\theta)/(h(\theta))^n, \qquad z = n, n+1, n+2, \ldots, \tag{11.18}$$

where

$$[h(\theta)]^n = \sum_z b(z, n)\, \phi^z(\theta). \tag{11.19}$$

By using the Lagrange expansion on $(h(\theta))^n$ under the transformation $\theta = ug(\theta) = u(1-\theta)^{-\beta+1}$ as $\phi(\theta) = \theta/g(\theta)$, we obtain

$$(h(\theta))^n = \sum_{z=n}^{\infty} \frac{n}{z!} \left\{ \left(\frac{d}{d\theta}\right)^{z-1} \left[(1-\theta)^{-\beta z+z-1} \{-\ln(1-\theta)\}^{n-1} \right] \right\}_{\theta=0} \left(\frac{\theta}{g(\theta)}\right)^z$$

$$= \sum_{z=n}^{\infty} \frac{n!}{z!} \left\{ \left(\frac{d}{d\theta}\right)^{z-1} \left[\sum_{k=n-1}^{\infty} \frac{1}{k!} |s(k, n-1)| \cdot \sum_{r=0}^{\infty} (-1)^r \theta^{r+k} \binom{-\beta z + z - 1}{r} \right] \right\}_{\theta=0}$$

$$\times \left(\frac{\theta}{g(\theta)}\right)^z,$$

where

$$\binom{a}{b} = \begin{cases} \frac{a(a-1)a-2)\cdots(a-b+1)}{1\cdot 2\cdot 3\cdots b}, & a > b \geq 0, \\ 0, & b < 0. \end{cases}$$

On further simplification, we get

$$(h(\theta))^n = \sum_{z=n}^{\infty} \frac{n!}{z} \left\{ \sum_{t=n-1}^{z} \frac{(-1)^{z-1-t}}{t!} |s(t, n-1)| \binom{-\beta z + z - 1}{z-1-t} \right\} \left(\frac{\theta}{g(\theta)}\right)^z, \tag{11.20}$$

where $s(t, n-1)$ are the Stirling numbers of the first kind. Comparing (11.19) with (11.20) and using (11.18), we obtain the distribution of Z as

$$P(Z = z) = \frac{n!}{z} \sum_{t=n-1}^{z-1} (-1)^{z-1-t} \frac{|s(t, n-1)|}{t!} \binom{-\beta z + z - 1}{z-1-t} \frac{\theta^z (1-\theta)^{\beta z - z}}{\{-\ln(1-\theta)\}^n}. \tag{11.21}$$

The above probability distribution has a complex form; however,

$$\mathrm{E}[Z] = \mathrm{E}\left[\sum_{i=1}^{n} X_i\right] = \sum_{i=1}^{n} \mathrm{E}[X_i] = n\alpha\theta(1-\theta\beta)^{-1}$$

and

$$\mathrm{Var}[Z] = \sum_{i=1}^{n} \mathrm{Var}[X_i] = n\alpha\theta(1-\theta)(1-\theta\beta)^{-3} - n\left[\alpha\theta(1-\theta\beta)^{-1}\right]^2.$$

Thus,

$$\mathrm{E}[\bar{X}] = \alpha\theta(1-\theta\beta)^{-1}, \tag{11.22}$$

$$\mathrm{Var}[\bar{X}] = n^{-1}\left[\alpha\theta(1-\theta)(1-\theta\beta)^{-3} - \alpha^2\theta^2(1-\theta\beta)^{-2}\right]. \tag{11.23}$$

Theorem 11.1. *The GLSD model (11.1) is not strongly unimodal but is unimodal for all values of θ in $0 < \theta < \beta^{-1}$ and for $\beta \geq 1$, and the mode is at the point $x = 1$ (Famoye, 1987).*

Proof. Keilson and Gerber (1971) showed that the sequence $\{P_x\}$ is strongly unimodal if and only if $P_x^2/(P_{x-1}\,P_{x+1}) \geq 1$ for all values of x.

For $x = 2$, the GLSD gives

$$\frac{P_2^2}{P_1 P_3} = \frac{3}{2} \frac{(2\beta-1)^2}{(3\beta-1)(3\beta-2)}. \tag{11.24}$$

By using logarithmic differentiation, it can be shown that the right-hand side of (11.24) is a decreasing function of β, and hence it takes the maximum value when $\beta = 1$. Therefore, (11.24) becomes

$$\frac{P_2^2}{P_1 P_3} < \frac{3}{4} < 1 .$$

Thus, the GLSD model (11.1) does not satisfy the property of strong unimodality for $x = 2$ and so the model is not strongly unimodal, or equivalently, not log-concave.

When $\beta = 1$, the GLSD reduces to the logarithmic series distribution whose unimodality is well established (see Johnson, Kotz, and Komp 1992, p. 290). We now consider the unimodality of GLSD for $\beta > 1$. For the mode to be at the point $x = 1$, it suffices to show that

$$P_{x+1} < P_x \quad \text{for all} \quad x = 1, 2, 3, \ldots .$$

Now,

$$\frac{P_{x+1}}{P_x} = \frac{x}{(x+1)^2} \frac{(\beta x + \beta)!(\beta x - x)!}{[(\beta - 1)(x+1)]!(\beta x)!} \cdot \theta(1-\theta)^{\beta-1} . \tag{11.25}$$

Since $0 < \theta < \beta^{-1}$, the expression $\theta(1-\theta)^{\beta-1}$ is an increasing function of θ and its maximum occurs at $\theta = \beta^{-1}$. Thus, (11.25) becomes

$$\begin{aligned}
\frac{P_{x+1}}{P_x} &< \frac{x}{(x+1)^2} \cdot \frac{1}{\beta} \cdot \left(\frac{\beta-1}{\beta}\right)^{\beta-1} \frac{(\beta x + \beta)(\beta x + \beta - 1) \cdots (\beta x + 1)}{(\beta x - x + \beta - 1)(\beta x - x + \beta - 2) \cdots (\beta x - x + 1)} \\
&= \frac{x}{x+1} \left(\frac{\beta-1}{\beta}\right)^{\beta-1} \prod_{i=1}^{\beta-1} \left(1 + \frac{x}{\beta x + \beta - x - i}\right) \\
&< \frac{x}{x+1} \left(\frac{\beta-1}{\beta}\right)^{\beta-1} \left(1 + \frac{x}{\beta x + \beta - x - \beta + 1}\right)^{\beta-1} \\
&= \frac{x}{x+1} \left(\frac{\beta(\beta x + 1) - \beta x - 1}{\beta(\beta x + 1) - \beta x}\right)^{\beta-1} \\
&< 1 \quad \text{for} \quad x = 1, 2, 3, \ldots .
\end{aligned}$$

Therefore, the GLSD is unimodal with its mode at the point $x = 1$. □

Hansen and Willekens (1990) showed that the GLSD in (11.1) is strictly log-convex and is infinitely divisible.

The mean μ_1' in (11.3) is a monotone increasing function of θ and β. When $\beta \geq 1$, the second moment of the GLSD is always more than the mean. Furthermore, the mean and variance tend to increase with an increase in the value of the parameter β. The variance increases faster than the mean.

For computation of probabilities, the following recurrence relation between the probabilities is useful:

$$P_{x+1}(\theta, \beta) = \left(\beta - \frac{x}{x+1}\right) \theta(1-\theta)^{\beta-1} \prod_{j=1}^{x-1} \left(1 + \frac{\beta}{\beta x - j}\right) P_x(\theta, \beta) \tag{11.26}$$

for $x = 1, 2, 3, \ldots$, where $P_1 = \theta(1-\theta)^{\beta-1}/\{-\ln(1-\theta)\}$ and $P_2 = (\beta - \frac{1}{2})\theta(1-\theta)^{\beta-1} P_1$.

11.5 Estimation

11.5.1 Point Estimation

Suppose a random sample of size n is taken from the GLSD model (11.1). Let the observed values be $1, 2, \ldots, k$ with corresponding frequencies $n_1, n_2, \ldots, n_k$, where k is the largest observed value of x in the sample, and let

$$n = \sum_{x=1}^{k} n_x .$$

Let the first two sample moments be

$$\bar{x} = \frac{1}{n} \sum_{x=1}^{k} x\, n_x \tag{11.27}$$

and

$$s^2 = \frac{1}{n-1} \sum_{x=1}^{k} (x - \bar{x})^2 n_x . \tag{11.28}$$

Moment Estimates

On equating the first two sample moments (11.27) and (11.28) to the corresponding population moments (11.3) and (11.7), we get

$$\bar{x} = \alpha\theta(1 - \theta\beta)^{-1} \tag{11.29}$$

and

$$s^2 = \alpha\theta(1 - \theta)(1 - \theta\beta)^{-3} - \bar{x}^2 . \tag{11.30}$$

From (11.29), we obtain

$$\beta = \theta^{-1} - \alpha(\bar{x})^{-1}, \tag{11.31}$$

and on using (11.31) in (11.30),

$$f(\theta) = (1 - \theta)\bar{x}^3 \alpha^{-2} - \theta^2 (s^2 + \bar{x}^2) = 0 . \tag{11.32}$$

We solve equation (11.32) for θ by using an iterative procedure.

As $\theta \to 0$, the function $f(\theta)$ in (11.32) tends to $1 - (s^2 + \bar{x}^2)(\bar{x})^{-3}$ and as $\theta \to 1$,

$$f(\theta) \to -(s^2 + \bar{x}^2)(\bar{x})^{-3} < 0.$$

When $1 - (s^2 + \bar{x}^2)(\bar{x})^{-3} > 0$, there exists at least one solution for equation (11.32) in the interval $(0, \beta^{-1})$. For $k = 2$ and $k = 3$, $1 - (s^2 + \bar{x}^2)(\bar{x})^{-3} > 0$, and by using mathematical induction

$$1 - (s^2 + \bar{x}^2)(\bar{x})^{-3} > 0 \qquad \text{for all values of } k.$$

Furthermore, $f'(\theta)$ first increases and then decreases in the interval $(0, \beta^{-1})$. Since $f''(\theta) < 0$, the function $f(\theta)$ in (11.32) is concave down in the interval $(0, \beta^{-1})$. Famoye (1995) has remarked that equation (11.32) gave one and only one solution to all simulation problems considered by him.

On getting the moment estimate $\tilde{\theta}$, the value of $\tilde{\beta}$, the moment estimate of β, can be obtained from (11.31).

Maximum Likelihood Estimates

The loglikelihood function of the GLSD is given by

$$\ell(\theta, \beta) = \ln\left[\prod_{x=1}^{k} \{P_x(\theta, \beta)\}^{n_x}\right]$$

$$= n\bar{x}\,\ln\theta + n\bar{x}(\beta - 1)\,\ln(1-\theta) - n\,\ln\{-\ln(1-\theta)\}$$

$$+ \sum_{x=2}^{k} n_x \left\{\sum_{i=1}^{x-1} \ln(\beta x - i) - \ln(x!)\right\}. \tag{11.33}$$

On differentiating (11.33) with respect to θ and β, we obtain the likelihood equations

$$\frac{\partial \ell(\theta, \beta)}{\partial \theta} = \frac{n\bar{x}}{\theta} - \frac{(\beta - 1)n\bar{x}}{1-\theta} + \frac{n}{(1-\theta)\ln(1-\theta)} = 0 \tag{11.34}$$

and

$$\frac{\partial \ell(\theta, \beta)}{\partial \beta} = n\bar{x}\,\ln(1-\theta) + \sum_{x=2}^{k}\sum_{i=1}^{x-1} \frac{x n_x}{\beta x - i} = 0. \tag{11.35}$$

Equations (11.34) and (11.35) are solved simultaneously to obtain $\hat{\theta}$ and $\hat{\beta}$, the ML estimates of parameters θ and β, respectively.

It has not yet been shown that the ML estimators are unique. However, in all simulation results involving the GLSD model (11.1), the ML estimating equations give a unique solution set for the parameters θ and β. It can easily be shown that (11.34) represents a curve where β is a monotonically decreasing function of θ. Similarly, the equation (11.35) represents a curve in which β is a monotonically decreasing function of θ. Thus, if the two curves intersect, it will be a unique point.

Proportion of Ones and Sample Mean Method

If P_1 denotes the probability of the class $X = 1$, then from (11.1),

$$P_1 = \alpha\theta(1-\theta)^{\beta-1}. \tag{11.36}$$

By equating (11.36) to the proportion of ones in the sample, we obtain

$$\frac{n_1}{n} = \alpha\,\theta(1-\theta)^{\beta-1}. \tag{11.37}$$

By using (11.31) in (11.37) and on taking the logarithms of both sides, we get

$$g(\theta) = \ln\theta + \left[\theta^{-1} - \bar{x}^{-1}\{-\ln(1-\theta)\}^{-1} - 1\right] \ln(1-\theta)$$

$$- \ln\{-\ln(1-\theta)\} - \ln(n_1/n) = 0. \tag{11.38}$$

When $\theta \to 0$, the function $g(\theta)$ in (11.38) tends to $\bar{x}^{-1} - \ln(n_1/n) - 1$, and as $\theta \to 1$, the function $g(\theta) \to +\infty$. By using mathematical induction as in the case of moment estimation,

$$\bar{x}^{-1} - \ln(n_1/n) - 1 > 0.$$

Furthermore, $g'(\theta) < 0$ for all values of θ. Thus, the function $g(\theta)$ is monotonically decreasing in the interval $\left(0, \beta^{-1}\right)$, and hence equation (11.38) has a unique solution θ^* of θ.

The corresponding estimate β^* of β can be obtained by substituting the value of θ^* in equation (11.31).

Minimum Variance Unbiased Estimation

Patel (1980) obtained the MVU estimators of some functions of parameters of the GLSD. Since the GLSD is a member of the MPSD class, $Z = \sum_{i=1}^{n} X_i$ is a complete and sufficient statistic for θ, and so the sample mean $\bar{X} = Z/n$ is a MVU estimator of the mean $\mu = \alpha\theta(1-\theta\beta)^{-1}$ of the GLSD, defined by (11.1).

The MVU estimators of other functions $\ell(\theta)$ of the parameter θ can be obtained by following the method given in subsection 7.5.2 and Theorem 7.4.

By formula (7.51) and the result (11.21), we get the value

$$b(n,z) = \frac{n!}{z} \sum_{t=n-1}^{z-1} (-1)^{z-1-t} \frac{|s(t,n-1)|}{t!} \binom{-\beta z + z - 1}{z-1-t}. \tag{11.39}$$

(i) *MVU estimator for* $\ell(\theta) = \{-\ln(1-\theta)\}^m$. By section 11.4, the Lagrange expansion of $\ell(\theta)\,(h(\theta))^n = \{-\ln(1-\theta)\}^{m+n}$, under the transformation $\theta = ug(\theta) = u(1-\theta)^{-\beta+1}$ and $\phi(\theta) = \theta/g(\theta)$, is given by

$$\{-\ln(1-\theta)\}^{m+n} = \sum_{z=m+n}^{\infty} \frac{m+n}{z!} \left\{ \left(\frac{d}{d\theta}\right)^{z-1} \left[(1-\theta)^{-\beta z+z-1}\{-\ln(1-\theta)\}^{m+n-1}\right]\right\}_{\theta=0} \times \left(\frac{\theta}{g(\theta)}\right)^z.$$

On simplification as in section 11.4 and using the result in (11.21), we get

$$\{-\ln(1-\theta)\}^{m+n} = \sum_{z=m+n}^{\infty} c(m+n,z) \left(\frac{\theta}{g(\theta)}\right)^z,$$

where

$$c(m+n,z) = \frac{(m+n)!}{z}\left[\sum_{t=m+n-1}^{z} \frac{(-1)^{z-1-t}}{t!} |s(t,m+n-1)| \binom{\beta z + z - 1}{z-1-t}\right].$$

Thus, the MVU estimator of $\{-\ln(1-\theta)\}^m$ is

$$f(z) = c(m+n,z)/b(n,z)$$

for $z = m+n, m+n+1, \ldots$ and zero otherwise.

(ii) *MVU estimator of* θ^m, *m is a positive integer.* Now,

$$\frac{d}{d\theta}\left[\theta^n\,(h(\theta))^n\right] = \frac{d}{d\theta}\left[\theta^m\{-\ln(1-\theta)\}^n\right]$$
$$= m\theta^{m-1}\{-\ln(1-\theta)\}^n + m(1-\theta)^{-1}\theta^m\{-\ln(1-\theta)\}^{n-1}.$$

By section 11.4, the Lagrange expansion of $\theta^m\{-\ln(1-\theta)\}^n$, under the transformation $\theta = ug(\theta) = u(1-\theta)^{-\beta+1}$ and $\phi(\theta) = \theta/g(\theta)$, is given by

$$\theta^m\{-\ln(1-\theta)\}^n = \sum_{z=m+n}^{\infty} \frac{1}{z!}\left\{\left(\frac{d}{d\theta}\right)^{z-1}(1-\theta)^{-\beta z+z}\Big[m\theta^{m-1}\{-\ln(1-\theta)\}^n\right.$$

$$\left. + \, n(1-\theta)^{-1}\theta^m\{-\ln(1-\theta)\}^{n-1}\Big]\right\}_{\theta=0}\left(\frac{\theta}{g(\theta)}\right)^z$$

$$= \sum_{z=m+n}^{\infty}\left[c_1(m+n,z)+c_2(m+n,z)\right]\left(\frac{\theta}{g(\theta)}\right)^z,$$

where

$$c_1(m+n,z) = \frac{m}{z!}\left\{\left(\frac{d}{d\theta}\right)^{z-1}\left[\theta^{m-1}(1-\theta)^{-\beta z+z}(-\ln(1-\theta))^n\right]\right\}_{\theta=0}$$

$$= \frac{m!}{(z-m)!z}\left\{\left(\frac{d}{d\theta}\right)^{z-m}\left[\sum_{k=n}^{\infty}\frac{1}{k!}|s(k,n)|\sum_{r=0}^{\infty}(-1)^r\theta^{r+k}\binom{-\beta z+z}{r}\right]\right\}_{\theta=0}$$

$$= \frac{m!}{z}\sum_{k=n}^{\infty}\frac{(-1)^{z-m-k}}{k!}|s(k,n)|\binom{-\beta z+z}{z-m-k}$$

and

$$c_2(m+n,z) = \frac{n}{z!}\left\{\left(\frac{d}{d\theta}\right)^{z-1}\left[\theta^{m}(1-\theta)^{-\beta z+z-1}(-\ln(1-\theta))^{n-1}\right]\right\}_{\theta=0}$$

$$= \frac{n}{(z-m-1)!z}\left\{\left(\frac{d}{d\theta}\right)^{z-m-1}\left[\sum_{k=n-1}^{\infty}\frac{1}{k!}|s(k,n-1)|\sum_{r=0}^{\infty}(-1)^r\theta^{r+k}\right.\right.$$

$$\left.\left.\times\binom{-\beta z+z-1}{r}\right]\right\}_{\theta=0}$$

$$= \frac{n}{z}\sum_{k=n-1}^{\infty}\frac{(-1)^{z-m-k-1}}{k!}|s(k,n-1)|\binom{-\beta z+z-1}{z-m-k-1}.$$

By Theorem 7.4 the MVU estimator of $\ell(\theta) = \theta^m$ for the GLSD becomes

$$f(z) = \left[c_1(m+n,z)+c_2(m+n,z)\right]/b(n,z)$$

for $z = m+n, m+n+1, m+n+2, \ldots$ and zero otherwise.

(iii) *MVU estimator of* $P_x(\theta,\beta) = \frac{1}{\beta x}\binom{\beta x}{x}\theta^x(1-\theta)^{\beta x-x}/[-\ln(1-\theta)]$. It can be obtained by determining the MVU estimator of

$$\ell(\theta) = \theta^x(1-\theta)^{\beta x-x}/[-\ln(1-\theta)]$$

in the same manner as (ii) above. However, the method will be longer and somewhat more complex.

11.5.2 Interval Estimation

The distribution of the sample sum $Y = \sum_{i=1}^{n} X_i$ is given by (11.18). When the sample size is small, a $100(1-\alpha)\%$ CI for θ when β is known can be obtained from the equations

$$\sum_{x=y}^{\infty} \frac{n!}{x} \sum_{k=n-1}^{x-1} (-1)^{x-1-k} \frac{|s(k,n-1)|}{k!} \binom{-\beta x + x - 1}{x-1-k} \frac{\left(\theta_\ell(1-\theta_\ell)^{\beta-1}\right)^x}{[-\ln(1-\theta_\ell)]^n} = \frac{\alpha}{2} \tag{11.40}$$

and

$$\sum_{x=0}^{y} \frac{n!}{x} \sum_{k=n-1}^{x-1} (-1)^{x-1-k} \frac{|s(k,n-1)|}{k!} \binom{-\beta x + x - 1}{x-1-k} \frac{\left(\theta_u(1-\theta_u)^{\beta-1}\right)^x}{[-\ln(1-\theta_u)]^n} = \frac{\alpha}{2}, \tag{11.41}$$

where $|s(k, n-1)|$ denotes the Stirling numbers of the first kind (see chapter 1).

Equations (11.40) and (11.41) can be solved iteratively with the help of a computer program to obtain θ_ℓ and θ_u, respectively. The quantities θ_ℓ and θ_u are the respective lower and upper $100(1-\alpha)\%$ confidence bounds for the parameter θ.

When the sample size is large, we apply the same method as used for the MPSD. By replacing the value of μ in expression (7.64), we obtain

$$1 - \alpha = P_r \left\{ \bar{X} - z_{\alpha/2}\, s/\sqrt{n} < \frac{\theta}{(1-\theta\beta)\{-\ln(1-\theta)\}} < \bar{X} + z_{\alpha/2}\, s/\sqrt{n} \right\}. \tag{11.42}$$

We obtain θ_ℓ and θ_u by solving the equations

$$\frac{\theta}{(1-\theta\beta)\{-\ln(1-\theta)\}} = h(\theta) = \bar{x} - z_{\alpha/2}\, s/\sqrt{n} \tag{11.43}$$

and

$$\frac{\theta}{(1-\theta\beta)\{-\ln(1-\theta)\}} = h(\theta) = \bar{x} + z_{\alpha/2}\, s/\sqrt{n} \tag{11.44}$$

numerically through an iterative procedure to yield θ_ℓ and θ_u, respectively, when the numerical values of n, β, $\bar{x}$, s, and $z_{\alpha/2}$ are substituted. Since $h(\theta)$ is an increasing function of θ, the equations in (11.43) and (11.44) have either no solutions or unique solutions. If $\theta \to 0$, $h(\theta) \to 1$ and if $\theta \to \beta^{-1}$, $h(\theta) \to \infty$. Therefore, equations (11.43), for example, will have a unique solution if and only if

$$1 - (\bar{x} - z_{\alpha/2}\, s/\sqrt{n}) < 0\,.$$

11.6 Statistical Testing

The goodness-of-fit test of the GLSD can be based on the chi-square statistic

$$\chi^2 = \sum_{x=1}^{k} (O_x - E_x)^2 / E_x, \tag{11.45}$$

where O_x and E_x are the observed and the expected frequencies for class x. The parameters θ and β are estimated by the ML technique. The expected value E_x is computed by

$$E_x = nP_x(\theta, \beta), \tag{11.46}$$

where n is the sample size.

The random variable χ^2 in (11.45) has a chi-square distribution with $k - 1 - r$ degrees of freedom, where r is the number of estimated parameters in the GLSD.

Famoye (2000) developed goodness-of-fit test statistics based on the EDF for the GLSD model. For small or moderate sample sizes, the tests are compared with respect to their simulated power of detecting some alternative hypotheses against a null hypothesis of a GLSD. The discrete version of the Cramer–von Mises test (defined in subsection 9.7.3) and the Anderson–Darling test (defined in subsection 9.7.3) are found to be the most powerful among the EDF tests.

11.7 Characterizations

Theorem 11.2. *Let X and Y be two independent discrete r.v.s. The conditional distribution of X, given $X + Y = z$, is*

$$P(X = x \mid X + Y = z) = \frac{\frac{1}{\beta x}\binom{\beta x}{x}\frac{1}{\beta(z-x)}\binom{\beta z - \beta x}{z-x}u^x}{\sum_{j=1}^{z-1}\frac{1}{\beta j}\binom{\beta j}{j}\frac{1}{\beta(z-j)}\binom{\beta(z-j)}{z-j}u^j}, \tag{11.47}$$

where $u = \frac{\theta_1}{\theta_2}\left(\frac{1-\theta_1}{1-\theta_2}\right)^{\beta-1}$ and $z = 2, 3, \ldots$, if and only if X and Y each has a GLSD with parameters (θ_1, β) and (θ_2, β), respectively.

Proof. Assume that X has a GLSD with parameters (θ_1, β) and that Y has a GLSD with parameters (θ_2, β). Since X and Y are independent

$$P(X + Y = z) = \sum_{j=1}^{z-1}\frac{1}{\beta j}\binom{\beta j}{j}\frac{1}{\beta(z-j)}\binom{\beta(z-j)}{z-j}\frac{\theta_2^z(1-\theta_2)^{\beta z - z}\,u^j}{\ln(1-\theta_1)\ln(1-\theta_2)}.$$

Now

$$\begin{aligned} P(X = x \mid X + Y = z) &= \frac{P(X = x) \cdot P(Y = z - x)}{P(X + Y = z)} \\ &= \frac{\frac{1}{\beta x}\binom{\beta x}{x}\frac{(\theta_1^x(1-\theta_1))^{\beta x - x}}{\{-\ln(1-\theta_1)\}}\;\frac{1}{\beta(z-x)}\binom{\beta(z-x)}{z-x}\frac{\theta_2^{z-x}(1-\theta_2)^{(\beta-1)(z-x)}}{\{-\ln(1-\theta_2)\}}}{P(X+Y=z)} \\ &= \frac{\frac{1}{\beta x}\binom{\beta x}{x}\frac{1}{\beta(z-x)}\binom{\beta(z-x)}{z-x}u^x}{\sum_{j=1}^{z-1}\frac{1}{\beta j}\binom{\beta j}{j}\frac{1}{\beta(z-j)}\binom{\beta(z-j)}{z-j}u^j}, \end{aligned}$$

which is the result in (11.47).

The second part of the proof is to assume that (11.47) holds. This part is straightforward. See Exercise 11.3. □

Theorem 11.3. *Suppose X and Y are two independent r.v.s such that*

$$P(X = 1 \mid X + Y = z) = \frac{\frac{1}{\beta(z-1)}\binom{\beta(z-1)}{z-1}u}{\sum_{j=1}^{z-1}\frac{1}{\beta j}\binom{\beta j}{j}\frac{1}{\beta(z-j)}\binom{\beta(z-j)}{z-j}u^j} \tag{11.48}$$

and

$$P(X=2 \mid X+Y=z) = \frac{\frac{1}{2\beta}\binom{2\beta}{2}\frac{1}{\beta(z-2)}\binom{\beta(z-2)}{z-2}u^2}{\sum_{j=1}^{z-1}\frac{1}{\beta j}\binom{\beta j}{j}\frac{1}{\beta(z-j)}\binom{\beta(z-j)}{z-j}u^j} \tag{11.49}$$

where $\beta \geq 1,\ 0<\theta_1,\ \theta_2<1,\ u=\frac{\theta_1}{\theta_2}\left(\frac{1-\theta_1}{1-\theta_2}\right)^{\beta-1}$. *Show that X and Y are GLS variates with parameters* (θ_1,β) *and* (θ_2,β), *respectively.*

Proof. Let $P(X=x)=f(x)$ with $\sum_{x=1}^{\infty} f(x)=1$ and $P(Y=y)=g(y)$ with $\sum_{y=1}^{\infty} g(y)=1$.

By condition (11.48),

$$\frac{f(1)g(z-1)}{\sum_{j=1}^{z-1} f(j)\,g(z-j)} = \frac{\frac{1}{\beta(z-1)}\binom{\beta(z-1)}{z-1}u}{\sum_{j=1}^{z-1}\frac{1}{\beta j}\binom{\beta j}{j}\frac{1}{\beta(z-j)}\binom{\beta(z-j)}{z-j}u^j}, \tag{11.50}$$

and by condition (11.49),

$$\frac{f(2)g(z-2)}{\sum_{j=1}^{z-1} f(j)\,g(z-j)} = \frac{\frac{1}{2\beta}\binom{2\beta}{2}\frac{1}{\beta(z-2)}\binom{\beta(z-2)}{z-2}u^2}{\sum_{j=1}^{z-1}\frac{1}{\beta j}\binom{\beta j}{j}\frac{1}{\beta(z-j)}\binom{\beta(z-j)}{z-j}u^j}. \tag{11.51}$$

Dividing (11.50) by (11.51) yields

$$\frac{f(1)g(z-1)}{f(2)\,g(z-2)} = \frac{\frac{1}{\beta(z-1)}\binom{\beta(z-1)}{z-1}}{u\left(\beta-\frac{1}{2}\right)\frac{1}{\beta(z-2)}\binom{\beta(z-2)}{z-2}},$$

which gives

$$g(z-1) = \frac{f(2)}{u\left(\beta-\frac{1}{2}\right)f(1)}\cdot\frac{\frac{1}{\beta(z-1)}\binom{\beta(z-1)}{z-1}}{\frac{1}{\beta(z-2)}\binom{\beta(z-2)}{z-2}}\,g(z-2).$$

When $z=3$,

$$g(2) = \frac{f(2)}{u\left(\beta-\frac{1}{2}\right)f(1)}\cdot\frac{1}{2\beta}\binom{2\beta}{2}g(1).$$

Also, when $z=4$,

$$g(3) = \left\{\frac{f(2)}{u\left(\beta-\frac{1}{2}\right)f(1)}\right\}^2\frac{1}{3\beta}\binom{3\beta}{3}g(1).$$

Hence, a recurrence relation is obtained as

$$g(z) = \left\{\frac{f(2)}{u\left(\beta-\frac{1}{2}\right)f(1)}\right\}^{z-1}\frac{1}{\beta z}\binom{\beta z}{z}g(1). \tag{11.52}$$

Now by assigning $\frac{f(2)}{\left(\beta-\frac{1}{2}\right)f(1)}=\theta_1(1-\theta_1)^{\beta-1}$, so that $\frac{f(2)}{u\left(\beta-\frac{1}{2}\right)f(1)}=\theta_2(1-\theta_2)^{\beta-1}$, the relation in (11.52) gives

$$g(y) = \left\{\theta_2(1-\theta_2)^{\beta-1}\right\}^y\frac{1}{\beta y}\binom{\beta y}{y}g(1). \tag{11.53}$$

By using the fact that $\sum_{y=1}^{\infty} g(y) = 1$ and the Lagrange expansion, we obtain from (11.53)

$$g(1) = \frac{\theta_2(1-\theta_2)^{\beta-1}}{\{-\ln(1-\theta_2)\}}.$$

Therefore,

$$g(y) = \frac{1}{\beta y}\binom{\beta y}{y}\frac{\theta_2^y(1-\theta_2)^{\beta y-y}}{[-\ln(1-\theta_2)]},$$

which is a GLSD with parameters (θ_2, β). Similarly, it can be shown that the r.v. X has a GLSD with parameters (θ_1, β). □

11.8 Applications

The GLSD has been found useful in research areas where the logarithmic series distribution has been applied. The GLSD can be used for the distribution of animal species and to represent population growth. Jain and Gupta (1973) used the GLSD to model the number of publications written by biologists.

Hansen and Willekens (1990) have shown that, at least asymptotically, the GLSD is a discretized version of the inverse Gaussian distribution and that it can be used as a lifetime distribution between two renewal points. They considered the application of GLSD in risk theory. Suppose an insurance company has a portfolio in which claims X_i $(i = 1, 2, \ldots)$ occur at consecutive time points Y_i $(i = 1, 2, \ldots)$. By assuming that $\{Y_i\}$ are independent negative exponentials with parameter λ and that $\{X_i\}$ is a sequence of independent r.v.s with the same distribution F, and independent of $\{Y_i\}$, the total claim size distribution up to time t is given by

$$F_t(x) = e^{-\lambda t}\sum_{k=0}^{\infty}(\lambda t)^k(k!)^{-1}\,F^{*k}(x), \qquad x \geq 0, \tag{11.54}$$

where F^{*k} is the kth convolution power of F. Hansen and Willekens (1990) assumed that F_t is GLSD and obtained the claim size distribution as

$$P(X_1 = n) = \frac{1}{\theta\beta}\left(\frac{\beta-1}{\beta(1-\theta)}\right)^{\beta-1}\frac{1}{\lambda t}\frac{\log(1-\theta)}{\log(1-\beta^{-1})}\,P_n(\theta,\beta), \tag{11.55}$$

where $P_n(\theta, \beta)$ is the GLSD in (11.1).

11.9 Related Distributions

Truncated GLSD

Since the GLSD is a modified power series distribution and since its truncated form is also a modified power series distribution, all the results in subsection 7.7.2 for truncated MPSD are applicable to the truncated GLSD.

The probability mass function of GLSD truncated on the left at the point $x = r - 1$ is given by

$$P_x(\theta,\beta,r) = \frac{1}{\beta x}\binom{\beta x}{x}\frac{\theta^x(1-\theta)^{\beta x-x}}{g_1(\theta,r)}, \qquad x = r,\ r+1,\ldots, \tag{11.56}$$

where $0 < \theta < 1,\ 1 \leq \beta < \theta^{-1}$, and

$$g_1(\theta, r) = -\ln(1-\theta) - \sum_{x=1}^{r-1} \frac{1}{\beta x}\binom{\beta x}{x}\theta^x(1-\theta)^{\beta x - x}. \tag{11.57}$$

Charalambides (1974) obtained the distribution of $Z = \sum_{i=1}^{n} X_i$ for truncated GLSD as

$$P_x(z, \beta, \theta) = \frac{\bar{s}(z, n, r)\theta^z(1-\theta)^{\beta z - z}}{g_n(\theta, r)\, z!}, \qquad z = nr,\ nr+1, \ldots, \tag{11.58}$$

where

$$g_n(\theta, r) = \frac{1}{n!}\left[-\ln(1-\theta) - \sum_{x=1}^{r-1}\frac{1}{\beta x}\binom{\beta x}{x}\theta^x(1-\theta)^{\beta x - x}\right]^n \tag{11.59}$$

and the coefficients

$$\bar{s}(z, n, r) = \frac{z!}{n!}\sum\prod_{i=1}^{n}\frac{1}{\beta x_i}\binom{\beta x_i}{x_i}, \tag{11.60}$$

where the summation is extended over all n-tuples $(x_1, x_2, \ldots, x_n)$ of integers $x_i \geq r$ such that $\sum_{i=1}^{n} x_i = z$.

Modified GLSD

Jani (1986) defined and studied the GLSD with zeroes. The pmf of generalized logarithmic series distribution with zeroes (GLSD$_0$) is given by

$$P_x(\theta, \beta, \gamma) = \begin{cases} 1-\gamma, & x = 0, \\ \frac{\gamma}{\beta x}\binom{\beta x}{x}\frac{\theta^x(1-\theta)^{\beta x - x}}{\{-\ln(1-\theta)\}}, & x > 0, \end{cases} \tag{11.61}$$

for $0 < \gamma < 1,\ 0 < \theta < 1$, and $\theta^{-1} > \beta \geq 1$. For $\gamma = 1$, the modified probability distribution in (11.61) reduces to the GLSD defined in (11.1). If the GLSD provides a good fit to a zero-truncated data set, then the GLSD with zeros must provide a satisfactory fit to the complete data set with zeros.

Hansen and Willekens (1990) showed that the GLSD with zeros (GLSD$_0$) is log-convex and hence it is infinitely divisible. The rth moment about the origin for the modified GLSD can be obtained easily by multiplying the rth moment about the origin for the GLSD in (11.1) by γ. Thus, one can obtain the first six moments for GLSD$_0$ by multiplying the results in (11.3) and (11.7)–(11.11) by γ.

Suppose a random sample of size n is taken from a population that has GLSD$_0$. Suppose the frequency at each x-value is denoted by n_x such that $n = \sum_{x=0}^{k} n_x$, where k is the largest observed value of x. The likelihood function is given by

$$L = \prod_{x=0}^{k}\{P_x(\theta, \beta, \gamma)\}^{n_x}. \tag{11.62}$$

On taking the logarithm of the likelihood function in (11.62) and differentiating partially with respect to the three parameters γ, θ, and β, the likelihood equations after simplification become

$$\hat{\gamma} = \frac{n - n_0}{n}, \tag{11.63}$$

$$\frac{\hat{\gamma}\,\hat{\alpha}\,\hat{\theta}}{1 - \hat{\theta}\hat{\beta}} = \bar{x}, \tag{11.64}$$

and

$$\hat{\gamma} \sum_{x=2}^{\infty} \sum_{j=1}^{x-1} x\, n_x (\hat{\beta}x - j)^{-1} = n\bar{x}, \tag{11.65}$$

where $\bar{x}$ is the sample mean and $\alpha = [-\ln(1-\theta)]^{-1}$. (See Exercise 11.5).
Equations (11.64) and (11.65) can be solved for the maximum likelihood estimators $\hat{\theta}$ and $\hat{\beta}$ by using the Newton–Raphson iterative method. It is quite possible that iteration may fail to converge.

Since the method of maximum likelihood may not yield a solution due to nonconvergence, one may use the method of zero-cell frequency and the first two sample moments. By equating the zero-class probability to the sample proportion for zeroclass and equating the first two population moments to the corresponding sample moments, we obtain the following three estimating equations:

$$1 - \tilde{\gamma} = \frac{n_0}{n}, \tag{11.66}$$

$$\mu_1' = \bar{x}, \tag{11.67}$$

and

$$\mu_2' - (\mu_1')^2 = s^2. \tag{11.68}$$

On solving the above simultaneously, we obtain estimators based on the zero-cell frequency and the first two moments (see Exercise 11.6).

11.10 Exercises

11.1 Show that the function $G_x(u)$ given in (11.4) is a pgf for a r.v. X with GLSD.
11.2 By using the recurrence relation in (11.16), obtain the third, fourth, and fifth central moments for GLSD.
11.3 Assuming that the result in equation (11.47) holds, prove that the r.v.s X and Y have GLSD with parameters (θ_1, β) and (θ_2, β), respectively.
11.4 Suppose a r.v. X has GLSD with zeros. Obtain a recurrence relation for the central moments. By using your relation, or otherwise, obtain the first four central moments for X.
11.5 Verify the likelihood equations in (11.63)–(11.65).
11.6 By substituting for the values of μ_1' and μ_2' in equations (11.67) and (11.68), simplify the estimating equations in (11.66)– (11.68).
11.7 A r.v. X has the logarithmic series distribution given by

$$P(X = x) = \frac{\theta^x}{x\{-\ln(1-\theta)\}}, \qquad x = 1, 2, 3, \ldots,$$

and zero otherwise and $0 < \theta < 1$. If θ is a continuous r.v. having the beta density function with parameters a and b, obtain the unconditional probability distribution of X with parameters a and b.

11.8 An insurance company has a portfolio in which the claims X_i, $i = 1, 2, 3, \ldots$, occur at consecutive time points Y_i, $i = 1, 2, 3, \ldots$, which are independent negative exponential r.v.s with parameter λ. If $\{X_i\}$ is a sequence of independent r.v.s with the same distribution function F, and independent of $\{Y_i\}$, show that the total claim size distribution up to time t is given by

$$F_t(x) = e^{-\lambda t} \sum_{k=0}^{\infty} (\lambda t)^k \, (k!)^{-1} \, F^{*k}(x), \quad x > 0,$$

where F^{*k} is the kth convolution power of F. What will be the form of the above distribution if F_i is GLSD? (Hint: See Hansen and Willekens, 1990.)

11.9 Prove that the GLSD in (11.1) is strictly log-concave and is infinitely divisible.

11.10 By using the mean of GLSD in (11.3) and the variance of GLSD in (11.17), determine the range of parameter values for which the mean is (a) smaller than, (b) equal to, and (c) greater than the variance. (Hint: Use computer programming and note that both the mean and variance of GLSD are increasing functions of β.)

12

Lagrangian Katz Distribution

12.1 Introduction and Definition

A discrete r.v. X is said to follow a Lagrangian Katz distribution (LKD) with parameters a, b, and β if its pmf is given by

$$P(X=x)=P_x(a,b,\beta)=\frac{a/\beta}{a/\beta+xb/\beta+x}\binom{a/\beta+xb/\beta+x}{x}\beta^x(1-\beta)^{a/\beta+xb/\beta}, \tag{12.1}$$

for $x=0,1,2,3,\ldots$ and zero otherwise, where $a>0$, $b>-\beta$, and $\beta<1$. Consul and Famoye (1996) formally defined and studied the above LKD. It is a member of the class of Lagrangian probability distributions $L(f;g;x)$ and is also a member of its subclass, the MPSDs. The probability model in (12.1) reduces to the Katz distribution (Katz, 1945, 1965) when $b=0$. The LKD reduces to the binomial distribution with parameters n and θ (i) when $0<\beta=\theta<1$, $a=n\theta$, $b=-\theta$ and n (positive integer); and (ii) when $b=0$, $\beta<0$, $\beta(\beta-1)^{-1}=\theta$, $-a/\beta=n$ (an integer). It reduces to the Poisson distribution when $b=0$ and $\beta\to 0$, and to the negative binomial distribution with parameters k and θ (i) when $b=0$, $0<\beta=\theta<1$, $a/\beta=k$; and (ii) when $b=\theta(1-\theta)^{-1}$, $a=k\theta(1-\theta)^{-1}$, and $\beta=-\theta(1-\theta)^{-1}$.

The GNBD in chapter 10 also contains three parameters just like the LKD; however, the GNBD lies in a narrow domain of the LKD as indicated below:

(i) When $0<\beta=\theta<1$, the values $a=n\theta$ and $b=(m-1)\theta$ change the LKD to the GNBD.
(ii) When $\beta<0$, the values $\beta=-\theta(1-\theta)^{-1}$, $a=n\theta(1-\theta)^{-1}$, and $b=m\theta(1-\theta)^{-1}$ change the LKD to the GNBD.

For other values of a, b, and β, the LKD is well defined outside the domain of the GNBD. Also, the LKD provides the generalized Poisson distribution (discussed in chapter 9) as a limiting form when $\beta\to 0$ under suitable conditions.

12.2 Generating Functions

The pgf of the LKD is given by

$$H(u)=f(z)=(1-\beta+\beta z)^{a/\beta}\quad\text{where}\quad z=u(1-\beta+\beta z)^{1+b/\beta} \tag{12.2}$$

for all values of the parameters in (12.1).

Another pgf for the LKD is

$$G(u) = f(z) = \left(\frac{1-\beta z}{1-\beta}\right)^{-a/\beta}, \tag{12.3}$$

where

$$z = u\left(\frac{1-\beta z}{1-\beta}\right)^{-b/\beta}. \tag{12.4}$$

Thus, the LKD is one of those few probability distributions which is generated by two sets of functions as given above. Note that for some values of a, b, and β in (12.1), the four functions in (12.2), (12.3), and (12.4) may not be pgfs in z; however, the functions are such that their values for $z = 1$ are unity.

By replacing β with c in the exponents of the functions $\left(\frac{1-\beta z}{1-\beta}\right)^{-a/\beta}$ and $(1-\beta+\beta z)^{1+b/\beta}$ in the two pgfs (12.3) with (12.4) and in (12.2), Janardan (1998) has generalized the two pgfs and their corresponding LKD in (12.1) to the form

$$P(X = x) = \frac{a/c}{a/c + xb/c + x}\binom{a/c + xb/c + x}{x}\beta^x(1-\beta)^{a/c+xb/c} \tag{12.5}$$

for $x = 0, 1, 2, \ldots$ and where $a > 0$, $c > 0$, $b \geq -c$, $0 < \beta < 1$, and has called it the generalized Pólya–Eggenberger distribution. As this model is not related to the Pólya–Eggenberger distribution and is a generalization of the LKD, we rename it the generalized Lagrangian Katz distribution I (GLKD_1).

Obviously, the special cases of the GLKD_1 are (i) the LKD (when c is replaced with β), (ii) the Katz distribution (when $b = 0$ and c is replaced with β), (iii) the GNBD (for $c = 1$), (iv) the negative binomial distribution (for $b = 0$), (v) the binomial distribution (for $c = 1$, $b = -1$), and (vi) the Poisson distribution (for $c \to 0$, $\beta \to 0$ such that $a\beta/c = \theta$).

12.3 Moments, Cumulants, and Recurrence Relations

Let $a = c\beta$ and $b = h\beta$ in the LKD defined in (12.1). Thus, the probability model in (12.1) becomes

$$P_x(c\beta, h\beta, \beta) = \frac{c}{c + hx + x}\binom{c + hx + x}{x}\beta^x(1-\beta)^{c+hx}, \tag{12.6}$$

which gives

$$(1-\beta)^{-c} = \sum_{x=0}^{\infty}\frac{c}{c + hx + x}\binom{c + hx + x}{x}\beta^x(1-\beta)^{hx}.$$

On differentiating with respect to β,

$$c(1-\beta)^{-c-1} = \sum_{x=0}^{\infty}\frac{c}{c + hx + x}\binom{c + hx + x}{x}\cdot x(1-\beta-h\beta)\cdot\beta^{x-1}(1-\beta)^{hx-1}.$$

The above gives

$$\mu_1' = \sum_{x=0}^{\infty} x \cdot P_x(c\beta, h\beta, \beta) = c\beta(1-\beta-h\beta)^{-1}. \tag{12.7}$$

Now, the rth noncentral moment is given by

$$\mu'_r = E(X^r) = \sum_{x=0}^{\infty} x^r P_x(c\beta, h\beta, \beta). \tag{12.8}$$

By using the method of differentiation on (12.8) with respect to β and on simplification, a recurrence relation between the noncentral moments of LKD becomes

$$\mu'_{r+1} = \frac{\beta(1-\beta)}{1-\beta-h\beta}\frac{d\mu'_r}{d\beta} + \mu'_1\mu'_r \tag{12.9}$$

for $r = 0, 1, 2, \ldots$, where $\mu'_0 = 1$, $\mu'_1 = \beta c(1-\beta-h\beta)^{-1} = a(1-b-\beta)^{-1}$. Note that one has to differentiate μ'_r with respect to β first and then substitute for the values of $c = a/\beta$ and $h = b/\beta$.

By putting $r = 1$ in (12.9),

$$\mu'_2 - (\mu'_1)^2 = \frac{\beta(1-\beta)}{1-\beta-h\beta)}\frac{d}{d\beta}\left[\beta c(1-\beta-h\beta)^{-1}\right] = \frac{c\beta(1-\beta)}{(1-\beta-h\beta)^3}.$$

Thus,

$$\sigma^2 = c\beta(1-\beta)(1-\beta-h\beta)^{-3} = a(1-\beta)(1-b-\beta)^{-3}. \tag{12.10}$$

By using the same method of differentiation with respect to β as in the noncentral moments, a recurrence relation between the central moments of LKD can be shown to be

$$\mu_{r+1} = \frac{\beta(1-\beta)}{1-\beta-h\beta}\frac{d\mu_r}{d\beta} + r\mu_2\mu_{r-1} \tag{12.11}$$

for $r = 1, 2, 3, \ldots$, where $\mu_1 = 0$, $\mu_2 = a(1-\beta)(1-b-\beta)^{-3}$.

Consul and Famoye (1996) obtained the first four cumulants of the LKD as

$$\left.\begin{aligned} L_1 &= a(1-b-\beta)^{-1} = \mu \\ L_2 &= a(1-\beta)(1-b-\beta)^{-3} = \sigma^2 \\ L_3 &= a(1-\beta^2)(1-b-\beta)^{-4} + 3ab(1-\beta)(1-b-\beta)^{-5} \end{aligned}\right\} \tag{12.12}$$

and

$$L_4 = a(1-\beta)\Big\{(1+4\beta+\beta^2)(1-b-\beta)^{-5} + 10b(1+\beta)(1-b-\beta)^{-6}$$
$$+15b^2(1-b-\beta)^{-7}\Big\}. \tag{12.13}$$

A recurrence relation between the LKD probabilities is

$$P_{x+1} = \frac{a+b(x+1)+\beta x}{x+1}(1-\beta)^{b/\beta}\prod_{i=1}^{x-1}\left(1+\frac{b}{a+bx+\beta i}\right)\cdot P_x \tag{12.14}$$

for $x = 1, 2, 3, 4, \ldots$, where $P_0 = (1-\beta)^{a/\beta}$ and $P_1 = a(1-\beta)^{b/\beta}P_0$. The recurrence relation in (12.14) is useful for computing LKD probabilities.

Janardan (1998) has also given a recurrence relation (similar to (12.14)) for the probabilities of GLKD_1. The mean and the variance of the GLKD_1 are

$$L_1 = \mu = a(\beta/c)(1-\beta-b\beta/c)^{-1}$$

and

$$L_2 = \sigma^2 = a(1-\beta)(\beta/c)(1-\beta-b\beta/c)^{-3}.$$

12.4 Other Important Properties

Some of the important properties of GLKD_1 and LKD are given in this section. The results for GLKD_1 reduce to the results for LKD when $c = \beta$.

Theorem 12.1. *If X_1 and X_2 are two mutually independent $GLKD_1$ variates with parameters (a_1, b, c, β) and (a_2, b, c, β), then their sum $Y = X_1 + X_2$ is a $GLKD_1$ variate with parameters $(a_1 + a_2, b, c, \beta)$.*

Proof. The joint probability distribution of X_1 and X_2 is given as

$$P(X_1 = x_1, X_2 = x_2) = J_{x_1}(a_1, b, c) J_{x_2}(a_2, b, c) \beta^{x_1+x_2} (1-\beta)^{(a_1+a_2+bx_1+bx_2)/c},$$

where

$$J_{x_i}(a_i, b, c) = \frac{a_i/c}{a_i/c + bx_i/c + x_i} \binom{a_i/c + bx_i/c + x_i}{x_i}, \quad i = 1, 2.$$

Therefore, the probability distribution of the sum $Y = X_1 + X_2$ is obtained by putting $x_2 = y - x_1$ and by summing over x_1 as

$$P(Y = y) = \beta^y (1-\beta)^{(a_1+a_2+by)/c} \sum_{x_1=0}^{y} J_{x_1}(a_1, b, c) J_{y-x_1}(a_2, b, c).$$

By using identity (1.85) on the above, it gives

$$P(Y = y) = \frac{(a_1 + a_2)/c}{(a_1 + a_2)/c + by/c + y} \beta^y (1-\beta)^{(a_1+a_2+by)/c} \tag{12.15}$$

for $y = 0, 1, 2, \ldots$, which is a GLKD_1 with parameters $(a_1 + a_2, b, c, \beta)$. □

Theorem 12.1 can be extended to any number of GLKD_1 variates. Hence, the GLKD_1 is closed under convolution.

Theorem 12.2. *Let X_1 and X_2 be two independent $GLKD_1$ variates with parameters (a_1, b, c, β) and $(a - a_1, b, c, \beta)$. Then the conditional distribution of $X_1 = x$, given that $X_1 + X_2 = z$, is a generalized negative hypergeometric distribution with parameters $(a/c,\ a_1/c,\ b/c)$.*

Proof. By definition and on cancellation of the common terms,

$$P(X_1 = x | X_1 + X_2 = z) = \frac{J_x(a_1, b, c) J_{z-x}(a - a_1, b, c)}{\sum_{x=0}^{z} J_x(a_1, b, c) J_{z-x}(a - a_1, b, c)}.$$

On using the identity (1.85) on the denominator, we get

$$P(X_1 = x | X_1 + X_2 = z) = \frac{J_x(a_1, b, c) J_{z-x}(a - a_1, b, c)}{J_z(a, b, c)}, \tag{12.16}$$

which is the generalized negative hypergeometric distribution with parameters $(a/c, a_1/c, b/c)$. □

Theorem 12.3. *The zero-truncated LKD with parameters a, b, and β approaches the generalized logarithmic series distribution with parameters b and b/β as $a \to 0$ and is given by*

$$P(X = x) = [-\ln(1-\beta)]^{-1} \frac{(xb/\beta + x - 1)!}{x!(xb/\beta)!} \beta^x (1-\beta)^{xb/\beta}$$

for $x = 1, 2, 3, \ldots$ and zero otherwise.

Proof. See Exercise 12.3

Theorem 12.4. *Let X_1 and X_2 be two independent r.v.s such that their sum $Z = X_1 + X_2$ with $z = 0, 1, 2, 3, \ldots$. If the conditional distribution of $X_1 = x$, given that $Z = z$ is the generalized negative hypergeometric distribution in Theorem 12.2, then each one of the r.v.s X_1 and X_2 has a LKD distribution.*

Proof. Let $p(x_1)$ and $q(x_2)$ denote the pmfs of X_1 and X_2, respectively. Then

$$\frac{p(x)q(z-x)}{\sum_{x=0}^{z} p(x)q(z-x)} = \frac{J_x(a_1, b, \beta) J_{z-x}(a_2, b, \beta)}{J_z(a_1 + a_2, b, \beta)}.$$

Therefore, for all integral values of x and $z \geq x$, we have

$$\frac{p(x)q(z-x)}{p(x-1)q(z-x+1)} = \frac{J_x(a_1, b, \beta) J_{z-x}(a_2, b, \beta)}{J_{x-1}(a_1, b, \beta) J_{z-x+1}(a_2, b, \beta)}.$$

By putting $z = x$ in the above, we have

$$\frac{p(x)}{p(x-1)} = \frac{q(1)}{q(0)} \cdot \frac{J_x(a_1, b, \beta)}{J_{x-1}(a_1, b, \beta)} \cdot \frac{\beta}{a_2}. \tag{12.17}$$

On using $x = 1, 2, 3, \ldots$ in (12.17) and multiplying them together, we obtain

$$p(x) = p(0)\beta^x \left(\frac{q(1)}{a_2 q(0)} \right)^x J_x(a_1, b, \beta). \tag{12.18}$$

Since $p(x) > 0$ for all values of x and $\sum_{x=0}^{\infty} p(x) = 1$, the summation of (12.18) over all values of x gives

$$1 = \sum_{x=0}^{\infty} \frac{a_1/\beta}{a_1/\beta + xb/\beta + x} \binom{a_1/\beta + xb/\beta + x}{x} \beta^x \left(\frac{q(1)}{a_2 q(0)} \right)^x \cdot p(0). \tag{12.19}$$

The right-hand-side of (12.19) is the sum of a power series similar to the sum of LKD in (12.1) over $x = 0$ to ∞. Thus, we have that

$$p(0) = (1-\beta)^{a_1/\beta}.$$

Therefore, the r.v. X_1 has a LKD with parameters (a_1, b, β). Similarly, it can be shown that X_2 has a LKD with parameters (a_2, b, β). □

12.5 Estimation

Let a random sample of size n be taken from the LKD model (12.1) and let the observed values be $x_1, x_2, \ldots, x_n$. Also, let n_i denote the frequency of the ith class, $i = 0, 1, 2, \ldots, k$, where k is the largest observed value. The sample sum y can be written in the form

$$y = \sum_{j=1}^{n} x_j = \sum_{i=0}^{k} i n_i, \tag{12.20}$$

where $\sum_{i=0}^{k} n_i = n$ and $\bar{x} = y/n$ is the sample mean. The sample variance is

$$S_2 = (n-1)^{-1} \sum_{i=0}^{k} n_i (i - \bar{x})^2 = (n-1)^{-1} \sum_{j=1}^{n} (x_j - \bar{x})^2. \tag{12.21}$$

The third central moment for the sample is given by

$$S_3 = (n-1)^{-1} \sum_{i=0}^{k} n_i (i - \bar{x})^3 = (n-1)^{-1} \sum_{j=1}^{n} (x_j - \bar{x})^3. \tag{12.22}$$

Moment Estimation

Consul and Famoye (1996) have given the moment estimators of a, b, and β of LKD in (12.1) as

$$\tilde{\beta} = 2 - \frac{1}{2}(A \pm \sqrt{A(A-4)}), \tag{12.23}$$

$$\tilde{a} = \frac{1}{2}(\bar{x})^{3/2}(S_2)^{-1/2}(\sqrt{A} \pm \sqrt{A-4}), \tag{12.24}$$

and

$$\tilde{b} = -1 + \frac{1}{2}(\sqrt{A} \pm \sqrt{A-4})(\sqrt{A} - \sqrt{\bar{x}/S_2}, \tag{12.25}$$

where

$$A = (3S_2^2 - S_3\bar{x})^2(\bar{x} - S_2^3)^{-1}. \tag{12.26}$$

From (12.23), the value of A has to be more than 4, otherwise β is not a real number. For a frequency distribution in which $A < 4$, the moment estimates for parameters a, b, and β do not exist.

Estimation Based on Moments and Zero-Class Frequency

By equating the zero-class probability with zero-class sample proportion $n_0/n = f_0$, it can be seen that

$$(1-\beta)^{a/\beta} = f_0 \quad \text{or} \quad a = \frac{\beta \log(f_0)}{\log(1-\beta)}. \tag{12.27}$$

The estimates based on zero-class frequency and moments are obtained from solving equation (12.27) along with the following two equations:

$$\mu_1' = a(1-b-\beta)^{-1} = \bar{x} \tag{12.28}$$

and

$$\mu_2' = a(1-\beta)(1-b-\beta)^{-3} = S_2. \tag{12.29}$$

On eliminating a and b between (12.27), (12.28), and (12.29),

$$(1-\beta)[\log(1-\beta)]^2 = \beta^2 S_2(\bar{x})^{-3}[\log(f_0)]^2, \tag{12.30}$$

which gives an estimate for β either graphically or by a numerical solution using the Newton–Raphson method. On getting the estimate $\bar{\beta}$ of β from (12.30), the estimates of a and b are given, respectively, by

$$\bar{a} = [\bar{x}^3(1-\bar{\beta})/S_2]^{1/2}$$

and

$$\bar{b} = 1 - \bar{\beta} - \bar{a}/\bar{x}.$$

Maximum Likelihood Estimation

The three unknown parameters in the LKD are a, b, and β. The log likelihood function of the LKD is

$$\begin{aligned}\ell &= \log L(a, b, \beta) \\ &= (n - n_0)\log(a) + \frac{n(a + b\bar{x})}{\beta}\log(1 - \beta) - \sum_{i=2}^{k} n_i \log(i!) + \sum_{i=2}^{k}\sum_{j=1}^{i-1} n_i \log(a + bi + \beta j).\end{aligned} \tag{12.31}$$

On differentiating (12.31) partially and setting the derivatives equal to zero, the likelihood equations become

$$\frac{\partial \ell}{\partial a} = \frac{n - n_0}{a} + \frac{n}{\beta}\log(1 - \beta) + \sum_{i=2}^{k}\sum_{j=1}^{i-1} \frac{n_i}{a + bi + \beta j} = 0, \tag{12.32}$$

$$\frac{\partial \ell}{\partial b} = \frac{n\bar{x}\log(1 - \beta)}{\beta} + \sum_{i=2}^{k}\sum_{j=1}^{i-1} \frac{i n_i}{a + bi + \beta j} = 0, \tag{12.33}$$

and

$$\frac{\partial \ell}{\partial \beta} = \frac{-n(a + b\bar{x})\log(1 - \beta)}{\beta^2} - \frac{n(a + b\bar{x})}{\beta(1 - \beta)} + \sum_{i=2}^{k}\sum_{j=1}^{i-1} \frac{j n_i}{a + bi + \beta j} = 0. \tag{12.34}$$

On multiplying equation (12.32) by a, equation (12.33) by b, equation (12.34) by β, and simplifying, it can be shown that

$$\hat{a} = \bar{x}(1 - b - \beta). \tag{12.35}$$

On using (12.35) in (12.33) and (12.34), the two ML equations become

$$\frac{n\bar{x}\log(1 - \beta)}{\beta} + \sum_{i=2}^{k}\sum_{j=1}^{i-1} \frac{i n_i}{\bar{x}(1 - b - \beta) + bi + \beta j} = 0 \tag{12.36}$$

and

$$\frac{-n\bar{x}(1 - \beta)\log(1 - \beta)}{\beta} - \frac{n\bar{x}}{\beta} + \sum_{i=2}^{k}\sum_{j=1}^{i-1} \frac{j n_i}{\bar{x}(1 - b - \beta) + bi + \beta j} = 0. \tag{12.37}$$

The maximum likelihood estimates $\hat{b}$ and $\hat{\beta}$ are obtained by solving equations (12.36) and (12.37) iteratively, starting with the moment estimates $\tilde{b}$ and $\tilde{\beta}$ (or the moment and zero-class estimates $\bar{b}$ and $\bar{\beta}$) as the initial values of b and β. The Newton–Raphson iterative technique or some other techniques can be used. Then, the ML estimate $\hat{a}$ of a is given by (12.35).

12.6 Applications

The LKD is a very versatile probability model. It was shown in section 12.1 that the domain of its three parameters a, b, and β is much larger than the domain of the parameters of the GNBD and that it reduces to the GNBD for some specific values of the parameters when β is positive as well as when β is negative. The limiting form of the LKD is the GPD. Thus, the LKD reduces to the binomial and the negative binomial distributions, which are special cases of the GNBD, and to the Poisson distribution, which is a special case of the GPD. The LKD is applicable to physical situations for which the binomial, the negative binomial, the Poisson, the GNB, and the GP distributions are found useful.

Consul and Famoye (1996) have applied the LKD to fit three sets of data from Beall and Rescia (1953). The fit of LKD to the data on Lespedeza capitata was found to be better (as judged by the chi-square goodness-of-fit test) than the fit from Neyman Type A distribution. For frequency data on tribes in Clearwater, Idaho, the LKD provided a better fit than the Neyman Type A distribution. The LKD was also applied to the frequency distribution of potato beetle data and the fit from LKD was better than that of Neyman Type A distribution.

12.7 Related Distributions

12.7.1 Basic LKD of Type I

A basic LKD of type I is denoted by basic LKD-I and is defined by the pmf

$$P(X=x) = \frac{1}{x}\binom{xb/\beta + x - 2}{x-1}\beta^{x-1}(1-\beta)^{xb/\beta} \tag{12.38}$$

for $x = 1, 2, 3, \ldots$ and zero otherwise.

The basic LKD-I is the limit of zero-truncated LKD as $a \to -\beta$. The basic LKD-I in (12.38) reduces to the

(i) Borel distribution when $\beta \to 0$;
(ii) Consul distribution (discussed in chapter 8) when $\beta < 0$, $\beta = -\alpha$, $b = m\alpha$, and $\alpha(1+\alpha)^{-1} = \theta$;
(iii) Geeta distribution (discussed in chapter 8) when $0 < \beta = \theta < 1$ and $b = (m-1)\theta$.

The pgf for the basic LKD-I is given by

$$H(u) = \varphi(z) = z, \quad \text{where} \quad z = u\left(\frac{1-\beta z}{1-\beta}\right)^{-b/\beta}. \tag{12.39}$$

Following the same method as for LKD, it can be shown that the mean and variance of basic LKD-I are given, respectively, by

$$\mu = (1-\beta)(1-b-\beta)^{-1} \quad \text{and} \quad \sigma^2 = b(1-\beta)(1-b-\beta)^{-3}. \tag{12.40}$$

Putting $b = h\beta$ in (12.38) and denoting the rth moment, of basic LKD-I, about the origin by ${}_1\mu'_r$,

$${}_1\mu'_r = \sum_{x=1}^{\infty} x^{r-1}\binom{xh + x - 2}{x-1}\beta^{x-1}(1-\beta)^{xh}.$$

On differentiation with respect to β and on simplification,

$$\frac{d}{d\beta}\left({}_1\mu'_r\right) = \sum_{x=1}^{\infty} x^{r-1}\binom{xh+x-2}{x-1}\beta^{x-1}(1-\beta)^{xh}\frac{x(1-\beta-h\beta)-(1-\beta)}{\beta(1-\beta)}$$

$$= \frac{1-\beta-h\beta}{\beta(1-\beta)}\left({}_1\mu'_{r+1}\right) - \left({}_1\mu'_r\right)\beta^{-1},$$

which gives the recurrence relation

$${}_1\mu'_{r+1} = \frac{\beta(1-\beta)}{1-\beta-h\beta}\frac{d}{d\beta}\left({}_1\mu'_r\right) + \left({}_1\mu'_r\right)\left({}_1\mu'_1\right).$$

Though the mean and the variance given in (12.40) of the basic LKD-I are different from those of the LKD, yet the above recurrence relation is exactly the same as (12.9) for the LKD.

The basic LKD-I is not closed under convolution even though the distribution of the sum of two independent basic LKD-I variates can be obtained. The distribution of the sum Y of n independent basic LKD-I variates is

$$P(Y=y) = \frac{n}{y}\binom{yb/\beta+y-n-1}{y-n}\beta^{y-n}(1-\beta)^{yb/\beta} \tag{12.41}$$

for $y = n, n+1, n+2, \ldots$ and zero otherwise. The probability distribution in (12.41) is sometimes called the delta-LKD.

12.7.2 Basic LKD of Type II

A basic LKD of type II (basic LKD-II) is defined by the probability mass

$$P(X=x) = \frac{1}{x}\binom{xb/\beta+x}{x-1}\beta^{x-1}(1-\beta)^{1+xb/\beta} \tag{12.42}$$

for $x = 1, 2, 3, \ldots$ and zero otherwise.

The basic LKD-II is the limit of zero-truncated LKD as $a \to \beta$. The basic LKD-II in (12.42) reduces to the

(i) Borel distribution when $\beta \to 0$;
(ii) Consul distribution (discussed in chapter 8) when $0 < \beta = \theta < 1$ and $b = (m-1)\theta$;
(iii) Geeta distribution (discussed in chapter 8) when $\beta < 0$, $\beta = -\alpha$, $b = m\alpha$ and $\alpha(1+\alpha)^{-1} = \theta$.

The pgf for the basic LKD-II is given by

$$H(u) = \varphi(z) = z, \quad \text{where} \quad z = u(1-\beta+\beta z)^{1+b/\beta}. \tag{12.43}$$

It can easily be shown that the mean and variance of the basic LKD-II are, respectively, given by

$$\mu = (1-b-\beta)^{-1} \quad \text{and} \quad \sigma^2 = (b+\beta)(1-\beta)(1-b-\beta)^{-3}. \tag{12.44}$$

Putting $b = h\beta$ in (12.42) and denoting the rth moment of the basic LKD-II, about the origin, by ${}_2\mu'_r$, we have

$${}_2\mu'_r = \sum_{x=1}^{\infty} x^{r-1} \binom{xh+x}{x} \beta^x (1-\beta)^{1+xh}.$$

Differentiating the above with respect to β and on simplification,

$$\frac{d}{d\beta}\left({}_2\mu'_r\right) = \sum_{x=1}^{\infty} x^{r-1} \binom{xh+x}{x} \beta^{x-1}(1-\beta)^{1+xh} \left[\frac{x(1-\beta-h\beta)-1}{\beta(1-\beta)}\right],$$

which gives the recurrence relation among the moments of the basic LKD-II as

$${}_2\mu'_{r+1} = \frac{\beta(1-\beta)}{1-\beta-h\beta)}\frac{d}{d\beta}\left({}_2\mu'_r\right) + \left({}_2\mu'_r\right)\left({}_2\mu'_1\right).$$

Again, the mean and variance in (12.44), of the basic LKD-II are different than the mean and the variance of the basic LKD-I. However, the recurrence relation between the moments about the origin is the same for both models.

12.7.3 Basic GLKD of Type II

By taking the Lagrangian pgf

$$h(u) = f(z) = \left(\frac{1-\beta z}{1-\beta}\right)^{-a/c}, \quad \text{where} \quad z = u\left(\frac{1-\beta z}{1-\beta}\right)^{-b/c},$$

and using the Lagrange expansion in (1.80), Janardan (1998) obtained a GLKD of type II ($GLKD_2$) as

$$P(X=x) = (1-\beta-b\beta/c)\binom{a/c+(1+b/c)x-1}{x}\beta^x(1-\beta)^{(a+bx)/c-1} \tag{12.45}$$

for $x = 0, 1, 2, \ldots$, and zero elsewhere, and $a > 0,\ c > 0,\ b \geq -c,\ 0 < \beta < 1$. Just like $GLKD_1$, the model $GLKD_2$ has many special cases:

(i) the negative binomial distribution (for $b = 0$),
(ii) the Katz distribution (for $b = 0,\ c = \beta$),
(iii) the binomial distribution (for $b = -1,\ c = 1,\ a$ an integer),
(iv) the linear function negative binomial distribution (for $c = 1$) with pmf

$$P(X=x) = [1-\beta(b+1)]\binom{a-1+(b+1)x}{x}\beta^x(1-\beta)^{a-1+bx}, \quad x = 0, 1, 2, \ldots,$$

with parameters $(a-1, b+1, \beta)$ (Janardan and Rao, 1983).
(v) When $c = \beta$ in (12.45), the $GLKD_2$ reduces to a weighted LKD with pmf

$$P(X=x) = (1-\beta-b)\binom{a/\beta+xb/\beta+x-1}{x}\beta^x(1-\beta)^{a/\beta+xb/\beta-1}, \quad x = 0, 1, 2, \ldots.$$

Denoting the combinatorial function in (12.45) by H and since $\sum_x P(X = x) = 1$, (12.45) gives

$$(1-\beta)^{-a/c} = \sum_x H(1-\beta-b\beta/c)\beta^x(1-\beta)^{bx/c-1}.$$

On differentiation of the above with respect to β and multiplying by $(1-\beta)^{a/c}$, on simplification, it gives the mean μ as

$$\mu = \frac{a\beta/c}{1-\beta-b\beta/c} + \frac{b\beta/c}{(1-\beta-b\beta/c)^2}.$$

In a similar manner, one can show that the variance is given by

$$\sigma^2 = \frac{a(1-\beta)\beta/c}{(1-\beta-b\beta/c)^3} + \frac{b(1+\beta+b\beta/c)(1-\beta)\beta/c}{(1-\beta-b\beta/c)^4}.$$

Theorem 12.5. *If a r.v. X has the model GLKD$_1$ and it is weighted by $\omega(X) = a + bX$, then the weighted r.v. X^* is GLKD$_2$ (Janardan, 1998).*

The theorem can be easily proved by getting $E[a + bX]$ and by multiplying the probability mass function of GLKD$_1$ by $(a + bx)\,\{E[a + bX]\}^{-1}$.

Theorem 12.6. *Let X be a GLKD$_2$ r.v. with parameters (a, b, c, β). If it is weighted by $\omega(X) = c(a + bX)^{-1}$, then the weighted r.v. X^* follows the GLKD$_1$ with parameters (a, b, c, β).*

Proof. $E[c(a+bX)^{-1}] = (1-\beta-b\beta/c)(c/a)(1-\beta)^{-1}$ by using the sum of the probability mass function of GLKD$_1$ over $x = 0$ to ∞. Then

$$P(X^* = x) = \frac{c(a+bx)^{-1}(1-\beta-b\beta/c)}{(1-\beta-b\beta/c)(c/a)(1-\beta)^{-1}} \frac{(a/c+xb/c+x-1)!}{x!(a/c+xb/c-1)!}\beta^x(1-\beta)^{a/c+xb/c-1}$$

$$= \frac{a/c}{a/c+xb/c+x} \frac{(a/c+xb/c+x)!}{x!(a/c+xb/c)!}\beta^x(1-\beta)^{a/c+xb/c}.$$

□

12.8 Exercises

12.1 By differentiating the result in (12.8) with respect to β, show that a recurrence relation between the noncentral moments of LKD is given by (12.9). Also, show that a recurrence relation between the central moments is given by (12.11).

12.2 Show that the basic LKD-II is not closed under convolution. Find the distribution of the sum Y of n independent basic LKD-II variates.

12.3 Prove that the zero-truncated LKD with parameters a, b, and β approaches the generalized logarithmic series distribution with parameters b and b/β as $a \to 0$.

12.4 Show that the basic LKD-I is the limit of zero-truncated LKD when the parameter $a \to -\beta$.

12.5 Obtain the moment estimates for the parameters of LKD-I and LKD-II.

12.6 Show that the function $G(u)$ given in (12.3) is a pgf for a r.v. X with LKD.

12.7 Suppose $x_1, x_2, \ldots, x_n$ is a random sample from a LKD (given by (12.6)) with parameter β as the only unknown. Under a prior distribution which is beta with parameters a and b, find the Bayes estimator of β.

12.8 Show that the LKD defined by

$$P(X = x) = \frac{a/\beta}{a/\beta + xb/\beta + x} \binom{a/\beta + xb/\beta + x}{x} \beta^x (1 - \beta)^{a/\beta + xb/\beta}$$

for $x = 0, 1, 2, 3, \ldots$ and zero otherwise, where $a > 0,\ b > -\beta$, and $\beta < 1$, provides the GPD as a limiting form when $\beta \to 0$ under suitable conditions.

12.9 A r.v. X is a GLKD$_1$ given by

$$P(X = x) = \frac{a/c}{a/c + xb/c + x} \binom{a/c + xb/c + x}{x} \beta^x (1 - \beta)^{a/c + xb/c}$$

for $x = 0, 1, 2, 3, \ldots$, zero otherwise, and where $a > 0,\ c > 0,\ b \geq -c,\ 0 < \beta < 1$. If each random variable X is weighted by $\omega(X) = a + bX$, obtain the probability distribution of the weighted r.v. X^*.

12.10 Show that the pgf of the GLKD$_1$ in Exercise 12.9 is

$$G(u) = f(z) = \left(\frac{1 - \beta z}{1 - \beta} \right)^{-a/c}, \quad \text{where} \quad z = u \left(\frac{1 - \beta z}{1 - \beta} \right)^{-b/c},$$

by expanding $G(u)$ in powers of u and by obtaining the coefficient of u^x.

12.11 Obtain the zero-truncated LKD. By using the method of differentiation, obtain a recurrence relation between its noncentral moments. Obtain the mean and variance of zero-truncated LKD.

13

Random Walks and Jump Models

13.1 Introduction

A particle is said to perform a simple random walk on a line; when starting from an initial position (an integer n) on the line, it moves each time from its position either a unit step $(+1)$ in the positive direction with some probability p $(0 < p < 1)$ or a unit step (-1) in the negative direction with probability $q = 1 - p$. In a general random walk problem, the particle may move each time from its position a unit step (-1) in the negative direction, or it may stay at its position or jump $1, 2, 3, \ldots, k$ steps in the positive direction on the line. Various probabilities are assigned for the various possible mutually exclusive moves or jumps of the particle such that the sum of the probabilities is unity. Also, a random walk may have either absorbing or reflecting barriers on one or both sides on the line. Thus, there can be numerous variations in the random walk problems.

The random walk models provide a first approximation to the theory of Brownian motion. The random walk problems, the classical gamblers' ruin problems, the queuing problems, the problems of the total size of an epidemic, and the ballot problems have been independently studied for a very long time by numerous researchers. The methods used for solving such problems are based on either the study of lattice paths, formulating difference equations, or the generating functions. All these techniques essentially use different combinatorial methods (see McKendrick, 1926; Feller, 1957, 1968; Takács, 1962; Spitzer, 1964; Prabhu, 1965; Mohanty, 1966, 1979). The probability models for these problems are often given in the form of pgfs and they appear to be different on account of some differences in the assumptions to the problems. Kemp and Kemp (1968) showed that the distribution of the number of steps to ruin in a gambler's ruin random walk with initial position n and the distribution of the number of customers served during a busy period of M|M|1, when the service starts with n customers, are the same.

Hill and Gulati (1981) considered a simple random walk with an absorbing barrier at zero associated with the game of roulette and used the method of difference equations and Lagrange expansion to obtain the probability of ruin at exactly the nth step. Recently, Consul (1994b) has given a general theorem for obtaining the pgf of the distribution of the probabilities of absorption of a particle at the origin when it starts from a point n and performs a polynomial walk or jumps in each step and has applied it to obtain the exact probability distributions for a number of specific walks.

13.2 Simplest Random Walk with Absorbing Barrier at the Origin

Let a particle be at a distance of unit space from the origin, which has an absorbing barrier. The barrier does not allow the particle to jump away from the origin. Let p be the probability of a unit jump towards the origin and $q = 1 - p$ be the probability of failing to jump. Let X denote the number of trials in which the particle reaches the origin. It is evident that X has a geometric distribution given by $P(X = x) = pq^{x-1}$, $x = 1, 2, 3, \ldots$, whose pgf is $l(u) = \sum_x u^x pq^{x-1}$.

Let us now apply the Lagrange transformation. For each trial made by the particle the pgf is $g(z) = p + qz$ and thus the Lagrange transformation becomes $z = u(p + qz)$, i.e., $z(1 - uq) = up$ or

$$z = l(u) = up(1 - uq)^{-1} = \sum_{x=1}^{\infty} u^x pq^{x-1}, \tag{13.1}$$

which is the pgf of the geometric distribution for the random variable X. Thus, the two pgfs are the same.

13.3 Gambler's Ruin Random Walk

The gambler's ruin problem (playing against an infinitely rich person) is a random walk on the points $0, 1, 2, 3, \ldots$, with an initial sum (position) of n and with an absorbing barrier at $x = 0$. We shall first consider $n = 1$. If p is the probability of one step towards the origin and $q = 1 - p$ is the probability of one step to the right, i.e., of reaching the sum of 2, then the pgf of these changes is $g(z) = p + qz^2$.

Let X denote the total number of steps to absorption (ruin). According to Feller (1957) and Kemp and Kemp (1968), the pgf of the distribution of X is

$$l(u) = \left[\left\{1 - (1 - 4pqu^2)^{\frac{1}{2}}\right\} \Big/ 2qu\right]. \tag{13.2}$$

By using the Lagrange transformation on the above $g(z) = p + qz^2$, we get

$$z = u(p + qz^2) \qquad \text{i.e.} \quad uqz^2 - z + up = 0.$$

On solving this quadratic equation for z, we have the pgf

$$z = l(u) = \left\{1 - \sqrt{(1 - 4pqu^2)}\right\} \Big/ 2qu,$$

which is the same as (13.2). Thus, the use of Lagrange transformation provides the pgf of the total number of trials before absorption in a random walk. By using the Lagrange expansion of z one can express the same pgf in the form

$$\begin{aligned} z = l(u) &= \sum_{x=1}^{\infty} \frac{u^x}{x!} D^{x-1} \left[(p + qz^2)^x\right]_{z=0} \\ &= \sum_{x=1}^{\infty} \frac{u^x}{x!} D^{x-1} \left[\sum_{k=0}^{x} \binom{x}{k} q^k z^{2k} p^{x-k}\right]_{z=0} \\ &= \sum_{k=0}^{\infty} \frac{u^{2k+1}}{(2k+1)!} p^{k+1} q^k \cdot \binom{2k+1}{k} (2k)!, \end{aligned} \tag{13.3}$$

which gives the probability distribution of X, the number of steps to ruin, as

$$P(X = 2k + 1) = \frac{(2k)!}{k!(k + 1)!} q^k p^{k+1} \tag{13.4}$$

for $k = 0, 1, 2, 3, \ldots$, and where the particle takes k steps to the right and $k+1$ steps to the left before absorption at the origin (i.e., ruin of gambler). The above probability distribution of X is a modified form of the GNBD defined by Jain and Consul (1971) as the r.v. X takes the odd integral values $1, 3, 5, 7, \ldots$ only. One can easily verify that (13.2) and (13.3) are equivalent to each other.

When the gambler starts with an initial capital of n (instead of 1), the pgf of X can be written by taking the nth power of $l(u)$ in (13.2) or (13.3), as done by Kemp and Kemp (1968); however, the computation of the absorption (ruin) probabilities is somewhat involved.

A simpler method of computing the absorption probabilities is to put $f(z) = z^n$ and $g(z) = p + qz^2$ in (2.3). The coefficients of the various powers of u in the Lagrange expansion denote the absorption probabilities. This result is proved for a very general polynomial random walk in the next section.

Thus the pgf of the distribution of X for the ruin of the gambler is

$$\begin{aligned} z^n = G(u) &= \sum_{x=n}^{\infty} \frac{u^x}{x!} D^{x-1}[nz^{n-1}(p + qz^2)^x]_{z=0} \\ &= \sum_{k=0}^{\infty} u^{n+2k} \frac{n}{n + 2k} \binom{n + 2k}{k} q^k p^{n+k} \end{aligned} \tag{13.5}$$

and the probability distribution of X, the number of steps to gambler's ruin, becomes

$$P(X = n + 2k) = \frac{n}{n + 2k} \binom{n + 2k}{k} q^k p^{n+k} \tag{13.6}$$

for $k = 0, 1, 2, 3, \ldots$, where k denotes the number of steps moved to the right before ruin. Obviously, $X = n, n + 2, n + 4, \ldots$, and the result (13.4) is a particular case of (13.6) given by $n = 1$.

It can easily be shown that the mean and variance of the probability distribution of the games played before the ruin of the gambler with capital n, given by (13.6), are

$$\mu = n(1 - 2q)^{-1} \quad \text{and} \quad \sigma^2 = 4npq(1 - 2q)^{-3} \text{ for } 0 < q < \frac{1}{2}. \tag{13.7}$$

13.4 Generating Function of Ruin Probabilities in a Polynomial Random Walk

Let the initial capital (position) of a player (particle) be n. Also, in each play (step) the player (particle) changes his capital (position) by $-1, 0, 1, 2, \ldots, k$ with probabilities p_{-1}, p_0, $p_1, \ldots, p_k$, respectively, where $p_{-1} + p_0 + p_1 + \cdots + p_k = 1$ and the pgf of the change in the capital (position) after each play (step) is

$$g(z) = p_{-1} + p_0 z + p_1 z^2 + \cdots + p_k z^{k+1}. \tag{13.8}$$

Since the adversary (casino) is infinitely rich, the model has a single absorbing barrier at the origin; i.e., absorption (ruin) occurs when the particle hits the origin. Let $p_{i,x}$ denote the probability that the particle is at position i after step x. Therefore, initially $p_{n,0} = 1$ and $p_{i,0} = 0$ for $i \neq n$; i.e., initially we have

$$p_{i,0} = \delta_{i,n}, \tag{13.9}$$

where $\delta_{i,n}$ is the usual Kronecker delta where $\delta_{n,n} = 1$ and zero otherwise.

The difference equations satisfied by $p_{i,x}$ for $x \geq 1$ are as follows:

$$p_{0,x} = p_{-1}\; p_{1,x-1}, \tag{13.10}$$

$$p_{i,x} = \sum_{h=-1}^{i-1} p_h\; p_{i-h,x-1} \qquad \text{for } i = 1, 2, 3, \ldots, k, \tag{13.11}$$

$$p_{i,x} = \sum_{h=-1}^{k} p_h\; p_{i-h,x-1} \qquad \text{for } i \geq k+1. \tag{13.12}$$

Let $P_i(u)$, $(i \geq 0)$ be a pgf defined by

$$P_i(u) = \sum_{x=1}^{\infty} p_{i,x}\, u^{x-1}, \tag{13.13}$$

so that $P_0(u)$ is the pgf of the ruin probabilities in this polynomial random walk problem.

By multiplying (13.10), (13.11), and (13.12) by u^x and summing over x from $x = 1$ to ∞, we obtain

$$P_0(u) = p_{-1}\, u P_1(u), \tag{13.14}$$

$$P_i(u) = \sum_{h=-1}^{i-1} [p_h\; u p_{i-h}(u) + p_h\; \delta_{i-h,n}], \qquad i = 1, 2, 3, \ldots, k, \tag{13.15}$$

$$P_i(u) = \sum_{h=-1}^{k} [p_h u P_{i-h}(u) + p_h\; \delta_{i-h,n}], \qquad i \geq k+1. \tag{13.16}$$

Defining a new generating function $P(u, z)$ by

$$P(u, z) = \sum_{i=0}^{\infty} P_i(u) \cdot z^i, \tag{13.17}$$

multiplying (13.15) and (13.16) by z^i, and adding these for $i = 1, 2, \ldots, k, k+1, \ldots$ with (13.14), we get

$$\begin{aligned} P(u, z) &= \left[p_{-1} u z^{-1} + p_0 u + p_1\, u z + \cdots + p_k\, u z^{k-1}\right] [P(u, z) - P_0(u)] \\ &\quad + z^{n-1}\left[p_{-1} + p_0 z + p_1 z^2 + \cdots + p_k z^{k+1}\right] \\ &= u z^{-1} g(z)\, [P(u, z) - P_0(u)] + z^{n-1} g(z). \end{aligned}$$

The above relation gives

$$P(u, z) = \frac{z^n - uP_0(u)}{z/g(z) - u}. \tag{13.18}$$

The pgf $P_0(u)$ can be determined from (13.18) by requiring that $P(u, z)$ be analytic and $u = z/g(z)$ so that

$$P_0(u) = z^n/u, \tag{13.19}$$

where z is defined in terms of u by the Lagrange transformation $z = ug(z)$. Thus the Lagrange expansion of z^n in powers of u, under the transformation $z = u\ g(z)$, gives the pgf of the ruin probabilities of the above polynomial random walk. By using the Lagrange expansion (2.3),

$$P_0(u) = u^{-1}z^n = \sum_{x=1}^{\infty} \frac{u^{x-1}}{x!} D^{x-1}\left[nz^{n-1}(g(z))^x\right]_{z=0}$$

$$= \sum_{x=n}^{\infty} u^{x-1} \frac{n}{(x-n)!x} D^{x-n}\left[(g(z))^x\right]_{z=0}, \tag{13.20}$$

where $D = \partial/\partial z$, and the probabilities of the player's ruin become

$$p_{0,x} = P(X = x) = \frac{n}{(x-n)!x} D^{x-n}\left[(p_{-1} + p_0 z + p_1 z^2 + \cdots + p_k z^{k+1})^x\right]_{z=0}$$

$$= \frac{n}{(x-n)!x} D^{x-n} \sum \frac{x!}{r_0!r_1!r_2!\ldots(r_{k+1})!} (p_{-1})^{r_0}(p_0)^{r_1}(p_1)^{r_2}\ldots(p_k)^{r_{k+1}}(z)^{\sum_{i=1}^{k+1} ir_i}\Big|_{z=0}$$

$$= \frac{n}{x} \sum \frac{x!}{r_0!r_1!r_2!\ldots(r_{k+1})!} (p_{-1})^{r_0}(p_0)^{r_1}(p_1)^{r_2}\ldots(p_k)^{r_{k+1}} \tag{13.21}$$

for $x = n, n+1, n+2, \ldots$ and zero otherwise, and where $r_i \geq 0,\ i = 0, 1, 2, \ldots, k+1$, and the summation is taken over all sets of nonnegative integers $r_0, r_1, r_2, \ldots, r_{k+1}$ from 0 to x such that $\sum_{i=0}^{k+1} r_i = x$ and $\sum_{i=0}^{k+1} ir_i = x - n$.

Some particular values of $p_{0,x}$, given by (13.21), are

$$p_{0,n} = (p_{-1})^n, \qquad p_{0,n+1} = np_0(p_{-1})^n,$$

$$p_{0,n+2} = np_1(p_{-1})^{n+1} + \binom{n+1}{2} p_0^2 (p_{-1})^n,$$

$$p_{0,n+3} = np_2(p_{-1})^{n+2} + n(n+2)p_1p_0(p_{-1})^{n+1} + \binom{n+2}{3} p_0^3 (p_{-1})^n,$$

$$p_{0,n+4} = np_3(p_{-1})^{n+3} + n(n+3)p_2p_0(p_{-1})^{n+2} + \frac{1}{2}n(n+3)p_1^2(p_{-1})^{n+2}$$

$$+ \frac{1}{2}n(n+2)(n+3)p_1p_0^2(p_{-1})^{n+1} + \binom{n+3}{4} p_0^4 (p_{-1})^n.$$

It is clear from the above that the actual determination of the ruin probabilities for this polynomial random walk is quite complex on account of the summations and the conditions on those summations.

A number of simpler random walks, for special values of $g(z)$ will be considered in sections 13.5 through 13.9 of this chapter.

The expression for $p_{0,x}$ in (13.21) represents the nonzero probabilities of the gambler's ruin (or the absorption of the particle) at the nth, $(n+1)$th,... step i.e., at the xth step, whatever values it may have in a particular problem.

However, the player of the game is more concerned with the probability that he or she is still playing after the Nth step (play). By (13.21) one can write

$$\Pr\{\text{still playing after the } N\text{-th step}\} = 1 - \sum_{x=n}^{N} p_{0,x}. \tag{13.22}$$

Probability of Capital i after the Nth Step

The generating function $P(u,z)$, obtained in (13.18), can also be written in the alternative form

$$P(u,z) = P_0(u) + \left[z^{n-1} g(z) - P_0(u)\right] (1 - u\, g(z)/z)^{-1}. \tag{13.23}$$

Since u and z are arbitrary variables, we assume that $0 < u\; g(z)/z < 1$, and we expand both sides in powers of u and z by (13.17), (13.13), and the binomial $(1 - u\; g(z)/z)^{-1}$. Thus

$$\sum_{i=0}^{\infty}\sum_{x=1}^{\infty} p_{i,x} u^{x-1} z^i \equiv \sum_{x=1}^{\infty} p_{o,x} u^{x-1} + z^{n-1} g(z) \sum_{j=0}^{\infty} u^j (g(z)/z)^j - \sum_{k=1}^{\infty} u^{k-1} \sum_{j=1}^{k} p_{o,j} (g(z)/z)^{k-j}.$$

By equating the coefficients of u^{N-1} on both sides,

$$\sum_{i=0}^{\infty} p_{i,N} z^i = p_{0,N} + z^{n-N} (g(z))^N - \sum_{j=1}^{N} p_{0,j} (g(z)/z)^{N-j}. \tag{13.24}$$

Since $p_{0,j} = 0$ for $j = 1, 2, \ldots, n-1$ and $p_{0,N}$ gets canceled with the last term in the summation, (13.24) reduces to

$$\sum_{i=0}^{\infty} p_{i,N} z^i = z^{n-N} (g(z))^N - \sum_{j=n}^{N-1} p_{0,j} (g(z)/z)^{N-j}. \tag{13.25}$$

Since (13.21) gives only nonzero probabilities of ruin after step x, the nonzero values of $p_{i,N}$ for $i \geq 1$ are given by (13.25) as the coefficient of z^i on the right-hand side. Therefore,

$$p_{i,N} = \text{coefficient of } z^{i-n+N} \text{ in } (g(z))^N - \sum_{j=n}^{N-1} p_{0,j} \times \text{ coefficient of } z^{N+i-j} \text{ in } (g(z))^{N-j}. \tag{13.26}$$

The actual computation of the values of $p_{i,N}$ for the general value of $g(z)$ in (13.8) is difficult. These will be determined for the particular cases in sections 13.5, 13.6, 13.7, and 13.8.

13.5 Trinomial Random Walks

Good (1958) has defined a "recurrent right-handed walk" as one that starts and ends at the same point and never goes left of this point and has obtained the probability of a particle starting at the origin and of returning to the origin for the first time. Let the probabilities of taking steps of $-1, 0, +1$ by a particle at any point be p_{-1}, p_0, p_1, respectively, so that the pgf for any one step is $g(z) = p_{-1} + p_0 z + p_1 z^2$, where $p_{-1} + p_0 + p_1 = 1$.

Let the particle start from a point n steps from the origin and let there be an absorbing barrier at the origin like that of a gambler who starts with a capital of n and who can't start again after getting ruined. Also, let X denote the total number of steps taken by the particle before absorption at the origin.

The pgf of the probability distribution of X will be given by the Lagrange expansion (2.3), where $f(z) = z^n$ or by (13.20). Then the pgf is

$$z^n = G(u) = \sum_{x=n}^{\infty} \frac{u^x}{x!} D^{x-1} \left[n z^{n-1} (p_{-1} + p_0 z + p_1 z^2)^x \right]_{z=0}$$

$$= \sum_{x=n}^{\infty} \frac{u^x}{x!} \binom{x-1}{n-1} n! D^{x-n} \left[\sum_{i,j=0}^{x} \frac{x! p_1^i p_0^j (p_{-1})^{x-i-j}}{i! j! (x-i-j)!} \cdot z^{2i+j} \right]_{z=0},$$

where $i + j \le x$. On differentiating $(x - n)$ times and putting $z = 0$, the pgf of r.v. X becomes

$$G(u) = \sum_{x=n}^{\infty} u^x \frac{n}{x} \sum_{i=0}^{[a]} \frac{x! p_1^i (p_{-1})^{n+i} (p_0)^{x-n-2i}}{i!(n+i)!(x-n-2i)!}, \tag{13.27}$$

where $a = \frac{1}{2}(x-n)$, and $[a]$ is the integral part of a.

Thus the probability distribution of X, the number of steps to absorption, is

$$p_{0,x} = P(X = x) = \frac{n}{x} \sum_{i=0}^{[a]} \frac{x!}{i!(n+i)!(x-n-2i)!} (p_{-1})^{n+i} p_1^i (1 - p_1 - p_{-1})^{x-n-2i}. \tag{13.28}$$

By using the results on the cumulants of the Lagrangian probability distributions, given in chapter 2, it can be shown that the mean and variance of the above probability distribution are

$$\mu = \frac{n}{1 - p_0 - 2p_1} \quad \text{and} \quad \sigma^2 = \frac{n[p_0 q_0 + 4p_1 q_1 - 4p_0 p_1]}{(1 - p_0 - 2p_1)^3}, \tag{13.29}$$

which exist when $0 < p_0 + 2p_1 < 1$, i.e., when $p_{-1} > p_1$. Thus if $p_{-1} = p_0 = p_1 = \frac{1}{3}$, the mean μ will not exist; i.e., the particle may never reach the origin and the player may never get ruined.

The probability of the particle not being absorbed after the Nth step can easily be written down by substituting the value of $p_{0,x}$ from (13.28) in (13.22). Thus, the probability of still playing after the Nth step is

$$1 - (p_{-1})^n \left[1 + np_0 + \binom{n+1}{2} p_0^2 + np_1 p_{-1} + \binom{n+2}{3} p_0^3 + n(n+2) p_{-1} p_0 p_1 + \dots \right.$$

$$\left. + \frac{n}{N} \sum_{i=0}^{(N-n)/2} \frac{N!}{i!(n+i)!(N-n-2i)!} (p_1 p_{-1})^i \, p_0^{N-n-2i} \right].$$

Since the expansion of $(p_{-1} + p_0 z + p_1 z^2)^N$ in powers of z can easily be written down by the trinomial theorem, the probability of the particle being at position i after the Nth step is given by (13.26) as

$$p_{i,N} = \sum_{s=0}^{[b]} \frac{N! p_1^s p_0^{i-n+N-2s} (p_{-1})^{n+s-i}}{s!(n+s-i)!(i-n+N-2s)!} - \sum_{j=n}^{N-1} p_{0,j} \sum_{s=0}^{[c]} \frac{(N-j)! p_1^s (p_{-1})^{i+s} p_0^{N-i-j-2s}}{s!(i+s)!(N-i-j-2s)!} \tag{13.30}$$

for $N = n+1, n+2, n+3, \ldots$ and zero otherwise, and where $b = \frac{1}{2}(i - n + N)$, $c = \frac{1}{2}(N - i - j)$, $[b]$ and $[c]$ denote the integer parts of b and c, and the second summation is zero if $N < n+1$.

Special Cases

Case I. When $n = 1$, the pgf of the recurrent random walk, with an absorbing barrier at zero, can also be expressed in another form, instead of (13.27), by the Lagrange transformation as given below:

$$z = u\,(p_{-1} + p_0 z + p_1 z^2)$$

$$\text{or} \quad u\, p_1 z^2 + (up_0 - 1)z + up_{-1} = 0$$

$$\text{or} \quad z = \frac{1 - up_0 - (1 - 2up_0 - u^2 B^2)^{\frac{1}{2}}}{2up_1}, \tag{13.31}$$

where $B^2 = 4p_1 p_{-1} - p_0^2$.

Case II. When $p_{-1} = p^2$, $p_0 = 2pq$, and $p_1 = q^2$, where $p + q = 1$, the probability distribution (13.28) gets reduced to the simple form

$$P(X = x) = \frac{n}{x} \binom{2x}{x-n} q^{x-n} p^{n+x} \tag{13.32}$$

for $x = n, n+1, n+2, \ldots$ and zero otherwise.

13.6 Quadrinomial Random Walks

Let a particle start from some point n (a positive integer) and let the particle be allowed to take steps of size $-1, 0, +1$, or $+2$ only from any given position in each trial. Thus the particle is like a gambler who can increase or decrease his capital by $-1, 0, +1, +2$ in each game and who keeps playing against an infinitely rich adversary till he gets ruined or becomes infinitely rich. Obviously, there is an absorbing barrier at the origin because he can't play after he loses all the money. There can be four different probabilities for these four possible alternatives. However, the general case does not provide reasonably nice results. Let $0 < p_0 = 1 - q_0 < 1$ and $0 < p = 1 - q < 1$. Also, let the probabilities of taking $-1, 0, +1$, or $+2$ steps by the particle be $pp_0, pq_0, p_0 q, qq_0$, respectively, so that the pgf of these moves is $g(z) = pp_0 + pq_0 z + p_0 q z^2 + qq_0 z^3 = (p_0 + q_0 z)(p + qz^2)$.

Let X be the total number of trials made by the particle before it reaches the origin for the first time and gets absorbed (i.e., the gambler gets ruined). Using the Lagrange expansion with $f(z) = z^n$, the pgf of the r.v. X is given by (13.20) in the following form:

$$z^n = G(u) = \sum_{x=n}^{\infty} \frac{u^x}{x!} D^{x-1} \left[n z^{n-1} (p_0 + q_0 z)^x (p + q z^2)^x \right]_{z=0}$$

$$= \sum_{x=n}^{\infty} \frac{u^x}{x!} \binom{x-1}{n-1} n! D^{x-n} \left[\sum_{i=0}^{x} \binom{x}{i} p_0^{x-i} q_0^i z^i \sum_{j=0}^{x} \binom{x}{j} q^j p^{x-j} z^{2j} \right]_{z=0}$$

$$= \sum_{x=n}^{\infty} \frac{u^x}{(x-n)!} \frac{n}{x} D^{x-n} \left[\sum_{k=0}^{3x} \sum_{j=0}^{[\frac{1}{2}k]} \binom{x}{j} \binom{x}{k-2j} q^j p^{x-j} q_0^{k-2j} p_0^{x+2j-k} z^k \right]_{z=0}$$

$$= \sum_{x=n}^{\infty} u^x \frac{n}{x} \sum_{j=0}^{[a]} \binom{x}{j} \binom{x}{x-n-2j} q^j q^{x-j} p_0^{n+2j} q_0^{x-n-2j}, \tag{13.33}$$

where $a = \frac{1}{2}(x-n)$, and which gives the probability distribution of X, the number of steps for absorption, as

$$p_{0,x} = P(X = x) = \frac{n}{x} p^x p_0^n q_0^{x-n} \sum_{j=0}^{[a]} \binom{x}{j} \binom{x}{x-n-2j} \left(\frac{q p_0^2}{p q_0^2} \right)^j \tag{13.34}$$

for $x = n, n+1, n+2, \ldots$ and zero otherwise.

By using the formulas for the cumulants of the Lagrangian probability distributions given in chapter 2, the mean and the variance of the above probability distribution (13.34) can be shown to be

$$\mu = n(1 - q_0 - 2q)^{-1}, \qquad 0 < q_0 + 2q < 1, \tag{13.35}$$

and

$$\sigma^2 = n(p_0 q_0 + 4pq)(1 - q_0 - 2q)^{-3}. \tag{13.36}$$

The particle will ultimately reach the origin if $0 < q_0 + 2q < 1$ or if $1 < p + \frac{1}{2}p_0 < \frac{3}{2}$. However, if $p + \frac{1}{2}p_0 < 1$ the particle may not get absorbed. The probability of the particle not getting absorbed after the Nth step can easily be written down by substituting the value of $p_{0,x}$ from (13.34) in (13.22).

Also, the expansion of $(g(z))^N = (p_0 + q_0 z)^N (q + q z^2)^N$ in powers of z is

$$(p_0 + q_0 z)^N (p + q z^2)^N = \sum_{k=0}^{3N} \sum_{s=0}^{[a]} z^k \binom{N}{s} \binom{N}{k-2s} q^s p^{N-s} q_0^{k-2s} p_0^{N-k+2s}, \tag{13.37}$$

where $a = \frac{1}{2}k$ and $[a]$ is the integer part of a. By using the expansion (13.37) in (13.26) twice, the probability that the particle is at position i (≥ 1) just after step N is

$$p_{i,N} = \sum_{s=0}^{[b]} \binom{N}{s} \binom{N}{i-n+N-2s} q^s p^{N-s} q_0^{i-n+N-2s} p_0^{n-i+2s}$$
$$- \sum_{j=n}^{N-1} p_{0,j} \sum_{s=0}^{[c]} \binom{N-j}{s} \binom{N-j}{N+i-j-2s} q^s p^{N-j-s} q_0^{N+i-2j-2s} p_0^{j+2s-i} \tag{13.38}$$

for $N = n+1, n+2, n+3, \dots$ and zero otherwise, and where $b = \frac{1}{2}(N+i-n)$, $c = \frac{1}{2}(N+i-j)$, $[b]$ and $[c]$ are the integer parts of b and c, and the second summation is zero if $N < n+1$.

13.7 Binomial Random Walk (Jumps) Model

Let the initial position of a particle be $+n$ and each time let the particle be allowed to take mutually exclusive steps of size $-1, +1, +3, +5, \dots, 2m-1$ only with probabilities of $\binom{m}{i} p^{m-i} q^i$, $i = 0, 1, 2, \dots, m$, respectively, so that the pgf of its mutually exclusive steps in this kind of walk (or jump) is $g(z) = \sum_{i=0}^{m} \binom{m}{i} p^{m-i} q^i z^{2i} = (p+qz^2)^m$ for each time (trial), where $p + q = 1$.

Let X be the total number of jumps taken by the particle before it reaches the origin for the first time, where it gets absorbed, as the origin has an absorbing barrier. If the particle is a gambler playing with an infinitely rich adversary, he or she gets ruined when the capital becomes zero.

Then, the pgf of the r.v. X is given by the Lagrange expansion in (13.20), under the transformation $z = u(p+qz^2)^m$, as

$$
\begin{aligned}
z^n = l(u) &= \sum_{x=n}^{\infty} \frac{u^{x-1}}{x!} D^{x-1}[nz^{n-1}(p+qz^2)^{mx}]_{z=0} \\
&= \sum_{x=n}^{\infty} \frac{u^{x-1}}{x!} \frac{(x-1)!n!}{(n-1)!(x-n)!} D^{x-n}[(p+qz^2)^{mx}]_{z=0} \\
&= \sum_{x=n}^{\infty} u^{x-1} \frac{n}{(x-n)!x} D^{x-n} \left[\sum_{k=0}^{mx} \binom{mx}{k} q^k z^{2k} p^{mx-k} \right]_{z=0} \\
&= \sum_{k=0}^{\infty} u^{n+2k-1} \frac{n}{n+2k} \binom{mn+2mk}{k} q^k p^{mn+2mk-k}.
\end{aligned}
\tag{13.39}
$$

The above pgf gives the probability distribution of X, the number of steps on which the particle gets absorbed at the origin, as

$$
p_{0,x} = P(X = x = n+2k) = \frac{n}{n+2k} \binom{mn+2mk}{k} q^k p^{mn+2mk-k} \tag{13.40}
$$

for $k = 0, 1, 2, 3, \dots$ and zero otherwise.

When $n = 1$ and $m = 1$, the model (13.40) reduces to the ordinary random walk distribution (13.4) of a gambler who starts with unit capital, which has been discussed by Feller (1957) and Kemp and Kemp (1968).

The model (13.40) is a special case of the GNBD, discussed in detail in chapter 10. The mean and the variance of X are

$$
\left.
\begin{aligned}
E(X) &= n(1-2mq)^{-1}, \quad 0 < q < (2m)^{-1} \\
\sigma_x^2 &= 4nmqp(1-2mq)^{-3}, \; 0 < q < (2m)^{-1}.
\end{aligned}
\right\}
\tag{13.41}
$$

The probability that the particle does not get absorbed at the origin until the end of the Nth step is given by (13.22) and can be written down easily by substituting the values of $p_{0,x}$ from (13.40) in (13.20) and becomes $1 - \sum_{x=n}^{N} p_{0,x}$.

Since

$$\left(g(z)^N\right) = \left(p + qz^2\right)^{mN} = \sum_{s=0}^{mN} \binom{mN}{s} q^s p^{mN-s} z^{2s},$$

the probability that the particle is at position i (≥ 1) after the Nth step (or jump) is given by (13.26) in the form

$$\begin{aligned} p_{i,N} = & \binom{mN}{\frac{1}{2}(i-n+N)} q^{(N+i-n)/2} p^{(2mN-N-i+n)/2} \\ & - \sum_{j=n}^{N-1} p_{0,j} \binom{mN-mj}{(N+i-j)/2} q^{(N+i-j)/2} p^{mN-mj-(N+i-j)/2} \end{aligned} \tag{13.42}$$

for $N = n+1, n+2, n+3, \ldots$ and zero otherwise, and where the second summation is zero if $N < n+1$.

13.8 Polynomial Random Jumps Model

Let the particle start from some given point n (a positive integer) and let us suppose that in each trial or game it can either move one step backwards or stay at its place, or it can move forward by either one of $1, 2, 3, \ldots, 3m$ steps. These movements are mutually exclusive. Also, let the pgf of these movements in each game or trial be given by $g(z) = (p_0 + q_0 z)^m (p + qz^2)^m$, where $p_0^m p^m$ is the probability of moving one step backward, $mp_0^{m-1} q_0 p^m$ is the probability of staying at its place, $\frac{1}{2}m(m-1)p^m p_0^{m-2} q_0^2 + mp_0^m p^{m-1} q$ is the probability of moving one step to the right, and so on. Also, $0 < p_0 = 1 - q_0 < 1$ and $0 < p = 1 - q < 1$.

If there is an absorbing barrier at the origin, the random walks terminate as soon as the particle reaches the origin, i.e., when the particle takes n steps backwards from its original position. Let X denote the total number of trials or games in which the random jumps of the particle terminate. The pgf of the probability distribution of X is given by the Lagrange expansion of $f(z) = z^n$, under the transformation $z = u\ (p_0 + q_0 z)^m (p + qz^2)^m$, by the formula (13.20) as

$$\begin{aligned} z^n = G(u) &= \sum_{x=n}^{\infty} \frac{u^{x-1}}{x!} D^{x-1} \left[nz^{n-1} (p_0 + q_0 z)^{mx} (p + qz^2)^{mx} \right]_{z=0} \\ &= \sum_{x=n}^{\infty} \frac{u^{x-1}}{(x-n)!x} \frac{n}{x} D^{x-n} \left[\sum_{i=0}^{mx} \binom{mx}{i} p_0^{mx-i} q_0^i z^i \sum_{j=0}^{mx} \binom{mx}{j} p^{mx-j} q^j z^{2j} \right]_{z=0} \\ &= \sum_{x=n}^{\infty} \frac{u^{x-1}}{(x-n)!} \frac{n}{x} D^{x-n} \left[\sum_{k=0}^{3x} \sum_{j=0}^{[\frac{1}{2}x]} \binom{mx}{j} \binom{mx}{k-2j} q^j p^{mx-j} q_0^{k-2j} p_0^{mx-k+2j} z^k \right]_{z=0} \\ &= \sum_{x=n}^{\infty} u^x \frac{n}{x} \sum_{j=0}^{[a]} \binom{mx}{j} \binom{mx}{x-n-2j} q^j p^{mx-j} q_0^{x-n-2j} p_0^{mx+n-x+2j}, \end{aligned} \tag{13.43}$$

where $a = \frac{1}{2}(x-n)$. Thus, the probability distribution of X, the number of trials made by the particle before absorption, becomes

$$p_{0,x} = P(X=x) = \frac{n}{x}\sum_{j=0}^{[a]}\binom{mx}{j}\binom{mx}{x-n-2j} q^j p^{mx-j} q_0^{x-n-2j} p_0^{mx+n-x+2j} \qquad (13.44)$$

for $x = n, n+1, n+2, \ldots$ and zero otherwise. The expression becomes a little smaller for the special case of $p_0 = p$.

The mean and the variance of the probability distribution of X can be obtained by using the formulas for the cumulants of Lagrangian probability distributions given in chapter 2 and are

$$\left.\begin{aligned} \mu &= n(1 - mq_0 - 2mq)^{-1} \\ \sigma_x^2 &= nm(p_0q_0 + 4pq)(1 - mq_0 - 2mq)^{-3}, \end{aligned}\right\} \qquad (13.45)$$

where $0 < mq_0 + 2mq < 1$. Thus, the particle will definitely get absorbed at the origin if $q_0 + 2q < m^{-1}$.

The probability of the particle not getting absorbed even after the Nth step is given by substituting the value of $p_{0,x}$ from (13.44) in (13.22).

Also,

$$(g(z))^N = (p_0 + q_0 z)^{mN}(p + qz^2)^{mN}$$

gives the power series in z as

$$(g(z))^N = \sum_{r,s=0}^{mN}\binom{mN}{s}\binom{mN}{r} q_0^r p_0^{mN-r} q^s p^{mN-s} z^{r+2s}.$$

By using the above expansion and the expression (13.26), the probability that the particle is at position i (≥ 1) after the Nth step can be written in the form

$$\begin{aligned} p_{i,N} &= \sum_{s=0}^{[a]}\binom{mN}{s}\binom{mN}{N+i-n-2s} q^s p^{mN-s} p_0^{mN} (p_0/q_0)^{n+2s-N-i} \\ &\quad - \sum_{j=n}^{N-1} p_{0,j}\binom{mN-mj}{s}\binom{mN-mj}{N+i-2j-2s}(pp_0)^{mN-mj}(q/p)^s (q_0/p_0)^{N+i-2s-2j} \end{aligned} \qquad (13.46)$$

for $N = n+1, n+2, n+3, \ldots$ and zero otherwise, and where $a = \frac{1}{2}(N+i-n)$, $[a]$ is the integer part of a, and the second summation is zero if $N < n+1$.

13.9 General Random Jumps Model

Let $+n$ be the initial position of a particle. Each time the particle jumps, it takes mutually exclusive steps of size $-1, a-1, 2a-1, 3a-1, \ldots, am-1$ only with probabilities $\binom{m}{i}p^{m-i}q^i$, $i = 0, 1, 2, \ldots, m$, respectively, so that the pgf of its jump each time is $g(z) = \sum_{i=0}^{m}\binom{m}{i}p^{m-i}q^i z^{ai} = (p+qz^a)^m$, where $0 < p = 1-q < 1$.

Let X denote the total number of jumps taken by the particle before it reaches the origin, which has an absorbing barrier, and where it gets absorbed. By section 13.4, the pgf of the probability distribution of X is given by the Lagrange expansion in (13.20) of z^n, under the transformation $z = u\,(p+qz^a)^m$, in the form

$$
\begin{aligned}
z^n = l(u) &= \sum_{x=n}^{\infty} \frac{u^{x-1}}{x!} D^{x-1}\left[nz^{n-1}(p+qz^a)^{mx}\right]_{z=0} \\
&= \sum_{x=n}^{\infty} u^{x-1} \frac{n}{(x-n)!x} D^{x-n}\left[\sum_{k=0}^{mx}\binom{mx}{k} q^k p^{mx-k} z^{ak}\right]_{z=0} \\
&= \sum_{k=0}^{\infty} u^{n+ak-1} \frac{n}{n+ak}\binom{mn+mak}{k} q^k p^{mn+mak-k},
\end{aligned}
\tag{13.47}
$$

which gives the probability distribution of X, the number of jumps made by the particle before it gets absorbed at the origin as

$$
p_{0,x} = P(X = x = n+ak) = \frac{n}{n+ak}\binom{mn+mak}{k} q^k p^{mn+mak-k} \tag{13.48}
$$

for $k = 0, 1, 2, 3, \ldots$ and zero otherwise. The above is the GNBD discussed in detail in chapter 10. The mean and the variance of X are

$$
\left.
\begin{aligned}
E[X] &= n(1-amq)^{-1} \\
\sigma_x^2 &= mna^2pq(1-amq)^{-3},
\end{aligned}
\right\} \tag{13.49}
$$

which exist when $q < (am)^{-1}$.

The probability of the particle not getting absorbed until after the Nth jump can be written by substituting the value of $p_{0,x}$ from (13.48) in (13.22). Also, the probability $p_{i,N}$ that the particle is at position i (≥ 1) after the Nth jump is given by (13.26). By using $g(z) = (p + qz^a)^m$, the probability becomes

$$
\begin{aligned}
p_{i,N} = &\binom{mN}{(N+i-n)/a} q^{(N+i-n)/a} p^{mN-(N+i-n)/a} \\
&- \sum_{j=n}^{N-1} p_{0,j}\binom{mN-mj}{(N+i-j)/a} q^{(N+i-j)/a} p^{mN-mj-(N+i-j)/a}
\end{aligned}
\tag{13.50}
$$

for $N = n+1, n+2, n+3, \ldots$ and zero otherwise, and where the second summation is zero if $N < n+1$.

13.10 Applications

It was shown in chapter 6 that the probability distributions of the first busy period in many queuing problems, of the branching processes, of the spread of various epidemics, and of the spread of the sales of a product, are all Lagrangian probability models, which are given by using different pgfs.

A large number of researchers have used many techniques, based on combinatorial methods, for solving the gambler's ruin problems, random walk problems, and the ballot problems. However, it has been shown in the previous sections that the Lagrange expansion can be used effectively to obtain the probability distributions for all these models and to compute the probabilities for many useful cases.

The main use of the random walks models and the random jumps models, discussed in this chapter, will be for applied problems in the different disciplines of study and in the industry. The researcher in this field of study has to give a suitable interpretation to the different variables, analyze the problem carefully, break it up systematically into small time periods, and reduce it to the form of a random walk, a random jump, or a branching process.

The manufacturer (company) of a new product has to invest capital (money) in the initial development of the product and its utility value, in its approval by government, in setting up the production line, in the actual cost of production, and in sales campaigns through various kinds of media like magazine advertisements, TV advertisements, radio, sales agents, etc., over a reasonable period of time. Thus, the company has the initial costs of development, of approval, and of manufacturing plant which have to be realized from the sales of the product. Then, the company has the basic cost of the product and the advertisement costs. Considering the retailers' and wholesalers' commissions the product is priced in such a manner that the initial costs are realized within the first three years together with the cost of the product, all the cost of advertisements, and a reasonable return on all the investments made by the company. The price must be competitive with the prices of other similar products. If this product is completely new and there is no similar product in the market, then the price must be affordable to the public.

The manufacturing company shall have a break-even point of minimum sales volume and the company suffers losses till the sales reach that break-even point. If all the initial capital amount gets used up and the company is not able to raise more capital, it will go bankrupt. As the company increases its monthly sales above the break-even point, it becomes more profitable. The researchers in the company have to determine a probability model for increase in sales after each advertisement campaign and have to prescribe some numbers for break-even point sales, for net 10% profit sales, for net 20% profit sales, ..., and then they have to formulate a suitable random walk model to determine the probabilities for reaching those levels. Every company has to collect data on the effect of the advertisement campaigns to determine the probabilities for the sales of the product and on the various costs of the product as well as the profit margins etc. to determine the various profit levels so that this research can be completed. This research is an ongoing process in every industry.

13.11 Exercises

13.1 Consider the trinomial random walk model in (13.32). Obtain a recurrence relation between the noncentral moments. Hence or otherwise, obtain the first three noncentral moments. Obtain the moment estimate for the parameter p.

13.2 (a) Show that the mean and variance of the random walk model (13.6) are given by (13.7).
  (b) Show that the mean and variance of the trinomial random walk model in (13.28) are given by (13.29).
  (c) Show that the mean and variance of the binomial random walk model (13.40) are given by (13.41).

13.3 Show that the mean and variance of the Katz distribution (model (6), Table 2.1) are

$$\mu = (1-\beta)(1-\beta-b)^{-1}, \quad \sigma^2 = b(1-\beta)(1-\beta-b)^{-3}.$$

13.4 Let the probabilities of taking steps $-1, 0, +1$ on a straight line by a particle at any point be p_{-1}, p_0, p_1, respectively, where $p_{-1}+p_0+p_1 = 1$. Let the particle start from a point n steps away from the origin, which has an absorbing barrier, and let X denote the total number of steps taken by the particle before its absorption at the origin. Show that the probability distribution of the r.v. X is

$$P(X = x) = \frac{n}{x}\sum_{i=0}^{[a]} \frac{x!}{i!(n+i)(x-n-2i)!} p_1^i (p_{-1})^{n+1} (p_0)^{x-n-2i},$$

where $a = \frac{1}{2}(x-n)$, $[a]$ is the integral part of a, and $x = n, n+1, n+2, \ldots$. Also, express the pgf $G(u)$ of the r.v. X for the above trinomial walk as an implicit function

$$G(u) = z^n, \quad \text{where} \quad z = u(p_{-1} + p_0 z + p_1 z^2).$$

13.5 Let a particle start from a point one step away from the origin which has an absorbing barrier and let the particle make random jumps to points $-1, +1, +3, \ldots, (2m-1)$ with the probabilities given by the pgf $g(z) = (p+qz^2)^m$, $0 < p = 1-q < 1$. If X denotes the total number of trials (jumps) made by the particle before it gets absorbed at the origin $(z = 0)$, show that the probability distribution of X is

$$P(X = 2k+1) = \frac{m}{m+2mk}\binom{m+2mk}{k} q^k p^{m+2mk-k}, \quad k = 0, 1, 2, \ldots.$$

Also, show that the pgf of the r.v. X is

$$G(u) = z, \quad \text{where} \quad z = u(p+qz^2)^m.$$

13.6 Let a particle start from some point n (a positive integer > 0) and let the particle be allowed to take mutually exclusive steps of size $-1,\ 0,\ +1,\ +2$, only from any given position in each trial. Let the pgf of these moves in each trial be $g(z) = p_0 + p_1 z + p_2 z^2 + p_3 z^3$, where $0 < p_i < 1,\ i = 0, 1, 2, 3$, and $p_0+p_1+p_2+p_3 = 1$. Let X denote the total number of steps taken by the particle before absorption at the origin. Obtain the probability distribution of X, the number of steps to absorption at the origin.

13.7 Obtain the values of $E[X]$ and $\text{Var}(X)$ in Exercise 13.6.

13.8 Obtain the probability that the particle in Exercise 13.6 will be at the point 20 after the nth step.

13.9 A company manufactures hair shampoo and supplies its product to the retailers in 10 oz. bottles. The market is very competitive and so they have to run TV advertisements every month. If they are able to sell N bottles every month, they make a reasonable profit. On account of the monthly advertisements and the strong competition, the monthly sales of the bottles increase or decrease by mutually exclusive numbers of $-1,\ +1,\ +3, \ldots, 2m-1$ with probabilities given by the pgf $g(z) = (p+qz^2)^m$ for each TV advertisement (trial), where $0 < q = 1-p < 1$. Find the probability that the monthly sales of the shampoo bottles are $\geq N$ after n months.

14

Bivariate Lagrangian Distributions

14.1 Definitions and Generating Functions

Univariate Lagrangian probability distributions, in general, and many of their special models have been discussed in chapters 1 through 13. Shenton and Consul (1973) studied the bivariate Lagrangian distributions (BLDs). Jain and Singh (1975) and Churchill and Jain (1976) studied some bivariate power series distributions associated with the Lagrange expansion. Shoukri and Consul (1982) and Shoukri (1982) considered the bivariate modified power series distributions associated with the Lagrange expansion.

Shenton and Consul (1973) applied Poincaré (1886) generalization of Lagrange expansion to develop the bivariate Lagrangian and Borel–Tanner distributions and showed that these probability distributions represent the number of customers served in a single server queuing system. When the service begins with a queue consisting of i customers of type I and j customers of type II with different arrival rates requiring separate kinds of service for each type of customer, the general probability distribution of the number of customers of type I and type II served in a busy period is a Lagrangian-type bivariate Borel–Tanner probability distribution.

Let $h(t_1, t_2) \equiv h$, $k(t_1, t_2) \equiv k$ and $f(t_1, t_2) \equiv f$ be any three nonnegative bivariate meromorphic functions such that $h(1,1) = k(1,1) = f(1,1) = 1$ and $h(0,0)$ and $k(0,0)$ are nonzero. As all these properties are satisfied by bivariate pgfs, the functions h, k, and f can be pgf but it is not necessary. Consider the transformations

$$\begin{aligned} t_1 &= u\, h(t_1, t_2), \\ t_2 &= v\, k(t_1, t_2), \end{aligned} \tag{14.1}$$

which give $u = v = 0$ for $t_1 = t_2 = 0$ and $u = v = 1$ for $t_1 = t_2 = 1$. Since u and v are bivariate analytic functions of t_1 and t_2 in the neighborhood of the origin, the smallest positive roots of the transformations in (14.1) give for $f(t_1, t_2)$, by Poincaré's bivariate generalization of Lagrange expansion, a power series expansion in terms of u and v. (See (1.93), (1.94), and (1.95)). Let X and Y be two discrete random variables and let the probability $P(X = x, Y = y)$ be denoted by $P(x, y)$. The bivariate function $f(t_1, t_2)$ in (1.93) provides the pgf $l(u, v)$, which can be written as

$$f(t_1, t_2) = l(u, v) = \sum_{x=0}^{\infty} \sum_{y=0}^{\infty} P(x, y) u^x v^y \tag{14.2}$$

for $x + y > 0,\ x, y = 0, 1, 2, \ldots$ and where

$$P(0,0) = f(0,0)$$

and

$$P(x,y) = \frac{1}{x!}\frac{1}{y!}\left(\frac{\partial}{\partial t_1}\right)^{x-1}\left(\frac{\partial}{\partial t_2}\right)^{y-1}\left\{h^x k^y \frac{\partial^2 f}{\partial t_1 \partial t_2} + h^x \frac{\partial k^y}{\partial t_1}\frac{\partial f}{\partial t_2} + k^y \frac{\partial h^x}{\partial t_2}\frac{\partial f}{\partial t_1}\right\}_{t_1=t_2=0}. \tag{14.3}$$

The values of $P(x, y)$ for particular integral values of x and y are nonnegative probabilities. When the bivariate functions $h(t_1, t_2)$, $k(t_1, t_2)$, and $f(t_1, t_2)$ are not pgfs, one has to impose the conditions that the values of $P(x, y)$ in (14.3) are nonnegative for all values of x and y. Since $l(u, v) = 1$ for $u = v = 1$, the sum of all $P(x, y),\ x, y = 0, 1, 2, \ldots$ is unity.

In general, each one of the three bivariate functions, $h,\ k$, and f, have three parameters and so the bivariate Lagrangian probability distribution $P(x, y)$ in (14.3) will have nine unknown parameters. In statistical applications of the probability models, the estimation of the unknown parameters is one of the main areas of study. The estimation becomes very difficult when there are too many parameters. In view of this problem, Consul (1994a) showed that when the three bivariate analytic functions f, h, and k are of the form $f(t_1, t_2) = [g(t_1, t_2)]^a$, $h(t_1, t_2) = [g(t_1, t_2)]^b$ and $k(t_1, t_2) = [g(t_1, t_2)]^c$, where a, b, c are real nonnegative constants, $g(1, 1) = 1$, and $g(0, 0)$ is nonzero, then the power series expansion in (14.2) provides the bivariate pgf $l(u, v)$, which can be simplified to any one of the two forms

$$f(t_1,t_2) = f(0,0) + \sum_{x=0}^{\infty}\sum_{y=0}^{\infty}\frac{u^x}{x!}\frac{v^y}{y!}\left\{\left(\frac{\partial}{\partial t_2}\right)^{y}\left(\frac{\partial}{\partial t_1}\right)^{x-1}\left[h^x k^y \frac{\partial f}{\partial t_1}\right]\right\}_{t_1=t_2=0}, \tag{14.4}$$

$$x + y > 0,$$

or

$$f(t_1,t_2) = f(0,0) + \sum_{x=0}^{\infty}\sum_{y=0}^{\infty}\frac{u^x}{x!}\frac{v^y}{y!}\left\{\left(\frac{\partial}{\partial t_1}\right)^{x}\left(\frac{\partial}{\partial t_2}\right)^{y-1}\left[h^x k^y \frac{\partial f}{\partial t_2}\right]\right\}_{t_1=t_2=0}, \tag{14.5}$$

$$x + y > 0.$$

The bivariate Lagrangian probability distribution in (14.3) will then be simplified to the form

$$P(x,y) = \frac{1}{x!y!}\left\{\left(\frac{\partial}{\partial t_1}\right)^{x}\left(\frac{\partial}{\partial t_2}\right)^{y-1}\left[h^x k^y \frac{\partial f}{\partial t_2}\right]\right\}_{t_1=t_2=0}, \tag{14.6}$$

for $x + y > 0, x, y = 0, 1, 2, \ldots$ and $P(0, 0) = f(0, 0)$.

Numerous bivariate Lagrangian probability models can be generated by assigning different sets of bivariate probability generating functions to $f(t_1, t_2)$, $h(t_1, t_2)$, and $k(t_1, t_2)$. More bivariate Lagrangian probability models can also be generated by choosing specific functions $h,\ k$, and f, which satisfy the conditions in paragraph three of this section. Many such bivariate Lagrangian probability models will be derived and considered in later sections of this chapter.

There is another method as well by which some specific bivariate Lagrangian probability distributions can be obtained. Let $f \equiv f(\theta_1, \theta_2)$, $\phi \equiv \phi(\theta_1, \theta_2)$, and $\psi \equiv \psi(\theta_1, \theta_2)$ be three nonzero bivariate analytic functions of θ_1 and θ_2. Consider the bivariate transformations

$$\theta_1 = u\,\phi(\theta_1, \theta_2) \quad \text{and} \quad \theta_2 = v\,\psi(\theta_1, \theta_2),$$

which provide $u = v = 0$ when $\theta_1 = \theta_2 = 0$ and

$$u = \theta_1\,[\phi(\theta_1, \theta_2)]^{-1} = g(\theta_1, \theta_2), \quad v = \theta_2\,[\psi(\theta_1, \theta_2)]^{-1} = h(\theta_1, \theta_2). \tag{14.7}$$

Since u and v are bivariate analytic functions of θ_1 and θ_2 in the neighborhood of the origin, the smallest positive roots of the transformations provide, by Poincaré's formula (1.93), a power series expansion in terms of u and v as

$$f(0, 0) = a(0, 0)$$

and

$$f(\theta_1, \theta_2) = \sum_{\substack{x=0 \\ x+y>0}}^{\infty} \sum_{y=0}^{\infty} a(x, y) u^x v^y = \sum_{\substack{x=0 \\ x+y>0}}^{\infty} \sum_{y=0}^{\infty} a(x, y)\,(g(\theta_1, \theta_2))^x\,(h(\theta_1, \theta_2))^y, \tag{14.8}$$

where

$$a(x, y) = \frac{1}{x!}\frac{1}{y!}\left(\frac{\partial}{\partial\theta_1}\right)^{x-1}\left(\frac{\partial}{\partial\theta_2}\right)^{y-1}\left[\phi^x\psi^y\frac{\partial^2 f}{\partial\theta_1\partial\theta_2} + \phi^x\frac{\partial\psi^y}{\partial\theta_1}\frac{\partial f}{\partial\theta_2} + \psi^y\frac{\partial\phi^x}{\partial\theta_2}\frac{\partial f}{\partial\theta_1}\right]_{\theta_1=\theta_2=0} \tag{14.9}$$

defined over $(x, y) \in S$, where S is a subset of the cartesian product of the set of nonnegative integers. If $a(x, y) > 0$ for all $(x, y) \in S$ and $\theta_1, \theta_2 \geq 0$, the relation (14.8) can be divided by the function $f(\theta_1, \theta_2)$ on both sides. On division, it provides the families of bivariate Lagrangian probability distribution, given by

$$P(X = x, Y = y) = P(x, y) = a(x, y)\,[g(\theta_1, \theta_2)]^x\,[h(\theta_1, \theta_2)]^y\,/f(\theta_1, \theta_2), \quad (x, y) \in S, \tag{14.10}$$

where $a(x, y)$ is given by (14.9).

By assigning suitable values of the functions $f(\theta_1, \theta_2)$, $g(\theta_1, \theta_2)$, and $h(\theta_1, \theta_2)$, subject to the conditions stated earlier, one can obtain a large number of bivariate Lagrangian probability models. This class of bivariate Lagrangian probability distribution (14.10) was given the name *bivariate modified power series distribution (BMPSD)* by Shoukri and Consul (1982) and was studied in detail. Many specific bivariate Lagrangian probability distributions, obtained from (14.3) or (14.6), belong to the class of BMPSD. The study of the BMPSD is much simpler than the study of the general class of bivariate Lagrangian probability distributions in (14.3) or (14.6).

14.2 Cumulants of Bivariate Lagrangian Distributions

Shenton and Consul (1973) obtained expressions for the first-order cumulants in terms of the cumulants of $h(t_1, t_2)$ and $k(t_1, t_2)$, for bivariate Lagrangian probability models. Let $H_{r,s}$ and $K_{r,s}$, $r, s, = 0, 1, 2, \ldots$, be the cumulants of the probability distributions generated by $h(t_1, t_2)$ and $k(t_1, t_2)$. Let $F_{r,s}$, $r, s = 1, 2, \ldots$, be the cumulants of the distribution generated by the pgf $f(t_1, t_2)$. Also, let $L_{r,s}$, $r, s = 0, 1, 2, \ldots$, be the cumulants of the bivariate Lagrangian probability distributions.

Now using e^{T_1}, e^{T_2}, e^{β_1}, and e^{β_2} in place of t_1, t_2, u, and v, respectively, in (14.1) and in (14.2), taking the logarithms of all three results, and by expanding them in power series, we obtain

$$\left.\begin{aligned} T_1 &= \beta_1 + (H_{10}T_1 + H_{01}T_2) \\ &\quad + \left(H_{20}T_1^2/2! + H_{11}T_1T_2 + H_{02}T_2^2/2!\right) + \dots \\ T_2 &= \beta_2 + (K_{10}T_1 + K_{01}T_2) \\ &\quad + \left(K_{20}T_1^2/2! + K_{11}T_1T_2 + K_{02}T_2^2/2!\right) + \dots \end{aligned}\right\} \tag{14.11}$$

$$\left.\begin{aligned} &F_{10}T_1 + F_{01}T_2 + F_{20}T_1^2/2! + F_{11}T_1T_2 + F_{02}T_2^2/2! + \dots \\ &\quad = L_{10}\beta_1 + L_{01}\beta_2 + L_{20}\beta_1^2/2! + L_{11}\beta_1\beta_2 + L_{02}\beta_2^2/2! + \dots \end{aligned}\right\}. \tag{14.12}$$

On substituting the values of β_1 and β_2 from (14.11) in (14.12) and by equating the coefficients of T_1 and T_2 with each other, we have

$$\left.\begin{aligned} F_{10} &= L_{10}(1 - H_{10}) + L_{01}(-K_{10}) \\ F_{01} &= L_{10}(-H_{01}) + L_{01}(1 - K_{01}) \end{aligned}\right\}, \tag{14.13}$$

which give the two first-order noncentral moments of BLD as

$$L_{10} = \frac{F_{10}(1 - K_{01}) + F_{01}K_{10}}{(1 - H_{10})(1 - K_{01}) - H_{01}K_{10}} \tag{14.14}$$

and

$$L_{01} = \frac{F_{01}(1 - H_{10}) + F_{10}H_{01}}{(1 - H_{10})(1 - K_{01}) - H_{01}K_{10}}. \tag{14.15}$$

By equating together the coefficients of T_1^2, T_1T_2, T_2^2 on both sides of (14.12) after eliminating β_1 and β_2 by (14.11), Shenton and Consul (1973) obtained the following relations among the second-order cumulants:

$$\left.\begin{aligned} &L_{20}(1 - H_{10})^2 - 2L_{11}K_{10}(1 - H_{10}) + L_{02}K_{10}^2 = Q_{20} \\ &L_{20}H_{01}(1 - H_{10}) - L_{11}\left\{(1 - H_{10})(1 - K_{01}) - H_{01}K_{10}\right\} + L_{02}K_{10}(1 - K_{01}) = Q_{11} \\ &L_{20}H_{01}^2 - 2L_{11}H_{01}(1 - K_{01}) + L_{02}(1 - K_{01})^2 = Q_{02} \end{aligned}\right\}, \tag{14.16}$$

where

$$\left.\begin{aligned} Q_{20} &= F_{20} + L_{10}H_{20} + L_{01}K_{20} \\ Q_{11} &= -(F_{11} + L_{10}H_{11} + L_{01}K_{11}) \\ Q_{02} &= F_{02} + L_{10}H_{02} + L_{01}K_{02} \end{aligned}\right\}. \tag{14.17}$$

On solving equations (14.16) by Cramer's rule, the second-order cumulants or covariances of BLD can easily be evaluated. The first- and second-order cumulants will exist if $H_{10} + H_{01} < 1$ and $K_{10} + K_{01} < 1$.

14.3 Bivariate Modified Power Series Distributions

14.3.1 Introduction

The class of bivariate modified power series distribution (BMPSD) is defined (Shoukri and Consul, 1982) by a bivariate discrete random variable (X, Y) as

$$P(X = x, Y = y) = a(x, y)\,[g(\theta_1, \theta_2)]^x\,[h(\theta_1, \theta_2)]^y\,/f(\theta_1, \theta_2), \quad (x, y) \in S, \tag{14.18}$$

and zero otherwise, where S is a subset of the cartesian product of the set of nonnegative integers, $a(x, y) > 0$, θ_1, $\theta_2 \geq 0$ and $g(\theta_1, \theta_2)$, $h(\theta_1, \theta_2)$ and $f(\theta_1, \theta_2)$ are finite positive and differentiable functions of θ_1 and θ_2 such that

$$f(\theta_1, \theta_2) = \sum_{(x,y)\in S} a(x, y) g^x(\theta_1, \theta_2) h^y(\theta_1, \theta_2).$$

Whenever $g(\theta_1, \theta_2) = \Phi_1$ and $h(\theta_1, \theta_2) = \Phi_2$ can be solved for θ_1 and θ_2, the BMPSD reduces to the class of bivariate generalized power series distribution.

The class of BMPSD includes, among others, the trinomial distribution, the bivariate negative binomial distribution, the bivariate Poisson distribution, the bivariate Borel–Tanner distribution, the bivariate GNBD, and the power series distributions associated with Lagrange expansion as defined by Jain and Singh (1975). A truncated BMPSD is also a BMPSD.

Many important families of the BMPSD can be generated by using the Lagrange expansion in (14.2). Jain and Singh (1975) used the expansion in (14.2) on $f(\theta_1, \theta_2)$ under the transformation $\theta_1 = u g(\theta_1, \theta_2)$ and $\theta_2 = v h(\theta_1, \theta_2)$ to obtain the BMPSD in (14.18). Shoukri and Consul (1982) provided a table of the sets of pgfs with their corresponding BMPSDs. Some of these are given in Table 14.1.

In subsequent subsections, we use the letters g, h, and f only for the functions $g(\theta_1, \theta_2)$, $h(\theta_1, \theta_2)$, and $f(\theta_1, \theta_2)$, respectively, and the following symbols:

$$D_i = \frac{\partial}{\partial \theta_i},$$

$$g_i = \frac{\partial \log g}{\partial \theta_i}, \quad h_i = \frac{\partial \log h}{\partial \theta_i}, \quad f_i = \frac{\partial \log f}{\partial \theta_i}, \tag{14.19}$$

$$g_{ij} = \frac{\partial^2 \log g}{\partial \theta_i \partial \theta_j} \quad \text{and so on, where} \quad i, j = 1, 2. \tag{14.20}$$

The sign $\sum$ will be used for the sum over all points $(x, y) \in S$, unless otherwise specified. We denote

$$\Delta = g_1 h_2 - g_2 h_1. \tag{14.21}$$

For all nonnegative integers r and s,

$$\mu'_{rs} = \mathrm{E}(X^r Y^s), \quad \mu_{rs} = \mathrm{E}\left[(X - \mu'_{10})^r (Y - \mu'_{01})^s\right]. \tag{14.22}$$

14.3.2 Moments of BMPSD

Since $\sum P(X = x, Y = y) = 1$, we have $f = \sum a(x, y) g^x h^y$. On differentiating this partially with respect to θ_1 and θ_2, respectively, and dividing by f, we obtain

$$f_1 = g_1 \mu'_{10} + h_1 \mu'_{01},$$

$$f_2 = g_2 \mu'_{10} + h_2 \mu'_{01}.$$

On solving the above, one obtains

$$\mu'_{10} = (f_1 h_2 - f_2 h_1)/\Delta \tag{14.23}$$

Table 14.1. Some members of BMPSDs

Name, $f(\theta_1, \theta_2)$	$g(\theta_1, \theta_2)$	$h(\theta_1, \theta_2)$	$a(x, y)$
Bivariate modified double Poisson, $f(\theta_1, \theta_2) = e^{\theta_1+\theta_2}$	$\theta_1 e^{-m_1(\theta_1+\theta_2)}$ $\theta_1,\ \theta_2 > 0$	$\theta_2 e^{-m_2(\theta_1+\theta_2)}$ $0 < m_1\theta_1 + m_2\theta_2 < 1$	$\frac{(1+m_1x+m_2y)^{x+y-1}}{x!y!}$ $m_1 > 0,\ m_2 > 0$ $x, y = 0, 1, 2, \ldots$
Bivariate modified negative binomial, $f(\theta_1, \theta_2) = (1-\theta_1-\theta_2)^{-n}$	$\theta_1(1-\theta_1-\theta_2)^{\beta_1-1}$ $0 < \theta_1,\ \theta_2 < 1$	$\theta_2(1-\theta_1-\theta_2)^{\beta_2-1}$ $0 < \beta_1\theta_1 + \beta_2\theta_2 < 1$	$\frac{n\Gamma(n+\beta_1x+\beta_2y)}{x!y!\Gamma(n+\beta_1x+\beta_2y-x-y+1)}$ $n > 0$ $x, y = 0, 1, 2, \ldots$
Bivariate modified logarithmic, $f(\theta_1, \theta_2) = -\ln(1-\theta_1-\theta_2)$	$\theta_1(1-\theta_1-\theta_2)^{\beta_1-1}$ $0 < \theta_1,\ \theta_2 < 1$	$\theta_2(1-\theta_1-\theta_2)^{\beta_2-1}$ $0 < \beta_1\theta_1 + \beta_2\theta_2 < 1$	$\frac{\Gamma(\beta_1x+\beta_2y)}{x!y!\Gamma(\beta_1x+\beta_2y-x-y+1)}$ $x, y = 1, 2, 3, \ldots$
Bivariate modified delta Poisson, $f(\theta_1, \theta_2) = \theta_1^m\theta_2^n$	$\theta_1 e^{-\theta_1-\theta_2}$ $\theta_1,\ \theta_2 > 0$	$\theta_2 e^{-\theta_1-\theta_2}$	$\frac{(m+n)(x+y)^{x+y-m-n-1}}{(x-m)!(y-n)!}$ $x = m(1),\ y = n(1)$
Bivariate modified delta binomial, $f(\theta_1, \theta_2) = \frac{\theta_1^m\theta_2^n}{(1-\theta_1-\theta_2)^{m+n}}$	$\theta_1(1-\theta_1-\theta_2)^{\beta_1-1}$ $0 < \theta_1,\ \theta_2 < 1$	$\theta_2(1-\theta_1-\theta_2)^{\beta_2-1}$ $0 < \beta_1\theta_1 + \beta_2\theta_2 < 1$	$\frac{(m\beta_1+n\beta_2)}{(x-m)!(y-n)!}$ $\times \frac{\Gamma(\beta_1x+\beta_2y)}{\Gamma(\beta_1x+\beta_2y+m+n-x-y+1)}$ $x = m(1),\ y = n(1)$

and

$$\mu'_{01} = (f_2g_1 - f_1g_2)/\Delta \tag{14.24}$$

In general, $\mu'_{r,s} = \sum x^r y^s a(x, y) g^x h^y / f$. On differentiating partially with respect to θ_1 and θ_2, respectively, we obtain

$$D_1\mu'_{r,s} = g_1\mu'_{r+1,s} + h_1\mu'_{r,s+1} - f_1\mu'_{r,s},$$

$$D_2\mu'_{r,s} = g_2\mu'_{r+1,s} + h_2\mu'_{r,s+1} - f_2\mu'_{r,s}.$$

On solving the two equations, the recurrence relations for the higher product moments are given by

$$\mu'_{r+1,s} = \left[h_2 \cdot D_1\mu'_{r,s} - h_1 \cdot D_2\mu'_{r,s}\right]/\Delta + \mu'_{r,s}\mu'_{10}, \tag{14.25}$$

$$\mu'_{r,s+1} = \left[g_1 \cdot D_2\mu'_{r,s} - g_2 \cdot D_1\mu'_{r,s}\right]/\Delta + \mu'_{r,s}\mu'_{01}. \tag{14.26}$$

Similarly, by differentiating $\mu_{r,s}$ partially with respect to θ_1 and θ_2, respectively, and on simplification, the two recurrence relations between the central product moments are

$$\mu_{r+1,s} = \Delta^{-1}\left\{(h_2 D_1 - h_1 D_2)\mu_{r,s} + r\mu_{r-1,s}(h_2 D_1 - h_1 D_2)\mu'_{10} + s\mu_{r,s-1}(h_2 D_1 - h_1 D_2)\mu'_{01}\right\} \quad (14.27)$$

and

$$\mu_{r,s+1} = \Delta^{-1}\left\{(g_1 D_2 - g_2 D_1)\mu_{r,s} + r\mu_{r-1,s}(g_1 D_2 - g_2 D_1)\mu'_{10} + s\mu_{r,s-1}(g_1 D_2 - g_2 D_1)\mu'_{01}\right\}. \quad (14.28)$$

The above two formulas will give not only the variances σ_x^2 and σ_y^2 by putting $r = 1, s = 0$ and $r = 0, s = 1$, but will also give the recurrence relations among higher central moments by putting $r = 0$ or $s = 0$. The coefficient of correlation ρ_{xy} can be obtained from the recurrence relation.

Recurrence relations among the factorial moments $\mu_x^{[r]} = \mathrm{E}\left(X^{[r]}\right)$ and $\mu_y^{[s]} = \mathrm{E}\left(Y^{[s]}\right)$ of the BMPSD can be obtained, and these are given by

$$\mu_x^{[r+1]} = \Delta^{-1}\left[h_2 D_1 - h_1 D_2\right]\mu_x^{[r]} + (\mu'_{10} - r)\mu_x^{[r]} \quad (14.29)$$

and

$$\mu_y^{[s+1]} = \Delta^{-1}\left[g_1 D_2 - g_2 D_1\right]\mu_y^{[s]} + (\mu'_{01} - s)\mu_y^{[s]}, \quad (14.30)$$

where $\mu_x^{[1]} = \mu'_{10}$ and $\mu_y^{[1]} = \mu'_{01}$. The cumulants can be obtained from the moments by using the relationship between central moments and cumulants or by using the results for cumulants in section 14.2.

14.3.3 Properties of BMPSD

The following properties are given by Shoukri and Consul (1982).

Let $(X_i, Y_i),\ i = 1, 2, 3, \ldots, N$, be a random sample of size N taken from the BMPSD given by (14.18) and let

$$Z_1 = \sum_{i=1}^{N} X_i \quad \text{and} \quad Z_2 = \sum_{i=1}^{N} Y_i.$$

When the functions g and h are zero at $\theta_1 = 0$ and $\theta_2 = 0$, due to the properties of the power series functions, the joint probability function of (Z_1, Z_2) can be written as

$$P(Z_1 = z_1, Z_2 = z_2) = b(z_1, z_2, N) g^{z_1} h^{z_2} / f^N, \quad (14.31)$$

where

$$b(z_1, z_2, N) = \sum \prod_{i=1}^{N} a(x_i, y_i),$$

and the summation extends over all ordered N-tuples $\{(x_1, y_1), \ \ldots, \ (x_N, y_N)\}$ of nonnegative integers of the set S under the conditions

$$\sum_{i=1}^{N} x_i = z_1 \quad \text{and} \quad \sum_{i=1}^{N} y_i = z_2.$$

The evaluation of $b(Z_1, Z_2, N)$ seems to be difficult on account of the summation over (x_i, y_i) on the products of $a(x_i, y_i)$, but it can be obtained more easily from (14.3) by replacing f, g, and k by f^N, $\theta_1 g^{-1}$, and $\theta_2 h^{-1}$, respectively, using θ_1 and θ_2 instead of t_1 and t_2 and putting Z_1 and Z_2 instead of x_1 and x_2, respectively. Thus, the BMPSD satisfies the convolution property.

The following two theorems, given by Shoukri and Consul (1982), can easily be proved for a BMPSD.

Theorem 14.1. *The means μ'_{10} and μ'_{01} of a BMPSD with $f(0,0) = 1$ are proportional to the parametric functions g and h, respectively, if and only if it is a double Poisson probability distribution.*

Theorem 14.2. *The means μ'_{10} and μ'_{01} of a BMPSD with $f(0,0) = 1$ are equal to $cg(1 - g - h)^{-1}$ and $ch(1 - g - h)^{-1}$, respectively, where c is any real number and g and h are the two parametric functions of the BMPSD, if and only if it is a bivariate negative binomial distribution with pmf*

$$P(X = x,\ Y = y) = \frac{(x + y + n - 1)!}{x!y!n!}\theta_1^x\theta_2^y(1 - \theta_1 - \theta_2)^n.$$

14.3.4 Estimation of BMPSD

Maximum Likelihood Estimation

If (X_i, Y_i), $i = 1, 2, 3, \ldots, N$, is a random sample of size N taken from the BMPSD given by (14.18), the logarithm of its likelihood function L becomes

$$\ln L = \text{constant} + \sum_{i=1}^{N} x_i \log g + \sum_{i=1}^{N} y_i \log h - N \log f. \tag{14.32}$$

On differentiating partially with respect to θ_1 and θ_2 and equating to zero, the ML equations become

$$\bar{x}g_1 + \bar{y}h_1 - f_1 = 0, \tag{14.33}$$

$$\bar{x}g_2 + \bar{y}h_2 - f_2 = 0, \tag{14.34}$$

where $(\bar{x}, \bar{y})$ is the sample mean. Assuming that other parameters in the BMPSD are known, the solution of the above two equations for θ_1 and θ_2 is not easy because these two parameters are involved in the functions g_1, g_2, h_1, h_2, f_1, and f_2. However, these equations can easily be solved for $\bar{x}$ and $\bar{y}$ to give

$$\bar{x} = \frac{f_1h_2 - f_2h_1}{g_1h_2 - g_2h_1} = \hat{\mu}'_{10} \tag{14.35}$$

and

$$\bar{y} = \frac{f_2g_1 - f_1g_2}{g_1h_2 - g_2h_1} = \hat{\mu}'_{01}. \tag{14.36}$$

Thus, $\bar{X}$ and $\bar{Y}$ are the ML estimators for the means μ'_{10} and μ'_{01}, respectively. This shows that the ML estimators of the means μ'_{10} and μ'_{01} are identical to their moment estimators. If equations (14.35) and (14.36) do not give an explicit solution for θ_1 and θ_2, an iterative method can be used to get a convergent solution starting with some values $(\theta_{10}, \theta_{20})$. The variances and covariance of the ML estimators for θ_1 and θ_2 are given by Shoukri and Consul (1982).

MVU Estimation

The BMPSD belongs to the exponential class. Accordingly, the sample sums Z_1 and Z_2 become jointly sufficient statistics for the parameters θ_1 and θ_2. Shoukri (1982) has shown that the statistic (Z_1, Z_2) is complete. Shoukri (1982) has also obtained the necessary and sufficient conditions for the existence of a minimum variance unbiased estimator for a parametric function of parameters θ_1 and θ_2.

Let the set of positive integers $\{(x, y) : x \geq r, \; y \geq s\}$ of a two-dimensional space be denoted by ${}_rI_s$, where r and s are nonnegative integers. A subset U_N of ${}_0I_0$ is said to be the index-set of the function f^N if

$$f^N = \sum b(Z_1, Z_2, N) g^{Z_1} h^{Z_2},$$

where $b(Z_1, Z_2, N) > 0$ for $(Z_1, Z_2) \in U_N \subseteq {}_0I_0$. Let $K(\theta_1, \theta_2)$ be a real valued parametric function of θ_1 and θ_2 such that, for a random sample of size N, we have

$$K(\theta_1, \theta_2) \cdot f^N = \sum c(Z_1, Z_2, N) g^{Z_1} h^{Z_2},$$

where $c(Z_1, Z_2, N) \neq 0$ for $(Z_1, Z_2) \in U_N^* \subseteq {}_0I_0$. U_N^* is the index-set of the function $K(\theta_1, \theta_2) \cdot f^N$.

Shoukri (1982) stated and proved the following theorem which provides a necessary and sufficient condition for the existence of a MVU estimator of $K(\theta_1, \theta_2)$.

Theorem 14.3. *Necessary and sufficient conditions for $K(\theta_1, \theta_2)$ to be MVU estimable on the basis of a random sample of size N taken from the BMPSD are that $K(\theta_1, \theta_2) f^N$ is analytic at the origin and that $U_N^* \subseteq U_N$, where U_N^* and U_N are the index-sets of the functions $K(\theta_1, \theta_2) f^N$ and f^N, respectively. Also, when $K(\theta_1, \theta_2)$ is MVU estimable, its MVU estimator $\ell(Z_1, Z_2, N)$ is given by*

$$\ell(Z_1, Z_2, N) = \begin{cases} \frac{c(Z_1, Z_2, N)}{b(Z_1, Z_2, N)}, & (Z_1, Z_2) \in U_N^*, \\ 0, & \textit{otherwise}. \end{cases} \tag{14.37}$$

Proof. Condition is necessary: Let $K(\theta_1, \theta_2)$ be MVU estimable for some N; i.e., there exists a function $\ell(Z_1, Z_2, N)$ such that $\text{E}\left[\ell(Z_1, Z_2, N)\right] = K(\theta_1, \theta_2)$. Thus,

$$\sum_{U_N} \ell(Z_1, Z_2, N) b(Z_1, Z_2, N) g^{Z_1} h^{Z_2} = K(\theta_1, \theta_2) f^N$$

and $K(\theta_1, \theta_2) \cdot f^N$ must possess an expansion in powers of g and h; i.e., it must be analytic at the origin. In view of the assumption of the expansion taken earlier, by equating the two summations, we have

$$\sum_{U_N} \ell(Z_1, Z_2, N) b(Z_1, Z_2, N) g^{Z_1} h^{Z_2} = \sum_{U_N^*} c(Z_1, Z_2, N) g^{Z_1} h^{Z_2}.$$

Now, for every $(Z_1, Z_2) \in U_N^*$, $b(Z_1, Z_2, N)$ must be > 0, i.e., $(Z_1, Z_2) \in U_N$, which implies that $U_N^* \subseteq U_N$.

By equating the coefficients of $g^{Z_1} h^{Z_2}$ on both sides for all $(Z_1, Z_2) \in U_N$, the MVU estimator of $K(\theta_1, \theta_2)$ becomes

$$\ell(Z_1, Z_2, N) = \begin{cases} \frac{c(Z_1, Z_2, N)}{b(Z_1, Z_2, N)}, & (Z_1, Z_2) \in U_N^*, \\ 0, & \text{otherwise.} \end{cases}$$

Condition is sufficient: Let $U_N^* \subseteq U_N$ and $K(\theta_1, \theta_2) \cdot f^N$ be analytic at the origin. Expanding $K(\theta_1, \theta_2) \cdot f^N$ in powers of g and h by bivariate Lagrange expansion,

$$K(\theta_1, \theta_2) \cdot f^N = \sum_{U_N^*} c(Z_1, Z_2, N) g^{Z_1} h^{Z_2}.$$

Thus,

$$K(\theta_1, \theta_2) = \sum_{U_N^*} \frac{c(Z_1, Z_2, N)}{b(Z_1, Z_2, N)} \cdot b(Z_1, Z_2, N) g^{Z_1} h^{Z_2} / f^N$$

$$= \sum_{U_N} \ell(Z_1, Z_2, N) P(Z_1 = z_1, Z_2 = z_2),$$

implying that $\ell(Z_1, Z_2, N)$ is an unbiased estimator for $K(\theta_1, \theta_2)$. Since $\ell(Z_1, Z_2, N)$ is a function of the joint complete sufficient statistic (Z_1, Z_2), it must be an MVU estimator for $K(\theta_1, \theta_2)$. □

Remark. The parametric function $g^a h^b / f^c$, where a, b are any nonnegative integers and c is a positive integer, is MVU estimable for all sample sizes $N \geq c$ if and only if $U_{N-c}^* \subseteq U_N$, and in that case the MVU estimator for $g^a h^b / f^c$ is

$$\ell(Z_1, Z_2, N) = \begin{cases} \frac{b(Z_1 - a, Z_2 - b, N - c)}{b(Z_1, Z_2, N)}, & (Z_1, Z_2) \in U_{N-c}^*, \\ 0, & \text{otherwise.} \end{cases} \tag{14.38}$$

Bivariate Modified Double Poisson Model

The ML equations (14.33) and (14.34) become

$$\bar{x}(\theta_1^{-1} - m_1) - m_2 \bar{y} - 1 = 0 \quad \text{and} \quad -\bar{x} m_1 + \bar{y}(\theta_2^{-1} - m_2) - 1 = 0,$$

which provide the ML estimators for the parameters θ_1 and θ_2 as

$$\hat{\theta}_1 = \frac{\bar{X}}{m_1 \bar{X} + m_2 \bar{Y} + 1} = \frac{Z_1}{N + m_1 Z_1 + m_2 Z_2}$$

and

$$\hat{\theta}_2 = \frac{Z_2}{N + m_1 Z_1 + m_2 Z_2}.$$

It can easily be shown that $\mathrm{E}[\hat{\theta}_1] = N\theta_1 (N + m_1)^{-1}$ and $\mathrm{E}[\hat{\theta}_2] = N\theta_2 (N + m_2)^{-1}$, so that $(N + m_1) Z_1 (N + m_1 Z_1 + m_2 Z_2)^{-1} N^{-1}$ is an unbiased estimator for the parameter θ_1. Thus, the MVU estimator for θ_1 becomes

$$(N + m_1) Z_1 (N + m_1 Z_1 + m_2 Z_2)^{-1} N^{-1}.$$

Similarly, the MVU estimator of θ_2 is

$$(N + m_2) Z_2 (N + m_1 Z_1 + m_2 Z_2)^{-1} N^{-1}.$$

Bivariate Modified Negative Binomial Model

The ML equations (14.33) and (14.34) become

$$\bar{x}\theta_1^{-1}(1-\theta_1-\theta_2)-\bar{x}(\beta_1-1)-\bar{y}(\beta_2-1)-n=0$$

and

$$-\bar{x}(\beta_1-1)+\bar{y}\theta_2^{-1}(1-\theta_1-\theta_2)-\bar{y}(\beta_2-1)-n=0,$$

which provide the ML estimators for the parameters θ_1 and θ_2 as

$$\hat{\theta}_1=\frac{\bar{X}}{N+\beta_1\bar{X}+\beta_2\bar{Y}}=\frac{Z_1}{N(N+\beta_1 Z_1+\beta_2 Z_2)}$$

and

$$\hat{\theta}_2=\frac{Z_2}{N(N+\beta_1 Z_1+\beta_2 Z_2)}.$$

One can easily find their expected values, which can be used to find the unbiased estimators for θ_1 and θ_2. These unbiased estimators will be the MVU estimators of θ_1 and θ_2.

Bivariate Modified Delta Poisson Model

The ML equations (14.33) and (14.34) become

$$\bar{x}(\theta_1^{-1}-1)-\bar{y}-m\theta_1^{-1}=0 \quad \text{and} \quad -\bar{x}-\bar{y}(\theta_2^{-1}-1)-n\theta_2^{-1}=0,$$

which provide the ML estimators for the parameters θ_1 and θ_2 as

$$\hat{\theta}_1=\frac{Z_1-mN}{Z_1+Z_2} \quad \text{and} \quad \hat{\theta}_2=\frac{Z_2-nN}{Z_1+Z_2}.$$

By taking the expected values of $\hat{\theta}_1$ and $\hat{\theta}_2$, one can easily find the MVU estimators of θ_1 and θ_2.

Bivariate Modified Delta Binomial Model

The ML equations (14.33) and (14.34) become

$$(\bar{x}-m)\theta_1^{-1}(1-\theta_1-\theta_2)=m+n+\beta_1\bar{x}+\beta_2\bar{y}-\bar{x}-\bar{y}$$

and

$$(\bar{y}-n)\theta_2^{-1}(1-\theta_1-\theta_2)=m+n+\beta_1\bar{x}+\beta_2\bar{y}-\bar{x}-\bar{y},$$

which provide the ML estimators for the parameters θ_1 and θ_2 as

$$\hat{\theta}_1=\frac{Z_1-mN}{\beta_1 Z_1+\beta_2 Z_2} \quad \text{and} \quad \hat{\theta}_2=\frac{Z_2-nN}{\beta_1 Z_1+\beta_2 Z_2}.$$

On taking the expected values of $\hat{\theta}_1$ and $\hat{\theta}_2$, one can determine the unbiased estimators for θ_1 and θ_2, which become their MVU estimators.

14.4 Some Bivariate Lagrangian Delta Distributions

These families of probability models are obtained by taking $f(t_1, t_2) = t_1^m t_2^n$ in the Lagrangian expansion in (14.2), as the other expansions in (14.4) and (14.5) are not applicable. When $h(t_1, t_2) = (g(t_1, t_2))^{c_1}$, $c_1 > 1$, and $k(t_1, t_2) = (g(t_1, t_2))^{c_2}$, $c_2 > 1$, the expansions in (14.2) and (14.3), under the transformation (14.1), provide the pgf of the bivariate Lagrangian delta distributions in the form

$$t_1^m t_2^n = \sum_{i=m}^{\infty}\sum_{j=n}^{\infty} \frac{u^i v^j}{(i-m)!(j-n)!} \frac{mc_1 + nc_2}{c_1 i + c_2 j} D_1^{i-m} D_2^{j-n} g^{c_1 i + c_2 j} \bigg|_{t_1=t_2=0}$$

as a power series in u and v and the family of bivariate probability distribution as

$$P(X = x, Y = y) = \frac{mc_1 + nc_2}{(x-m)!(y-n)!Q} D_1^{x-m} D_2^{y-n} \left[g(t_1, t_2)\right]^Q \bigg|_{t_1=t_2=0}$$

for $x = m, m+1, m+2, \ldots,\ y = n, n+1, n+2, \ldots,$ and zero otherwise, and where $Q = c_1 x + c_2 y$.

Three important families of the bivariate Lagrangian delta probability distributions are given in Table 14.2, where $a = min(x-m, y-n)$:

Table 14.2. Some members of bivariate Lagrangian delta distributions

No.	$g(t_1, t_2)$	$P(X = x, Y = y)$
1.	$e^{\theta_1(t_1-1)+\theta_2(t_2-1)+\theta_3(t_1t_2-1)}$ $\theta_i \geq 0,\ i = 1, 2, 3$	$(c_1 m + c_2 n) Q^{x+y-m-n-1} e^{-Q(\theta_1+\theta_2+\theta_3)}$ $\times \sum_{k=0}^{a} \frac{\theta_1^{x-m-k}\theta_2^{y-n-k}\theta_3^k Q^{-k}}{(x-m-k)!(y-n-k)!k!}$
2.	$(\theta_0 + \theta_1 t_1 + \theta_2 t_2 + \theta_3 t_1 t_2)^r$ $0 \leq \theta_i \leq 1,\ i = 1, 2, 3$ $\theta_0 = 1 - \theta_1 - \theta_2 - \theta_3 > 0$	$(c_1 m + c_2 n) r$ $\times \sum_{k=0}^{a} \frac{(Q-1)!\theta_1^{x-m-k}\theta_2^{y-n-k}\theta_3^k\theta_0^{Q-x-y+m+n+k}}{(x-m-k)!(y-n-k)!k!(Q-x-y+m+n+k)!}$
3.	$\theta_0^{-1}(1 - \theta_1 t_1 - \theta_2 t_2 - \theta_3 t_1 t_2)$ $0 \leq \theta_i \leq 1,\ i = 1, 2, 3$ $\theta_0 = 1 - \theta_1 - \theta_2 - \theta_3 > 0$ $h = \left[g(t_1, t_2)\right]^{1-c_1},\ c_1 > 1$ $k = \left[g(t_1, t_2)\right]^{1-c_2},\ c_2 > 1$	$\frac{(c_1 m + c_2 n)\theta_0^{Q-x-y}}{Q}$ $\times \sum_{k=0}^{a} \frac{\theta_1^{x-m-k}\theta_2^{y-n-k}\theta_3^k \Gamma(Q-m-n-k)}{(x-m-k)!(y-n-k)!k!\Gamma(Q-x-y)}$

Notes:

(i) The model (1) is called the bivariate Lagrangian delta-Poisson probability distribution. Its particular case, given by $\theta_3 = 0,\ c_1\theta_1 = m_1,\ c_2\theta_2 = m_2,\ c_2\theta_1 = M_1$, and $c_2\theta_1 = M_1$, was studied by Shenton and Consul (1973).

(ii) The model (2) can be called the bivariate Lagrangian delta-binomial probability distribution. Model (1) can be obtained as a limiting form of this model.

(iii) The model (3) can be called the bivariate Lagrangian delta-negative binomial probability distribution. One can obtain model (1) as a limiting distribution of this model as well.

14.5 Bivariate Lagrangian Poisson Distribution

14.5.1 Introduction

The pgf of the bivariate Poisson probability model, given by Holgate (1964), is

$$g(t_1, t_2) = \exp\left[\theta_1(t_1 - 1) + \theta_2(t_2 - 1) + \theta_3(t_1t_2 - 1)\right].$$

Now, let

$$f(t_1, t_2) = e^{n[\theta_1(t_1-1)+\theta_2(t_2-1)+\theta_3(t_1t_2-1)]} \tag{14.39}$$

and

$$\left.\begin{aligned} h(t_1, t_2) &= e^{\beta_1[\theta_1(t_1-1)+\theta_2(t_2-1)+\theta_3(t_1t_2-1)]} \\ k(t_1, t_2) &= e^{\beta_2[\theta_1(t_1-1)+\theta_2(t_2-1)+\theta_3(t_1t_2-1)]} \end{aligned}\right\}. \tag{14.40}$$

Under the transformations (14.1), Consul (1994a) used the formula (14.4) on (14.39) and (14.40) and obtained the bivariate Lagrangian Poisson distribution (BLPD) as

$$P(0,0) = f(0,0) = e^{-n(\theta_1+\theta_2+\theta_3)}$$

and

$$P(x, y) = \sum_{u=0}^{\min(x,y)} \frac{\theta_1^{x-u}}{(x-u)!}\frac{\theta_2^{y-u}}{(y-u)!}\frac{\theta_3^{u}}{u!} n \cdot Q^{x+y-u-1} e^{-Q(\theta_1+\theta_2+\theta_3)} \tag{14.41}$$

for $x,\ y = 0, 1, 2, \ldots,\ x + y > 0$ and $Q = n + \beta_1 x + \beta_2 y$. The values of $P(x, y)$ in (14.41) reduce to the bivariate Poisson distribution, defined by Holgate (1964), when $\beta_1 = \beta_2 = 0$ and $n = 1$. When $\theta_3 = 0$ and $n = 1$, the BLPD in (14.41) reduces to the bivariate modified double Poisson model in Table 14.1.

Ambagaspitiya (1998) derived recurrence formulas for the bivariate Lagrangian Poisson distribution. Using the definition

$$S = \begin{bmatrix} X_1, X_2, \ldots, X_M \\ \\ Y_1, Y_2, \ldots, Y_N \end{bmatrix}$$

for modeling a book of business containing two classes of insurance policies where claim frequency among two classes are correlated but claim severities are independent of frequencies and assuming that (M, N) have a BLPD, he obtained the pmf of S as the compound BLPD. He obtained some recurrence formulas for this model as well.

14.5.2 Moments and Properties

Let $M_{j,k} = \mathrm{E}[X^j Y^k]$ denote the bivariate jth and kth noncentral moment for the BLPD. Using the BLPD model probabilities given in (14.41) to evaluate $\mathrm{E}[X^j Y^k]$, Consul (1994a) obtained the following two recurrence relations among the bivariate moments

$$\begin{aligned} aM_{j+1,k} &= n(\theta_1+\theta_3)M_{j,k} + \beta_2(\theta_1+\theta_3)(\theta_2 D_2 + \theta_3 D_3)M_{j,k} \\ &\quad + (1-\beta_2\theta_2 - \beta_2\theta_3)(\theta_1 D_1 + \theta_3 D_3)M_{j,k} \end{aligned} \tag{14.42}$$

and

$$\begin{aligned} aM_{j,k+1} &= n(\theta_2+\theta_3)M_{j,k} + \beta_1(\theta_2+\theta_3)(\theta_1 D_1 + \theta_3 D_3)M_{j,k} \\ &\quad + (1-\beta_1\theta_1 - \beta_1\theta_3)(\theta_2 D_2 + \theta_3 D_3)M_{j,k}, \end{aligned} \tag{14.43}$$

where

$$a = 1 - \beta_1(\theta_1+\theta_3) - \beta_2(\theta_2+\theta_3),$$

$$D_i = \frac{\partial}{\partial\theta_i}, \quad i = 1, 2, 3,$$

and $M_{0,0} = 1$.

By using $j = 0 = k$ in (14.42) and (14.43), the two means become

$$\mu'_{10} = n(\theta_1+\theta_3)a^{-1} \tag{14.44}$$

and

$$\mu'_{01} = n(\theta_2+\theta_3)a^{-1}. \tag{14.45}$$

On finding the partial derivatives of μ'_{10} and μ'_{01}, with respect to θ_1, θ_2, and θ_3 and on further simplifications, the variances μ_{20} and μ_{02} become

$$\mu_{20} = n(\theta_1+\theta_3)\left\{1 - 2\beta_2\theta_2 + (\theta_1+\theta_2)(\theta_2+\theta_3)\beta_2^2\right\}a^{-3} \tag{14.46}$$

and

$$\mu_{02} = n(\theta_2+\theta_3)\left\{1 - 2\beta_1\theta_1 + (\theta_1+\theta_2)(\theta_1+\theta_3)\beta_1^2\right\}a^{-3}. \tag{14.47}$$

Also, the covariance of the BLPD is given by

$$\mu_{11} = n\left[\theta_3 + \beta_1\theta_2(\theta_1+\theta_3) + \beta_2\theta_1(\theta_2+\theta_3) - \beta_1\beta_2(\theta_1+\theta_2)(\theta_2+\theta_3)(\theta_1+\theta_3)\right]a^{-3}. \tag{14.48}$$

Consul (1994a) showed that the BLPD possesses the convolution property. If (X_1, X_2) and (Y_1, Y_2) are two BLPD r.v.s with the six parameters

$$(n_1, \theta_1, \theta_2, \theta_3, \beta_1, \beta_2) \quad \text{and} \quad (n_2, \theta_1, \theta_2, \theta_3, \beta_1, \beta_2),$$

respectively. Then the r.v.s $X_1 + Y_1 = Z_1$ and $X_2 + Y_2 = Z_2$ have a BLPD with parameters $(n_1+n_2, \theta_1, \theta_2, \theta_3, \beta_1, \beta_2)$.

14.5.3 Special BLPD

The special case of the BLPD for $\theta_3 = 0$ and $n = 1$ is the *bivariate modified double Poisson model* of the BMPSD class discussed in section 14.3 with the functions defined as

$$g = \theta_1 e^{-\beta_1(\theta_1+\theta_2)}, \; h = \theta_2 e^{-\beta_2(\theta_1+\theta_2)}, \; \text{and } f = e^{\theta_1+\theta_2}.$$

Its means, variances and covariance can be written from the above values. However, they can be computed from (14.22), (14.23), and (14.24) as well. The functions give $f_1 = f_2 = 1$, $g_1 = \theta_1^{-1} - \beta_1$, $g_2 = -\beta_1$, $h_1 = -\beta_2$, and $h_2 = \theta_2^{-1} - \beta_2$.
By the formulas given in section 14.3 for BMPSD, we obtain

$$\Delta = g_1 h_2 - g_2 h_1 = (1 - \theta_1\beta_1 - \theta_2\beta_2)/\theta_1\theta_2,$$

$$\mu'_{10} = (f_1 h_2 - f_2 h_1)/\Delta = \theta_1/(1 - \theta_1\beta_1 - \theta_2\beta_2), \tag{14.49}$$

and

$$\mu_{20} = (h_2 D_1 \mu'_{10} - h_1 D_2 \mu'_{10})/\Delta = \frac{\theta_1\left[1 - 2\beta_2\theta_2 + \beta_2^2\theta_2(\theta_1 + \theta_2)\right]}{(1 - \theta_1\beta_1 - \theta_2\beta_2)^3}, \tag{14.50}$$

where $\theta_1\beta_1 + \theta_2\beta_2 < 1$.

The values of μ'_{01} and μ_{02} can be written down by symmetry. It can also be proved that the coefficient of correlation between X and Y is given by (Shoukri, 1982)

$$\rho = \frac{\theta_1\theta_2[\beta_1(1 - \theta_2\beta_2) + \beta_2(1 - \theta_1\beta_1)]}{[\theta_1\theta_2(1 - 2\theta_1\beta_1 + \theta_1^2\beta_1^2 + \theta_1\theta_2\beta_1^2)(1 - 2\theta_2\beta_2 + \theta_2^2\beta_2^2 + \theta_1\theta_2\beta_2^2)]^{1/2}}. \tag{14.51}$$

The conditional expectation and variance of Y for a given x become

$$\mathrm{E}(Y|x) = \frac{\theta_2(1 + \beta_1 x)}{1 - \theta_2\beta_2}, \qquad x = 0, 1, 2, \ldots, \tag{14.52}$$

and

$$\mathrm{Var}(Y|x) = \frac{\theta_2(1 + \beta_1 x)}{(1 - \theta_2\beta_2)^2}, \qquad x = 0, 1, 2, \ldots. \tag{14.53}$$

14.6 Other Bivariate Lagrangian Distributions

14.6.1 Bivariate Lagrangian Binomial Distribution

A bivariate discrete r.v. (X, Y) is said to follow a bivariate Lagrangian binomial distribution (BLBD) if the joint probability distribution of X and Y is given by

$$P(x, y) = \sum_{u=0}^{\min(x,y)} \frac{n\Gamma(n + m_1 x + m_2 y)}{(n + m_1 x + m_2 y - x - y + u)!} \frac{\theta_1^{x-u}}{(x-u)!} \frac{\theta_2^{y-u}}{(y-u)!} \frac{\theta_3^u}{u!}$$
$$\times (1 - \theta_1 - \theta_2 - \theta_3)^{n + m_1 x + m_2 y - x - y + u + 1} \tag{14.54}$$

for $x, \; y = 0, 1, 2, 3, \ldots$, and where $\theta_i \geq 0, \; i = 1, 2, 3$, such that $0 < \theta_1 + \theta_2 + \theta_3 < 1$ and m_1, m_2, and n are nonnegative integers. Consul (1994a) obtained the model (14.54) by defining its bivariate pgf as

$$\ell(u, v) = f(t_1, t_2) = [1 + \theta_1(t_1 - 1) + \theta_2(t_2 - 1) + \theta_3(t_1 t_2 - 1)]^n, \tag{14.55}$$

under the transformations

$$t_1 = u\, h(t_1, t_2) \quad \text{and} \quad t_2 = v\, k(t_1, t_2),$$

where

$$h(t_1, t_2) = [1 + \theta_1(t_1 - 1) + \theta_2(t_2 - 1) + \theta_3(t_1 t_2 - 1)]^{m_1}, \tag{14.56}$$

$$k(t_1, t_2) = [1 + \theta_1(t_1 - 1) + \theta_2(t_2 - 1) + \theta_3(t_1 t_2 - 1)]^{m_2}. \tag{14.57}$$

To obtain the means, variances, and covariance of the BLBD in (14.54), the bivariate cumulants $F_{10},\ F_{01},\ F_{20},\ F_{11},\ F_{02},\ H_{10},\ H_{01},\ H_{20},\ H_{11},\ H_{02},\ K_{10},\ K_{01},\ K_{20},\ K_{11}$, and K_{02} can be derived for the bivariate generating functions $f(t_1, t_2),\ h(t_1, t_2)$, and $k(t_1, t_2)$ and then substitute their values in the formulas (14.14), (14.15), (14.16), and (14.17) to get the means L_{10} and L_{01} and the variances L_{20} and L_{02}. The process is time consuming.

Another alternative is to define $M_{j,k} = \mathrm{E}[X^j Y^k]$ and then obtain recurrence relations for $M_{j+1,k}$ and $M_{j,k+1}$ by differentiation of $M_{j,k}$ with respect to $\theta_1, \theta_2, \theta_3$, respectively, and by solving them. This method will also be long.

The BLBD (14.54) reduces to the *bivariate modified negative binomial model*, when $\theta_3 = 0$, which belongs to the BMPSD class defined by Shoukri and Consul (1982) and for which the relevant functions are

$$g = \theta_1(1 - \theta_1 - \theta_2)^{m_1 - 1}, \quad h = \theta_2(1 - \theta_1 - \theta_2)^{m_2 - 1}, \quad \text{and} \quad f = (1 - \theta_1 - \theta_2)^{-n-1}.$$

From the above, we get

$$f_1 = (n + 1)(1 - \theta_1 - \theta_2)^{-1} = f_2,$$

$$g_1 = \theta_1^{-1} - (m_1 - 1)(1 - \theta_1 - \theta_2)^{-1}, \quad g_2 = -(m_1 - 1)(1 - \theta_1 - \theta_2)^{-1},$$

$$h_1 = -(m_2 - 1)(1 - \theta_1 - \theta_2)^{-1}, \quad h_2 = \theta_2^{-1} - (m_2 - 1)(1 - \theta_1 - \theta_2)^{-1},$$

and

$$\Delta = g_1 h_2 - g_2 h_1 = (1 - m_1\theta_1 - m_2\theta_2)/\left[(1 - \theta_1 - \theta_2)\theta_1\theta_2\right].$$

By substituting their values in (14.23), (14.24), (14.25), and (14.26) the means and the variances of the bivariate modified negative binomial model become

$$\mu'_{10} = (n + 1)\theta_1(1 - m_1\theta_1 - m_2\theta_2)^{-1}, \quad \mu'_{01} = (n + 1)\theta_2(1 - m_1\theta_1 - m_2\theta_2)^{-1},$$

$$\mu_{20} = (n + 1)\theta_1[1 - \theta_1 - 2m_2\theta_2 + m_2^2\theta_2(\theta_1 + \theta_2)]/(1 - m_1\theta_1 - m_2\theta_2)^3,$$

$$\mu_{02} = (n + 1)\theta_2[1 - \theta_2 - 2m_1\theta_1 + m_1^2\theta_1(\theta_1 + \theta_2)]/(1 - m_1\theta_1 - m_2\theta_2)^3,$$

where $0 < m_1\theta_1 + m_2\theta_2 < 1$. The ML estimators and the MVU estimators for θ_1 and θ_2 are in section 14.3.

14.6.2 Bivariate Lagrangian Negative Binomial Distribution

The bivariate Lagrangian negative binomial distribution (BLNBD) is defined by Consul (1994a) as

$$P(x,y) \tag{14.58}$$
$$= \sum_{u=0}^{\min(x,y)} \frac{n\Gamma(n+\beta_1 x+\beta_2 y-u)}{(n+\beta_1 x+\beta_2 y-x-y)!} \frac{\theta_1^{x-u}}{(x-u)!} \frac{\theta_2^{y-u}}{(y-u)!} \frac{\theta_3^u}{u!} (1-\theta_1-\theta_2-\theta_3)^{n+\beta_1 x+\beta_2 y-x-y}$$

for $x, y = 0, 1, 2, 3, \ldots$ and where $\theta_i \geq 0,\ i = 1, 2, 3$, such that $0 < \theta_1+\theta_2+\theta_3 < 1, n > 0$, and $\beta_i \geq 1,\ i = 1, 2$. The pgf of the joint probability distribution in (14.58) is given by

$$\ell(u,v) = f(t_1,t_2) = \left(\frac{1-\theta_1 t_1-\theta_2 t_2-\theta_3 t_1 t_2}{1-\theta_1-\theta_2-\theta_3}\right)^{-n} \tag{14.59}$$

under the transformations

$$t_1 = u\, h(t_1,t_2) \quad \text{and} \quad t_2 = v\, k(t_1,t_2),$$

where

$$h(t_1,t_2) = \left(\frac{1-\theta_1 t_1-\theta_2 t_2-\theta_3 t_1 t_2}{1-\theta_1-\theta_2-\theta_3}\right)^{1-\beta_1}, \tag{14.60}$$

$$k(t_1,t_2) = \left(\frac{1-\theta_1 t_1-\theta_2 t_2-\theta_3 t_1 t_2}{1-\theta_1-\theta_2-\theta_3}\right)^{1-\beta_2}. \tag{14.61}$$

When $\theta_3 = 0$ the probability model BLNBD, defined by (14.58), reduces to the *bivariate modified negative binomial model* which belongs to the BMPSD class. Jain and Singh (1975) obtained this particular model by a different method and had called it generalized bivariate negative binomial distribution and obtained a recurrence relation between the probabilities and the first two means and the second factorial moments for the model. Shoukri (1982) obtained the MVU estimator for $\theta_1^{\gamma_1}\theta_2^{\gamma_2}$ when $\theta_3 = 0$ in (14.58) as

$$\ell(Z_1,Z_2,N) = \begin{cases} \frac{Z_1!Z_2!\Gamma(nN+\beta_1 Z_1+\beta_2 Z_2-\gamma_1-\gamma_2)}{nN(Z_1-\gamma_1)!(Z_2-\gamma_2)!\Gamma(nN+\beta_1 Z_1+\beta_2 Z_2)} \\ \times[nN+\gamma_1(\beta_1-1)+\gamma_2(\beta_2-1)],\ Z_1 \geq \gamma_1,\ Z_2 \geq \gamma_2, \\ 0, \qquad \text{otherwise,} \end{cases} \tag{14.62}$$

where Z_1, Z_2, and N are defined in section 14.3.

For a probabilistic interpretation of the BLNBD model, defined by (14.58), consider a sequence of independent trials with four possible outcomes:

(i) occurrences (successes) of events of Type I with probability θ_1,
(ii) occurrences of events of Type II with probability θ_2,
(iii) occurrences of some events of Type I and some events of Type II together with probability θ_3, and
(iv) the nonoccurrences (failures) of any event with probability $1-\theta_1-\theta_2-\theta_3$.

Then the probability that x events of type I, y events of type II, and u events of both type I and type II together do take place with exactly $n+\beta_1 x+\beta_2 y-x-y$ failures in $n+\beta_1 x+\beta_2 y$ trials will provide the model (14.58).

The means, variances, and covariance of the model in (14.58) can be obtained from the formulas (14.14), (14.15), (14.16), and (14.17) by substituting the values of the bivariate cumulants F_{ij}, H_{ij}, and K_{ij}, $i, j = 0, 1, 2$, which can be determined from the pgfs in (14.59), (14.60), and (14.61), respectively, by replacing t_1 and t_2 by e^{T_1} and e^{T_2} in f, g, and h, taking the logarithms of f, g, and h and by expanding them in powers of T_1 and T_2. The cumulants are the coefficients of the various powers of T_1 and T_2.

14.6.3 Bivariate Lagrangian Logarithmic Series Distribution

Consul (1994a) defined bivariate Lagrangian logarithmic series distribution (BLLSD) as

$$P(x, y) = \frac{(1-\theta_1-\theta_2-\theta_3)^{\beta_1 x+\beta_2 y-x-y}}{[-\ln(1-\theta_1-\theta_2-\theta_3)]} \sum_{u=0}^{\min(x,y)} \frac{\Gamma(\beta_1 x+\beta_2 y-u)}{(\beta_1 x+\beta_2 y-x-y)!} \frac{\theta_1^{x-u}}{(x-u)!} \frac{\theta_2^{y-u}}{(y-u)!} \frac{\theta_3^u}{u!} \tag{14.63}$$

for $x, y = 1, 2, 3, \ldots\ x + y \geq 1$, where $\theta_i \geq 0,\ i = 1, 2, 3$, such that $0 < \theta_1 + \theta_2 + \theta_3 < 1$ and $\beta_i \geq 1,\ i = 1, 2$. The pgf of the bivariate distribution in (14.63) is given by

$$\ell(u, v) = f(t_1, t_2) = \frac{\ln(1-\theta_1 t_1-\theta_2 t_2-\theta_3 t_1 t_2)}{\ln(1-\theta_1-\theta_2-\theta_3)} \tag{14.64}$$

under the transformations

$$t_1 = u\, h(t_1, t_2) \quad \text{and} \quad t_2 = v\, k(t_1, t_2),$$

where

$$h(t_1, t_2) = \left(\frac{1-\theta_1 t_1-\theta_2 t_2-\theta_3 t_1 t_2}{1-\theta_1-\theta_2-\theta_3}\right)^{1-\beta_1}, \tag{14.65}$$

$$k(t_1, t_2) = \left(\frac{1-\theta_1 t_1-\theta_2 t_2-\theta_3 t_1 t_2}{1-\theta_1-\theta_2-\theta_3}\right)^{1-\beta_2}. \tag{14.66}$$

The means, variances, and covariance of the BLLSD, defined above, can be obtained from the formulas (14.14), (14.15), (14.16), and (14.17) by substituting the values of the cumulants $F_{ij},\ H_{ij}$, and $K_{ij},\ i, j = 0, 1, 2$, which can be determined from the pgfs (14.64), (14.65), and (14.66), respectively.

When $\theta_3 = 0$, the above probability model in (14.63) reduces to the *bivariate modified logarithmic series distribution* (Shoukri and Consul, 1982), which belongs to the BMPSD class with the functions

$$f = -\ln(1-\theta_1-\theta_2), \quad g = \theta_1(1-\theta_1-\theta_2)^{\beta_1-1}, \quad \text{and} \quad h = \theta_2(1-\theta_1-\theta_2)^{\beta_2-1}.$$

According to the definitions in section 14.3, for this special case of $\theta_3 = 0$, we get

$$f_1 = \frac{(1-\theta_1-\theta_2)^{-1}}{-\ln(1-\theta_1-\theta_2)} = f_2,$$

$$g_1 = \theta_1^{-1} - (\beta_1-1)(1-\theta_1-\theta_2)^{-1}, \quad g_2 = -(\beta_1-1)(1-\theta_1-\theta_2)^{-1},$$

$$h_1 = -(\beta_2-1)(1-\theta_1-\theta_2)^{-1}, \quad h_2 = \theta_2^{-1} - (\beta_2-1)(1-\theta_1-\theta_2)^{-1},$$

and

$$\Delta = g_1 h_2 - g_2 h_1 = (1-\beta_1\theta_1-\beta_2\theta_2)/[(1-\theta_1-\theta_2)\theta_1\theta_2].$$

Therefore,

$$\mu'_{10} = (f_1 h_2 - f_2 h_1)/\Delta = \frac{\theta_1(1 - \beta_1\theta_1 - \beta_2\theta_2)^{-1}}{[-\ln(1 - \theta_1 - \theta_2)]},$$

$$\mu'_{01} = (f_2 g_1 - f_1 g_2)/\Delta = \frac{\theta_2(1 - \beta_1\theta_1 - \beta_2\theta_2)^{-1}}{[-\ln(1 - \theta_1 - \theta_2)]}.$$

The values of μ_{20} and μ_{02} can be similarly derived from (14.27) and (14.28).

14.6.4 Bivariate Lagrangian Borel–Tanner Distribution

Shenton and Consul (1973) defined the bivariate Lagrangian Borel–Tanner distribution (BLBTD) by using the bivariate Lagrange expansion and gave its pgf as

$$\ell(u, v) = f(t_1, t_2) = t_1^i t_2^j \tag{14.67}$$

under the transformations in (14.1), where

$$h(t_1, t_2) = \exp[m_1(t_1 - 1) + m_2(t_2 - 1)], \tag{14.68}$$

$$k(t_1, t_2) = \exp[M_1(t_1 - 1) + M_2(t_2 - 1)]. \tag{14.69}$$

The expansion gives the bivariate joint probability for the BLBTD as

$$P(x, y) = e^{-(m_1+m_2)x-(M_1+M_2)y} \frac{(m_1 x + M_1 y)^{x-i}}{(x - i)!} \frac{(m_2 x + M_2 y)^{y-j}}{(y - j)!}$$

$$\times \left[\frac{ij}{xy} + \frac{x - i}{x} \frac{M_1 j}{m_1 x + M_1 y} + \frac{y - j}{y} \frac{m_2 i}{m_2 x + M_2 y} \right]. \tag{14.70}$$

The cumulants of the probability models represented by (14.67), (14.68), and (14.69) are

$$F_{10} = i, \quad F_{01} = j, \quad F_{20} = F_{02} = F_{11} = 0,$$

$$H_{10} = H_{20} = m_1, \quad H_{01} = H_{02} = m_2, \quad H_{11} = 0,$$

$$K_{10} = K_{20} = M_1, \quad K_{01} = K_{02} = M_2, \quad K_{11} = 0.$$

Substituting their values in (14.14) and (14.15), the means of the BLBTD become

$$L_{10} = [i(1 - M_2) + jM_1] \,/\, [(1 - m_1)(1 - M_2) - m_2 M_1]$$

and

$$L_{01} = [j(1 - m_1) + im_2] \,/\, [(1 - m_1)(1 - M_2) - m_2 M_1].$$

The variances and the covariance of the BLBTD are given by

$$L_{20} = \frac{Q_{20}(1 - M_2)^2 + Q_{02}M_1^2}{[(1 - m_1)(1 - M_2) - m_2 M_1]^2}, \quad L_{02} = \frac{Q_{20}(1 - m_2)^2 + Q_{02}m_1^2}{[(1 - m_1)(1 - M_2) - m_2 M_1]^2},$$

$$L_{11} = \frac{Q_{20}m_2(1 - M_2) + Q_{02}M_1(1 - m_1)}{[(1 - m_1)(1 - M_2) - m_2 M_1]^2},$$

where

$$Q_{20} = [(i+j)(M_1+m_1)+i(M_1m_2-M_2m_1)]\,/\,[(1-m_1)(1-M_2)-m_2M_1],$$
$$Q_{02} = [(i+j)(M_2+m_2)+j(M_1m_2-M_2m_1)]\,/\,[(1-m_1)(1-M_2)-m_2M_1].$$

We now consider the following three cases.

Case 1. If $m_1 = M_1$ and $m_2 = M_2$, in the above BLBTD one obtains the BLBTD defined by Shoukri and Consul (1982), and the probability function in (14.70) reduces to

$$P(x,y) = (i+j)\frac{m_1^{x-i}}{(x-i)!}\frac{m_2^{y-j}}{(y-j)!}(x+y)^{x+y-i-j-1}e^{-(m_1+m_2)(x+y)} \tag{14.71}$$

for $x \geq i$ and $y \geq j$.

Case 2. If $m_1 = 0$ and $M_2 = 0$, the BLBTD in (14.70) reduces to

$$P(x,y) = \frac{M_1(M_1y)^{x-i-1}}{(x-i)!}\frac{m_2(m_2x)^{y-j-1}}{(y-j)!}(xj+yi-ij)e^{-m_2x-M_1y} \tag{14.72}$$

for $x \geq i$ and $y \geq j$.

Case 3. If $m_2 = 0$ and $M_1 = 0$, the BLBTD in (14.70) reduces to the probability function

$$P(x,y) = \frac{(m_1x)^{x-i}}{(x-i)!}\frac{(M_2y)^{y-j}}{(y-j)!}\frac{ij}{xy}e^{-m_1x-M_2y} \tag{14.73}$$

for $x \geq i$ and $y \geq j$.

Jain and Singh (1975) have defined a bivariate Borel–Tanner distribution as a limiting form of their bivariate generalized negative binomial distribution and have obtained its first two noncentral moments. In fact, what they obtained is the bivariate modified double Poisson model, given in Table 14.1, and it is a particular case of the BLPD for $\theta_3 = 0$.

14.6.5 Bivariate Inverse Trinomial Distribution

Shimizu, Nishii, and Minami (1997) defined the bivariate inverse trinomial distribution (BITD). Let

$$g_1(t_1,t_2) = g_2(t_1,t_2) = p + \sum_{j=1}^{2}(q_jt_j + r_jt_j^2),$$

and $f(t_1,t_2) = t_1^{k_1}t_2^{k_2}$ be three pgfs where $p > 0$, q_j, $r_j \geq 0$, $p + \sum_{j=1}^{2}(q_j + r_j) = 1$, and k_j $(j = 1, 2)$ are nonnegative integers. By using the Lagrange expansion (14.2) with $h(t_1,t_2) = g_1(t_1,t_2)$ and $k(t_1,t_2) = g_2(t_1,t_2)$, the pmf of BITD becomes

$$P(x_1,x_2) = \frac{k_1+k_2}{x_1+x_2}\sum_{i_1=0}^{[\frac{x_1-k_1}{2}]}\sum_{i_2=0}^{[\frac{x_2-k_2}{2}]} p^{\sum_{j=1}^{2}(i_j+k_j)}$$
$$\times \prod_{j=1}^{2}\left\{q_j^{x_j-2i_j-k_j}r_j^{i_j}\left(\begin{matrix}x_1+x_2\\ i_1, i_2, x_1-2i_1-k_1, x_2-2i_2-k_2, \sum(i_j+k_j)\end{matrix}\right)\right\} \tag{14.74}$$

for $x_j = k_j, k_j+1, k_j+2, \ldots$ and $k_1 + k_2 > 0$.

On using the transformation $X_j = Y_j + k_j$ for $j = 1, 2$, and putting $\lambda = \sum_{j=1}^{2} k_j$, we obtain the probability function of $Y = (Y_1, Y_2)$ as

$$P(y_1, y_2) = \lambda p^{\lambda} \prod_{j=1}^{2} q_j^{y_j} \sum_{i_1=0}^{[\frac{y_1}{2}]} \sum_{i_2=0}^{[\frac{y_2}{2}]} \prod_{j=1}^{2} \left(\frac{pr_j}{q_j^2} \right)^{i_j}$$

$$\times \binom{\lambda + y_1 + y_2}{i_1, i_2, y_1 - 2i_1, y_2 - 2i_2, \lambda + i_1 + i_2} \tag{14.75}$$

for $y_j = 0, 1, 2, \ldots$ and $p \geq r_1 + r_2$.

As long as $\lambda > 0$, the probability function in (14.75) has the pgf

$$\phi(u_1, u_2) = \left(\frac{2p}{1 - \sum_{j=1}^{2} q_j u_j + \sqrt{\left(1 - \sum_{j=1}^{2} q_j u_j\right)^2 - 4p \sum_{j=1}^{2} r_j u_j^2}} \right)^{\lambda} \tag{14.76}$$

for $0 < u_j \leq 1$, $j = 1, 2$ and $\phi(1, 1) = 1$ if $p \geq r_1 + r_2$ and $\phi(1, 1) = \left(\frac{p}{r_1+r_2}\right)^{\lambda}$ if $p \leq r_1 + r_2$.

14.6.6 Bivariate Quasi-Binomial Distribution

Mishra (1996) considered a five-urn model with different colored balls and described successive drawings of balls by a player from the different urns under a number of conditions and calculated the probability of the player being a winner. Based on the numbers of balls in the various urns and the number of balls being added, the probability had six integer parameters. Mishra (1996) replaced some ratios with different symbols and obtained the probability of the player being a winner in a nice form as

$$P(x, y) = \sum_{u=0}^{\min(x,y)} \frac{n!}{u!(x-u)!(y-u)!(n-x-y+u)!} \alpha_1 \beta_1 (\alpha_1 + \alpha_2 u)^{u-1} (p - \alpha_1 - \alpha_2 u)^{x-u}$$

$$\times [\beta_1 + (y-u)\beta_2]^{y-u-1} [1 - p - \beta_1 - (y-u)\beta_2]^{n-x-y+u}, \tag{14.77}$$

where $x, y = 0, 1, 2, \ldots, n$ and $0 < p < 1$, $0 < \alpha_i, \beta_i < 1$ for $i = 1, 2$, and n is a positive integer. Mishra (1996) proved that it was a true bivariate probability distribution and named it the *bivariate quasi-binomial distribution* (BQBD) because its conditional distribution, for given x, is the quasi-binomial distribution. Also, Mishra (1996) obtained the first- and the second-order moments of BQBD in (14.77) in the form

$$\text{E}[X] = np,$$

$$\text{E}[Y] = n \sum_{i=0}^{n-1} (n-1)^{(i)} \left(\alpha_1 \alpha_2^i + \beta_1 \beta_2^i \right), \tag{14.78}$$

$$\text{E}\left[X^2\right] = n(n-1)p^2 + np,$$

$$\mathrm{E}\left[Y^2\right] = n(n-1)\left[\sum_{i=0}^{n-2}(i+1)(n-2)^{(i)}\left(\alpha_1^2\alpha_2^i + \beta_1^2\beta_2^i\right)\right.$$
$$+ 2\alpha_1\beta_1\sum_{r=0}^{n-2}(n-2)^{(r)}\alpha_2^r\sum_{j=0}^{n-2-r}(n-2-r)^{(j)}\beta_2^j$$
$$\left. + \sum_{j=0}^{n-1}\frac{1}{2}(j+1)(j+2)(n-1)^{(j)}\left(\alpha_1\alpha_2^j + \beta_1\beta_2^j\right)\right] \tag{14.79}$$

and

$$\mathrm{Cov}(X, Y) = -np\sum_{i=0}^{n-1}(i+1)(n-1)^{(i)}\left(\alpha_1\alpha_2^i + \beta_1\beta_2^i\right). \tag{14.80}$$

Mishra (1996) fitted the model (14.77) to a demographic sample survey data of 515 families of Patna (India) by estimating the parameters $p,\ \alpha_1,\ \alpha_2,\ \beta_1$, and β_2 by ML method.

14.7 Exercises

14.1 Obtain the variances and covariance of the ML estimates of parameters θ_1 and θ_2 in a BMPSD.

14.2 For a BMPSD, state and prove the theorem that provides a necessary and sufficient condition for existence of an MVU estimator of $K(\theta_1, \theta_2)$.

14.3 For a BMPSD, derive the results in (14.27) and (14.28).

14.4 For a BLPD, show the results in (14.42) and (14.43).

14.5 Let $\theta_3 = 0$ in the BLPD defined in (14.41). Compute μ'_{01} and μ_{02}.

14.6 Consider the BLBD in (14.54). Show that the distribution is a BMPSD when $\theta_3 = 0$. When $\theta_3 = 0$, find the moment estimates of the parameters θ_1, θ_2, m_1, and m_2 if n is a known constant.

14.7 Let $X_i,\ i = 1, 2, \ldots, M$, and $Y_j,\ j = 1, 2, \ldots, N$, be the claim severities in two classes of insurance policies, which are independent of each other and are also independent of the frequencies. Assume that (M, N) have a BLPD. If a book of business consists of

$$S = \begin{bmatrix} X_1, X_2, \ldots, X_M \\ Y_1, Y_2, \ldots, Y_N \end{bmatrix},$$

show that the pmf of S is a compound BLPD (Ambagaspitiya, 1998).

14.8 If the service begins with a queue consisting of i customers of type I and j customers of type II with different arrival rates for each type of customer and requiring separate kinds of service for each type of customer, show that the probability distribution of the number of customers of type I and type II served in a busy period is a Lagrangian-type bivariate Borel–Tanner probability distribution (Shenton and Consul, 1973).

14.9 Consider a bivariate branching process similar to the univariate branching process given in section 6.2. Define the total number of objects in the zeroth, first, second, ..., nth generations. Assume that the probability distribution of the number of objects produced by each object remains unaltered over successive generations and that the branching process started with two objects, one of each kind I and kind II. Show that the probability distribution of the total progeny is the basic BLD.

14.10 Find the variance and covariance for the bivariate logarithmic series distribution.

14.11 Let $f = f(\theta_1, \theta_2) = e^{n(\theta_1+\theta_2)}$, $g = g(\theta_1, \theta_2) = \theta_1 e^{-\beta_1(\theta_1+\theta_2)}$, and $h = h(\theta_1, \theta_2) = \theta_2 e^{-\beta_2(\theta_1+\theta_2)}$. Find $a(x, y)$ such that f, g, and h give a BMPSD. Show that the means μ'_{10} and μ'_{01} are equal to the corresponding means for the bivariate modified negative binomial distribution.

15

Multivariate Lagrangian Distributions

15.1 Introduction

The multivariate generalizations of important discrete distributions, their properties, and some of their applications have been discussed by many researchers, including Neyman (1965), Olkin and Sobel (1965), Patil and Bildikar (1967), Teicher (1954), and Wishart (1949). A systematic account of these generalizations of discrete distributions has been given by Johnson and Kotz (1969) and Johnson, Kotz, and Balakrishnan (1997). The vast scope and importance of the multivariate generalizations of discrete distributions is abundantly clear from the works cited above.

However, very little work has been done on the multivariate generalizations of Lagrangian distributions, their importance, and their applications. The pioneering work in this direction was done by Good (1960, 1965), who generalized the Lagrange expansion to an arbitrary number of independent variables and applied it to the enumeration of trees and to stochastic processes. Consul and Shenton (1973b) used one of the expansions, given by Good (1960), to derive another form and to define the multivariate Lagrangian probability distributions. They described a number of possible applications, obtained its mean vector and variance-covariance matrix, and developed a multivariate generalization of the Borel–Tanner distribution. Khatri (1982) presented the complex form of the class of multivariate Lagrangian distributions (MLDs) in a more systematic form and explicitly gave the probability density functions for multivariate Lagrangian Poisson, multinomial, and quasi-Pólya distributions and obtained expressions for their moments.

Nandi and Das (1996) defined the multivariate Abel series distributions. They derived the quasi-multinomial distribution of type I and the multiple GPD as examples. Kvam and Day (2001) considered the multivariate Pólya distribution for application in combat models. Goodness-of-fit tests for the model are derived.

15.2 Notation and Multivariate Lagrangian Distributions

The letters $\mathbf{t},\ \mathbf{T},\ \mathbf{u},\ \boldsymbol{\beta},\ \mathbf{0}, \mathbf{x}$ represent k-variate column vectors so that

$$\left.\begin{array}{ll} \mathbf{t}' = (t_1, t_2, \ldots, t_k), & \mathbf{u}' = (u_1, u_2, \ldots, u_k) \\ \mathbf{T}' = (T_1, T_2, \ldots, T_k), & \boldsymbol{\beta}' = (\beta_1, \beta_2, \ldots, \beta_k) \\ \mathbf{0}' = (0, 0, \ldots, 0), & \mathbf{x}' = (x_1, x_2, \ldots, x_k) \\ \mathbf{T}^{-1} = (T_1^{-1}, T_2^{-1}, \ldots, T_k^{-1})' & \boldsymbol{\beta}^{-1} = (\beta_1^{-1}, \beta_2^{-1}, \ldots, \beta_k^{-1})' \end{array}\right\}. \tag{15.1}$$

We shall use the following symbols:

$$g_i(\mathbf{t}') = g_i, \qquad (g_i)^{x_\nu} = g_i^{x_\nu},$$

$$\mathbf{x}! = \prod_{i=1}^{k}(x_i)!, \quad \text{and} \quad D_i^r = \left(\frac{\partial}{\partial t_i}\right)^r \tag{15.2}$$

for $i = 1, 2, 3, \ldots, k$ and where D_i^r is a differential operator that will be used on the left-hand side of all functions and will operate on all of the functions on the right-hand side of the operator.

Let the multivariate functions $g_i(\mathbf{t}')$ of $\mathbf{t}'$ be analytic in the neighborhood of the origin, such that $g_i(\mathbf{0}') \neq \mathbf{0}'$ $(i = 1, 2, \ldots, k)$. Note that the transformations

$$t_i = u_i g_i(\mathbf{t}'), \quad i = 1, 2, \ldots, k, \tag{15.3}$$

imply that $\mathbf{u} = \mathbf{0} \Longleftrightarrow \mathbf{t} = \mathbf{0}$.

Let $f(\mathbf{t}')$ be another meromorphic function in $\mathbf{t}'$. Good (1960) proved that

$$f(\mathbf{t}'(\mathbf{u}')) = \sum \frac{u_1^{x_1} u_2^{x_2} \ldots u_k^{x_k}}{x_1! x_2! \ldots x_k!} D_1^{x_1} \ldots D_k^{x_k} \left\{ f(\mathbf{t}') g_1^{x_1} \ldots g_k^{x_k} \| \delta_i^\nu - u_i g_i^\nu \| \right\} \Big|_{\mathbf{t}'=\mathbf{0}'}, \tag{15.4}$$

where the summation is taken over all nonnegative integers $x_1, x_2, \ldots, x_k$. Good (1960) pointed out that the factor multiplying $u_1^{x_1} \ldots u_k^{x_k}$ in (15.4) is not a proper "coefficient." Consul and Shenton (1973b) modified Good's general result in (15.4) so that the factor multiplying $u_1^{x_1} \ldots u_k^{x_k}$ becomes a true coefficient. They gave a new form of Good's multivariate Lagrange-type expansion as

$$f(\mathbf{t}'(\mathbf{u}')) = \sum_{\mathbf{x}} \frac{u_1^{x_1} u_2^{x_2} \ldots u_k^{x_k}}{x_1! x_2! \ldots x_k!} \left[D_1^{x_1-1} \ldots D_k^{x_k-1} \| D_\nu (g_\nu)^{x_\nu} \mathbf{I} - \mathbf{G} \| f(\mathbf{t}') \right]_{\mathbf{t}'=\mathbf{0}'}, \tag{15.5}$$

where $\mathbf{I}$ is the $k \times k$ identity unit matrix and $\mathbf{G}$ is the $k \times k$ matrix $\frac{\partial}{\partial t_j}(g_i)^{x_i}$. The result in (15.5) is a generalization of Poincaré's result, as the factor multiplying $u_1^{x_1} \ldots u_k^{x_k}$ is a true coefficient. When the operational determinant is positive for all $\mathbf{x}$, then the coefficient of $\prod_{i=1}^k u_i^{x_i}$ provides the multivariate Lagrangian probability distribution. To show that (15.5) is a generalization of Poincaré's expansion, we consider the value of the operational determinant in (15.5) for $k = 2$ as

$$\begin{vmatrix} D_1 g_1^{x_1} - \partial_1 g_1^{x_1} & -\partial_2 g_1^{x_1} \\ -\partial_1 g_2^{x_2} & D_2 g_2^{x_2} - \partial_2 g_2^{x_2} \end{vmatrix} f(t_1, t_2), \quad \partial_i = \frac{\partial}{\partial t_i}.$$

On evaluating the determinant and simplifying it by the use of the operators D_1 and D_2 on the functions, one obtains

$$g_1^{x_1} g_2^{x_2} \partial_1 \partial_2 f + g_1^{x_1} \partial_1 g_2^{x_2} \partial_2 f + g_2^{x_2} \partial_2 g_1^{x_1} \partial_1 f,$$

which is the expression in the bivariate Lagrange expansion in section 14.1.

Considering

$$\mathbf{u}^{\mathbf{x}} = \prod_{i=1}^{k} u_i^{x_i} = \prod_{i=1}^{k} t_i^{x_i} \Big/ \prod_{i=1}^{k} g_i^{x_i}$$

and

$$f(\mathbf{t}) = \ell(\mathbf{u}) = \sum_{\mathbf{x}} b(\mathbf{x}')\mathbf{u}^{\mathbf{x}}/(\mathbf{x}!), \tag{15.6}$$

Khatri (1982) combined (15.4) and (15.5) and expressed the coefficient of $\mathbf{u}^{\mathbf{x}}$ in a very compact form as

$$b(\mathbf{x}') = \left[\prod_{i=1}^{k} D_i^{x_i} f(\mathbf{t}') \prod_{i=1}^{k} \left(g_i(\mathbf{t}')\right)^{x_i} \mid \mathbf{I} - \mathbf{A}(\mathbf{t}') \mid \right]_{\mathbf{t}'=\mathbf{0}'}, \tag{15.7}$$

where $\mathbf{A}(\mathbf{t}') = \left(a_{ij}(\mathbf{t})\right)$ and $a_{ij}(\mathbf{t}) = t_i D_j \left\{\ln g_i(\mathbf{t}')\right\}$ for all $i,\ j$. If $b(\mathbf{x}') \geq \mathbf{0}'$ for all $\mathbf{x}$, the multivariate Lagrangian probability distribution becomes

$$P(X_i = x_i,\ i = 1, 2, \ldots, k) = b(\mathbf{x}')/(\mathbf{x}!) \tag{15.8}$$

for all $\mathbf{x}$ and for all $\mathbf{t}$ within the radius of convergence. If $g_i(\mathbf{t}') = \mathbf{1}'$ for all $i = 1, 2, \ldots, k$, then (15.7) together with (15.8) reduces to the multivariate power series distribution.

Since (15.7) is in a very compact form and its importance may not be realized, Khatri (1982) considered its value for particular cases $k = 1, 2, 3$, simplified (15.7), and gave different expressions. For brevity, let

$$\left(g_i(\mathbf{t}')\right)^{x_i} = g_i^{x_i} = h_i \quad \text{for} \quad i = 1, 2, 3, \ldots, k.$$

For $k = 1$, (15.7) gives

$$b(x_1) = \left[D_1^{x_1} \left\{ f(h_1 - x_1^{-1} t_1 D_1 h_1) \right\} \right]_{t_1=0}$$

$$= \left[\sum_{i=1}^{x_1} \binom{x_1}{i} \left(D_1^i f\right) \left(D_1^{x_1-i} h_1 - (x_1 - i) x_1^{-1} D_1^{x_1-i} h_1\right) \right]_{t_1=0}$$

$$= \sum_{i=1}^{x_1} \binom{x_1}{i} \left(D_1^i f\right) \left(\frac{i}{x_1}\right) D_1^{x_1-i} h_1 \,|_{t_1=0} = D_1^{x_1-1} \left(h_1 D_1 f\right) |_{t_1=0},$$

which is the term of the univariate Lagrange expansion.

For $k = 2$, (15.7) gives

$$b(x_1, x_2) = D_1^{x_1} D_2^{x_2} f(\mathbf{t}') g_1^{x_1} g_2^{x_2} \begin{vmatrix} 1 - t_1 D_1(\ln g_1) & -t_1 D_2(\ln g_1) \\ -t_2 D_1(\ln g_2) & 1 - t_2 D_2(\ln g_2) \end{vmatrix}_{t_1=t_2=0}$$

$$= D_1^{x_1} D_2^{x_2} f \cdot \begin{vmatrix} h_1 - x_1^{-1} t_1 D_1 h_1 & -x_1^{-1} t_1 D_2 h_1 \\ -x_2^{-1} t_2 D_1 h_2 & h_2 - x_2^{-1} t_2 D_2 h_2 \end{vmatrix}_{t_1=t_2=0}$$

$$= D_1^{x_1} D_2^{x_2} \Big[f \Big\{ h_1 h_2 - t_1 x_1^{-1} h_2 D_1 h_1 - t_2 x_2^{-1} h_1 D_2 h_2$$

$$+ t_1 t_2 x_1^{-1} x_2^{-1} (D_1 h_1 D_2 h_2 - D_1 h_2 D_2 h_1) \Big\} \Big]_{t_1=t_2=0}$$

$$= D_1^{x_1} D_2^{x_2} (f h_1 h_2) - D_1^{x_1-1} D_2^{x_2} f h_2 D_1 h_1 - D_1^{x_1} D_2^{x_2-1} f h_1 D_2 h_2$$

$$+ D_1^{x_1-1} D_2^{x_2-1} (D_1 h_1 D_2 h_2 - D_1 h_2 D_2 h_1) \,|_{t_1=t_2=0} \,.$$

On simplification, the above expression gives

$$b(x_1, x_2) = D_1^{x_1-1} D_2^{x_2-1} \left[h_1 h_2 D_1 D_2 f + h_1 (D_1 h_2)(D_2 f) + h_2 (D_2 h_1)(D_1 f)\right]_{t_1=t_2=0},$$

which is the term of the bivariate Lagrange expansion in chapter 14.

Similarly, for $k = 3$, the expression (15.7) can finally be simplified to the form

$$\begin{aligned} b(x_1, x_2, x_3) = D_1^{x_1-1} D_2^{x_2-1} D_3^{x_3-1} & [h_1 h_2 h_3 D_1 D_2 D_3 f + h_1 (D_1 h_2 h_3) D_2 D_3 f \\ & + h_2 (D_2 h_1 h_3) D_1 D_3 f + h_3 (D_3 h_1 h_2) D_1 D_2 f \\ & + (\Delta_{12} h_1 h_2 h_3) D_3 f + (\Delta_{13} h_1 h_3 h_2) D_2 f \\ & + (\Delta_{23} h_2 h_3 h_1) D_1 f]_{t_1=t_2=t_3=0}, \end{aligned} \tag{15.9}$$

where

$$\Delta_{12} h_1 h_2 h_3 = h_1 h_2 h_3 + h_1 (D_1 h_2)(D_2 h_3) + h_2 (D_2 h_1)(D_1 h_3). \tag{15.10}$$

The expressions (15.9) and (15.10), together with (15.8), define the trivariate Lagrangian probability distributions as

$$P(X_1 = x_1, X_2 = x_2, X_3 = x_3) = b(x_1, x_2, x_3) \big/ [x_1! x_2! x_3!] \tag{15.11}$$

for all nonnegative integral values of x_1, x_2, x_3 if $b(x_1, x_2, x_3) \geq 0$.

By assigning suitable values for the functions $f(\mathbf{t}')$, $g_1(\mathbf{t}')$, $g_2(\mathbf{t}')$, $g_3(\mathbf{t}')$ for $\mathbf{t}' = (t_1, t_2, t_3)$, one can determine numerous trivariate Lagrangian probability models, like the bivariate Lagrangian probability models in chapter 14.

Thus, the multivariate Lagrangian probability distributions are given by the expression in (15.8) together with the expression (15.7) for all nonnegative integral values of x_i, $i = 1, 2, \ldots, k$. Another form of the class of multivariate Lagrangian probability distributions is given by the generating function (15.5) as

$$P(\mathbf{X}' = \mathbf{0}') = f(\mathbf{0}'),$$

$$P(\mathbf{X}' = \mathbf{x}') = \frac{1}{x_1! \ldots x_k!} D_1^{x_1-1} \ldots D_k^{x_k-1} \| D_\nu \, (g_\nu)^{x_\nu} \, \mathbf{I} - \mathbf{G} \| f(\mathbf{t}')|_{\mathbf{t}'=\mathbf{0}'} \tag{15.12}$$

for $(x_1, \ldots, x_k) \neq (0, \ldots, 0)$ and where $\mathbf{G}$ is a $k \times k$ matrix $\frac{\partial}{\partial t_j} (g_i)^{x_i}$, $j = 1, 2, , \ldots, k$.

To get many families of discrete multivariate Lagrangian probability distributions easily, the functions $g_i(\mathbf{t}')$, $i = 1, 2, \ldots, k$, and $f(\mathbf{t}')$ are replaced by particular sets of multivariate pgfs. When one chooses some other nonnegative multivariate meromorphic functions for $g_i(\mathbf{t}')$, $i = 1, 2, \ldots, k$, and $f(\mathbf{t}')$, which are not pgfs, satisfying the conditions $g_i(\mathbf{0}') \neq \mathbf{0}'$ and $f(\mathbf{1}') = \mathbf{1}'$, they must be such that all the terms in (15.12) are nonnegative.

Khatri (1982) has also shown, by using Jacobians of some transformations, that the expression given by him for $b(\mathbf{x}')$ can be transformed to the form given by Good (1960). Also, he has stated that the actual computation of $b(\mathbf{x}')$ for the various multivariate Lagrangian probability models is quite difficult even when specific values are chosen for the pgfs f, g_i, $i = 1, 2, \ldots, k$. The expression (15.9), together with (15.10), for $b(x_1, x_2, x_3)$ makes it clear that the expressions for different trivariate Lagrangian probability models will be quite long and complicated. Even Good (1975) has stated that one need not try to open up the compact form of the model and that the properties be studied by other ingenious methods as given by Consul and Shenton (1972) for computing cumulants. Good (1975) used the same methods to get the cumulants for the multivariate Lagrangian distributions because he did not realize that it had already been done by Consul and Shenton (1973b). Good (1975) has stated that every probability model is trivially Lagrangian, as $g_i(\mathbf{t}')$ can be taken to be unity.

Since the general form of the MLD, defined by (5.8) with (15.7) or by (15.12) are not easy to use, we shall consider some special cases of the MLD in sections 15.4 and 15.5.

15.3 Means and Variance-Covariance

Consul and Shenton (1973b) obtained the mean vector and the $k \times k$ variance-covariance matrix for the MLD by a very ingenious method, which is described here. Let the mean vector of the MLDs, defined by (15.12) or by (15.8), together with (15.7), be denoted by $\mathbf{L}_{(1)} = (L_1, L_2, \ldots, L_k)$. Also, let the mean row vectors of the discrete distributions given by $g_i(\mathbf{t}')$ and $f(\mathbf{t}')$ be denoted by $\mathbf{G}^i_{(1)} = (G^i_1, G^i_2, \ldots, G^i_k)$ and $\mathbf{F}_{(1)} = (F_1, F_2, \ldots, F_k)$. If some of these $(k+1)$ multivariate probability distributions do not have the same number (k) of variates, the particular means, corresponding to the variates absent, are taken to be zero and the corresponding t's in the $g_i(\mathbf{t}')$ and $f(\mathbf{t}')$ can be replaced by unity. Since $\mathbf{G}^i_{(1)}$ denotes the mean vector, let $(\mathbf{G}^i_{(1)})$ be a $k \times k$ matrix of mean values given by

$$\left(\mathbf{G}^i_{(1)}\right) = \begin{bmatrix} G^1_1 \; G^1_2 \; \ldots \; G^1_k \\ G^2_1 \; G^2_2 \; \ldots \; G^2_k \\ \ldots \\ G^k_1 \; G^k_2 \; \ldots \; G^k_k \end{bmatrix}. \tag{15.13}$$

Also, let the variance-covariance matrix of the multivariate probability distributions, given by the pgfs $g_i(\mathbf{t}')$, $i = 1, 2, \ldots, k$, be represented by

$$\mathbf{G}^i_{(2)} = \begin{bmatrix} G^i_{11} \; G^i_{12} \; \cdots \; G^i_{1k} \\ G^i_{21} \; G^i_{22} \; \cdots \; G^i_{2k} \\ \cdots \\ G^i_{k1} \; G^i_{k2} \; \cdots \; G^i_{kk} \end{bmatrix} \quad \text{for} \quad i = 1, 2, 3, \ldots, k. \tag{15.14}$$

Let the variance-covariance matrix of the probability distribution, given by $f(\mathbf{t}')$, be denoted by $\mathbf{F}_{(2)}$ and the variance-covariance matrix of the MLD (15.12) be denoted by $\mathbf{L}_{(2)}$. Since $\mathbf{G}^i_{(2)}$ is a $k \times k$ matrix, let $(\mathbf{G}^i_{(2)})$ represent a three-dimensional $k \times k \times k$ matrix whose element in the jth row and lth column is the column vector $(G^1_{jl}, G^2_{jl}, \ldots, G^k_{jl})$ for $j,\ l = 1, 2, 3, \ldots, k$.

By replacing each t_i and u_i by e^{T_i} and e^{β_i}, respectively, in the k transformations (15.3), taking the logarithms of each one of them, and on expanding them, Consul and Shenton (1973b) got the relations (given in the vector form)

$$\mathbf{T} = \boldsymbol{\beta} + \left(\mathbf{G}^i_{(1)}\right)\mathbf{T} + \frac{1}{2}\mathbf{T}'\left(\mathbf{G}^i_{(2)}\right)\mathbf{T} + \cdots, \tag{15.15}$$

which gives the vector relation

$$\boldsymbol{\beta} = \left\{ \mathbf{I} - \left(\mathbf{G}^{i}_{(1)} \right) - \frac{1}{2} \mathbf{T}' \left(\mathbf{G}^{i}_{(2)} \right) - \cdots \right\} \mathbf{T}. \tag{15.16}$$

Similarly, by replacing each t_i and u_i by e^{T_i} and e^{β_i}, respectively, in the relation (15.5), which can be expressed as

$$f(\mathbf{t}') = \sum u_1^{x_1} u_2^{x_2} \ldots u_k^{x_k} \mathbf{L} \left(\mathbf{g}',\ f;\ \mathbf{x}' \right) = \ell(\mathbf{u}') \tag{15.17}$$

also, and by taking logarithms on both sides and by expanding in powers of T_i and β_i, we get the vector relation

$$\mathbf{F}_{(1)}\mathbf{T} + \frac{1}{2}\mathbf{T}'\mathbf{F}_{(2)}\mathbf{T} + \cdots = \mathbf{L}_{(1)}\boldsymbol{\beta} + \frac{1}{2}\boldsymbol{\beta}'\mathbf{L}_{(2)}\boldsymbol{\beta} + \cdots . \tag{15.18}$$

On eliminating the column vector $\boldsymbol{\beta}$ between (15.18) and (15.16), one obtains

$$\left.\begin{aligned} \mathbf{F}_{(1)}\mathbf{T} + \tfrac{1}{2}\mathbf{T}'\mathbf{F}_{(2)}\mathbf{T} + \cdots = \mathbf{L}_{(1)} & \left\{ \mathbf{I} - \left(\mathbf{G}^{i}_{(1)} \right) - \tfrac{1}{2} \mathbf{T}' \left(\mathbf{G}^{i}_{(2)} \right) - \cdots \right\} \mathbf{T} \\ & + \tfrac{1}{2}\mathbf{T}' \left\{ \mathbf{I} - \left(\mathbf{G}^{i}_{(1)} \right) - \tfrac{1}{2} \mathbf{T}' \left(\mathbf{G}^{i}_{(2)} \right) - \cdots \right\}' \mathbf{L}_{(2)} \\ & \times \left\{ \mathbf{I} - \left(\mathbf{G}^{i}_{(1)} \right) - \tfrac{1}{2} \mathbf{T}' \left(\mathbf{G}^{i}_{(2)} \right) - \cdots \right\} \mathbf{T} + \cdots \end{aligned}\right\} . \tag{15.19}$$

Since both sides of (15.19) must be identical, a simple comparison of the terms on the two sides gives the following two relations together with many others for higher values:

$$\mathbf{F}_{(1)} = \mathbf{L}_{(1)} \left(\mathbf{I} - \left(\mathbf{G}^{i}_{(1)} \right) \right)$$

and

$$\mathbf{F}_{(2)} = -\mathbf{L}_{(1)} \left(\mathbf{G}^{i}_{(2)} \right) + \left(\mathbf{I} - \left(\mathbf{G}^{i}_{(1)} \right) \right)' \mathbf{L}_{(2)} \left(\mathbf{I} - \left(\mathbf{G}^{i}_{(1)} \right) \right).$$

Hence the mean-vector of the MLD becomes

$$\mathbf{L}_{(1)} = \mathbf{F}_{(1)} \left(\mathbf{I} - \left(\mathbf{G}^{i}_{(1)} \right) \right)^{-1}, \tag{15.20}$$

and the corresponding variance-covariance matrix, when $(\mathbf{I} - (\mathbf{G}^{i}_{(1)}))$ is a nonsingular $k \times k$ matrix, is given by

$$\mathbf{L}_{(2)} = \left[\left(\mathbf{I} - \left(\mathbf{G}^{i}_{(1)} \right) \right)' \right]^{-1} \left(\mathbf{F}_{(2)} + \mathbf{L}_{(1)} \left(\mathbf{G}^{i}_{(2)} \right) \right) \left(\mathbf{I} - \left(\mathbf{G}^{i}_{(1)} \right) \right)^{-1}, \tag{15.21}$$

where the elements of the middle matrix are of the form

$$F_{jl} + L_1 G^1_{jl} + L_2 G^2_{jl} + L_3 G^3_{jl} + \cdots + L_k G^k_{jl} \tag{15.22}$$

for $j, l = 1, 2, 3, \ldots, k$.

Minami (1998) has obtained the above variance-covariance matrix $\mathbf{L}_{(2)}$, but the same had been obtained twenty-five years earlier by Consul and Shenton (1973b).

15.4 Multivariate Lagrangian Distributions (Special Form)

The multivariate Lagrange expansions (15.4), given by Good (1960), and (15.5), given by Consul and Shenton (1973b), or the form (15.6) with (15.7) given by Khatri (1982), are in very nice compact form. However, each one of these forms involves the determinant of a $k \times k$ matrix, whose evaluation becomes quite difficult. In view of this we shall consider some special forms of the generating functions to get a simpler form for the multivariate Lagrange expansion.

Let the functions $g_i(\mathbf{t}') = (g(\mathbf{t}'))^{c_i}$, $i = 1, 2, 3, \ldots, k$, and let the function $f(\mathbf{t}') = g(\mathbf{t}')^m$ or the derivative of $f(\mathbf{t}')$ with respect to any one of the variables t_i be of that form. For the present, we assume that $\frac{\partial}{\partial t_i} f(\mathbf{t}')$ is of that form. When these multivariate functions are of these forms, the determinant of the $k \times k$ matrix in the Lagrange expansion (15.5) can be opened up in a systematic manner, and it is found that almost all the terms, except one, cancel out with each other. The whole expression gets simplified to a form similar to the bivariate simplified form (14.4) or (14.5) and gives the new multivariate Lagrange expansion as

$$f(\mathbf{t}') = f(\mathbf{0}') + \sum \frac{u_1^{x_1} u_2^{x_2} \ldots u_k^{x_k}}{x_1! x_2! \ldots x_k!} \left[D_1^{x_1 - 1} D_2^{x_2} \ldots D_k^{x_k} \left\{ g_1^{x_1} g_2^{x_2} \ldots g_k^{x_k} \frac{\partial f}{\partial t_1} \right\} \right]_{\mathbf{t}'=\mathbf{0}'} . \quad (15.23)$$

The above multivariate Lagrange expansion provides the multivariate Lagrangian probability distributions in the nice form

$$P(\mathbf{X}' = \mathbf{0}') = f(\mathbf{0}'),$$

$$P(\mathbf{X}' = \mathbf{x}') = \frac{1}{x_1! x_2! \ldots x_k!} D_1^{x_1 - 1} D_2^{x_2} \ldots D_k^{x_k} \left\{ g_1^{x_1} g_2^{x_2} \ldots g_k^{x_k} \frac{\partial f}{\partial t_1} \right\} \Bigg|_{\mathbf{t}'=\mathbf{0}'}$$

$$= \frac{1}{x_1! x_2! \ldots x_k!} D_1^{x_1 - 1} D_2^{x_2} \ldots D_k^{x_k} \left\{ (g(\mathbf{t}'))^{\sum c_i x_i} \frac{\partial f}{\partial t_1} \right\} \Bigg|_{\mathbf{t}'=\mathbf{0}'} \quad (15.24)$$

for $i = 1, 2, \ldots, k$, and the summation $\sum c_i x_i$ is on all values of i and $x_i = 0, 1, 2, \ldots$ for each value of i.

15.4.1 Multivariate Lagrangian Poisson Distribution

Let the multivariate functions $g_i(\mathbf{t}')$ and $f(\mathbf{t}')$ be given by

$$g_i(\mathbf{t}') = \exp\left\{ m_i \sum_{j=1}^{k} \theta_j (t_j - 1) \right\}, \quad i = 1, 2, \ldots, k, \quad \text{and} \quad f(\mathbf{t}') = \exp\left\{ \alpha \sum_{j=1}^{k} \theta_j (t_j - 1) \right\} \quad (15.25)$$

for $\theta_j > 0$, $m_i > 0$, for $i, j = 1, 2, 3, \ldots, k$ and $\alpha > 0$.

Substituting these values in (15.24) and on taking the derivatives successively and simplifying, we get the multivariate Lagrangian Poisson probability distribution in the form

$$P(\mathbf{X}' = \mathbf{x}') = \left[\prod_{i=1}^{k} \frac{\theta_i^{x_i}}{x_i!} \right] \alpha \left(\alpha + \sum_{i=1}^{k} m_i x_i \right)^{\sum_{i=1}^{k} x_i - 1} \exp\left\{ - \left(\alpha + \sum_{i=1}^{k} m_i x_i \right) \sum_{i=1}^{k} \theta_i \right\} \quad (15.26)$$

for $x_i = 0, 1, 2, 3 \ldots$, for $i = 1, 2, \ldots, k$. Some further conditions on the parameters may have to be imposed for the existence of cumulants. Khatri (1982) had obtained a more complex form of the multivariate Lagrangian Poisson distribution, but it contains a $k \times k$ determinant for evaluation with k derivatives.

15.4.2 Multivariate Lagrangian Negative Binomial Distribution

Let the multivariate probability generating functions $g_i(\mathbf{t}')$ and $f(\mathbf{t}')$ be given by

$$g_i(\mathbf{t}') = \left(\frac{1 - \sum_{j=1}^{k} \theta_j t_j}{1 - \sum_{j=1}^{k} \theta_j} \right)^{1-\beta_i}, \quad i = 1, 2, \ldots, k, \tag{15.27}$$

and

$$f(\mathbf{t}') = \left(\frac{1 - \sum_{j=1}^{k} \theta_j t_j}{1 - \sum_{j=1}^{k} \theta_j} \right)^{-\alpha}, \tag{15.28}$$

where $\theta_j > 0$, such that $\sum_{j=1}^{k} \theta_j < 1,\ 1 < \beta < \left(\max \theta_j\right)^{-1}$ for $i = 1, 2, 3, \ldots, k$ and $\alpha > 0$.

Substituting the above values of the functions $g_i(\mathbf{t}')$ and $f(\mathbf{t}')$ in (15.24), differentiating successively with respect to $t_i,\ i = 1, 2, \ldots, k$, and simplifying the expressions, the multivariate generalized negative binomial distribution becomes

$$P(\mathbf{X}' = \mathbf{x}') = \frac{\alpha \left(\alpha + \sum_{i=1}^{k} \beta_i x_i\right)!}{\left(\alpha + \sum_{i=1}^{k} (\beta_i - 1) x_i\right)!} \prod_{i=1}^{k} \frac{\theta_i^{x_i}}{x_i!} \left(1 - \sum_{i=1}^{k} \theta_i\right)^{\alpha + \sum_{i=1}^{k} (\beta_i - 1) x_i} \tag{15.29}$$

for $x_i = 0, 1, 2, 3, \ldots,$ for $i = 1, 2, \ldots, k$.

By using combinatorial methods, Mohanty (1966, 1979) obtained the model (15.29) and proved that it represents the probability of a particle from the origin to the point

$$\left(\alpha + \sum (\beta_i - 1) x_i;\ x_1, x_2, \ldots, x_k\right),$$

not touching the hyperplane

$$z = \alpha + \sum (\beta_i - 1) z_i$$

except at the end.

When $\beta_i = 1$ for $i = 1, 2, \ldots, k$, the model (15.29) reduces to the multivariate negative binomial distribution. When $\alpha \to \infty,\ \beta_i \to \infty$, and $\theta_i \to 0$ for $i = 1, 2, \ldots, k$ such that $\alpha\theta_i = a_i,\ \beta_i\theta_j = b_{ij}$, it can be proved that the limiting form of the multivariate Lagrangian negative binomial model (15.29) approaches another multivariate generalization of Poisson distribution given by

$$P(\mathbf{X}' = \mathbf{x}') = \prod_{j=1}^{k} \frac{a_j \left(a_j + \sum b_{ji} x_i\right)^{x_j - 1}}{x_j!} \exp\left[- \sum_{j=1}^{k} \left(a_j + \sum_{i=1}^{k} b_{ij} x_i \right) \right]. \tag{15.30}$$

The expression in (15.30) can be changed to the same form as (15.26) by putting $a_i = c d_i$ and $b_{ji} = d_j m_i$ for $i = 1, 2, 3, \ldots, k$ and $j = 1, 2, 3, \ldots, k$.

15.4.3 Multivariate Lagrangian Logarithmic Series Distribution

Let the multivariate probability generating functions $g_i(\mathbf{t}')$ be given by (15.27) and the function $f(\mathbf{t}')$ be

$$f(\mathbf{t}') = \frac{\ln\left(1 - \sum_{j=1}^{k} \theta_j t_j\right)}{\ln\left(1 - \sum_{j=1}^{k} \theta_j\right)}, \tag{15.31}$$

where $\theta_j > 0$ such that $\sum_{j=1}^{k} \theta_j < 1,\ 1 < \beta_i < \left(\max \theta_j\right)^{-1}$ for $i = 1, 2, \ldots, k$.

Substituting the values of the functions $g_i(\mathbf{t}')$ and $f(\mathbf{t}')$ in the expression (15.24), differentiating with respect to $t_1, t_2, \ldots, t_k$ successively, and simplifying the expression, the multivariate Lagrangian logarithmic series distribution is given in the form

$$P(\mathbf{X}' = \mathbf{x}') = \frac{\Gamma\left(\sum_{i=1}^{k} \beta_i x_i\right)}{\Gamma\left(1 + \sum_{i=1}^{n} (\beta_i - 1) x_i\right) \left\{-\ln(1 - \sum_{i=1}^{k} \theta_i)\right\}} \prod_{i=1}^{k} \frac{\theta_i^{x_i}}{x_i!} \left(1 - \sum_{i=1}^{k} \theta_i\right)^{\sum_{i=1}^{k} (\beta_i - 1) x_i} \tag{15.32}$$

for $x_i = 1, 2, 3, \ldots$, for $i = 1, 2, \ldots, k$, and the summations are on $i = 1$ to k.

15.4.4 Multivariate Lagrangian Delta Distributions

Let the multivariate binomial analytic probability generating functions be

$$g_i(\mathbf{t}') = \left[1 + \sum_{j=1}^{k} \theta_j (t_j - 1)\right]^{\beta_i}, \quad i = 1, 2, \ldots, k, \tag{15.33}$$

where $\theta_j > 0$ such that $\sum_j^k \theta_j < 1$ and β_i are positive integers and let

$$f(\mathbf{t}') = t_1^{m_1} t_2^{m_2} \ldots t_k^{m_k}$$

be another multivariate analytic function, where $m_i,\ i = 1, 2, \ldots, k$ are positive integers. The use of these multivariate functions in (15.5) provides the multivariate Lagrangian delta binomial probability model in the form

$$P(\mathbf{X}' = \mathbf{x}') = \frac{\sum_{i=1}^{k} \beta_i m_i \Gamma\left(\sum_{i=1}^{k} \beta_i x_i\right)}{\Gamma\left(\sum_{i=1}^{k} [(\beta_i - 1) x_i + m_i] + 1\right)} \prod_{i=1}^{k} \frac{\theta_i^{x_i - m_i}}{(x_i - m_i)!} \left(1 - \sum_{i=1}^{k} \theta_i\right)^{\sum_{i=1}^{k} [(\beta_i - 1) x_i + m_i]} \tag{15.34}$$

for $x_i = m_i, m_i + 1, m_i + 2, \ldots$, for $i = 1, 2, \ldots, k$, and all summations are on $i = 1$ to k.

Similarly, by considering the multivariate Poisson probability generating functions

$$g_i(\mathbf{t}') = \exp\left[m_i \sum_{j=1}^{k} \theta_j (t_j - 1)\right], \quad i = 1, 2, \ldots, k, \tag{15.35}$$

where $m_i > 0,\ \theta_j > 0$ for all values of i and by taking

$$f(\mathbf{t}') = t_1^{n_1} t_2^{n_2} \ldots t_k^{n_k},$$

where $n_1, n_2, \ldots, n_k$ are positive integers, and using them in the multivariate Lagrange expansion (15.5) gives the multivariate Lagrangian delta Poisson probability model as

$$P(\mathbf{X}'=\mathbf{x}') = \prod_{i=1}^{k} \frac{\theta_i^{x_i-n_i}}{x_i-n_i} \left(\sum_{i=1}^{k} m_i n_i\right)\left(\sum_{i=1}^{k} m_i x_i\right)^{\sum_{i=1}^{k}(x_i-n_i)-1} \exp\left[-\left(\sum_{i=1}^{k} m_i x_i\right)\sum_{i=1}^{k}\theta_i\right], \tag{15.36}$$

where $x_i = n_i, n_i+1, n_i+2, \ldots$ for $i = 1, 2, \ldots, k$ and all the summations are for $i = 1$ to k.

Consul and Shenton (1973b) had taken the parameter a_{ij}, instead of the product $m_i\theta_j$ used in (15.35), and had obtained another version of multivariate Lagrangian delta Poisson probability distribution as

$$P(\mathbf{X}'=\mathbf{x}') = \left[\prod_{i=1}^{k}\left\{\frac{e^{-x_i\left(\sum_j a_{ij}\right)}}{(x_i-n_i)!}\left(\sum_j a_{ij}x_j\right)^{x_i-n_i}\right\}\right]\left|\mathbf{I} - \frac{a_{rj}(x_r-n_r)}{\sum_i x_i a_{ir}}\right| \tag{15.37}$$

for $x_i = n_i, n_i+1, n_i+2, \ldots,\ i = 1, 2, \ldots, k$

Shimizu, Nishii, and Minami (1997) extended the bivariate inverse trinomial distribution to the multivariate Lagrangian inverse trinomial distribution and obtained its pgf and conjectured the form of the distribution.

Minami (1998) has derived a number of multivariate Lagrangian distributions, which are either particular cases or the same as those obtained earlier by Consul and Shenton (1973b) and by Khatri (1982). Possibly, Minami did not know about these earlier works.

15.5 Multivariate Modified Power Series Distributions

It has been shown earlier that the univariate MPSD and the bivariate MPSD belong to the class of Lagrangian probability distributions and form a separate subclass for which many results can be obtained more easily. The multivariate Lagrange expansion will now be used to obtain the multivariate MPSD.

Let $g_i = g_i(a_1, a_2, \ldots, a_k),\ i = 1, 2, \ldots, k$, be a set of nonnegative multivariate analytic functions of the parameters $a_1, a_2, \ldots, a_k$ such that $g_i(\mathbf{0}') \neq \mathbf{0}'$ for $i = 1, 2, \ldots, k$. Also, let $a_i/g_i(\mathbf{a}') = h_i(\mathbf{a}') = h_i$ for $i = 1, 2, \ldots, k$.

Let $f = f(a_1, a_2, \ldots, a_k)$ be another nonnegative multivariate analytic function of the same parameters such that $f(\mathbf{0}') \neq \mathbf{0}'$. Also, let there be k multivariate Lagrange transformations given by

$$a_i = u_i g_i(a_1, a_2, \ldots, a_k), \quad i = 1, 2, 3, \ldots, k,$$

which give $u_i = a_i/g_i$ and $u_i = 0$ when $a_i = 0$. These k transformations imply that each a_i is a function of the k variables $(u_1, u_2, \ldots, u_k)$ and that the function $f = f(\mathbf{a}')$ can be expanded near the smallest root into a power series in $u_i,\ i = 1, 2, \ldots, k$, which is given by the multivariate Lagrange expansion (15.4), given by Good (1960), or the expansion (15.5) given by Consul and Shenton (1973b). Thus,

$$f(\mathbf{a}') = \sum_{\mathbf{x}} \frac{u_1^{x_1}u_2^{x_2}\ldots u_k^{x_k}}{x_1!x_2!\ldots x_k!}\left[D_1^{x_1-1}\ldots D_k^{x_k-1}\|D_\nu\,(g_\nu)^{x_\nu} - \mathbf{G}\|\,f(\mathbf{a}')\right]_{\mathbf{a}'=\mathbf{0}'}, \tag{15.38}$$

where $\mathbf{G}$ is the $k\times k$ matrix $\left(\frac{\partial}{\partial a_\nu} g_i^{x_i}\right)$ and the summations are over all $x_i = 0, 1, 2, 3, \ldots,\ i = 1, 2, \ldots, k$.

By replacing $u_i = a_i/g_i(\mathbf{a}') = h_i(\mathbf{a}') = h_i$, the above becomes

$$f(\mathbf{a}') = \sum_{\mathbf{x}} b(x_1, x_2, \ldots, x_k) \frac{h_1^{x_1} h_2^{x_2} \ldots h_k^{x_k}}{x_1! x_2! \ldots x_k!}. \tag{15.39}$$

where

$$b(x_1, x_2, \ldots, x_k) = D_1^{x_1-1} D_2^{x_2-1} \ldots D_k^{x_k-1} \| D_\nu (g_\nu)^{x_\nu} - \mathbf{G} \| f(\mathbf{a}') \mid_{\mathbf{a}'=\mathbf{0}'} . \tag{15.40}$$

When $b(x_1, x_2, \ldots, x_k) \geq 0$ for all values of x_i, $i = 1, 2, \ldots, k$, the multivariate series sum (15.39), on division by $f(\mathbf{a}')$ on both sides, becomes unity and provides the multivariate MPSD as

$$P(\mathbf{X}' = \mathbf{x}') = b(x_1, x_2, \ldots, x_k) \frac{h_1^{x_1} h_2^{x_2} \ldots h_k^{x_k}}{x_1! x_2! \ldots x_k!} \left[f(\mathbf{a}') \right]^{-1} \tag{15.41}$$

for $x_i = 0, 1, 2, 3, \ldots,$ $i = 1, 2, \ldots, k$, and where $b(\mathbf{x}')$ is given by (15.40).

In the general case, as above, the computation of $b(x_1, x_2, \ldots, x_k)$ is very time consuming. When the multivariate functions $g_i(\mathbf{a}')$ are of the form $\left(g(\mathbf{a}')\right)^{c_i}$ for $i = 1, 2, 3, \ldots, k$ and the multivariate function $f(\mathbf{a}') = \left(g(\mathbf{a}')\right)^m$ or the derivative of $f(\mathbf{a}')$ with respect to any one of the parameters a_i is of that form, as in section 15.4, then the form of (15.40) becomes much simpler. Assuming $\frac{\partial}{\partial a_i} f(\mathbf{a}')$ to be of that form, the expression (15.40) gives

$$b(x_1, x_2, \ldots, x_k) = D_1^{x_1-1} D_2^{x_2} \ldots D_k^{x_k} \left\{ g_1^{x_1} g_2^{x_2} \ldots g_k^{x_k} \frac{\partial f}{\partial a_i} \right\} \Bigg|_{\mathbf{a}'=\mathbf{0}'} . \tag{15.42}$$

The following three subsections provide three multivariate Lagrangian models as applications of the multivariate MPSDs.

15.5.1 Multivariate Lagrangian Poisson Distribution

Let $f(\mathbf{a}') = e^{\alpha \sum_i a_i}$ and $g_i = e^{m_i \sum_j a_j}$, $i = 1, 2, \ldots, k$, $j = 1, 2, \ldots, k$, and $\alpha > 0$ and $m_i > 0$, $i = 1, 2, \ldots, k$, so that $h_i = a_i/g_i$. On substitution in (15.42), we get

$$b(x_1, x_2, \ldots, x_k) = D_1^{x_1-1} D_2^{x_2} \ldots D_k^{x_k} \left\{ \alpha \exp \left[\left(\alpha + \sum_i m_i x_i \right) \sum_i a_i \right] \right\} \Bigg|_{\mathbf{a}'=\mathbf{0}'}$$

$$= \alpha \left(\alpha + \sum_i m_i x_i \right)^{\sum_i x_i - 1} .$$

Thus, the MPSD formula (15.41) gives the multivariate Lagrangian Poisson distribution as

$$P(\mathbf{X} = \mathbf{x}) = \alpha \left(\alpha + \sum_i m_i x_i \right)^{\sum_i x_i - 1} \left[\prod_{i=1}^{k} \frac{a_i^{x_i}}{x_i!} \right] \exp \left\{ - \left(\alpha + \sum_i m_i x_i \right) \sum_i a_i \right\}$$

for $x_i = 0, 1, 2, 3, \ldots,$ $i = 1, 2, \ldots, k$, and where all summations are for $i = 1$ to k. The above expressions for the probabilities are the same as (15.26), where a_i is replaced by θ_i.

15.5.2 Multivariate Lagrangian Negative Binomial Distribution

The multivariate Lagrangian negative binomial distribution is also a multivariate MPSD for

$$f(\mathbf{a}') = \left(1 - \sum_i a_i\right)^{-\alpha} \quad \text{and} \quad g_j(\mathbf{a}') = \left(1 - \sum_i a_i\right)^{-\beta_j+1},$$

where $\alpha > 0,\ \beta_j > 1,\ j = 1, 2, 3, \ldots, k$, and $a_i > 0$ such that $\sum_{i=1}^{k} a_i < 1$ and $h_i(\mathbf{a}') = a_i/g_i(\mathbf{a}')$.

Substitution of the values in (15.42) provides the value of $b(\mathbf{x}')$ and the substitution in (15.41) gives the probabilities, which are similar to (15.29).

15.5.3 Multivariate Lagrangian Logarithmic Series Distribution

The Multivariate Lagrangian logarithmic series distribution is also a multivariate MPSD which is given by taking

$$f(\mathbf{a}') = -\ln\left(1 - \sum_i a_i\right) \quad \text{and} \quad g_j(\mathbf{a}') = \left(1 - \sum_i a_i\right)^{-\beta_j+1},$$

where $a_i > 0$ such that $\sum_{i=1}^{k} a_i < 1$ for $i = 1, 2, \ldots, k$ and $\beta_j > 1,\ j = 1, 2, 3, \ldots, k$, and $h_i(\mathbf{a}') = a_i/g_i(\mathbf{a}')$. By using these values in (15.42) one gets $b(x_1, x_2, \ldots, x_k)$, and then substituting them in (15.41), where $h_i = a_i/g_i(\mathbf{a}')$, provides the same probabilities as in (15.32).

15.5.4 Moments of the General Multivariate MPSD

The following symbols will be used in this section:

$$\frac{\partial}{\partial a_j} \ln h_i(\mathbf{a}') = h_{ij}, \qquad \frac{\partial}{\partial a_j} \ln f(\mathbf{a}') = f_j$$

for $i = 1, 2, 3, \ldots, k$ and $j = 1, 2, 3, \ldots, k$. Also, the mean vector of the general multivariate MPSD, defined by (15.41), will be denoted by $\mathrm{E}(X_1), \mathrm{E}(X_2), \ldots, \mathrm{E}(X_k)$.

Since $\sum_{\mathbf{x}'} P(\mathbf{X}' = \mathbf{x}') = 1$, the probabilities (15.41) give

$$f(\mathbf{a}') = f(\mathbf{a}') = \sum_{\mathbf{x}} \frac{b(x_1, x_2, \ldots, x_k)}{x_1! x_2! \ldots x_k!} h_1^{x_1} h_2^{x_2} \ldots h_k^{x_k}. \tag{15.43}$$

By differentiating the above successively with respect to $a_1, a_2, \ldots, a_k$ and multiplying each one of them by $\left[f(\mathbf{a}')\right]^{-1}$, we get the k relations

$$\begin{aligned}
f_1 &= \mathrm{E}(X_1)h_{11} + \mathrm{E}(X_2)h_{21} + \mathrm{E}(X_3)h_{31} + \cdots + \mathrm{E}(X_k)h_{k1},\\
f_2 &= \mathrm{E}(X_1)h_{12} + \mathrm{E}(X_2)h_{22} + \mathrm{E}(X_3)h_{32} + \cdots + \mathrm{E}(X_k)h_{k2},\\
f_3 &= \mathrm{E}(X_1)h_{13} + \mathrm{E}(X_2)h_{23} + \mathrm{E}(X_3)h_{33} + \cdots + \mathrm{E}(X_k)h_{k3},\\
&\ \ \vdots \qquad\quad \vdots \qquad\qquad \vdots \qquad\qquad \vdots \qquad \vdots\\
f_k &= \mathrm{E}(X_1)h_{1k} + \mathrm{E}(X_2)h_{2k} + \mathrm{E}(X_3)h_{3k} + \cdots + \mathrm{E}(X_k)h_{kk}.
\end{aligned} \tag{15.44}$$

By solving the above k relations for the k means of the probability model (15.41), we have

$$\mathrm{E}(X_1)\Delta_1^{-1} = -\mathrm{E}(X_2)\Delta_2^{-1} = \mathrm{E}(X_3)\Delta_3^{-1} = \cdots = (-1)^{k-1}\mathrm{E}(X_k)\Delta_k^{-1} = \Delta^{-1}, \tag{15.45}$$

where

$$\Delta_1 = \begin{vmatrix} f_1 & h_{21} & h_{31} & \cdots & h_{k1} \\ f_2 & h_{22} & h_{32} & \cdots & h_{k2} \\ \vdots & \vdots & \vdots & \vdots & \vdots \\ f_k & h_{2k} & h_{3k} & \cdots & h_{kk} \end{vmatrix}, \quad \Delta_2 = \begin{vmatrix} f_1 & h_{11} & h_{31} & \cdots & h_{k1} \\ f_2 & h_{12} & h_{32} & \cdots & h_{k2} \\ \vdots & \vdots & \vdots & \vdots & \vdots \\ f_k & h_{1k} & h_{3k} & \cdots & h_{kk} \end{vmatrix},$$

$$\Delta_3 = \begin{vmatrix} f_1 & h_{11} & h_{21} & \cdots & h_{k1} \\ f_2 & h_{12} & h_{22} & \cdots & h_{k2} \\ \vdots & \vdots & \vdots & \vdots & \vdots \\ f_k & h_{1k} & h_{2k} & \cdots & h_{kk} \end{vmatrix}, \quad \Delta_k = \begin{vmatrix} f_1 & h_{11} & h_{21} & \cdots & h_{k-1,1} \\ f_2 & h_{12} & h_{22} & \cdots & h_{k-1,2} \\ \vdots & \vdots & \vdots & \vdots & \vdots \\ f_k & h_{1k} & h_{2k} & \cdots & h_{k-1,k} \end{vmatrix},$$

and

$$\Delta = \begin{vmatrix} h_{11} & h_{21} & h_{31} & \cdots & h_{k1} \\ h_{12} & h_{22} & h_{31} & \cdots & h_{k2} \\ \vdots & \vdots & \vdots & \vdots & \vdots \\ h_{1k} & h_{2k} & h_{3k} & \cdots & h_{kk} \end{vmatrix}.$$

The determination of the matrix of the second moments for the general multivariate MPSD becomes much more complex. Theoretically, one can obtain these quantities by taking the second derivatives of (15.43) with respect to $a_1, a_2, \ldots, a_k$ successively and then simplifying them to get another set of k relations, which have to be used in conjunction with the k relations (15.44) to get the k second moments for the probability model (15.41)

15.5.5 Moments of Multivariate Lagrangian Poisson Distribution

Khatri (1982) had obtained the expressions for the means and variances of the multivariate Lagrangian Poisson distributions but these contained $k \times k$ determinants and matrices with k derivatives and need complex computation for reduction into simple forms. Independent methods for their evaluations are being given here.

Since $\sum_{\mathbf{x}'} P(\mathbf{X}' = \mathbf{x}') = 1$, by (15.26) we have

$$e^{-\alpha \sum_i \theta_i} = \sum_{\mathbf{x}'} a(x_1, x_2, \ldots, x_k)\theta_1^{x_1}\theta_2^{x_2} \ldots \theta_k^{x_k} e^{-\sum_i m_i x_i \sum_j \theta_j}, \tag{15.46}$$

where

$$a(x_1, x_2, \ldots, x_k) = \frac{\alpha\left(\alpha + \sum_i m_i x_i\right)^{\sum_i x_i - 1}}{x_1! x_2! \ldots x_k!},$$

and $i, j = 1, 2, 3, \ldots, k$, and the summations are on all values of i and j. Differentiating (15.46) with respect to θ_i, multiplying both sides by $e^{-\alpha \sum_i \theta_i}$, and summing over all $\mathbf{x}' = (x_1, x_2, \ldots, x_k)$, we get

$$\alpha = \left(\theta_i^{-1} - m_i\right) \mathrm{E}\left[X_i\right] - \sum_{j \neq i}^{k} m_j \mathrm{E}\left[X_j\right], \quad i = 1, 2, \ldots, k. \tag{15.47}$$

By subtracting the relation for $i = 1$ in the above from each of the other $k - 1$ relations in (15.47),

$$\theta_1^{-1}\mathrm{E}[X_1] = \theta_2^{-1}\mathrm{E}[X_2] = \cdots = \theta_k^{-1}\mathrm{E}[X_k].$$

Then, by adding together the k relations in (15.47) and by using the above equalities, we obtain

$$\mathrm{E}[X_i] = \alpha\theta_i \left(1 - \sum_{j=1}^{k} m_j\theta_j\right)^{-1}, \quad i = 1, 2, 3, \ldots, k. \tag{15.48}$$

To obtain the variance $\mathrm{V}(X_1)$, we differentiate $\mathrm{E}[X_1]e^{\alpha \sum_i \theta_i}$ with respect to θ_1 and get

$$\frac{\partial}{\partial\theta_1} \frac{\alpha\theta_1 e^{\alpha \sum_i \theta_i}}{1 - \sum_i m_i\theta_i} = \frac{\partial}{\partial\theta_1} \left[\sum_{\mathbf{x}'} a(\mathbf{x}') x_1 \theta_1^{x_1} \theta_2^{x_2} \ldots \theta_k^{x_k} e^{-\sum_i m_i x_i \sum_j \theta_j} \right],$$

which gives, on division by $e^{\alpha \sum_i \theta_i}$,

$$\begin{aligned}
\frac{\alpha + \alpha^2\theta_1}{1 - \sum_i m_i\theta_i} + \frac{\alpha m_1\theta_1}{\left(1 - \sum_i m_i\theta_i\right)^2} &= \theta_1^{-1}\mathrm{E}[X_1^2] - \sum_{i=1}^{k} m_i \mathrm{E}[X_1 X_i] \\
&= \left(\theta_1^{-1} - m_1\right)\mathrm{E}[X_1^2] - \sum_{i=2}^{k} \frac{m_i\alpha\theta_1 \cdot \alpha\theta_i}{\left(1 - \sum_i m_i\theta_i\right)^2},
\end{aligned}$$

$$\begin{aligned}
\therefore \frac{1 - m_1\theta_1}{\theta_1}\mathrm{E}[X_1^2] &= \frac{\alpha + \alpha^2\theta_1}{1 - \sum_i m_i\theta_i} + \frac{\alpha m_1\theta_1}{\left(1 - \sum_i m_i\theta_i\right)^2} \\
&\quad + \frac{\alpha^2\theta_1\left[\left(\sum_i m_i\theta_i - 1\right) + 1 - m_1\theta_1\right]}{\left(1 - \sum_i m_i\theta_i\right)^2} \\
&= \frac{\alpha}{1 - \sum_i m_i\theta_i} + \frac{\alpha m_1\theta_1}{\left(1 - \sum_i m_i\theta_i\right)^2} + \frac{\alpha^2\theta_1(1 - m_1\theta_1)}{\left(1 - \sum_i m_i\theta_i\right)^2},
\end{aligned}$$

$$\therefore \mathrm{V}(X_1) = \mathrm{E}[X_1^2] - (\mathrm{E}[X_1])^2 = \frac{\alpha\theta_1(1 - m_1\theta_1)^{-1}}{1 - \sum_i m_i\theta_i} + \frac{\alpha m_1\theta_1^2(1 - m_1\theta_1)^{-1}}{\left(1 - \sum_i m_i\theta_i\right)^2}. \tag{15.49}$$

By symmetry,

$$\mathrm{V}(X_i) = \frac{\alpha\theta_i(1 - m_i\theta_i)^{-1}}{1 - \sum_i m_i\theta_i} + \frac{\alpha m_i\theta_i^2(1 - m_i\theta_i)^{-1}}{\left(1 - \sum_i m_i\theta_i\right)^2}, \quad i = 1, 2, \ldots, k, \tag{15.50}$$

and $\mathrm{Cov}(X_i, X_j) = 0$ for $i, j = 1, 2, \ldots, k$ and $i \neq j$.

15.5.6 Moments of Multivariate Lagrangian Negative Binomial Distribution

Since the probability model is given by (15.29) and $\sum_{\mathbf{x}'} P(\mathbf{X}' = \mathbf{x}') = 1$, one can write

$$\left(1 - \sum_i \theta_i\right)^{-\alpha} = \sum_{\mathbf{x}'} a(x_1, x_2, \ldots, x_k)\theta_1^{x_1}\theta_2^{x_2}\ldots\theta_k^{x_k}\left(1 - \sum_i \theta_i\right)^{\sum_i(\beta_i-1)x_i}, \quad (15.51)$$

where

$$a(x_1, x_2, \ldots, x_k) = \frac{\left(\alpha + \sum_i \beta_i x_i - 1\right)!\alpha}{x_1!x_2!\ldots x_k!\left(\alpha + \sum_i(\beta_i - 1)x_i\right)!}.$$

Differentiating (15.51) with respect to θ_j, multiplying both sides by $\left(1 - \sum_i \theta_i\right)^{\alpha+1}$, and simplifying, we have

$$\alpha = \left(1 - \sum_i \theta_i\right)\theta_j^{-1}\mathrm{E}[X_j] - \sum_i(\beta_i - 1)\mathrm{E}[X_i] \quad (15.52)$$

for $j = 1, 2, 3, \ldots, k$. The above k relations show that

$$\theta_1^{-1}\mathrm{E}[X_1] = \theta_2^{-1}\mathrm{E}[X_2] = \cdots = \theta_k^{-1}\mathrm{E}[X_k] = L \text{ (say)}.$$

Substituting these values in (15.52),

$$\alpha = L\left(1 - \sum_i \theta_i\right) - L\sum_i(\beta_i - 1)\theta_i = L\left(1 - \sum_i \beta_i\theta_i\right)$$

$$\therefore L = \alpha\left(1 - \sum_i \beta_i\theta_i\right)^{-1},$$

and thus

$$\mathrm{E}[X_j] = \alpha\theta_j\left(1 - \sum_i \beta_i\theta_i\right)^{-1}, \quad j = 1, 2, \ldots, k, \quad (15.53)$$

where $\sum_{i=1}^k \beta_i\theta_i < 1$.

To obtain the $\mathrm{V}(X_1)$, we differentiate $\mathrm{E}[X_1]\left(1 - \sum_i \theta_i\right)^{-\alpha}$ with respect to θ_1. Thus,

$$\frac{\partial}{\partial\theta_1}\left[\frac{\alpha\theta_1\left(1 - \sum_i \theta_i\right)^{-\alpha}}{1 - \sum_i \beta_i\theta_i}\right] = \sum_{\mathbf{x}'} a(\mathbf{x}')\theta_1^{x_1}\theta_2^{x_2}\ldots\theta_k^{x_k}\left\{x_1^2\theta_1^{-1}\left(1 - \sum_i \theta_i\right)^{\sum_i(\beta_i-1)x_i}\right.$$

$$\left. - \frac{x_1\sum_{i=1}^k(\beta_i - 1)x_i\left(1 - \sum_i \theta_i\right)^{\sum_i(\beta_i-1)x_i}}{1 - \sum_i \theta_i}\right\}.$$

Therefore, on division by $\left(1-\sum_i \theta_i\right)^{-\alpha}$,

$$\frac{\alpha^2\theta_1\left(1-\sum_i \theta_i\right)^{-1}+\alpha}{1-\sum_i \beta_i\theta_i}+\frac{\alpha\beta_1\theta_1}{\left(1-\sum_i \beta_i\theta_i\right)^2}$$

$$=\theta_1^{-1}\mathrm{E}[X_1^2]-\left(1-\sum_{i=1}^{k}\theta_i\right)^{-1}\sum_{i=1}^{k}(\beta_i-1)\mathrm{E}[X_iX_1]$$

$$=\left(\theta_1^{-1}-\frac{\beta_1-1}{1-\sum_i \theta_i}\right)\mathrm{E}[X_1^2]-\frac{\alpha^2\theta_1\sum_{i=2}^{k}(\beta_i-1)\theta_i}{\left(1-\sum_i \theta_i\right)\left(1-\sum_i \beta_i\theta_i\right)^2}$$

or

$$\left(1-\sum_i \theta_i-\beta_1\theta_1+\theta_1\right)\mathrm{E}[X_1^2]=\frac{\alpha\theta_1\left(1-\sum_i \theta_i\right)}{1-\sum_i \beta_i\theta_i}+\frac{\alpha\theta_1^2\beta_1\left(1-\sum_i \theta_i\right)}{\left(1-\sum_i \beta_i\theta_i\right)^2}$$

$$+\frac{\alpha^2\theta_1^2\left[1-\sum_i \theta_i-\beta_1\theta_1+\theta_1\right]}{\left(1-\sum_i \beta_i\theta_i\right)^2}.$$

$$\therefore \mathrm{E}[X_1^2]=\frac{\alpha^2\theta_1^2}{\left(1-\sum_i \beta_i\theta_i\right)^2}+\left[\frac{\alpha\theta_1}{1-\sum_i \beta_i\theta_i}+\frac{\alpha\beta_1\theta_1^2}{\left(1-\sum_i \beta_i\theta_i\right)^2}\right]\frac{1-\sum_i \theta_i}{1-\sum_i \theta_i-(\beta_1-1)\theta_1}$$

$$\therefore \mathrm{V}(X_1)=\mathrm{E}[X_1^2]-(\mathrm{E}[X_1])^2=\left[\frac{\alpha\theta_1}{1-\sum_i \beta_i\theta_i}+\frac{\alpha\beta_1\theta_1^2}{\left(1-\sum_i \beta_i\theta_i\right)^2}\right]\frac{1-\sum_i \theta_i}{1-\sum_i \theta_i-(\beta_1-1)\theta_1}.$$

By symmetry,

$$\mathrm{V}(X_j)=\left[\frac{\alpha\theta_j}{1-\sum_i \beta_i\theta_i}+\frac{\alpha\beta_j\theta_j^2}{\left(1-\sum_i \beta_i\theta_i\right)^2}\right]\frac{\left(1-\sum_i \theta_i\right)}{1-\sum_i \theta_i-(\beta_j-1)\theta_j} \tag{15.54}$$

for $j=1,2,\ldots,k$ and $\mathrm{Cov}(X_i,\ X_j)=0$.

15.6 Multivariate MPSDs in Another Form

A special case of the multivariate MPSDs, derived in (15.41), can also be written as

$$P(x_1,x_2,\ldots,x_k)=B(b_1,b_2,\ldots,b_k)f(x_1+x_2+\cdots+x_k)\frac{b_1^{x_1}b_2^{x_2}\ldots b_k^{x_k}}{x_1!x_2!\ldots x_k!}, \tag{15.55}$$

with the condition

$$B(b_1,b_2,\ldots,b_k)\sum_{\mathbf{x}=\mathbf{0}}^{\infty}f(x_1+x_2+\cdots+x_k)\frac{b_1^{x_1}b_2^{x_2}\ldots b_k^{x_k}}{x_1!x_2!\ldots x_k!}=1. \tag{15.56}$$

Kapur (1982) stated that these k summations can be done in two stages. First, the summations over all the values of $x_1,x_2,\ldots,x_k$ can be taken such that the sum $x_1+x_2+\cdots+x_k=x$ is an integer and then the sum over x can be taken from 0 to ∞ and such that the above expression will give

$$A(b) \sum_{x=0}^{\infty} f(x) \cdot \frac{b^x}{x!} = 1, \tag{15.57}$$

where $b = b_1 + b_2 + \cdots + b_k$ and $A(b) = B(b_1, b_2, \ldots, b_k)$, so that

$$P(x_1, x_2, \ldots, x_k) = A(b) f(x_1 + x_2 + \cdots + x_k) \frac{b_1^{x_1} b_2^{x_2} \ldots b_k^{x_k}}{x_1! x_2! \ldots x_k!}. \tag{15.58}$$

The mgf for (15.55) becomes

$$\begin{aligned} M(\mathbf{t}) &= B(\mathbf{b}) \sum_{\mathbf{x}=\mathbf{0}}^{\infty} f(x_1 + x_2 + \cdots + x_k) \frac{(b_1 e^{t_1})^{x_1} (b_2 e^{t_2})^{x_2} \ldots (b_k e^{t_k})^{x_k}}{x_1! x_2! \ldots x_k!} \\ &= \frac{A(b_1 + b_2 + \cdots + b_k)}{A(b_1 e^{t_1} + b_2 e^{t_2} + \cdots + b_k e^{t_k})}, \end{aligned} \tag{15.59}$$

from which all the moments can be determined.

Kapur (1982) provided the pmfs for the following thirteen families of multivariate MPSDs corresponding to the univariate Lagrangian probability models in chapter 2. Note that $x_1 + x_2 + \cdots + x_k = x$ and $b_1 + b_2 + \cdots + b_k = b$ in all models given below.

(i) *Multivariate generalized negative binomial distribution:*

$$f(t) = (q + pt)^n, \quad g(t) = (q + pt)^m, \quad 0 < 1 - q = p < m^{-1},$$

$$L(g, f; \mathbf{x}) = \frac{nq^n}{n + mx} \frac{(n + mx)!}{[n + (m-1)x]!} \frac{b_1^{x_1} b_2^{x_2} \ldots b_k^{x_k}}{x_1! x_2! \ldots x_k!}$$

for $x_i = 0, 1, 2, \ldots$ and $i = 1, 2, \ldots, k$, where $b = pq^{-m+1} = (1-q)q^{-m+1}$; the parameters are $m, n, b_1, b_2, \ldots, b_k, q$.

(ii) *Multivariate delta-binomial distribution:*

$$f(t) = t^n, \quad g(t) = (q + pt)^m, \quad 0 < 1 - q = p < m^{-1},$$

$$L(g, f; \mathbf{x}) = \frac{nq^{mn}}{n + x} \frac{[m(n + x)]!}{[mn + (m-1)x]!} \frac{b_1^{x_1} b_2^{x_2} \ldots b_k^{x_k}}{x_1! x_2! \ldots x_k!}$$

for $x_i = 0, 1, 2, \ldots,\ i = 1, 2, \ldots, k$, and $b = pq^{m+1} = (1-q)q^{m+1}$.

(iii) *Multivariate binomial-Poisson distribution:*

$$f(t) = e^{M(t-1)}, \quad g(t) = (q + pt)^m, \quad 0 < 1 - q = p < m^{-1},$$

$$L(g, f; \mathbf{x}) = e^{-M} {}_2F_0\left(1 - x, -mx; ; -\frac{p}{Mq}\right) \frac{b_1^{x_1} b_2^{x_2} \ldots b_k^{x_k}}{x_1! x_2! \ldots x_k!},$$

for $x_i = 0, 1, 2, \ldots,\ i = 1, 2, \ldots, k, \quad b = Mq^m, \quad q = (b/M)^{1/m}$.

(iv) *Multivariate binomial-negative binomial distribution:*

$$f(t) = (1-p')^k(1-p't)^{-k}, \quad g(t) = (q+pt)^m, \quad 0 < 1-q = p < m^{-1},$$
$$0 < p' = 1-q' < 1,$$

$$L(g, f; \mathbf{x})$$
$$= \frac{(1-p't)^k}{\Gamma(k)}\Gamma(k+x)\,{}_2F_1\left(1-x, -mx; 1-x-k; -\frac{p'(1-p')^{-2}}{qp}\right)\frac{b_1^{x_1}b_2^{x_2}\dots b_k^{x_k}}{x_1!x_2!\dots x_k!}$$

for $x_i = 0, 1, 2, \dots,\ i = 1, 2, \dots, k$, and $b = q^m p'$.

(v) *Multivariate delta-Poisson distribution:*

$$f(t) = t^n, \quad g(t) = e^{\theta(t-1)}, \quad 0 < \theta < 1,$$

$$L(g, f; \mathbf{x}) = \frac{ne^{-n\theta}(n+x)^x}{n+x}\frac{b_1^{x_1}b_2^{x_2}\dots b_k^{x_k}}{x_1!x_2!\dots x_k!},$$

where $b = \theta e^{-\theta}$.

(vi) *Multivariate generalized Poisson distribution:*

$$f(t) = e^{\theta(t-1)}, \quad g(t) = e^{\lambda(t-1)}, \quad 0 < \lambda < 1,$$

$$L(g, f; \mathbf{x}) = \theta e^{-\theta}(\theta + \lambda x)^x \frac{b_1^{x_1}b_2^{x_2}\dots b_k^{x_k}}{x_1!x_2!\dots x_k!},$$

where $b = e^{-\lambda}$.

(vii) *Multivariate Poisson-binomial distribution:*

$$f(t) = (q+pt)^n, \quad g(t) = e^{\theta(t-1)}, \quad 0 < \theta < 1,$$
$$0 < 1-q = p < 1,$$

$$L(g, f; \mathbf{x}) = \frac{npq^{n-1}}{\theta}(x)^{x-1}\,{}_2F_0\left(1-x, 1-n; ; \frac{p}{\theta q^x}\right)\frac{b_1^{x_1}b_2^{x_2}\dots b_k^{x_k}}{x_1!x_2!\dots x_k!}$$

for $x_i \geq 0,\ i = 1, 2, \dots, k;\ x \neq 0$ and $b = \theta e^{-\theta}$ and $L(g, f; \mathbf{x}) = 0$ for $x = 0$.

(viii) *Multivariate Poisson-negative binomial distribution:*

$$f(t) = (1-p')^k(1-p't)^{-k}, \quad g(t) = e^{\theta(t-1)}, \quad 0 < \theta < 1,$$
$$0 < p' = 1-q' < 1,$$

$$L(g, f; \mathbf{x}) = \frac{kp'(1-p')^k}{\theta}(x)^{x-1}\,{}_2F_0\left(1-x, 1+k; ; -\frac{p'}{\theta x}\right)\frac{b_1^{x_1}b_2^{x_2}\dots b_k^{x_k}}{x_1!x_2!\dots x_k!}$$

for $x_i \geq 0,\ i = 1, 2, \dots, k;\ x \neq 0$ and $L(g, f; \mathbf{x}) = (1-p')^k$ for $x = 0$.

(ix) *Multivariate delta-negative binomial distribution:*

$$f(t) = t^n, \quad g(t) = (1-p')^k(1-p't)^{-k}, \quad kp'(1-p')^{-1} < 1,$$

$$L(g, f; \mathbf{x}) = \frac{n}{n+x}\frac{\Gamma[(k+1)x+kn]}{\Gamma(kx+n)}e^{-nk}\frac{b_1^{x_1}b_2^{x_2}\dots b_k^{x_k}}{x_1!x_2!\dots x_k!},$$

where $b = p'(1-p')^k$.

(x) *Multivariate negative binomial-binomial distribution:*

$$f(t) = (q+pt)^n, \quad g(t) = (1-p')^k(1-p't)^{-k}, \quad kp'(1-p')^{-1} < 1,$$

$$0 < p = 1-q < 1,$$

$$L(g, f; \mathbf{x}) = npq^{n-1}(p')^{-1}\,{}_2F_1\left(1-x, 1-n; 2-(k+1)x; -\frac{p}{qp'}\right)\frac{b_1^{x_1}b_2^{x_2}\dots b_k^{x_k}}{x_1!x_2!\dots x_k!}$$

for $x_i \geq 0,\ i = 1, 2, \dots, k;\ x \neq 0$ and $L(g, f; \mathbf{x}) = q^n$ for $x = 0$.

(xi) *Multivariate negative binomial-negative binomial distribution:*

$$f(t) = (1-p)^M(1-pt)^{-M}, \quad g(t) = (1-p')^k(1-p't)^{-k},$$

$$kp'(1-p')^{-k} < 1,$$

$$L(g, f; \mathbf{x}) = \frac{M(1-p')^M}{M+(k+1)x}\frac{\Gamma[(k+1)x+M+1]}{\Gamma[kx+M+1]}\frac{b_1^{x_1}b_2^{x_2}\dots b_k^{x_k}}{x_1!x_2!\dots x_k!},$$

where $b = p'(1-p')^k$.

(xii) *Multivariate Poisson-rectangular distribution:*

$$f(t) = \frac{1-t^n}{n(1-t)}, \quad n \geq 2, \quad g(t) = e^{m(t-1)}, \quad 0 < m < 1,$$

$$L(g, f; \mathbf{x}) = \frac{1}{n}\sum_{r=0}^{n-2}(r+1)\frac{(x-1)}{(x-r-1)}(mx)^{x-1}\frac{b_1^{x_1}b_2^{x_2}\dots b_k^{x_k}}{x_1!x_2!\dots x_k!},$$

where $b = e^{-m}$.

(xiii) *Multivariate Poisson logarithmic series distribution:*

$$f(t) = \frac{\ln(1-pt)}{\ln(1-p)}, \quad g(t) = e^{m(t-1)}, \quad 0 < m < 1,$$

$$L(g, f; \mathbf{x}) = \frac{p}{-m\ln(1-p)}\,{}_2F_0\left(1-x, 1; ; -\frac{p}{mx}\right)x^{x-1}\frac{b_1^{x_1}b_2^{x_2}\dots b_k^{x_k}}{x_1!x_2!\dots x_k!},$$

where $b = me^{-m}$.

None of the above thirteen models resembles any one of the models in Table 14.1 (in the bivariate case) or any one of the multivariate models given in sections 15.4 and 15.5. It seems that Kapur (1982) did not realize that the multivariate Lagrangian probability models would not be similar to the univariate probability models.

15.7 Multivariate Lagrangian Quasi-Pólya Distribution

Khatri (1982) developed the multivariate Lagrangian quasi-Pólya distributions by taking the Lagrange expansion of

$$f^{-a_0} = \left(1 - \sum_{i=1}^{k} \theta_i\right)^{-a_0}$$

in powers of $\theta_1 f^{-a_1}, \theta_2 f^{-a_2}, \ldots, \theta_k f^{-a_k}$ and then considering the product

$$f^{-(a_0+b_0)} = f^{-a_0} \cdot f^{-b_0}. \tag{15.60}$$

By equating the coefficients of $\prod_{i=1}^{k} \theta_i^{\nu_i} f^{-a_i \nu_i}$ in the expansion on the left-hand side of (15.60) with the coefficients in the product of the two expansions on the right-hand side of (15.60), Khatri (1982) obtained the formula

$$(a_0 + b_0)\left(a_0 + b_0 + \sum_{i=1}^{k} a_i \nu_i + 1\right)^{(\sum_i \nu_i - 1)} \Big/ \prod_{i=1}^{k} \nu_i!$$

$$= \sum_{x_1=0}^{\nu_1} \cdots \sum_{x_k=0}^{\nu_k} \frac{A(x_i, a_0) A(\nu_i - x_i, b_0)}{x_1! x_2! \ldots x_k! (\nu_1 - x_1)! (\nu_2 - x_2)! \ldots (\nu_k - x_k)!}, \tag{15.61}$$

where

$$A(x_i, a_0) = a_0 \left(a_0 + \sum_{i=1}^{k} a_i x_i + 1\right)^{(\sum_i x_i - 1)}.$$

By dividing both sides with the expression on the left-hand side of (15.61), Khatri (1982) defined the multivariate Lagrangian quasi-Pólya distribution as

$$P(\mathbf{X}' = \mathbf{x}') = \left[\prod_{i=1}^{k} \binom{\nu_i}{x_i}\right] \cdot \frac{A(x_i, a_0) A(\nu_i - x_i, b_0)}{A(\nu_i, a_0 + b_0)} \tag{15.62}$$

for $x_i = 0, 1, 2, \ldots, \nu_i,\ i = 1, 2, \ldots, k$, and where $a_0, b_0, a_1, a_2, \ldots, a_k$ are all positive.

Considering the other multivariate Lagrange expansion

$$\frac{f^{-a_0+1}}{f - \sum_{i=1}^{k} a_i \theta_i} = \sum_{\mathbf{x}'=\mathbf{0}'}^{\infty} \frac{\left(a_0 + \sum_{i=1}^{k} a_i x_i\right)^{\left(\sum_{i=1}^{k} x_i\right)}}{x_1! x_2! \ldots x_k!} \prod_{i=1}^{k} \left(\theta_i^{x_i} f^{a_i x_i}\right), \tag{15.63}$$

Khatri (1982) obtained the first moments of the distribution as

$$E[X_j] = a_0 \nu_j / (a_0 + b_0) \quad \text{for} \quad j = 1, 2, \ldots, k. \tag{15.64}$$

15.8 Applications of Multivariate Lagrangian Distributions

Good (1960, 1965) not only did the pioneering work of the multivariate Lagrange expansions of multivariate functions but also showed that these expansions could be usefully applied to random branching processes, to the enumeration of rooted ordered trees, and to some stochastic processes. Consul and Shenton (1973b) showed that the total number of customers of different kinds served in a busy period was given by a multivariate Lagrangian distribution and that the theory could easily be extended for use in different areas. We shall now consider some of these applications.

15.8.1 Queuing Processes

A busy airport is served everyday by a large number of airplanes of different types and different sizes which arrive at different rates and at different times on account of the various types of flights (passenger, commercial freight, consumer freight and mail, charter, training, etc.). Each type of flight service has its own multivariate probability distribution for the arrival of different types of airplanes. Each plane needs many kinds of services (like fuel supply, drinking water supply, nondrinking water, unloading, cleaning, checking of equipment, etc.) at the same airport. The service crew may consist of many persons but they have to work in sequence and in collaboration with each other and so they may all be regarded as one set of crews which looks after all the service needs of the different airplanes. Thus, the set of crews can be regarded as a single server giving multivariate services and the different types of airplanes may be supposed to form queues for the needed services.

Let the $(r_1, r_2, \ldots, r_k)$ denote the different types of airplanes waiting for services when the service begins, so that $f(\mathbf{t}') = t_1^{r_1} t_2^{r_2} \ldots t_k^{r_k}$. Also, let the average input vector of the ith type of flight service for unit time element be $(\lambda_1^{(i)}, \lambda_2^{(i)}, \ldots, \lambda_k^{(i)})$ and the average service vector be $(\mu_1, \mu_2, \ldots, \mu_k)$. Then the ratio vector $(\lambda_1^{(i)}/\mu_1, \lambda_2^{(i)}/\mu_2, \ldots, \lambda_k^{(i)}/\mu_k)$ will denote the average rates of change per unit time element. Let the mean vector of the multivariate probability distribution, with pgf $g^{(i)}(\mathbf{t}')$, be

$$G_{(1)}^{(i)} = \left(\lambda_1^{(i)}/\mu_1, \lambda_2^{(i)}/\mu_2, \ldots, \lambda_k^{(i)}/\mu_k\right).$$

The probability distribution of the number of airplanes of different types served in a busy period will be given by

$$P(\mathbf{X}' = \mathbf{x}') = \frac{1}{x_1! \ldots x_k!} \left[D_1^{x_1 - 1} \ldots D_k^{x_k - 1} \| D_\nu t_\nu^{r_\nu} g_x^{(\nu)} \mathbf{I} - t_\nu^{r_\nu} \mathbf{G} \| \right]_{\mathbf{t}' = \mathbf{0}'} \tag{15.65}$$

for $x_i \geq r_i$, $i = 1, 2, \ldots, k$, and G represents the $k \times k$ matrix $\left(g_{x_i j}^{(i)}\right)$ and

$$g_{x_i j}^{(i)} = \frac{\partial}{\partial t_j} \left(g^{(i)}\right)^{x_i}.$$

The above is a multivariate Lagrangian delta distribution. According to the values of the multivariate pgfs

$$g^{(i)}(\mathbf{t}'),$$

these probability distributions (15.65) will change. When the numbers $(r_1, r_2, \ldots, r_k)$ of the different types of airplanes, waiting in queues for service at the initial time, become random variables, given by pgf $f(\mathbf{t}')$, then the probability distribution of the number of planes of different types served in a busy period will be given by the general multivariate Lagrangian probability distribution (15.8) together with (15.7).

This queuing process plays its part in many different industries. Some simple examples are given below:

(i) Let each atomic fission generate k different types of reactions and have its own multivariate probability distribution of producing such reactions. If the number of different kinds of fissions is also k, then the pgfs of the different types of reactions may be given by $g^{(i)}(\mathbf{t}')$, $i = 1, 2, \ldots, k$. When the different kinds of atomic fissions are generated according to the pgf $f(\mathbf{t}')$, the probability that the whole process started by such atomic fissions will contain x_1 reactions of type 1, x_2 reactions of type 2, ..., x_k reactions of type k will be given by a multivariate discrete probability distribution of the Lagrangian type pgf (15.4) with suitable choice of $f(\mathbf{t}')$ and $g^{(i)}(\mathbf{t}')$, $i = 1, 2, \ldots, k$.

(ii) In the highly developed modern world, we are being constantly bombarded with various types of radioactive particles and cosmic rays from the computers, TVs, microwaves, etc., which slowly weaken the human immune system. Also, everyone is subjected to thousands of very mild shocks every day by static electricity generated by modern clothes and carpets, etc. Then we go through numerous medical tests which involve x-rays and other types of rays. When the number of hits by each of such sources exceeds certain levels $k_1, k_2, k_3, \ldots, k_n$, then their adverse effects in the form of different types of new diseases may be visible. Let $f(\mathbf{t}')$ denote the pgf of the generation of different types of such radioactive rays. If $g^{(i)}(\mathbf{t}')$ represents the pgf of the multivariate probability distributions of the number of persons who contacted such visible diseases on account of the ith source, $i = 1, 2, \ldots, k$, then the probability distribution of the number of such attacks by different diseases in the whole process will be given by one of the multivariate Lagrangian distributions with pgf (15.4) and some particular values of $f(\mathbf{t}')$ and $g^{(i)}(\mathbf{t}')$.

(iii) Numerous diseases are caused by (a) air pollution, (b) water pollution, (c) food pollution or contamination due to the use of insecticides, (d) bad sanitation, (e) consumption of too many preservatives, and (f) unhygienic conditions. All these sources have their own multivariate distributions of generating different types of diseases when each source exceeds some specific limits. If $g^{(i)}(\mathbf{t}')$ denotes the pgf of the multivariate probability distribution of the number of persons getting different diseases on account of the ith source, $i = 1, 2, \ldots, k$, and $f(\mathbf{t}')$ is the pgf of the multivariate distribution of the different types of susceptible persons who are exposed to these sources for diseases, then the pgf of the total number of persons affected by these diseases will be of the form given by (15.4).

15.8.2 Random Branching Processes with k Types of Females

Suppose a female of type i, $i = 1, 2, \ldots, k$, has a probability of giving birth in any generation (fixed discrete time interval) to $m_1, m_2, \ldots, m_k$ individuals of type 1, 2, ..., k (the original female being included as one of the children if she survives). Let the probability be the coefficient of $t_1^{m_1}, t_2^{m_2}, \ldots, t_k^{m_k}$ in a probability generating function $g_i(\mathbf{t}')$. Good (1955) has shown that the pgfs, $t_i(\mathbf{u}')$, of the size of the whole tree, including the original female (of type i), are given by

$$t_i = u_i g_i(\mathbf{t}').$$

The conditions $g_i(\mathbf{0}') \neq \mathbf{0}'$, $i = 1, 2, \ldots, k$, have a physical significance that every type of female must have a nonzero probability of being barren. Let the branching process start with r_1 females of type 1, r_2 females of type 2, ..., r_k females of type k. Good (1960) has shown that the probability that the whole process will contain precisely m_1 individuals of type 1, m_2 individuals of type 2, ..., m_k individuals of type k, is equal to the coefficient of $u_1^{m_1-r_1} u_2^{m_2-r_2} \ldots u_k^{m_k-r_k}$ in

$$g_1^{m_1} g_2^{m_2} \dots g_k^{m_k} \left\| \delta_i^\nu - \frac{u_i}{g_i} \frac{\partial g_i}{\partial u_\nu} \right\|.$$

As an example, Good (1960) stated that if the distributions of children were Poissonian, given by $g_i(\mathbf{t}') = \exp\left[\sum_\nu a_{i\nu}(t_\nu - 1)\right]$, then the probability of the whole process containing m_1 individuals of type 1, m_2 individuals of type 2, ..., m_k individuals of type k will be given by obtaining the coefficients from

$$\exp\left\{\sum_i m_i \sum_\nu a_{i\nu}(t_\nu - 1)\right\} \|\delta_i^\nu - t_i a_{i\nu}\|.$$

Good (1960) had also applied this theory to the enumeration of rooted trees. Good (1965) gave a very detailed description of the different kinds of trees, their colors and generations, and their ordered forms within colors, etc., and applied the theory of multivariate Lagrange expansions to obtain numerous results. Those persons who wish to work in this area should definitely read this important paper.

15.9 Exercises

15.1 Let the multivariate probability generating functions $g_i(\mathbf{t}')$ and $f(\mathbf{t}')$ be given by

$$g_i(\mathbf{t}') = \left[1 + \sum_{j=1}^k \theta_j(t_j - 1)\right]^{m_i}, \quad i = 1, 2, \dots, k, \quad \text{and} \quad f(\mathbf{t}') = \left[1 + \sum_{j=1}^k \theta_j(t_j - 1)\right]^n,$$

where $\theta_j > 0$ such that $\sum_j^k \theta_j < 1$; $n > 0$ and $m_i > 0$, $i = 1, 2, \dots, k$, are positive integers. By using (15.24), derive the multivariate Lagrangian binomial probability distribution.

15.2 For the multivariate Lagrangian quasi-Pólya distribution, show the result in (15.64).

15.3 Let the trivariate functions $g_i(\mathbf{t}')$ and $f(\mathbf{t}')$ be given by

$$g_i(\mathbf{t}') = \exp\left\{m_i \sum_{j=1}^3 \theta_j(t_j - 1)\right\}, \quad i = 1, 2, 3, \quad \text{and} \quad f(\mathbf{t}') = \exp\left\{\alpha \sum_{j=1}^3 \theta_j(t_j - 1)\right\}$$

for $\theta_j > 0$, $m_i > 0$, for $i, j = 1, 2, 3$ and $\alpha > 0$. By using the expressions (15.9) and (15.10), together with (15.8) define the trivariate Lagrangian Poisson probability distribution.

15.4 Let the multivariate functions $f(\mathbf{a}')$ and $g_i(\mathbf{a}')$ be given by

$$f(\mathbf{a}') = \left(1 - \sum_i a_i\right)^{-\alpha} \quad \text{and} \quad g_i(\mathbf{a}') = \left(1 - \sum_i a_i\right)^{-\beta_i + 1},$$

where $\alpha > 0$, $\beta_i > 1$, $i = 1, 2, 3, \dots, k$, and $a_i > 0$ such that $\sum_i a_i < 1$ and $h_i(\mathbf{a}') = a_i / g_i(\mathbf{a}')$. Using the above functions in (15.42), obtain a multivariate modified power series distribution which is similar to the multivariate Lagrangian negative binomial distribution in (15.29).

15.5 Consider the atomic fissions example (i) of subsection 15.8.1 and assign suitable values to the functions $f(\mathbf{t}')$ and $g^{(i)}(\mathbf{t}')$, $i = 1, 2, 3, 4$, so that the whole process started by such atomic fissions contains x_1 reactions of type 1, x_2 reactions of type 2, x_3 reactions of type 3, and x_4 reactions of type 4, and will generate a multivariate Lagrangian Poisson probability distribution. Prove the result in a systematic manner.

15.6 Describe a specific example with full details based on the material described in (ii) of subsection 15.8.1. Assign suitable values to the multivariate functions $f(\mathbf{t}')$ and $g^{(i)}(\mathbf{t}')$, $i = 1, 2, 3$, so that the probability distribution of the number of attacks by different diseases in the whole process will be given by the multivariate Lagrangian negative binomial model. Prove the result.

15.7 Consider the general example (iii) discussed in subsection 15.8.1 and formulate a specific example with full details so that the probability distribution of the total number of persons affected by various diseases will have a trinomial Lagrangian logarithmic series distribution. Prove the result.

15.8 Let the multivariate binomial pgfs be

$$g_i(\mathbf{t}') = \left(1 + \sum_{j=1}^{k} \theta_j (t_j - 1)\right)^{\beta_i}, \quad i = 1, 2, \ldots, k,$$

where $\theta_j > 0$ such that $\sum_{j=1}^{k} \theta_j < 1$ and β_i are positive integers. Let $f(\mathbf{t}') = t_1 t_2 \ldots t_k$ be another multivariate analytic function. By using the multivariate Lagrange expansion in (15.23), derive the multivariate basic Lagrangian binomial probability distribution. By using

$$g_i(\mathbf{t}') = \exp\left[m_i \sum_{j=1}^{k} \theta_j (t_j - 1)\right], \quad i = 1, 2, \ldots, k,$$

where $m_i > 0$, $\theta_j > 0$ for all values of i, derive the multivariate basic Lagrangian Poisson probability distribution.

16

Computer Generation of Lagrangian Variables

16.1 Introduction and Generation Procedures

The use of general (nonspecific) methods for generating pseudo-random variables from Lagrangian probability distributions will now be considered. The general methods include the inversion method and the alias method. Then, some specific method to certain Lagrangian probability distributions will be given for generating pseudo-random variables. The nonspecific generation methods are particularly suitable when the parameters of a Lagrangian distribution remain constant from call to call in the algorithm. When the parameters of a distribution change from call to call, the distribution-specific generation methods are recommended. These methods employ the structural properties of the distributions.

16.1.1 Inversion Method

Suppose X is a Lagrangian random variate with probability mass function

$$P(X = x_j) = P_{x_j}, \quad x_j \in T, \tag{16.1}$$

where T is a subset of nonnegative integers. The cumulative distribution function is given by

$$P(X \le x_j) = f(x_j) = \sum_{i \le x_j} P_i. \tag{16.2}$$

Let U be a random variate from uniform distribution on $(0, 1)$. Then

$$P\left(F(x_j - 1) < U \le F(x_j)\right) = \int_{F(x_j-1)}^{F(x_j)} du = F(x_j) - F(x_j - 1) = P_{x_j}.$$

A Lagrangian variate x is obtained by setting $x = x_j$ if

$$F(x_j - 1) < U \le F(x_j). \tag{16.3}$$

The inversion method consists of generating a uniform random variate U on $(0, 1)$ and obtaining X by monotone transformation of U into a variate from the Lagrangian distribution. Thus, a uniform $(0, 1)$ cdf is transformed into a Lagrangian cdf.

The inversion method is also called the "table look-up" method. One obtains a table containing the cumulative probabilities for the Lagrangian distribution and these probabilities are stored in computer memory. A uniform variate on (0, 1) is then generated and compared with the cumulative probabilities. The interval in which the uniform (0, 1) lies gives the random variate X. One can use one of the many search procedures (see Devroye, 1986, Chapter 3) that are now available to make this a very fast method.

The standard method is the sequential search method which compares the generated uniform (0, 1) variate with the list of $F(x_j)$. The search starts at the first value of $F(x_j)$ and compares the uniform (0, 1) with each $F(x_j)$ until the inequality in (16.3) is satisfied.

Chen and Asau (1974) introduced the "index table" search method. In this method, an index table is constructed and the value stored in the table is used to indicate which subset of $F(x_j)$ contains the value of x_j that satisfies the inequality in (16.3) for a generated variate U from uniform (0, 1). This method requires less computation time than the "sequential method" or the "bisection method." A modification of the "index table" search method was proposed by Ahrens and Kohrt (1981). This modification was designed for discrete distributions with long tails.

In the algorithms for generating Lagrangian pseudo-random variates, a sequential search method will be used. These algorithms are quite ideal for situations in which one or a few random variates are required. If several random variates are required, one can set up a table look-up algorithm with a sequential search.

16.1.2 Alias Method

Walker's (1974a, 1974b, 1977) alias method is a general nonspecific method for generating pseudo-random variates from discrete distributions. The method requires only one comparison for each generated variable. The alias method is related to "rejection" methods but differs from "rejection" methods because all rejected numbers are replaced by "aliases" that are used. Thus, a generated number is either accepted or replaced with an alias number. To generate a pseudo-random variate X with probability distribution (16.1) for $j = 1, 2, 3, \ldots, n$, we use a random number W which is uniformly distributed over the range $1, 2, 3, \ldots, n$.

Hence

$$P(W = j) = \frac{1}{n}.$$

The alias method consists of generating a pair (W, U) where U is uniform on (0, 1) and this is independent of W. The random variate is then defined by

$$X = \begin{cases} W & \text{if} \quad U \le E(W), \\ A(W) & \text{if} \quad U > E(W), \end{cases}$$

where $A(W)$ is an alias and $E(W)$ is a cutoff value. The functions $A(W)$ and $E(W)$ are chosen as indicated in the following algorithm.

Algorithm for Alias Method

1. [Initialize]
   $N \leftarrow X$ if $P_X > 1.0E-5$ and $P_{X+1} < 1.0E-5$, where P_X is the probability for X.
   $G_X \leftarrow P_X$ for $X = 0, 1, 2, \ldots, N$
   For $I = 0, N$, Do

$A_I \leftarrow I$
$E_I \leftarrow 0.0$
$B_I \leftarrow G_I - 1.0/N$

2. [Assign the alias and cutoff values]
   $S \leftarrow 0.0$
   For $I = 0, N$, Do
   $S \leftarrow S + B_I$
   while $S > 1.0E - 5$
   For $I = 0, N$, Do
   $C \leftarrow B_j$, the largest negative value of B
   $K \leftarrow J$, the corresponding position in array
   $D \leftarrow B_j$, the largest positive value of B
   $L \leftarrow J$, the corresponding position in array
   $A_k \leftarrow L$
   $E_k \leftarrow 1.0 + C \cdot N$
   $B_k \leftarrow 0.0$
   $B_L \leftarrow C + D$
3. [Generate the random variate]

Generate two independent variates (W, U) from uniform distribution on $(0, 1)$:

$$X \leftarrow \int (W \cdot N) + 1$$

$$\text{If} \quad (U > E_x) \quad \text{then} \quad X \leftarrow A$$

The inversion method algorithms for some members of Lagrangian probability distributions are described in subsequent sections. When the pseudo-random variate has been obtained, we say "Return X." The routine terminates if only one random variate is required, or loops back to the beginning if another random variate is desired.

The computation of quantities such as $(1 - \theta)^m$ may cause underflow when m is large. Also, the recurrence relation between the Lagrangian probabilities may be a potential source of roundoff errors. These errors may accumulate as x increases. To guard against these problems, implementation of all algorithms under double precision is recommended.

In addition to the inversion method, some distribution-specific generation algorithms have also been included in the following sections.

16.2 Basic Lagrangian Random Variables

The basic Lagrangian probability distributions are discussed in chapter 8. The pmf for the Borel distribution is given by

$$P_x = \frac{(x\lambda)^{x-1} e^{-x\lambda}}{x!}, \quad x = 1, 2, 3, \ldots.$$

A recurrence relation between the Borel probabilities is

$$P_x = \lambda e^{-\lambda} \left(1 + \frac{1}{x-1}\right)^{x-2} P_{x-1}, \quad x = 2, 3, \ldots, \tag{16.4}$$

with $P_1 = e^{-\lambda}$. The pmf for the Consul distribution is

$$P_x = \frac{1}{x}\binom{mx}{x-1}\theta^{x-1}(1-\theta)^{mx-x+1}, \quad x = 1, 2, 3, \ldots,$$

with a recurrence relation

$$P_x = \frac{(m-1)(x-1)+1}{x-1}\theta(1-\theta)^{m-1}\prod_{i=1}^{x-2}\left(1+\frac{m}{mx-m-i}\right)P_{x-1} \tag{16.5}$$

for $x = 2, 3, \ldots$ and $P_1 = (1-\theta)^m$. The pmf for the Geeta distribution is given by

$$P_x = \frac{1}{mx-1}\binom{mx-1}{x}\theta^{x-1}(1-\theta)^{mx-x}, \quad x = 1, 2, 3, \ldots.$$

A recurrence relation between the probabilities of the Geeta distribution is

$$P_x = \frac{m(x-1)-x}{x}\theta(1-\theta)^{m-1}\prod_{i=2}^{x}\left(1+\frac{m}{mx-m-1}\right)P_{x-1} \tag{16.6}$$

for $x = 2, 3, 4, \ldots$ with $P_1 = (1-\theta)^{m-1}$. The Haight distribution is a special case of the Geeta distribution when $m = 2$.

Borel Random Variate Generator based on Inversion Method
[Initialize: $\omega = \lambda e^{-\lambda}$]

1. $X \leftarrow 1$
2. $S \leftarrow e^{-\lambda}$ and $P \leftarrow S$
3. Generate U from uniform distribution on (0, 1).
4. While $U > S$, do
   $X \leftarrow X + 1$
   $C \leftarrow (1 + 1/(X-1))^{X-2}$
   $P \leftarrow \omega C P$
   $S \leftarrow S + P$
5. Return X

Consul Random Variate Generator based on Inversion Method
[Initialize: $\omega = \theta(1-\theta)^{m-1}$]

1. $X \leftarrow 1$
2. $S \leftarrow (1-\theta)^m$ and $P \leftarrow S$
3. Generate U from uniform distribution on (0, 1).
4. If $U \le S$, Return X
5. $X \leftarrow X + 1$
6. $P \leftarrow \omega m P$ and $S \leftarrow S + P$
7. If $U \le S$, Return X
8. While $U > S$, do
   $X \leftarrow X + 1$
   $C \leftarrow ((m-1)(X-1)+1)\prod_{i=1}^{X-2}(1 + m/(mX - m - i))$
   $P \leftarrow \omega C P/(X-1)$
   $S \leftarrow S + P$
9. Return X

Geeta Random Variate Generator based on Inversion Method
[Initialize: $\omega = \theta(1-\theta)^{m-1}$]

1. $X \leftarrow 1$
2. $S \leftarrow (1-\theta)^{m-1}$ and $P \leftarrow S$
3. Generate U from uniform distribution on (0, 1).
4. While $U > S$, do
   $X \leftarrow X + 1$
   $C \leftarrow ((mX - m - X)\prod_{i=2}^{X}(1 + m/(mX - m - i))$
   $P \leftarrow \omega C P/X$
   $S \leftarrow S + P$
5. Return X

16.3 Simple Delta Lagrangian Random Variables

The simple delta Lagrangian probability distributions are given in chapter 2. The pmf for the delta-Poisson distribution is given as

$$P_x = \frac{n}{x(x-n)!}(\lambda x)^{x-n}e^{-x\lambda} \quad \text{for} \quad x \geq n.$$

A recurrence relation between the delta-Poisson probabilities is

$$P_x = \left(\frac{x-1}{x-n}\right)\lambda e^{-\lambda}\left(1 + \frac{1}{x-1}\right)^{x-n-1} P_{x-1} \tag{16.7}$$

for $x = n+1, n+2, n+3, \ldots$ with $P_n = e^{-n\lambda}$. The delta-Poisson distribution reduces to the Borel distribution when $n = 1$.

The pmf for the delta-binomial distribution is given by

$$P_x = \frac{n}{x}\binom{mx}{x-n}\theta^{x-n}(1-\theta)^{n+mx-x} \quad \text{for } x \geq n.$$

A recurrence relation between the delta-binomial probabilities is

$$P_x = \frac{(m-1)(x-1)+n}{x-n}\theta(1-\theta)^{m-1}\prod_{i=1}^{x-n-1}\left(1 + \frac{m}{(m-1)x-i}\right)P_{x-1} \tag{16.8}$$

for $x \geq n+1$ with $P_n = (1-\theta)^{nm}$.

The delta-binomial distribution reduces to the Consul distribution when $n = 1$. The pmf for the delta-negative binomial distribution is

$$P_x = \frac{n}{x}\binom{mx-n-1}{x-n}\theta^{x-n}(1-\theta)^{mx-x}, \quad x \geq n,$$

with recurrence relation

$$P_x = \frac{m(x-1)-x}{x-n}\cdot\frac{x-1}{x}\cdot\theta(1\theta)^{m-1}\prod_{i=n+1}^{x}\left(1 + \frac{m}{x(x-1)-i}\right)P_{x-1} \tag{16.9}$$

for $x \geq n + 1$ with $P_n = (1 - \theta)^{n(m-1)}$. The delta-negative binomial distribution reduces to the Geeta distribution when $n = 1$.

The inversion method for generating simple delta Lagrangian random variables is similar to the inversion method for generating basic Lagrangian random variables as provided in section 16.2. We now give the inversion method for generating the delta-Poisson random variables. Similar algorithms for the delta-binomial and delta-negative binomial can be obtained.

Delta-Poisson Random Variate Generator based on Inversion Method
[Initialize: $\omega = \lambda e^{-\lambda}$]

1. $X \leftarrow n$
2. $S \leftarrow e^{-\lambda n}$ and $P \leftarrow S$
3. Generate U from uniform distribution on (0, 1).
4. While $U < S$, do
   $X \leftarrow X + 1$
   $C \leftarrow (X - 1)(1 + 1/(X - 1))^{X-n-1}$
   $P \leftarrow \omega CP/(X - n)$
   $S \leftarrow S + P$
5. Return X

Devroye (1992) introduced two universally applicable random variate generators for the family of delta Lagrangian distributions. The branching process method and the uniform bounding method were used to generate pseudo-random variates from the delta-Poisson, delta-binomial, and delta-negative binomial distributions.

The Branching Process Method
A pseudo-random variate X is generated based upon partial recreation of a certain branching process. In this method, there are no heavy numerical computations. The delta Lagrangian distribution was defined in chapter 2 as a Lagrangian distribution with generating functions (g, f), where $f(z) = z^n$ and g is a pgf. The pgf $f(z) = z^n$ puts mass 1 at point $x = n$. In a Galton–Watson branching process started with one individual, let every individual produce children independently in accordance with the distribution defined by the pgf $g(z)$. Let the size X be the number of elements that ever live in such a finite population. The pgf for X is $q(z)$, which is equal to the unique solution u of the equation

$$u = z\,g(u), \quad z \in [0, 1]. \tag{16.10}$$

If the branching process is started with Y individuals, where Y is a random variable having pgf $f(z)$, then the pgf of X is

$$E\left\{[q(z)]^Y\right\} = f(q(z)), \quad \text{where } q(z) \text{ is a solution of (16.10).}$$

Since the distribution of X is delta Lagrangian distribution with generating function (g, f), by using this property, one obtains the branching process method to generate the delta Lagrangian distribution.

The Branching Process Algorithm

1. Generate Y with pgf $f(z)$.
2. $X \leftarrow Y$

3. While $Y > 0$, do
   Generate W with pgf $g^Y(z)$.
   $(X, Y) \leftarrow (X + W, W)$
4. Return X

Uniform Bounding Method

Certain delta Lagrangian distributions have the property that

$$\sup_{\theta} P_{x,\theta} \leq q_x \tag{16.11}$$

for all x, where $P_{x,\theta} = P(X = x)$, $\sum_x q_x < \infty$, and θ is the collection of parameters. If P_x is a probability distribution such that $P_x \leq q_x$ for all x, then a pseudo-random variate X with probability distribution P_x can be generated by the rejection method. One generates pairs of independent random variates (X, U), where X has distribution cq_x for some constant c and U is from uniform distribution on (0, 1). When $Uq_x < P_x$, one returns X. This technique, called the uniform bounding method, has good speed even when θ changes from call to call.

Theorem 16.1. *The delta-Poisson, delta-binomial, and delta-negative binomial distributions with fixed n satisfy the condition where c is a constant depending on n and*

$$\sup_{\theta} P_{x,\theta} \leq \begin{cases} \frac{nc}{x\sqrt{x-n}} & \text{if } x > n, \\ 1 & \text{if } x = n, \end{cases} \tag{16.12}$$

where $c = 1/\sqrt{2\pi}$ *for the delta-Poisson,* $c = e^{\frac{1}{24n}}/\sqrt{n}$ *for the delta-binomial, and* $c = e^{\frac{1}{12}}/\sqrt{2\pi}$ *for the delta-negative binomial distribution (Devroye, 1992).*

Proof. We shall show the result for the delta-Poisson distribution. The results for the delta-binomial and delta-negative binomial distributions can be proved in a similar manner.

By using differentiation, it is not hard to show that

$$\sup_{u>0} u^a e^{-bu} = \left(\frac{a}{be}\right)^a \quad \text{for } a, b > 0. \tag{16.13}$$

By using (16.13), the delta-Poisson distribution can be written as

$$\sup_{0<\lambda\leq 1} P_{x,\lambda} \leq \frac{n[(x-n)/e]^{x-n}}{x(x-n)!}, \quad x \geq n. \tag{16.14}$$

By applying the Stirling's approximation

$$x! \geq \left(\frac{x}{e}\right)^x \sqrt{2\pi x}$$

to the inequality in (16.14), we obtain

$$\sup_{0<\lambda\leq 1} P_{x,\lambda} \leq \frac{n}{x\sqrt{2\pi(x-n)}}$$

$$= \frac{nc}{x\sqrt{x-n}}, \quad \text{where} \quad c = 1/\sqrt{2\pi}.$$

□

By using Theorem 16.1, the uniform bounding algorithm is developed for the delta-Poisson, delta-binomial, and delta-negative binomial distributions.

The Uniform Bounding Algorithm

1. [Initialize]
   Set $c \leftarrow 1/\sqrt{2\pi}$ for delta-Poisson distribution.
   Set $c \leftarrow e^{\frac{1}{24n}}/\sqrt{\pi}$ for delta-binomial distribution.
   Set $c \leftarrow e^{\frac{1}{12}}/\sqrt{2\pi}$ for delta-negative binomial distribution.
2. Repeat
   Generate U from uniform distribution on (0, 1).
   $V \leftarrow (1 + 4c\sqrt{n})U$
   Case I
   $V \leq 1$: Return $X \leftarrow n$
   $1 < V \leq 1 + 2c\sqrt{n}$: $Y \leftarrow n + 1 + (V-1)^2/4c^2$
   $T \leftarrow 2c^2/(V-1)$
   Case II
   $V > 1 + 2c\sqrt{n}$: $Y \leftarrow n + 1 + \left(\frac{2nc}{1+4c\sqrt{n}-V}\right)^2$
   $T \leftarrow nc/(Y-1-n)^{\frac{3}{2}}$
   Generate W from uniform distribution on (0, 1).
   Until $WT < P_{\lfloor Y \rfloor}$
   Return $X \leftarrow \lfloor Y \rfloor$, where $\lfloor Y \rfloor$ denotes rounding to the nearest smaller integer.

16.4 Generalized Poisson Random Variables

The pmf for the GPD is given in chapter 9 as

$$P_x = \theta(\theta + \lambda x)^{x-1} e^{-\theta - \lambda x}/x! \quad \text{for} \quad x = 0, 1, 2, 3, \ldots.$$

A recurrence relation between the generalized Poisson probabilities is given by

$$P_x = \frac{\theta - \lambda + \lambda x}{x} \left(1 + \frac{\lambda}{\theta - \lambda + \lambda x}\right)^{x-1} e^{-\lambda} P_{x-1} \tag{16.15}$$

for $x \geq 1$, with $P_0 = e^{-\theta}$.

Inversion Method (Famoye, 1997b)
[Initialize: $\omega = e^{-\lambda}$]

1. $X \leftarrow 0$
2. $S \leftarrow e^{-\theta}$ and $P \leftarrow S$
3. Generate U from uniform distribution on (0, 1).
4. While $U > S$, do
   $X \leftarrow X + 1$
   $c \leftarrow \theta - \lambda + \lambda X$
   $P \leftarrow \omega c(1 + \lambda/c)^{X-1} P/X$
   $S \leftarrow S + P$
5. Return X

Under the inversion method, one needs to guide against underflow and roundoff errors. The recursion formula in (16.15) is a potential source of roundoff errors. Using double precision arithmetic in the computer program can provide some protection for the roundoff errors.

Devroye (1989) worked out uniformly fast algorithms for generating the generalized Poisson random variables. Three regions in the parameter space (θ, λ) are considered in developing the algorithms. The regions are the Abel side of the parameter space, the Poisson side of the parameter space, and the region of monotonicity.

In the Abel side of the parameter space, $\theta \geq 1 + \lambda, \ \theta \geq 2\lambda(1-\lambda)^{-1}$. For this case, a two-tier rejection algorithm was developed. On the right tail, a polynomially decreasing distribution is used as a bounding function, while on the left tail, a geometrically increasing dominating function is used.

In the Poisson side of the parameter space, $\theta \geq \max(3, 2\lambda(1-\lambda)^{-1})$. The generalized Poisson probability P_x is bounded uniformly by a function $g(x)$. A rejection algorithm based on bounding continuous distributions is developed. The bounding distribution is the normal density with exponential tails added on both ends. The idea is to generate pairs (Y, U), where Y is a pseudo-random variate with density proportional to g and U is a uniform variate on $(0, 1)$ until $Ug(Y) \leq P_x$. When this inequality is satisfied, we set $X \leftarrow \lfloor Y \rfloor$, where X has the desired generalized Poisson distribution.

In the region of monotonicity, $\theta \leq 1 + \lambda$, the algorithm is a rejection method based upon the inequality

$$P_x \leq \theta e^{2-\lambda-\min(\lambda,\theta)} \sqrt{\frac{s}{\pi}} \left(\frac{1}{\sqrt{x}} - \frac{1}{\sqrt{x+1}} \right). \tag{16.16}$$

The algorithm can be applied to any region of the parameter space (θ, λ); however, it is recommended for parameter values of $\theta \leq 1 + \lambda$.

Rejection Algorithm with Polynomial Bound

1. [Initialize]
   $P_0 \leftarrow e^{-\theta}, b \leftarrow \theta e^{2-\lambda-\min(\lambda,\theta)} \sqrt{2/\pi}$
2. [Generator]
   Generate U from uniform distribution on $(0, 1)$.
   Repeat
   If $U \leq \frac{P_0}{P_0+b}$
   Then $X \leftarrow 0$ and Accept $\leftarrow$ True
   Else
   Generate V, W independent and identically distributed variables from uniform $(0, 1)$.
   Set $X \leftarrow \lfloor \frac{1}{W^2} \rfloor$
   Set Accept $\leftarrow \left[Vb \left(\frac{1}{\sqrt{X}} - \frac{1}{\sqrt{X+1}} \right) \leq P_X \right]$
   Until Accept
3. Return X

The evaluation of P_x for $x \geq 1$ can be done efficiently by using the recurrence relation in (16.15).

Famoye (1997b) developed a branching algorithm to generate generalized Poisson variates when the parameter λ is positive.

Branching Algorithm

1. Generate Y from Poisson distribution with mean θ.
2. $X \leftarrow Y$
3. While ($Y > 0$), do
   $k \leftarrow \lambda Y$
   Generate Z from Poisson distribution with mean k.
   $(X, Y) \leftarrow (X + Z, Z)$
4. Return X

An algorithm based on a normal approximation was developed by Famoye (1997b).

Normal Approximation Algorithm
[Initialize: $m \leftarrow \theta(1-\lambda)^{-1}$; $\nu \leftarrow \sqrt{\theta(1-\lambda)^{-3}}$]

1. Generate Y from a standard normal distribution.
2. $X \leftarrow \max(0, \lfloor m + \nu Y + 0.5 \rfloor)$
3. Return X

The above algorithm is recommended for $\theta \geq 10.0$ when $\lambda < 0$ and also for $\theta \geq 30.0$ when $0 < \lambda < 0.2$.

Famoye (1997b) compared the generalized Poisson variates generating algorithms. For low parameter values, the inversion method is faster than both the rejection and the branching methods. As the value of θ increases, the branching method performs faster than the inversion method.

16.5 Generalized Negative Binomial Random Variables

The pmf for the GNBD is given in chapter 10 as

$$P_x = \frac{m}{m+\beta x}\binom{m+\beta x}{x}\theta^x(1-\theta)^{m+\beta x-x}, \quad x = 0, 1, 2, 3, \ldots.$$

A recurrence relation between the generalized negative binomial probabilities is given by

$$P_x = \frac{m+(\beta-1)(x-1)}{x}\theta(1-\theta)^{\beta-1}\prod_{i=1}^{x-1}\left(1+\frac{\beta}{m+\beta x-\beta-i}\right)P_{x-1} \tag{16.17}$$

for $x \geq 1$, with $P_0 = (1-\theta)^m$.

Famoye (1998b) provided some of the theoretical bases for some of the generating algorithms developed in this section.

Theorem 16.2. *The GNBD in (10.1) is nonincreasing for all values of θ, β, and m such that $m\theta(1-\theta)^{\beta-1} < 1$ and satisfies the inequality*

$$P_x(\theta, \beta, m) \leq m(1+\sqrt{2})\left(\frac{1}{\sqrt{x}} - \frac{1}{\sqrt{x+1}}\right)[\pi\beta(\beta-1)]^{-1/2} \tag{16.18}$$

for all values of $x = 1, 2, 3, \ldots$.

Proof.

$$P_x(\theta, \beta, m) = \frac{m}{m + \beta x} \binom{m + \beta x}{x} \theta^x (1 - \theta)^{m+\beta x - x}$$

$$\leq \frac{m(\theta\beta)^x (1 - \theta)^{m+\beta x - x}}{\beta \sqrt{2\pi} x^{x+1/2} e^{-x}} \prod_{i=1}^{x-1} \left(x - \frac{i - m}{\beta} \right) \tag{16.19}$$

since by Stirling's approximation, $x! \geq \sqrt{2\pi} x^{x+1/2} e^{-x}$. The inequality in (16.19) can be rewritten as

$$P_x(\theta, \beta, m) \leq m \left(\beta \sqrt{2\pi} x^{3/2} \right)^{-1} e^R, \tag{16.20}$$

where

$$R = x \log(\theta\beta) + x + (m + \beta x - x) \log(1 - \theta) + \sum_{i=1}^{x-1} \log \left(1 - \frac{i - m}{\beta x} \right).$$

On differentiating R with respect to θ, we obtain

$$\frac{dR}{d\theta} = \frac{x - \theta(m + \beta x)}{\theta(1 - \theta)} \tag{16.21}$$

and this is zero when $\theta = x/(m + \beta x)$. When x is very large, the value of θ goes to β^{-1}.

When x is large: The result in (16.21) is always positive. Here, R is an increasing function of θ and R is maximum when $\theta = \beta^{-1}$. Thus,

$$R \leq x + (m + \beta x - x) \log(1 - \beta^{-1}) + \sum_{i=1}^{x-1} \log \left(1 - \frac{i - m}{\beta x} \right)$$

$$\leq -\frac{1}{2} \log(1 - \beta^{-1}).$$

When x is small: The result in (16.21) may change sign over the values of θ. If (16.21) is always negative, $R \leq -\frac{1}{2} \log(1 - \beta^{-1})$. On taking the second derivative of R with respect to θ, we obtain

$$\frac{d^2R}{d\theta^2} = -x\theta^{-2} - (m + \beta x - x)(1 - \theta)^{-2} < 0.$$

Hence, R has a maximum value at $\theta = x/(m + \beta x)$, and using this value, we obtain

$$R \leq -\frac{1}{2} \log(1 - \beta^{-1}).$$

Therefore,

$$P_x(\theta, \beta, m) \leq m \left[\beta \sqrt{2\pi} x^{3/2} \right]^{-1} \exp \left[-\frac{1}{2} \log(1 - 1/\beta) \right],$$

$$= m x^{-1} \left[2x\pi\beta(\beta - 1) \right]^{-1/2}.$$

But $x^{-3/2} \leq (2+\sqrt{2})\left(\frac{1}{\sqrt{x}} - \frac{1}{\sqrt{x+1}}\right)$ for all $x \geq 1$, and so

$$P_x(\theta, \beta, m) \leq m(1+\sqrt{2})\left(\frac{1}{\sqrt{x}} - \frac{1}{\sqrt{x+1}}\right)[\pi\beta(\beta-1)]^{-1/2},$$

which completes the proof. □

The GPD in (9.1) can be rewritten as

$$Q_x(p, \lambda) = p(p+\lambda x)^{x-1}e^{-p-\lambda x}/x!. \tag{16.22}$$

The GPD model in (16.22) is a limiting form of the GNBD when $m \to \infty$ and $\beta \to \infty$ such that $m\theta = p$ and $\beta\theta = \lambda$.

Theorem 16.3. *The GNBD in (10.1) satisfies the inequality*

$$P_x(\theta, \beta, m) \leq cQ_x(p, \lambda)$$

for all values of $x = 0, 1, 2, \ldots$, where m and β are large, $p = \theta m$, $\lambda = \theta\beta$, $Q_x(p, \lambda)$ is the GPD model, and the constant $c = e^{\theta}$.

Proof. We need to show that

$$\frac{m}{m+\beta x}\binom{m+\beta x}{x}\theta^x(1-\theta)^{m+\beta x-x} \leq \theta m \frac{(\theta m+\theta\beta x)^{x-1}}{x!}e^{-\theta m-\theta\beta x+\theta}.$$

That is, we need to show that

$$(1-\theta)^{m+\beta x-x}e^{\theta(m+\beta x-1)}\prod_{i=1}^{x-1}\left(1-\frac{i}{m+\beta x}\right) \leq 1. \tag{16.23}$$

When $x = 0$, (16.23) reduces to $(1-\theta)e^{\theta(m-1)} \leq 1$, which holds for all parameter values. The inequality in (16.23) also holds for $x = 1$. When $x = 2$, the left-hand side of (16.23) can be written as $e^{R(2)}$, where

$$R(2) = \log\left(1-\frac{1}{m+2\beta}\right) + (m+2\beta-2)\log(1-\theta) + \theta(m+2\beta-1).$$

$R(2)$ is maximum when $\theta = 1/(m+2\beta-1)$. Therefore,

$$R(2) \leq \log\left(1-\frac{1}{m+2\beta}\right) + (m+2\beta-2)\log(1-\frac{1}{m+2\beta-1}) + 1 \leq 0$$

when $m+2\beta > 2.27288$, which holds for large values of m and β. Therefore, $e^{R(2)} \leq 1$ and hence (16.23) holds for $x = 2$.

For $x = k$, the left-hand side of (16.23) becomes $e^{R(k)}$, where

$$R(k) = \sum_{i=1}^{k-1}\log\left(1-\frac{i}{m+\beta k}\right) + (m+\beta k-k)\log(1-\theta) + \theta(m+\beta k-1).$$

$R(k)$ is maximum at the point $\theta = (k-1)/(m+\beta k-1)$ and so

$$R(k) \leq \sum_{i=1}^{k-1}\log\left(1-\frac{i}{m+\beta k}\right) + (m+\beta k-k)\log\left(1-\frac{k-1}{m+\beta k-1}\right) + k - 1 \leq 0$$

for large values of m and β. Hence, (16.23) is satisfied for all values of x. □

Inversion Method (Famoye, 1998b)
[Initialize: $\omega = \theta(1-\theta)^{\beta-1}$]

1. $X \leftarrow 0$
2. $S \leftarrow (1-\theta)^m$ and $P \leftarrow S$
3. Generate U from uniform distribution on (0, 1).
4. If $U \leq S$, Return X
5. $X \leftarrow X + 1$
6. $P \leftarrow \omega m P$ and $S \leftarrow S + P$
7. If $U \leq S$, Return X
8. While $U > S$, do
   $X \leftarrow X + 1$
   $C \leftarrow \prod_{i=1}^{X-1}(1 + \beta/(m + \beta X - \beta - i)$
   $P \leftarrow \omega C(m + (\beta - 1)(X - 1))P/X$
   $S \leftarrow S + P$
9. Return X

Famoye (1998b) applied the inequality in (16.18) to develop a rejection algorithm. If U is a uniform random variate, the random variate $X \leftarrow \lfloor U^{-2} \rfloor$, where $\lfloor . \rfloor$ is the integer part of the value, satisfies

$$P(X \geq x) = P\left(1 \geq xU^2\right) = P\left(U \leq \frac{1}{\sqrt{x}}\right) = \frac{1}{\sqrt{x}} \quad \text{for all} \quad x \geq 1.$$

Therefore, $P(X = x) = 1/\sqrt{x} - 1/\sqrt{x+1}$. The rejection algorithm may be used for all values of θ, β, and m; however, the method is recommended for the case where $m\theta(1-\theta)^{\beta-1} \leq 1$.

Rejection Algorithm

1. [Initialize]
   Set $c \leftarrow m(1 + \sqrt{2})\,[\pi\beta(\beta - 1)]^{-1/2}$
   Set $P_0 \leftarrow (1-\theta)^m$
2. [Generator]
   Generate U from uniform distribution on (0, 1).
   Repeat
   If $U \leq P_0\,(c + P_0)^{-1}$
   Then $X \leftarrow 0$ and Accept $\leftarrow$ True
   Else
   Generate V, W independent and identically distributed from uniform (0, 1).
   Set $X \leftarrow \lfloor W^{-2} \rfloor$
   Set Accept $\leftarrow \left[Vc\left(\frac{1}{\sqrt{X}} - \frac{1}{\sqrt{X+1}}\right) \leq P_X\right]$
   Until Accept
3. Return X

Suppose the probability distribution Q_x, $x \geq 0$ is easy to generate. A rejection algorithm can be used to generate P_x, $x \geq 0$ if

$$P_x \leq cQ_x, \quad x \geq 0$$

where the rejection constant $c \geq 1$. Using Theorem 16.3, Famoye (1998b) developed a rejection algorithm based on the GPD.

Rejection Algorithm based on GPD

1. [Initialize]
   Set $\omega \leftarrow e^{\theta}$
2. [Generator]
   Generate U from uniform distribution on (0, 1).
   While $(\omega U Q_X > P_X)$, do
   Generate X from GPD with parameters $p = \theta m$ and $\lambda = \theta\beta$.
3. Return X

Famoye (1998b) developed a branching algorithm for the GNBD.

Branching Algorithm

1. Generate Y from the negative binomial distribution with parameters θ and m.
2. While $(Y > 0)$, do
   $k \leftarrow (\beta - 1)Y$
   Generate Z from the negative binomial distribution with parameters θ and k.
   $(X, Y) \leftarrow (Y + Z, Z)$
3. Return X

When parameters β and m are integers, the negative binomial distribution in the above algorithm can be replaced with binomial distribution.

In comparing the inversion, branching, and rejection methods, the rejection method is the slowest. For small values of θ, the inversion method is faster than the branching method. However, as parameter θ becomes large, the branching method becomes faster. In general, the inversion method tends to be the fastest when $\theta\beta \leq 0.6$. Famoye (1998b) recommended the inversion method when $\theta\beta \leq 0.6$ and the branching method when $\theta\beta > 0.6$.

16.6 Generalized Logarithmic Series Random Variables

The pmf for the GLSD is given in chapter 11 as

$$P_x = \frac{1}{\beta x}\binom{\beta x}{x}\frac{\theta^x(1-\theta)^{\beta x - x}}{-\ln(1-\theta)}, \quad x = 1, 2, 3, \ldots .$$

A recurrence relation between generalized logarithmic series probabilities is given by

$$P_x = \frac{(\beta-1)(x-1)}{x}\theta(1-\theta)^{\beta-1}\prod_{i=1}^{x-1}\left(1 + \frac{\beta}{\beta x - \beta - i}\right) \tag{16.24}$$

for $x \geq 2$, with $P_1 = \theta(1-\theta)^{\beta-1}[-\ln(1-\theta)]^{-1}$.

Famoye (1997d) presented some theoretical justification for some generating algorithms for the GLSD.

Theorem 16.4. *The GLSD in (11.1) is nonincreasing for all values of θ in $0 < \theta < \beta^{-1}$ and $\beta \geq 1$, and for all values of $x = 1, 2, 3, \ldots$ the GLSD satisfies the condition*

$$P_x(\theta, \beta) \leq \frac{1}{x}. \tag{16.25}$$

Proof.

$$P_x(\theta,\beta) = \frac{1}{\beta x}\binom{\beta x}{x}\frac{\theta^x(1-\theta)^{\beta x - x}}{[-\ln(1-\theta)]}$$

$$\leq \theta^{x-1}(1-\theta)^{(\beta-1)(x-1)}\prod_{i=1}^{x-1}\frac{\beta x - i}{x!} \tag{16.26}$$

since

$$P_1(\theta,\beta) = \theta(1-\theta)^{\beta-1}[-\ln(1-\theta)]^{-1} \leq 1.$$

The function $\theta^{x-1}(1-\theta)^{(\beta-1)(x-1)}$ is an increasing function of θ and it is maximum at $\theta = \beta^{-1}$. Thus, we can write (16.26) as

$$P_x(\theta,\beta) \leq \left(\frac{1}{\beta}\right)^{x-1}\left(1-\frac{1}{\beta}\right)^{(\beta-1)(x-1)}\prod_{i=1}^{x-1}\frac{\beta x - i}{x!}$$

$$= \left(1-\frac{1}{\beta}\right)^{(\beta-1)(x-1)}\prod_{i=1}^{x-1}\left(x-\frac{i}{\beta}\right)\frac{1}{x!}. \tag{16.27}$$

The right-hand side of (16.27) is a decreasing function of β and it is maximum when $\beta = 1$. Hence

$$P_x(\theta,\beta) \leq \prod_{i=1}^{x-1}\frac{x-i}{x!} = \frac{1}{x}.$$

□

By using the unimodality property and mathematical induction, one can also show that the inequality in (16.25) holds.

Theorem 16.5. *The GLSD in (11.1) is nonincreasing for all values of* θ *in* $0 < \theta < \beta^{-1}$ *and* $\beta > 1$ *and satisfies the inequality*

$$P_x(\theta,\beta) \leq (1+\sqrt{2})\left(\frac{1}{\sqrt{x}} - \frac{1}{\sqrt{x+1}}\right)\left[(-\ln(1-\theta))\sqrt{\pi\beta(\beta-1)}\right]^{-1} \tag{16.28}$$

for all values of $x = 1, 2, 3, \ldots$

Proof.

$$P_x(\theta,\beta) = \frac{1}{\beta x}\binom{\beta x}{x}\frac{\theta^x(1-\theta)^{\beta x - x}}{[-\ln(1-\theta)]}$$

$$\leq \frac{(\theta\beta)^x(1-\theta)^{\beta x - x}\prod_{i=1}^{x-1}(x - i/\beta)}{\beta[-\ln(1-\theta)]\sqrt{2\pi}x^{x+1/2}e^{-x}}, \tag{16.29}$$

since by Stirling's approximation, $x! \geq \sqrt{2\pi}x^{x+1/2}e^{-x}$. The inequality in (16.29) can be rewritten as

$$P_x(\theta,\beta) \leq \left\{\beta[-\ln(1-\theta)]\sqrt{2\pi}x^{3/2}\right\}^{-1}e^{Q}, \tag{16.30}$$

where

$$Q = x\log(\theta\beta) + x(\beta-1)\log(1-\theta) + x + \sum_{i=1}^{x-1}\log\left(1-\frac{i}{\beta x}\right).$$

Q is an increasing function of θ and it is maximum when $\theta = \beta^{-1}$. Hence

$$Q \le x(\beta-1)\log\left(1-\frac{1}{\beta}\right) + x + \sum_{i=1}^{x-1}\log\left(1-\frac{i}{\beta x}\right). \tag{16.31}$$

By using the results for $\sum_{i=1}^{n} i$, $\sum_{i=1}^{n} i^2$, $\sum_{i=1}^{n} i^3$, and $\sum_{i=1}^{n} i^4$ in (16.31), we obtain

$$Q \le -\frac{1}{2}\log\left(1-\frac{1}{\beta}\right). \tag{16.32}$$

On using (16.32) in (16.30), we obtain

$$P_x(\theta,\beta) \le \frac{\exp\left[-\frac{1}{2}\log\left(1-\frac{1}{\beta}\right)\right]}{\beta[-\log(1-\theta)]\sqrt{2\pi}\,x^{3/2}}. \tag{16.33}$$

But $x^{-3/2} \le (2+\sqrt{2})\left(\frac{1}{\sqrt{x}} - \frac{1}{\sqrt{x+1}}\right)$, and hence

$$P_x(\theta,\beta) \le (1+\sqrt{2})\left(\frac{1}{\sqrt{x}} - \frac{1}{\sqrt{x+1}}\right)\left[\{-\log(1-\theta)\}\sqrt{\pi\beta(\beta-1)}\right]^{-1}.$$

□

Inversion Method (Famoye, 1997d)
[Initialize: $\omega = \theta(1-\theta)^{\beta-1}$]

1. $X \leftarrow 1$
2. $S \leftarrow \omega/(-\ln(1-\theta))$ and $P \leftarrow S$
3. Generate U from uniform distribution on (0, 1).
4. While $U > S$, do
   $X \leftarrow X + 1$
   $C \leftarrow \prod_{i=1}^{X-1}(1+\beta/(\beta X - \beta - i))$
   $P \leftarrow \omega C(\beta-1)(X-1)P/X$
   $S \leftarrow S + P$
5. Return X

Famoye (1997d) used the upper bound of the inequality in (16.28) to develop an algorithm based on the rejection method. Consider the uniform (0, 1) random variate U. The random variate $X \leftarrow \lfloor U^{-2} \rfloor$, where $\lfloor . \rfloor$ is the integer part of the value, satisfies

$$P(X \ge x) = P(1 \ge xU^2) = P\left(U \le \frac{1}{\sqrt{x}}\right) = \frac{1}{\sqrt{x}}$$

for all values of $x \ge 1$. Therefore,

$$P(X = x) = \frac{1}{\sqrt{x}} - \frac{1}{\sqrt{x+1}},$$

and so a rejection algorithm for generating a generalized logarithmic series variate can be based on (16.28).

Rejection Algorithm

1. [Initialize]
   Set $c \leftarrow (1+\sqrt{2})\left[(-\log(1-\theta))\sqrt{\pi\beta(\beta-1)}\right]^{-1}$
2. [Generator]
   Repeat
   Generate $U,\ V$ i.i.d. from uniform distribution on (0,1).
   Set $X \leftarrow \lfloor V^{-2} \rfloor$
   Set Accept $\leftarrow Uc\left(\frac{1}{\sqrt{X}} - \frac{1}{\sqrt{X+1}}\right) \le P_X$
   Until Accept
3. Return X

The GLSD probabilities satisfy the inequality in (16.25). Devroye (1986) suggested a rejection algorithm for any distribution that satisfies the inequality. For the GLSD, the probability vector $P_1, P_2, \ldots, P_n$ is nonincreasing and the choice of $n = n(\theta, \beta)$ is given by the first integer n for which $P_n(\theta, \beta) < 10^{-5}$. Based on this inequality, Famoye (1997d) developed a monotone property algorithm for the GLSD.

Monotone Property Algorithm

1. [Initialize] Determine the first n such that $P_n(\theta, \beta) < 10^{-5}$.
2. Generate a random variate X with probability vector proportional to $1, \frac{1}{2}, \frac{1}{3}, \ldots, \frac{1}{n}$.
3. Repeat
   Generate U from uniform distribution on (0, 1).
   Set Accept $\leftarrow U \le XP_X$
   Until Accept
4. Return X

Famoye (1997d) developed a branching algorithm to generate GLS variates.

Branching Algorithm

1. Generate Y from logarithmic series distribution with parameter θ.
2. $X \leftarrow Y$
3. While ($Y > 0$), do
   $k \leftarrow (\beta - 1)Y$
   Generate Z from negative binomial distribution with parameters θ and k.
   $(X, Y) \leftarrow (X + Z, Z)$
4. Return X

Famoye (1997d) compared all the algorithms and found that the inversion method was the fastest for small values of θ and β. However, the branching method appeared to be the fastest for large values of θ and β. Famoye (1997d) recommended a modified algorithm which used an inversion method when $\theta\beta \le 0.45$ and used a branching method when $\theta\beta > 0.45$.

16.7 Some Quasi-Type Random Variables

The pmf for the QBD-I is given in chapter 4 as

$$P_x = \binom{m}{x} p(p + x\phi)^{x-1}(1 - p - x\phi)^{x-m}, \quad x = 1, 2, 3, \ldots.$$

A recurrence relation between the above probabilities is given by

$$P_x = \frac{m - x + 1}{x} \cdot \frac{p - \phi + x\phi}{1 - p + \phi - x\phi} \left(1 + \frac{\phi}{p - \phi + x\phi}\right)^{x-1} \left(1 - \frac{\phi}{1 - p + \phi - x\phi}\right)^{m-x} P_{x-1} \tag{16.34}$$

for $x \geq 1$ with $P_1 = (1 - p^m)$.

Quasi-Binomial I Random Variate Generator based on Inversion Method
[Initialize: $\omega = p - \phi$]

1. $X \leftarrow 0$
2. $S \leftarrow (1 - p)^m$ and $P \leftarrow S$
3. Generate U from uniform distribution on (0, 1).
4. While $U > S$, do
   $X \leftarrow X + 1$
   $C \leftarrow (1 + \phi/(\omega + X\phi))^{X-1}(1 - \phi/(1 - \omega - X\phi))^{m-X}$
   $P \leftarrow C(m - X + 1)(\omega + X\phi)P/(X(1 - \omega - X\phi))$
   $S \leftarrow S + P$
5. Return X

The pmf for the QHD-I is given by

$$P_x = \frac{\binom{n}{x} a(a + xr - 1)_{(x-1)}(b + nr - xr)_{(n-x)}}{(a + b + nr)_{(n)}}, \quad x = 0, 1, 2, 3, \ldots,$$

and a recurrence relation between its probabilities is given as

$$P_x = \frac{n - x + 1}{x} \cdot \frac{a + (x - 1)(r - 1)}{b + r + (r - 1)(n - x)} \prod_{i=1}^{x-1} \left(1 + \frac{r}{1 - r + xr - i}\right)$$

$$\times \prod_{i=0}^{n-x-1} \left(1 - \frac{r}{b + nr - r(x - 1) - i}\right) P_{x-1} \tag{16.35}$$

for $x > 1$, with $P_0 = \binom{b+nr}{n} \Big/ \binom{a+b+nr}{n}$.

Quasi-Hypergeometric I Random Variate Generator based on Inversion Method
[Initialize: $\omega_1 = a - r$ and $\omega_2 = b + nr + r$]

1. $X \leftarrow 0$
2. $S \leftarrow \binom{b+nr}{n} \Big/ \binom{a+b+nr}{n}$ and $P \leftarrow S$
3. Generate U from uniform distribution on (0, 1).

4. If $U \leq S$, Return X
5. $X \leftarrow X + 1$
6. $P \leftarrow a\left[\binom{b+nr-r}{n-1}/\binom{b+nr}{n}\right] P$ and $S \leftarrow S + P$
7. If $U \leq S$, Return X
8. While $U > S$, do

   $X \leftarrow X + 1$

   $c \leftarrow \prod_{i=1}^{X-1}(1 + r/(\omega_1 + Xr - i)) \prod_{j=0}^{n-X-1}(1 - r/(\omega_2 - Xr - j))$

   $P \leftarrow c(n - X + 1)(\omega_1 + X(r-1) + 1)P/(X(\omega_2 - n - X(r-1)))$

   $S \leftarrow S + P$
9. Return X

The pmf for the QPD-I is given by

$$P_x = \frac{\binom{n}{x} a(a + xr + c)^{[x-1,c]}(b + nr - xr)^{[n-x,c]}}{(a + b + nr)^{[n,c]}}, \quad x = 0, 1, 2, 3, \ldots$$

A recurrence relation between the above probabilities is given by

$$P_x = \frac{(n - x + 1)}{x} \frac{(a - r + xr + (x-1)c)}{b + nr + r + nc - x(r + c)} \prod_{i=1}^{x-1} \left(1 + \frac{r}{a - r + xr + ic}\right) \times \prod_{j=0}^{n-x-1} \left(1 - \frac{r}{b + nr + r - xr + ic}\right) P_{x-1} \tag{16.36}$$

for $x \geq 1$, with $P_0 = \binom{\frac{b+nr}{c}+n-1}{n} \Big/ \binom{\frac{a+b+nr}{c}+n-1}{n}$.

Quasi-Pólya I Random Variate Generator based on Inversion Method
[Initialize: $\omega_1 = a - r$ and $\omega_2 = b + nr + r$]

1. $X \leftarrow 0$
2. $S \leftarrow \binom{(b+nr)/c+n-1}{n} \Big/ \binom{(a+b+nr)/c+n-1}{n}$ and $P \leftarrow S$
3. Generate U from uniform distribution on (0, 1).
4. If $U \leq S$, Return X
5. $X \leftarrow X + 1$
6. $P \leftarrow a\left[\binom{(b+nr-r)/c+n-2}{n-1} \Big/ \binom{(b+nr)/c+n-1}{n}\right] P$ and $S \leftarrow S + P$
7. If $U \leq S$, Return X
8. While $U > S$, do

   $X \leftarrow X + 1$

   $c \leftarrow \prod_{i=1}^{X-1}(1 + r/(\omega_1 + Xr + ic)) \prod_{j=0}^{n-X-1}(1 - r/(\omega_2 - Xr - jc))$

   $P \leftarrow c(\omega_1 + Xr + (X-1)c)P/[X(\omega_2 + nc - X(r + c))]$

   $S \leftarrow S + P$
9. Return X

References

1. M. Abramowitz and I. A. Stegun. 1965. *Handbook of Mathematical Functions*. Dover Publications, Inc., New York.
2. M.S. Abu-Salih. 1980. On the resolution of a mixture of observations from two modified power series distributions. *Revista Colombiana de Matemáticas*, XIV:197–208.
3. J.H. Ahrens and K.D. Kohrt. 1981. Computer methods for efficient sampling from largely arbitrary statistical distributions. *Computing*, 26:19–31.
4. M. Ahsanullah. 1991a. Two characterizations of the generalized Poisson distribution. *Pakistan Journal of Statistics*, 7:15–19.
5. M. Ahsanullah. 1991b. Two characteristic properties of the generalized negative binomial distribution. *Biometrical Journal*, 33:861–864.
6. Ali-Amidi. 1978. A note on the moments of the generalized negative binomial distribution and on certain properties of this distribution. *SIAM Journal of Applied Mathematics*, 34:223–224.
7. R.S. Ambagaspitiya. 1995. A family of discrete distributions. *Insurance: Mathematics and Economics*, 16:107–127.
8. R.S. Ambagaspitiya. 1998. Compound bivariate Lagrangian Poisson distributions. *Insurance: Mathematics and Economics*, 23:21–31.
9. R.S. Ambagaspitiya and N. Balakrishnan. 1994. On the compound generalized Poisson distributions. *ASTIN Bulletin*, 24:255–263.
10. J. Angers and A. Biswas. 2003. A Bayesian analysis of zero-inflated generalized Poisson model. *Computational Statistics and Data Analysis*, 142:37–46.
11. N. Balakrishnan and V.B. Nevzorov. 2003. *A Primer on Statistical Distributions*. John Wiley & Sons, Inc., New York.
12. G. Beall and R.R. Rescia. 1953. A generalization of Neyman's contagious distributions. *Biometrics*, 9:354–386.
13. S. Berg. 1974. Factorial series distributions, with applications to capture-recapture problems. *Scandinavian Journal of Statistics*, 1:145–152.
14. S. Berg and L. Mutafchiev. 1990. Random mappings with an attracting center: Lagrangian distributions and a regression function. *Journal of Applied Probability*, 27:622–636.
15. S. Berg and K. Nowicki. 1991. Statistical inference for a class of modified power series distributions with applications to random mapping theory. *Journal of Statistical Planning and Inference*, 28:247–261.
16. E. Borel. 1942. Sur l'emploi du théorème de Bernoulli pour faciliter le calcul d'un infinité de coefficients. Application au problème de l'attente à un guichet. *Comptes Rendus, Académie des Sciences, Paris, Series A*, 214:452–456.
17. K.O. Bowman and L.R. Shenton. 1985. The distribution of a moment estimator for a parameter of the generalized Poisson distribution. *Communications in Statistics—Simulation and Computation*, 14(4):867–893.
18. A. Cayley. 1857. On the theory of the analytical forms called trees (and other titles). *Collected mathematical papers* (Cambridge, 1889–1897): **2**, 1–7; **3**, 242–246; **4**, 112–115; **9**, 202–204, 427–460, 544–545; **10**, 598–600; **11**, 365–367; **13**, 26–28. These articles were originally published in the period 1857–1889.

19. R.L. Chaddha. 1965. A case of contagion in binomial distribution. In G.P. Patil, editor, *Classical and Contagious Discrete Distributions*, pages 273–290. Statistical Publishing Society, Calcutta, India.
20. C.A. Charalambides. 1974. Minimum variance unbiased estimation for a class of left-truncated discrete distributions. *Sankhyā, Series A*, 36:397–418.
21. C.A. Charalambides. 1986. Gould series distributions with applications to fluctuations of sums of random variables. *Journal of Statistical Planning and Inference*, 14:15–28.
22. C.A. Charalambides. 1987. On some properties of the linear function Poisson, binomial and negative binomial distributions. *Metrika*, 33:203–216.
23. C.A. Charalambides. 1990. Abel series distributions with applications to fluctuations of sample functions of stochastic processes. *Communications in Statistics—Theory and Methods*, 19:317–335.
24. C.A. Charalambides and J. Singh. 1988. A review of the Stirling numbers, their generalizations and statistical applications. *Communications in Statistics—Theory and Methods*, 17:2533–2595,
25. H.C. Chen and Y. Asau. 1974. On generating random variates from an empirical distribution. *AIIE Transactions*, 6:163–166.
26. K.E. Churchill and G.C. Jain. 1976. Further bivariate distributions associated with Lagrange expansion. *Biometrical Journal*, 8:639–649.
27. P.C. Consul. 1974. A simple urn model dependent upon predetermined strategy. *Sankhyā, Series B*, 36:391–399.
28. P.C. Consul. 1975. Some new characterizations of discrete Lagrangian distributions. In G.P. Patil, S. Kotz, and J.K. Ord, editors, *Statistical Distributions in Scientific Work, 3: Characterizations and Applications*, pages 279–290. D. Reidel Publishing Company, Boston, MA.
29. P.C. Consul. 1981. Relation of modified power series distributions to Lagrangian probability distributions. *Communications in Statistics—Theory and Methods*, 10:2039–2046.
30. P.C. Consul. 1984. On the distributions of order statistics for a random sample size. *Statistica Neerlandica*, 38:249–256.
31. P.C. Consul. 1986. On the differences of two generalized Poisson variates. *Communications in Statistics—Theory and Methods*, 15:761–767.
32. P.C. Consul. 1988. On some models leading to the generalized Poisson distribution. *Communications in Statistics—Theory and Methods*, 17:423–4424.
33. P.C. Consul. 1989a. *Generalized Poisson Distributions: Applications and Properties*. Marcel Dekker, Inc., New York.
34. P.C. Consul. 1989b. On the differences of two generalized negative binomial variates. *Communications in Statistics—Theory and Methods*, 18:673–690.
35. P.C. Consul. 1990a. Two stochastic models for the Geeta distribution. *Communications in Statistics—Theory and Methods*, 19:3699–3706.
36. P.C. Consul. 1990b. Geeta distribution and its properties. *Communications in Statistics—Theory and Methods*, 19:3051–3068.
37. P.C. Consul. 1990c. New class of location-parameter discrete probability distributions and their characterizations. *Communications in Statistics—Theory and Methods*, 19:4653–4666.
38. P.C. Consul. 1990d. On some properties and applications of quasi-binomial distribution. *Communications in Statistics—Theory and Methods*, 19:477–504.
39. P.C. Consul. 1993. On the class of discrete Lagrangian distributions. Unpublished Research Work.
40. P.C. Consul. 1994a. Some bivariate families of Lagrangian probability distributions. *Communications in Statistics—Theory and Methods*, 23:2895–2906.
41. P.C. Consul. 1994b. On some probability distributions associated with random walks. *Communications in Statistics—Theory and Methods*, 23:3241–3255.
42. P.C. Consul. 1994c. A generalized stochastic urn model. Unpublished Research Work.
43. P.C. Consul and F. Famoye. 1985. Type I error in estimation and its control. Unpublished Research Work.
44. P.C. Consul and F. Famoye. 1986a. On the unimodality of the generalized Poisson distribution. *Statistica Neerlandica*, 40:117–122.
45. P.C. Consul and F. Famoye. 1986b. On the unimodality of the generalized negative binomial distribution. *Statistica Neerlandica*, 40:141–144.
46. P.C. Consul and F. Famoye. 1988. Maximum likelihood estimation for the generalized Poisson distribution when sample mean is larger than sample variance. *Communications in Statistics—Theory and Methods*, 17:299–309.

47. P.C. Consul and F. Famoye. 1989a. Minimum variance unbiased estimation for the Lagrange power series distributions. *Statistics*, 20:407–415.
48. P.C. Consul and F. Famoye. 1989b. The truncated generalized Poisson distribution and its estimation. *Communications in Statistics—Theory and Methods*, 18:3635–3648.
49. P.C. Consul and F. Famoye. 1995. On the generalized negative binomial distribution. *Communications in Statistics—Theory and Methods*, 24:459–472.
50. P.C. Consul and F. Famoye. 1996. Lagrangian Katz family of distributions. *Communications in Statistics—Theory and Methods*, 25:415–434.
51. P.C. Consul and F. Famoye. 2001. On Lagrangian distributions of the second kind. *Communications in Statistics—Theory and Methods*, 30:165–178.
52. P.C. Consul and F. Famoye. 2005. Equivalence of both classes of Lagrangian probability distributions. *Far East Journal of Theoretical Statistics*, 15:43–51.
53. P.C. Consul and H.C. Gupta. 1975. Generalized Poisson and generalized logarithmic distributions and their characterizations by zero regression. Unpublished Research Work.
54. P.C. Consul and H.C. Gupta. 1980. The generalized negative binomial distribution and its characterization by zero regression. *SIAM Journal of Applied Mathematics*, 39:231–237.
55. P.C. Consul and G.C. Jain. 1973a. A generalization of the Poisson distribution. *Technometrics*, 15:791–799.
56. P.C. Consul and G.C. Jain. 1973b. On some interesting properties of the generalized Poisson distribution. *Biometrische Zeitschrift*, 15:495–500.
57. P.C. Consul and S.P. Mittal. 1975. A new urn model with predetermined strategy. *Biometrische Zeitschrift*, 17:67–75.
58. P.C. Consul and S.P. Mittal. 1977. Some discrete multinomial probability models with predetermined strategy. *Biometrical Journal*, 19:161–173.
59. P.C. Consul and L.R. Shenton. 1972. Use of Lagrange expansion for generating generalized probability distributions. *SIAM Journal of Applied Mathematics*, 23:239–248.
60. P.C. Consul and L.R. Shenton. 1973a. Some interesting properties of Lagrangian distributions. *Communications in Statistics*, 2:263–272.
61. P.C. Consul and L.R. Shenton. 1973b. On the multivariate generalization of the family of discrete Lagrange distributions. In D.G. Kabe and R.P. Gupta, editors, *Multivariate Statistical Inference*, pages 13–23. North-Holland, New York.
62. P.C. Consul and L.R. Shenton. 1975. On the probabilistic structure and properties of discrete Lagrangian distributions. In G.P. Patil, S. Kotz, and J.K. Ord, editors, *Statistical Distributions in Scientific Work, 1: Models and Structures*, pages 41–57. D. Reidel Publishing Company, Boston, MA.
63. P.C. Consul and M.M. Shoukri. 1984. Maximum likelihood estimation for the generalized Poisson distribution. *Communications in Statistics—Theory and Methods*, 13:1533–1547.
64. P.C. Consul and M.M. Shoukri. 1985. The generalized Poisson distribution when the sample mean is larger than the sample variance. *Communications in Statistics—Theory and Methods*, 14:667–681.
65. P.C. Consul and M.M. Shoukri. 1988. Some chance mechanisms related to a generalized Poisson probability model. *American Journal of Mathematical and Management Sciences*, 8:181–202.
66. E.L Crow and G.E. Bardwell. 1965. Estimation of the parameters of the hyper-Poisson distributions. In G.P. Patil, editor, *Classical and Contagious Discrete Distributions*, pages 127–140. Statistical Publishing Society, Calcutta, India.
67. M.J. Crowder. 1978. Beta-binomial anova for proportions. *Applied Statistics*, 27:34–37.
68. L. Devroye. 1986. *Non-Uniform Random Variate Generation*. Springer-Verlag, New York.
69. L. Devroye. 1989. Random variate generators of the Poisson-Poisson and related distributions. *Computational Statistics and Data Analysis*, 8:247–278.
70. L. Devroye. 1992. The branching process method in Lagrange random variate generation. *Communications in Statistics—Simulation and Computation*, 21:1–14.
71. F. Eggenberger and G. Pólya. 1923. Über die Statistik verketteter Vorgänge. *Zeitschrift für angewandte Mathematik und Mechanik*, 3:279–289.
72. A. Erdélyi, W. Magnus, F. Oberhettinger, and F.G. Tricomi. 1953. *Higher Transcendental Functions*, volume I. McGraw-Hill, New York.
73. F. Faà di Bruno. 1855. Note sur une nouvelle formulae de calcul différential. *Quart. Jour. Pure and Applied Math.*, 1:359–360.

74. F. Famoye. 1987. A short note on the generalized logarithmic series distribution. *Statistics and Probability Letters*, 5:315–316.
75. F. Famoye. 1993. Testing for homogeneity: The generalized Poisson distribution. *Communications in Statistics—Theory and Methods*, 22:705–715.
76. F. Famoye. 1994. Characterization of generalized negative binomial distribution. *Journal of Mathematical Sciences*, 5:71–81.
77. F. Famoye. 1995. On certain methods of estimation for the generalized logarithmic series distribution. *Journal of Applied Statistical Sciences*, 2:103–117.
78. F. Famoye. 1997a. Generalized geometric distribution and some of its applications. *Journal of Mathematical Sciences*, 8:1–13.
79. F. Famoye. 1997b. Generalized Poisson random variate generation. *American Journal of Mathematical and Management Sciences*, 17:219–237.
80. F. Famoye. 1997c. Parameter estimation of generalized negative binomial distribution. *Communications in Statistics—Simulation and Computation*, 26:269–279.
81. F. Famoye. 1997d. Sampling from the generalized logarithmic series distribution. *Computing*, 58:365–375.
82. F. Famoye. 1998a. Bootstrap based tests for generalized negative binomial distribution. *Computing*, 61:359–369.
83. F. Famoye. 1998b. Computer generation of generalized negative binomial deviates. *Journal of Statistical Computation and Simulation*, 60:107–122.
84. F. Famoye. 1999. EDF tests for generalized Poisson distribution. *Journal of Statistical Computation and Simulation*, 63:159–168.
85. F. Famoye. 2000. Goodness of fit tests for generalized logarithmic series distribution. *Journal of Computational Statistics and Data Analysis*, 33:59–67.
86. F. Famoye and P.C. Consul. 1989a. Confidence interval estimation in the class of modified power series distributions. *Statistics*, 20:141–148.
87. F. Famoye and P.C. Consul. 1989b. A stochastic urn model for the generalized negative binomial distribution. *Statistics*, 20:607–613.
88. F. Famoye and P.C. Consul. 1990. Interval estimation and hypothesis testing for the generalized Poisson distribution. *American Journal of Mathematical and Management Sciences*, 10:127–158.
89. F. Famoye and P.C. Consul. 1993. The truncated generalized negative binomial distribution. *Journal of Applied Statistical Sciences*, 1:141–157.
90. F. Famoye and C.M.-S. Lee. 1992. Estimation of generalized Poisson distribution. *Communications in Statistics—Simulation and Computation*, 21:173–188.
91. W.S. Fazal. 1977. A test for a generalized Poisson distribution. *Biometrical Journal*, 19:245–251.
92. W. Feller. 1957. *An Introduction to Probability Theory and Its Applications*, volume 1. John Wiley & Sons, Inc., New York, second edition.
93. W. Feller. 1968. *An Introduction to Probability Theory and Its Applications*, volume 1. John Wiley & Sons, Inc., New York, third edition.
94. D.A.S. Fraser. 1952. Sufficient statistics and selection depending on the parameter. *Annals of Mathematical Statistics*, 23:417–425.
95. B. Friedman. 1949. A simple urn model. *Communications in Pure and Applied Mathematics*, 2:59–70.
96. I.J. Good. 1949. The number of individuals in a cascade process. *Proceedings of Cambridge Philosophical Society*, 45:360–363.
97. I.J. Good. 1955. The joint distribution of the sizes of the generations in a cascade process. *Proceedings of Cambridge Philosophical Society*, 51:240–242.
98. I.J. Good. 1958. Legendre polynomials and trinomial random walks. *Proceedings of Cambridge Philosophical Society*, 54:39–42.
99. I.J. Good. 1960. Generalizations to several variables of Lagrange's expansion, with applications to stochastic process. *Proceedings of Cambridge Philosophical Society*, 56:367–380.
100. I.J. Good. 1965. The generalization of Lagrange's expansion, and enumeration of trees. *Proceedings of Cambridge Philosophical Society*, 61:499–517.
101. I.J. Good. 1975. The Lagrange distributions and branching processes. *SIAM Journal of Applied Mathematics*, 28:270–275.

102. M.J. Goovaerts and R. Kaas. 1991. Evaluating compound generalized Poisson distributions recursively. *ASTIN Bulletin*, 21:193–197.
103. M. Gordon. 1962. Good's theory of cascade processes applied to the statistics of polymer distributions. *Proceedings of the Royal Statistical Society of London, Series A*, 268:240–256.
104. H.W. Gould. 1962. Congruences involving sums of binomial coefficients and a formula of Jensen. *Mathematical Notes*, pages 400–402, May.
105. H.W. Gould. 1966. Evaluation of a class of binomial coefficient summation. *Journal of Combinatorial Theory*, 1:233–247.
106. H.W. Gould. 1972. *Combinatorial Identities*. Morgantown Printing and Binding Company, Morgantown, WV.
107. P.L. Gupta. 1982. Probability generating functions of a MPSD with applications. *Mathematische Operationforschung und Statistik, series Statistics*, 13:99–103.
108. P.L. Gupta, R.C. Gupta, and R.C. Tripathi. 1995. Inflated modified power series distributions with applications. *Communications in Statistics—Theory and Methods*, 24:2355–2374.
109. P.L. Gupta, R.C. Gupta, and R.C. Tripathi. 1996. Analysis of zero-adjusted count data. *Computational Statistics and Data Analysis*, 23:207–218.
110. P.L. Gupta and J. Singh. 1981. On the moments and factorial moments of a MPSD. In C. Taillie, G.P. Patil, and B.A. Baldessari, editors, *Statistical Distributions in Scientific Work, 4: Models, Structures and Characterizations*, pages 189–195. Reidel Publishing Company, Dordrecht.
111. R.C. Gupta. 1974. Modified power series distributions and some of its applications. *Sankhyā, Series B*, 35:288–298.
112. R.C. Gupta. 1975a. Maximum-likelihood estimation of a modified power series distribution and some of its applications. *Communications in Statistics*, 4:687–697.
113. R.C. Gupta. 1975b. Some characterizations of discrete distributions by properties of their moment distributions. *Communications in Statistics*, 4:761–765.
114. R.C. Gupta. 1976. Distribution of the sum of independent generalized logarithmic series variables. *Communications in Statistics—Theory and Methods*, 5:45–48.
115. R.C. Gupta. 1977. Minimum variance unbiased estimation in modified power series distribution and some of its applications. *Communications in Statistics—Theory and Methods*, 6:977–991.
116. R.C. Gupta. 1984. Estimating the probability of winning (losing) in a gambler's ruin problem with applications. *Journal of Statistical Planning and Inference*, 9:55–62.
117. R.C. Gupta and J. Singh. 1982. Estimation of probabilities in the class of modified power series distributions. *Mathematische Operationforschung und Statistik, series Statistics*, 13:71–77.
118. R.C. Gupta and R.C. Tripathi. 1985. Modified power series distributions. In S. Kotz, N.L. Johnson, and C.B. Read, editors, *Encyclopedia of Statistical Sciences*, volume 5, pages 593–599. Wiley & Sons Inc., New York.
119. R.C. Gupta and R.C. Tripathi. 1992. Statistical inference based on the length-biased data for the modified power series distributions. *Communications in Statistics—Theory and Methods*, 21:519–537.
120. J. Hagen. 1891. Synopsis der Höheren Mathematik. *Berlin*, 1:64–68.
121. F.A. Haight. 1961. A distribution analogous to the Borel-Tanner. *Biometrika*, 48:167–173.
122. F.A. Haight and M.A. Breuer. 1960. The Borel-Tanner distribution. *Biometrika*, 47:145–150.
123. J.B.S. Haldane and S.M. Smith. 1956. The sampling distribution of a maximum likelihood estimate. *Biometrika*, 43:96–103.
124. B.B. Hansen and E. Willekens. 1990. The generalized logarithmic series distribution. *Statistics and Probability Letters*, 9:311–316.
125. T.E. Harris. 1947. *Some Theorems on the Bernoullian Multiplicative Process*. Ph.D. thesis, Princeton University, Princeton, NJ.
126. H.R. Henze and C.M. Blair. 1931. The number of structurally isomeric alcohols of the methanol series. *Journal of the American Chemical Society*, 53:3042–3946.
127. J.M. Hill and C.M. Gulati. 1981. The random walk associated with the game of roulette. *Journal of Applied Probability*, 18:931–936.
128. P. Holgate. 1964. Estimation of the bivariate Poisson distribution. *Biometrika*, 51:241–245.
129. R. Hoover and J.F. Fraumeni. 1975. Cancer mortality in U.S. counties with chemical industries. *Environmental Research*, 9:196–207.

130. J.O. Irwin. 1965. Inverse factorial series as frequency distributions. In G.P. Patil, editor, *Classical and Contagious Discrete Distributions*, pages 159–174. Statistical Publishing Society, Calcutta, India.
131. M. Itoh, S. Inagaki, and W.C. Saslaw. 1993. Gravitational clustering of galaxies: Comparison between thermodynamic theory and n-body simulations. iv. The effects of continuous mass spectra. *The Astrophysical Journal*, 403:476–496.
132. G.C. Jain. 1975a. A linear function Poisson distribution. *Biometrical Journal*, 17:501–506.
133. G.C. Jain. 1975b. On power series distributions associated with Lagrange expansion. *Biometrische Zeitschrift*, 17:85–97.
134. G.C. Jain and P.C. Consul. 1971. A generalized negative binomial distribution. *SIAM Journal of Applied Mathematics*, 21:501–513.
135. G.C. Jain and R.P. Gupta. 1973. A logarithmic series type distribution. *Trabajos de Estadística*, 24:99–105.
136. G.C. Jain and N. Singh. 1975. On bivariate power series distributions associated with Lagrange expansion. *Journal of the American Statistical Association*, 70:951–954.
137. K.G. Janardan. 1975. Markov-polya urn model with predetermined strategies. *Gujarat Statistical Review*, 2:17–32.
138. K.G. Janardan. 1978. On generalized Markov-Polya distribution. *Gujarat Statistical Review*, 5:16–32.
139. K.G. Janardan. 1987. Weighted Lagrange distributions and their characterizations. *SIAM Journal of Applied Mathematics*, 47:411–415.
140. K.G. Janardan. 1997. A wider class of Lagrange distributions of the second kind. *Communications in Statistics—Theory and Methods*, 26:2087–2091.
141. K.G. Janardan. 1998. Generalized Polya Eggenberger family of distributions and its relation to Lagrangian Katz family. *Communications in Statistics—Theory and Methods*, 27:2423–2443.
142. K.G. Janardan, H.W. Kerster, and D.J Schaeffer. 1979. Biological applications of the Lagrangian Poisson distribution. *Bioscience*, 29:599–602.
143. K.G. Janardan and B.R. Rao. 1983. Lagrange distributions of the second kind and weighted distributions. *SIAM Journal of Applied Mathematics*, 43:302–313.
144. K.G. Janardan and D.J. Schaeffer. 1977. Models for the analysis of chromosomal aberrations in human leukocytes. *Biometrical Journal*, 19:599–612.
145. K.G. Janardan, D.J Schaeffer, and R.J. DuFrain. 1981. A stochastic model for the study of the distribution of chromosome aberrations in human and animal cells exposed to radiation or chemicals. In C. Taillie, G.P. Patil, and B.A. Baldessari, editors, *Statistical Distributions in Scientific Work, 6: Applications in Physical, Social, and Life Sciences*, pages 265–277. Reidel Publishing Company, Dordrecht.
146. P.N. Jani. 1977. Minimum variance unbiased estimation for some left-truncated power series distributions. *Sankhyā, Series B*, 39:258–278.
147. P.N. Jani. 1978a. New numbers appearing in minimum variance unbiased estimation for decapitated negative binomial and Poisson distributions. *Journal of the Indian Statistical Association*, 16:41–48.
148. P.N. Jani. 1978b. On modified power series distributions. *Metron*, 36:173–186.
149. P.N. Jani. 1985. A characterization of modified power series and multivariate modified power series distributions. *Metron*, 43:219–229.
150. P.N. Jani. 1986. The generalized logarithmic series distribution with zeroes. *Journal of the Indian Society for Agricultural Statistics*, 38:345–351.
151. P.N. Jani and S.M. Shah. 1979a. Integral expressions for the tail probabilities of the modified power series distributions. *Metron*, 37:75–79.
152. P.N. Jani and S.M. Shah. 1979b. Misclassifications in modified power series distribution in which the value one is sometimes reported as zero, and some of its applications. *Metron*, 37:121–136.
153. J.L.W. Jensen. 1902. Sur une identité d'Abel et sur d'autres formules analogues. *Acta Mathematica*, 26:307–318.
154. Normal L. Johnson and Samuel Kotz. 1969. *Discrete Distributions*. Houghton Miffin, Boston, MA, first edition.
155. Normal L. Johnson and Samuel Kotz. 1977. *Urn Models and Their Applications*. John Wiley & Sons, Inc., New York.
156. N.L. Johnson, Samuel Kotz, and N. Balakrishnan. 1997. *Discrete Multivariate Distributions*. John Wiley & Sons, Inc., New York.

157. N.L. Johnson, Samuel Kotz, and Adriene W. Kemp. 1992. *Univariate Discrete Distributions*. John Wiley & Sons, Inc., New York. second edition.
158. J.N. Kapur. 1982. Maximum-entropy formalism for some univariate & multivariate Lagrangian distributions. *Aligarh Journal of Statistics*, 2:1–16.
159. S.K. Katti. 1967. Infinite divisibility of integer valued random variables. *Annals of Mathematical Statistics*, 38:1306–1308.
160. S.K. Katti and J. Gurland. 1961. The Poisson pascal distribution. *Biometrics*, 17:527–538.
161. S.K. Katti and L.E. Sly. 1965. Analysis of contagious data through behavioristic models. In G.P. Patil, editor, *Classical and Contagious Discrete Distributions*, pages 303–319. Statistical Publishing Society, Calcutta, India.
162. L. Katz. 1945. *Characteristics of frequency functions defined by first order difference equations*. Ph.D. thesis, University of Michigan, Ann Arbor, MI.
163. L. Katz. 1965. Unified treatment of a broad class of discrete probability distributions. In G.P. Patil, editor, *Classical and Contagious Discrete Distributions*, pages 175–182. Statistical Publishing Society, Calcutta, India.
164. J. Keilson and H. Gerber. 1971. Some results for discrete unimodality. *Journal of the American Statistical Association*, 66:386–389.
165. A.W. Kemp and C.D. Kemp. 1968. On a distribution associated with certain stochastic processes. *Journal of the Royal Statistical Society, Series B*, 30:160–163.
166. C.D. Kemp and A.W. Kemp. 1956. Generalized hypergeometric distributions. *Journal of the Royal Statistical Society, Series B*, 18:202–211.
167. C.G. Khatri. 1982. Multivariate Lagrangian Poisson and multinomial distributions. *Sankhyā, Series B*, 44:259–269.
168. J.C. Kleinman. 1973. Proportions with extraneous variance: Single and independent sample. *Journal of the American Statistical Association*, 68:46–54.
169. L. Kleinrock. 1975. *Queueing Systems*, volume 1. John Wiley & Sons, Inc., New York.
170. A. Kumar. 1981. Some applications of Lagrangian distributions in queueing theory and epidemiology. *Communications in Statistics—Theory and Methods*, 10:1429–1436.
171. A. Kumar and P.C. Consul. 1979. Negative moments of a modified power series distribution and bias of the maximum likelihood estimator. *Communications in Statistics—Theory and Methods*, 8:151–166.
172. A. Kumar and P.C. Consul. 1980. Minimum variance unbiased estimation for modified power series distribution. *Communications in Statistics—Theory and Methods*, 9:1261–1275.
173. P. Kvam and D. Day. 2001. The multivariate Polya distribution in combat modeling. *Naval Research Logistics*, 48:1–17.
174. J.L. Lagrange. 1788. *Mécanique Analytique*. Jacques Gabay, Paris, France.
175. C.M.-S. Lee and F. Famoye. 1996. A comparison of generalized Poisson distribution for modeling chromosome aberrations. *Biometrical Journal*, 38:299–313.
176. E.L. Lehmann. 1997. *Testing Statistical Hypotheses*. Springer-Verlag New York, Inc., New York. second edition.
177. B. Lerner, A. Lone, and M. Rao. 1997. On generalized Poisson distributions. *Probability and Mathematical Statistics*, 17:377–385.
178. G.S. Lingappaiah. 1986. On the generalized Poisson and linear function Poisson distributions. *Statistica*, 46:343–352.
179. G.S. Lingappaiah. 1987. Lagrange-negative binomial distribution in its simple and weighted forms. *Journal of Statistical Research*, 21:45–54.
180. K.V. Mardia. 1970. *Families of Bivariate Distributions*. Charles Griffin and Co., London.
181. A.G. McKendrick. 1926. Applications of mathematics to medical problems. *Proceedings of the Edinburgh Mathematical Society*, 44:98–130.
182. M. Minami. 1998. Multivariate Lagrange distributions and a subfamily. *Communications in Statistics—Theory and Methods*, 27:2185–2197.
183. M. Minami. 1999. Inverse relationship and limiting forms of Lagrange distributions. *Communications in Statistics—Theory and Methods*, 28:409–429.
184. A. Mishra. 1996. A generalized bivariate binomial distribution applicable in four-fold sampling. *Communications in Statistics—Theory and Methods*, 25:1943–1956.

185. A. Mishra and S.K. Singh. 1996. Moments of a quasi-binomial distribution. *Progress of Mathematics*, 30:59–67.
186. A. Mishra and D. Tiwary. 1985. On generalized logarithmic series distribution. *Journal of the Indian Society of Agricultural Statistics*, 37:219–225.
187. A. Mishra, D. Tiwary, and S.K. Singh. 1992. A class of quasi-binomial distributions. *Sankhyā, Series B*, 54:67–76.
188. S.G. Mohanty. 1966. On a generalized two-coin tossing problem. *Biometrische Zeitschrift*, 8:266–272.
189. S.G. Mohanty. 1979. *Lattice Path Counting and Applications*. Academic Press Inc., New York.
190. M. Murat and D. Szynal. 1998. Non-zero inflated modified power series distributions. *Communications in Statistics—Theory and Methods*, 27:3047–3064.
191. L. Mutafchiev. 1995. Local limit approximations for Lagrangian distributions. *Aequationes Mathematicae*, 49:57–85.
192. S.B. Nandi and K.K. Das. 1994. A family of the Abel series distributions. *Sankhyā, Series B*, 56:147–164.
193. S.B. Nandi and K.K. Das. 1996. A family of the multivariate Abel series distributions. *Sankhyā, Series A*, 58:252–263.
194. J. Neyman. 1965. Certain chance mechanisms involving discrete distributions. In G.P. Patil, editor, *Classical and Contagious Discrete Distributions*, pages 4–14. Statistical Publishing Society, Calcutta, India.
195. J. Neyman and E.L. Scott. 1964. A stochastic model of epidemics. In J. Gurland, editor, *Stochastic Models in Medicine and Biology*, pages 45–83. University of Wisconsin Press, Madison, WI.
196. M.S. Nikulin and V.G. Voinov. 1994. A chi-square goodness-of-fit test for modified power series distributions. *Sankhyā, Series B*, 56:86–94.
197. A. Noack. 1950. A class of random variables with discrete distributions. *Annals of Mathematical Statistics*, 21:127–132.
198. I. Olkin and M. Sobel. 1965. Integral expressions for tail probabilities of the multinomial and negative multinomial distributions. *Biometrika*, 52:167–179.
199. J.K. Ord. 1972. *Families of Frequency Distributions*. Charles Griffin and Co., London.
200. R. Otter. 1948. The number of trees. *Annals of Mathematics*, 49:583–599.
201. R. Otter. 1949. The multiplicative process. *Annals of Mathematical Statistics*, 20:206–224.
202. A.G. Pakes and T.P. Speed. 1977. Lagrange distributions and their limit theorems. *SIAM Journal of Applied Mathematics*, 32:745–754.
203. I.D. Patel. 1981. A generalized logarithmic series distribution. *Journal of the Indian Statistical Association*, 19:129–132.
204. S.R. Patel. 1980. Minimum variance unbiased estimation of generalized logarithmic series distribution. *Pure and Applied Mathematika Sciences*, 11:79–82.
205. S.R. Patel and P.N. Jani. 1977. On minimum variance unbiased estimation of generalized Poisson distributions and decapitated generalized Poisson distributions. *Journal of the Indian Statistical Association*, 15:157–159.
206. G.P. Patil. 1961. *Contributions to Estimation in a Class of Discrete Distributions*. Ph.D. thesis, University of Michigan, Ann Arbor, MI.
207. G.P. Patil. 1962. Certain properties of the generalized power series distribution. *Annals of the Institute of Statistical Mathematics, Tokyo*, 14:179–182.
208. G.P. Patil and S. Bildikar. 1967. Multivariate logarithmic series distribution as a probability model in population and community ecology and some of its statistical properties. *Journal of the American Statistical Association*, 62:655–674.
209. G.P. Patil, M.T. Boswell, S.W. Joshi, and M.V. Ratnaparkhi. 1984. *Dictionary and Bibliography of Statistical Distributions in Scientific Work, 1, Discrete Models*. International Co-operative Publishing House, Fairland, MD.
210. G.P. Patil and S.W. Joshi. 1968. *A Dictionary and Bibliography of Statistical Distributions*. Oliver and Boyd, Edinburgh.
211. G.P. Patil and C.R. Rao. 1978. Weighted distributions and size biased sampling with application to wild life populations and human families. *Biometrics*, 34:179–189.
212. H. Poincaré. 1886. Sur les résidus des intégrales doubles. *Acta Mathematica*, 9:321–380.
213. G. Pólya. 1930. Sur quelques points de la théorie des probabilités. *Annales de l'Institut Henri Poincaré*, 1:117–161.

214. G.Pólya. 1937. Kombinatorische anzahlbestimmungen für gruppen, graphen, und chesche verbindungen. *Acta Mathematica*, 68:145–253.
215. N.U. Prabhu. 1965. *Queues and Inventories*. John Wiley & Sons, Inc., New York.
216. D. Raikov. 1937a. A characteristic property of the Poisson laws. *C.R. Acad. Sci. (U.S.S.R.)*, 14:8–11.
217. D. Raikov. 1937b. A characteristic property of the Poisson laws. *Dokl. Akad. Nauk. (U.S.S.R.)*, 14:9–12.
218. C.R. Rao. 1965. On some discrete distributions arising out of methods of ascertainment. In G.P. Patil, editor, *Classical and Contagious Discrete Distributions*, pages 320–332. Statistical Publishing Society, Calcutta, India.
219. J. Riordan. 1968. *Combinatorial Identities*. John Wiley & Sons, Inc., New York.
220. S.M. Roman and G.C. Rota. 1978. The umbral calculus. *Advances in Mathematics*, 27:95–188.
221. Z. Rychlik and D. Zynal. 1973. On the limit behaviour of sums of a random number of independent random variables. *Colloquium Mathematicum*, 28:147–159.
222. K. Sarkadi. 1957. Generalized hypergeometric distributions. *Magyar Tudományos Akadémia Matematikai Kutató Intézetének Közlenényei*, 2:59–68.
223. D.J. Schaeffer, K.G. Janardan, H.W. Kerster, and A.C.M. Clarke. 1983. Statistically estimated free energies of chromosome aberration production from frequency data. *Biometrical Journal*, 25:275–282.
224. K. Sen and R. Jain. 1996. Generalized Makov-Polya urn models with predetermined strategies. *Statistics and Probability Letters*, 54:119–133.
225. R. Shanmugam. 1984. A test for homogeneity when the sample is a positive Lagrangian Poisson distribution. *Communications in Statistics—Theory and Methods*, 13:3243–3250.
226. L.R. Shenton and K.O. Bowman. 1977. *Maximum Likelihood Estimation in Small Samples*. Griffin, London.
227. L.R. Shenton and P.C. Consul. 1973. On bivariate Lagrange and Borel-Tanner distributions and their use in queueing theory. *Sankhyā, Series A*, 35:229–236.
228. R.K. Sheth. 1998. The generalized Poisson distribution and a model of clustering from Poisson initial conditions. *Monthly Notices of Royal Astronomical Society*, 299:207–217.
229. K. Shimizu, N. Nishii, and M. Minami. 1997. The multivariate inverse trinomial distribution as a Lagrangian probability model. *Communications in Statistics—Theory and Methods*, 26:1585–1598.
230. M.M. Shoukri. 1980. *Estimation Problems for Some Generalized Discrete Distributions*. Ph.D. thesis, University of Calgary, Calgary, Canada.
231. M.M. Shoukri. 1982. Minimum variance unbiased estimation in a bivariate modified power series distribution. *Biometrical Journal*, 24:97–101.
232. M.M. Shoukri and P.C. Consul. 1982. Bivariate modified power series distribution: Some properties, estimation and applications. *Biometrical Journal*, 24:787–799.
233. M.M. Shoukri and P.C. Consul. 1987. Some chance mechanisms generating the generalized Poisson probability models. In I.B. MacNeill and G.J. Umphrey, editors, *Advances in the Statistical Sciences V: Biostatistics*, volume 24, pages 259–268. D. Reidel Publishing Company, Boston, MA.
234. M. Sibuya, N. Miyawaki, and U. Sumita. 1994. Aspects of Lagrangian probability distributions. *Journal of Applied Probability*, 31A:185–197.
235. F. Spitzer. 1964. *Principles of Random Walk*. Van Nostrand, Princeton, NJ.
236. F.W. Steutel and K. van Harn. 1979. Discrete analogues of self-decomposability and stability. *Annals of Probability*, 7:893–899.
237. L. Takács. 1962. A generalization of the ballot problem and its applications in the theory of queues. *Journal of the American Statistical Association*, 57:327–337.
238. J.C. Tanner. 1953. A problem of interference between two queues. *Biometrika*, 40:58–69.
239. H. Teicher. 1954. On the multivariate Poisson distribution. *Skandinavisk Aktuarietidskrift*, 37:1–9.
240. R.C. Tripathi, P.L. Gupta, and R.C. Gupta. 1986. Incomplete moments of modified power series distributions with applications. *Communications in Statistics—Theory and Methods*, 15:999–1015.
241. R.C. Tripathi and R.C. Gupta. 1985. A generalization of the log-series distributions. *Communications in Statistics—Theory and Methods*, 14:1779–1799.
242. R.C. Tripathi and R.C. Gupta. 1988. Another generalization of the logarithmic series and the geometric distributions. *Communications in Statistics—Theory and Methods*, 17:1541–1547.
243. R.C. Tripathi and J. Gurland. 1977. A general family of discrete distributions with hypergeometric probabilities. *Journal of the Royal Statistical Society, Series B*, 39:349–356.

244. H.J.H. Tuenter. 2000. On the generalized Poisson distribution. *Statistica Neerlandica*, 54:374–376.
245. A.J. Walker. 1974a. New fast method for generating discrete random numbers with arbitrary frequency distributions. *Electronic Letter*, 10:127–128.
246. A.J. Walker. 1974b. Fast generation of uniformly distributed pseudorandom numbers with floating point representation. *Electronic Letter*, 10:553–554.
247. A.J. Walker. 1977. An efficient method for generating discrete random variables with general distributions. *ACM Transactions on Mathematical Software*, 3:253–256.
248. E.T. Whittaker and G.N. Watson. 1990. *A Course in Modern Analysis*. Cambridge University Press, Cambridge, England, 4th edition.
249. D.A. Williams. 1975. The analysis of binary responses from toxicological experiments involving reproduction and teratogenicity. *Biometrics*, 31:949–952.
250. J. Wishart. 1949. Cummulants of multivariate multinomial distributions. *Biometrika*, 36:47–58.
251. J.F. Yan. 1978. A new derivation of molecular size distribution in nonlinear polymers. *Macromolecules*, 11:648–649.
252. J.F. Yan. 1979. Cross-linking of polymers with a primary size distribution. *Macromolecules*, 12:260–264.

Index